Matrix Methods

Matrix Methods
Applied Linear Algebra and Sabermetrics

Fourth Edition

Richard Bronson

Gabriel B. Costa

ELSEVIER

ACADEMIC PRESS
An imprint of Elsevier

Academic Press is an imprint of Elsevier
125 London Wall, London EC2Y 5AS, United Kingdom
525 B Street, Suite 1650, San Diego, CA 92101, United States
50 Hampshire Street, 5th Floor, Cambridge, MA 02139, United States
The Boulevard, Langford Lane, Kidlington, Oxford OX5 1GB, United Kingdom

Notices
Knowledge and best practice in this field are constantly changing. As new research and experience broaden our understanding, changes in research methods, professional practices, or medical treatment may become necessary.

Practitioners and researchers must always rely on their own experience and knowledge in evaluating and using any information, methods, compounds, or experiments described herein. In using such information or methods they should be mindful of their own safety and the safety of others, including parties for whom they have a professional responsibility.

To the fullest extent of the law, neither the Publisher nor the authors, contributors, or editors assume any liability for any injury and/or damage to persons or property as a matter of products liability, negligence or otherwise, or from any use or operation of any methods, products, instructions, or ideas contained in the material herein.

Library of Congress Cataloging-in-Publication Data
A catalog record for this book is available from the Library of Congress

British Library Cataloguing-in-Publication Data
A catalogue record for this book is available from the British Library

ISBN: 978-0-12-818419-6

For information on all Academic Press publications visit our website at
https://www.elsevier.com/books-and-journals

Acquisition Editor: Katey Birtcher
Editorial Project Manager: Naomi Robertson
Production Project Manager: Kamatchi Madhavan
Cover Designer: Patrick C. Ferguson

Typeset by TNQ Technologies

To Evy … again.
R.B.

To my brother priests … and to the memory of
Archbishop Peter L. Gerety, the prelate who ordained me to
the priesthood 40 years ago.
G.B.C.

Contents

Preface to the third edition

It is no secret that matrices are used in many fields. They are naturally present in all branches of mathematics, as well as in many engineering and science fields. Additionally, this simple but powerful concept is readily applied to many other disciplines, such as economics, sociology, political science, nursing, and psychology.

The matrix is a dynamic construct. New applications of matrices are still evolving, and our third edition of **Matrix Methods: Applied Linear Algebra** (previously *An Introduction*) reflects important changes that have transpired since the publication of the previous edition.

In this third edition, we added material on optimization and probability theory. Chapter 4 is new and covers an introduction to the simplex method, one of the major applied advances in the last half of the 20th century. Chapter 9 is also new and introduces Markov chains, a primary use of matrices to probability applications. To ensure that the book remains appropriate in length for a one semester course, we deleted some of the subject matter that is more advanced; specifically, chapters on the Jordan canonical form and on special matrices (e.g., Hermitian and unitary matrices). We also included an Appendix dealing with technological support, such as computer algebra systems. The reader will also find that the text contains a considerable "modeling flavor".

This edition remains a textbook for the *student* not the instructor. It remains a book on methodology rather than theory. And, as in all past editions, proofs are given in the main body of the text only if they are easy to follow and revealing.

For most of this book, a firm understanding of basic algebra and a smattering of trigonometry are the only prerequisites; any references to calculus are few and far between. Calculus *is* required for Chapters 7 and 8; however, these chapters may be omitted with no loss of continuity, should the instructor wish to do so. The instructor will also find that he/she can "mix and match" chapters depending on the particular course requirements and the needs of the students.

Richard Bronson
Gabriel B. Costa

Preface to the fourth edition

We are very enthusiastic about this edition of **Matrix Methods**. Our publisher has given us the green light to blend the beauty and versatility of matrices and linear algebra with the timely topic of *sabermetrics*, defined as the search for objective knowledge about baseball.

For the past two generations, if not longer, sports analytics has dominated the world of professional athletics. Using myriad statistical measures, bolstered by computer technology, tactics and strategies have been influenced "by the numbers".

In Chapter 11 of this book, we present an introduction to the development of sabermetrics and how it is firmly entrenched in the culture of baseball, both in and out of the classroom. We give an overview of the topic and present some limitations dealing with some of the traditional metrics, which were used in the past. The reader will be able to access myriad data, primarily provided by websites, and he/she will appreciate the "matrix format" of programs, such as Excel, which are of great value in the field of sabermetrics. We also provide an introductory informal bibliography of this emerging field of study.

In Chapter 12, we delve into the "heart" of sabermetrics. How does one *do* sabermetrics? The reader will no doubt appreciate the applicability of such topics as "least squares" straight lines and probability, which are covered in prior chapters. This "module chapter" can serve as a springboard for educators considering offering a course on sabermetrics. Regarding this, readers are encouraged to contact the junior author, Gabe Costa, if any questions arise or assistance is sought regarding the teaching of sabermetrics. Gabe can be reached at the following e-mail address: Gabriel.costa@shu.edu.

It is assumed that, for these last two chapters, the reader is not only familiar with the basics of baseball, but also has a passion for the sport.

Before closing, the aforementioned junior author would like to make an acknowledgment. I would like to thank the senior author for years of friendship and for being a first-rate mentor: *Professor Richard Bronson*. I met Richard during my freshman year at Stevens Institute of Technology in 1967 and I've been learning from him ever since.

Richard Bronson
Gabriel B. Costa

About the authors

Richard Bronson is emeritus professor of mathematics at Fairleigh Dickinson University, where he served as department chairman, acting dean of his college, interim provost of his campus, university director of Government Affairs, and senior executive assistant to the President. He has authored or coauthored 11 books in mathematics, over 30 articles, primarily in mathematical modeling, and a novel.

Gabriel B. Costa is a Catholic priest, presently on an extended academic leave from Seton Hall University. He is currently serving at the United States Military Academy at West Point, where he is a professor of mathematics and an associate chaplain. His research interests are closed formed solutions of ordinary differential equations, mathematical education, and sabermetrics. He has coauthored several books with Richard Bronson.

Matrices

Chapter outline

1.1 Basic concepts

Definition 1. A *matrix* is a rectangular array of elements arranged in horizontal rows and vertical columns. Thus,

$$\begin{bmatrix} 1 & 3 & 5 \\ 2 & 0 & -1 \end{bmatrix}, \tag{1}$$

$$\begin{bmatrix} 4 & 1 & 1 \\ 3 & 2 & 1 \\ 0 & 4 & 2 \end{bmatrix}, \tag{2}$$

and

$$\begin{bmatrix} \sqrt{2} \\ \pi \\ 19.5 \end{bmatrix} \tag{3}$$

are all examples of a matrix.

Matrix Methods. https://doi.org/10.1016/B978-0-12-818419-6.00001-0

The matrix given in (1) has two rows and three columns; it is said to have *order* (or size) 2×3 (read two by three). By convention, the row index is always given first. The matrix in (2) has order 3×3, while that in (3) has order 3×1. The entries of a matrix are called *elements*.

In general, a matrix **A** (matrices will always be designated by uppercase boldface letters) of order $p \times n$ is given by

$$\mathbf{A} = \begin{bmatrix} a_{11} & a_{12} & a_{13} & \cdots & a_{1n} \\ a_{21} & a_{22} & a_{23} & \cdots & a_{2n} \\ a_{31} & a_{32} & a_{33} & \cdots & a_{3n} \\ \vdots & \vdots & \vdots & & \vdots \\ a_{p1} & a_{p2} & a_{p3} & \cdots & a_{pn} \end{bmatrix}, \tag{4}$$

which is often abbreviated to $[a_{ij}]_{p \times n}$ or just $[a_{ij}]$. In this notation, a_{ij} represents the general element of the matrix and appears in the *i*th row and the *j*th column. The subscript *i*, which represents the row, can have any value 1 through *p*, while the subscript *j*, which represents the column, runs 1 through *n*. Thus, if $i = 2$ and $j = 3$, a_{ij} becomes a_{23} and designates the element in the second row and third column. If $i = 1$ and $j = 5$, a_{ij} becomes a_{15} and signifies the element in the first row, fifth column. Note again that the row index is always given before the column index.

Any element having its row index equal to its column index is a *diagonal element*. Thus, the diagonal elements of a matrix are the elements in the $1-1$ position, $2-2$ position, $3-3$ position, and so on, for as many elements of this type that exist. Matrix (1) has 1 and 0 as its diagonal elements, while matrix (2) has 4, 2, and 2 as its diagonal elements.

If the matrix has as many rows as columns, $p = n$, it is called a *square matrix*; in general it is written as

$$\begin{bmatrix} a_{11} & a_{12} & a_{13} & \cdots & a_{1n} \\ a_{21} & a_{22} & a_{23} & \cdots & a_{2n} \\ a_{31} & a_{32} & a_{33} & \cdots & a_{3n} \\ \vdots & \vdots & \vdots & & \vdots \\ a_{n1} & a_{n2} & a_{n3} & \cdots & a_{nn} \end{bmatrix}. \tag{5}$$

In this case, the elements $a_{11}, a_{22}, a_{33}, \ldots, a_{nn}$ lie on and form the *main* (or principal) *diagonal*.

It should be noted that the elements of a matrix need not be numbers; they can be, and quite often arise physically as, functions, operators or, as we shall see later, matrices themselves. Hence,

$$\begin{bmatrix} \int_0^1 (t^2 + 1)\,dt & t^2 \sqrt{3t} \ 2 \end{bmatrix}, \quad \begin{bmatrix} \sin \theta & \cos \theta \\ -\cos \theta & \sin \theta \end{bmatrix},$$

and

$$\begin{bmatrix} x^2 & x \\ e^x & \dfrac{d}{dx}\ln x \\ 5 & x+2 \end{bmatrix}$$

are good examples of matrices. Finally, it must be noted that a matrix is an entity unto itself; it is not a number. If the reader is familiar with determinants, he or she will undoubtedly recognize the similarity in form between the two. *Warning*: the similarity ends there. Whereas a determinant (see Chapter 5) can be evaluated to yield a number, a matrix cannot. A matrix is a rectangular array, period.

Problems 1.1

1. Determine the orders of the following matrices:

$$A = \begin{bmatrix} 3 & 1 & -2 & 4 & 7 \\ 2 & 5 & -6 & 5 & 7 \\ 0 & 3 & 1 & 2 & 0 \\ -3 & -5 & 2 & 2 & 2 \end{bmatrix}, \quad B = \begin{bmatrix} 1 & 2 & 3 \\ 0 & 0 & 0 \\ 4 & 3 & 2 \end{bmatrix},$$

$$C = \begin{bmatrix} 1 & 2 & 3 & 4 \\ 5 & 6 & -7 & 8 \\ 10 & 11 & 12 & 12 \end{bmatrix}, \quad D = \begin{bmatrix} 3 & t & t^2 & 0 \\ t-2 & t^4 & 6t & 5 \\ t+2 & 3t & 1 & 2 \\ 2t-3 & -5t^2 & 2t^5 & 3t^2 \end{bmatrix},$$

$$E = \begin{bmatrix} \dfrac{1}{2} & \dfrac{1}{3} & \dfrac{1}{4} \\ \dfrac{2}{3} & \dfrac{3}{5} & -\dfrac{5}{6} \end{bmatrix}, \quad F = \begin{bmatrix} 1 \\ 5 \\ 10 \\ 0 \\ -4 \end{bmatrix}, \quad G = \begin{bmatrix} \sqrt{313} & -505 \\ 2\pi & 18 \\ 46.3 & 1.043 \\ 2\sqrt{5} & -\sqrt{5} \end{bmatrix},$$

$$H = \begin{bmatrix} 0 & 0 \\ 0 & 0 \end{bmatrix}, \quad J = \begin{bmatrix} 1 & 5 & -30 \end{bmatrix}.$$

2. Find, if they exist, the elements in the $1-3$ and the $2-1$ positions for each of the matrices defined in Problem 1.

3. Find, if they exist, a_{23}, a_{32}, b_{31}, b_{32}, c_{11}, d_{22}, e_{13}, g_{22}, g_{23}, and h_{32} for the matrices defined in Problem 1.

4. Construct the 2×2 matrix $\mathbf{A}$ having $a_{ij} = (-1)^{i+j}$.

5. Construct the 3×3 matrix $\mathbf{A}$ having $a_{ij} = i/j$.

6. Construct the $n \times n$ matrix $\mathbf{B}$ having $b_{ij} = n - i - j$. What will this matrix be when specialized to the 3×3 case?

7. Construct the 2×4 matrix $\mathbf{C}$ having

$$c_{ij} = \begin{cases} i & \text{when } i = 1, \\ j & \text{when } i = 2. \end{cases}$$

8. Construct the 3×4 matrix $\mathbf{D}$ having

$$d_{ij} = \begin{cases} i+j & \text{when } i > j, \\ 0 & \text{when } i = j, \\ i-j & \text{when } i < j. \end{cases}$$

9. Express the following times as matrices: (a) A quarter after nine in the morning. (b) Noon. (c) One thirty in the afternoon. (d) A quarter after nine in the evening.

10. Express the following dates as matrices:
   (a) July 4, 1776
   (b) December 7, 1941
   (c) April 23, 1809
   (d) October 31, 1688

11. A gasoline station currently has in inventory 950 gallons of regular unleaded gasoline, 1253 gallons of premium and 98 gallons of super. Express this inventory as a matrix.

12. Store 1 of a three-store chain has three refrigerators, five stoves, three washing machines, and four dryers in stock. Store 2 has in stock no refrigerators, two stoves, nine washing machines, and five dryers, while Store 3 has in stock four refrigerators, two stoves, and no washing machines or dryers. Present the inventory of the entire chain as a matrix.

13. The number of damaged items delivered by the SleepTight Mattress Company from its various plants during the past year is given by the matrix

$$\begin{bmatrix} 80 & 12 & 16 \\ 50 & 40 & 16 \\ 90 & 10 & 50 \end{bmatrix}.$$

The rows pertain to its three plants in Michigan, Texas, and Utah. The columns pertain to its regular model, its firm model, and its extra firm model, respectively. The company's goal for next year is to reduce by 10% the number of damaged regular mattresses shipped by each plant, to reduce by 20% the number of damaged firm mattresses shipped by its Texas plant, to reduce by 30% the number of damaged extra firm mattresses shipped by its Utah plant, and to keep all other entries the same as last year. What will next year's matrix be if all goals are realized?

14. A person purchased 100 shares of AT&T at $27 per share, 150 shares of Exxon at $45 per share, 50 shares of IBM at $116 per share, and 500 shares of PanAm at $2 per share. The current price of each stock is $29, $41, $116, and $3, respectively. Represent in a matrix all the relevant information regarding this person's portfolio.

15. On January 1, a person buys three certificates of deposit from different institutions, all maturing in 1 year. The first is for $1000 at 7%, the second is for $2000 at 7.5%, and the third is for $3000 at 7.25%. All interest rates are effective on an annual basis.

   (a) Represent in a matrix all the relevant information regarding this person's holdings.

   (b) What will the matrix be 1 year later if each certificate of deposit is renewed for the current face amount and accrued interest at rates one-half a percent higher than the present?

16. (**Markov Chains,** see Chapter 9) A finite Markov chain is a set of objects, a set of consecutive time periods, and a finite set of different states such that

   (a) during any given time period, each object is in only state (although different objects can be in different states), and

   (b) the probability that an object will move from one state to another state (or remain in the same state) over a time period depends only on the beginning and ending states.

A Markov chain can be represented by a matrix $\mathbf{P} = [p_{ij}]$ where p_{ij} represents the probability of an object moving from state i to state j in one time period. Such a matrix is called a *transition matrix*.

Start a transition matrix for the following Markov chain: Census figures show a population shift away from a large midwestern metropolitan city to its suburbs. Each year, 5% of all families living in the city move to the suburbs, while during the same time period only 1% of those living in the suburbs move into the city. *Hint*: Take state 1 to represent families living in the city, state 2 to represent families living in the suburbs, and one time period to equal a year.

17. Construct a transition matrix for the following Markov chain: Every 4 years, voters in a New England town elect a new mayor because a town ordinance prohibits mayors from succeeding themselves. Past data indicate that a Democratic mayor is succeeded by another Democrat 30% of the time and by a Republican 70% of the time. A Republican mayor, however, is succeeded by another Republican 60% of the time and by a Democrat 40% of the time. *Hint*: Take state 1 to represent a Republican mayor in office, state 2 to represent a Democratic mayor in office, and one time period to be 4 years.

18. Construct a transition matrix for the following Markov chain: The apple harvest in New York orchards is classified as poor, average, or good. Historical data indicates that if the harvest is poor 1 year then there is a 40% chance of having a good harvest the next year, a 50% chance of having an average harvest, and a 10% chance of having another poor harvest. If a harvest is average 1 year, the chance of a poor, average, or good harvest the next year is 20%, 60%, and 20%, respectively. If a harvest is good, then the chance of a poor, average, or good harvest the next year is 25%, 65%, and 10%, respectively. *Hint*: Take state 1 to be a poor harvest, state 2 to be an average harvest, state 3 to be a good harvest, and one time period to equal 1 year.

19. Construct a transition matrix for the following Markov chain. Brand X and brand Y control the majority of the soap powder market in a particular region and each has promoted its own product extensively. As a result of past advertising campaigns, it is known that over a 2-year period of time 10% of brand Y customers change to brand X and 25% of all other customers change to brand X. Furthermore, 15% of brand X customers change to brand Y and 30% of all other customers change to brand Y. The major brands also lose customers to smaller competitors, with 5% of brand X customers switching to a minor brand during a 2-year time period and 2% of brand Y customers doing likewise. All other customers remain loyal to their past brand of soap powder. *Hint*: Take state 1 to be a brand X customer, state 2 a brand Y customer, state 3 another brand customer, and one time period to be 2 years.

1.2 Operations

The simplest relationship between two matrices is equality. Intuitively one feels that two matrices should be equal if their corresponding elements are equal. This is the case, providing the matrices are of the same order.

Definition 1. Two matrices $\mathbf{A} = [a_{ij}]_{p \times n}$ and $\mathbf{B} = [b_{ij}]_{p \times n}$ are equal if they have the same order and if $a_{ij} = b_{ij}$ ($i = 1, 2, 3, ..., p; j = 1, 2, 3, ..., n$). Thus, the equality

$$\begin{bmatrix} 5x + 2y \\ x - 3y \end{bmatrix} = \begin{bmatrix} 7 \\ 1 \end{bmatrix}$$

implies that $5x + 2y = 7$ and $x - 3y = 1$.

The intuitive definition for matrix addition is also the correct one.

Definition 2. If $\mathbf{A} = [a_{ij}]$ and $\mathbf{B} = [b_{ij}]$ are both of order $p \times n$, then $\mathbf{A} + \mathbf{B}$ is a $p \times n$ matrix $\mathbf{C} = [c_{ij}]$, where $c_{ij} = a_{ij} + b_{ij}$ ($i = 1, 2, 3, ..., p; j = 1, 2, 3, ..., n$). Thus,

$$\begin{bmatrix} 5 & 1 \\ 7 & 3 \\ -2 & -1 \end{bmatrix} + \begin{bmatrix} -6 & 3 \\ 2 & -1 \\ 4 & 1 \end{bmatrix} = \begin{bmatrix} 5+(-6) & 1+3 \\ 7+2 & 3+(-1) \\ (-2)+4 & (-1)+1 \end{bmatrix} = \begin{bmatrix} -1 & 4 \\ 9 & 2 \\ 2 & 0 \end{bmatrix}$$

and

$$\begin{bmatrix} t^2 & 5 \\ 3t & 0 \end{bmatrix} + \begin{bmatrix} 1 & -6 \\ t & -t \end{bmatrix} = \begin{bmatrix} t^2+1 & -1 \\ 4t & -t \end{bmatrix};$$

but the matrices

$$\begin{bmatrix} 5 & 0 \\ -1 & 0 \\ 2 & 1 \end{bmatrix} \quad \text{and} \quad \begin{bmatrix} -6 & 2 \\ 1 & 1 \end{bmatrix}$$

cannot be added since they are not of the same order.

It is not difficult to show that the addition of matrices is both commutative and associative: that is, if **A**, **B**, **C** represent matrices of the same order, then

(A1) $\mathbf{A} + \mathbf{B} = \mathbf{B} + \mathbf{A}$.

(A2) $\mathbf{A} + (\mathbf{B} + \mathbf{C}) = (\mathbf{A} + \mathbf{B}) + \mathbf{C}$.

We define a zero matrix **0** to be a matrix consisting of only zero elements. Zero matrices of every order exist, and when one has the same order as another matrix **A**, we then have the additional property

(A3) $\mathbf{A} + \mathbf{0} = \mathbf{A}$.

Subtraction of matrices is defined in a manner analogous to addition: the orders of the matrices involved must be identical and the operation is performed elementwise.

Thus,

$$\begin{bmatrix} 5 & 1 \\ -3 & 2 \end{bmatrix} - \begin{bmatrix} 6 & -1 \\ 4 & -1 \end{bmatrix} = \begin{bmatrix} -1 & 2 \\ -7 & 3 \end{bmatrix}.$$

Another simple operation is that of multiplying a scalar times a matrix. Intuition guides one to perform the operation elementwise and once again intuition is correct. Thus, for example,

$$7\begin{bmatrix} 1 & 2 \\ -3 & 4 \end{bmatrix} = \begin{bmatrix} 7 & 14 \\ -21 & 28 \end{bmatrix} \quad \text{and} \quad t\begin{bmatrix} 1 & 0 \\ 3 & 2 \end{bmatrix} = \begin{bmatrix} t & 0 \\ 3t & 2t \end{bmatrix}.$$

Definition 3. If $\mathbf{A} = [a_{ij}]$ is a $p \times n$ matrix and if λ is a scalar, then $\lambda \mathbf{A}$ is a $p \times n$ matrix $\mathbf{B} = [b_{ij}]$, where $b_{ij} = \lambda a_{ij}$ $(i = 1, 2, 3, ..., p; j = 1, 2, 3, ..., n)$.

Example 1.   Find $5\mathbf{A} - \frac{1}{2}\mathbf{B}$ if

$$\mathbf{A} = \begin{bmatrix} 4 & 1 \\ 0 & 3 \end{bmatrix} \quad \text{and} \quad \mathbf{B} = \begin{bmatrix} 6 & -20 \\ 18 & 8 \end{bmatrix}$$

Solution

$$5\mathbf{A} - \frac{1}{2}\mathbf{B} = 5\begin{bmatrix} 4 & 1 \\ 0 & 3 \end{bmatrix} - \frac{1}{2}\begin{bmatrix} 6 & -20 \\ 18 & 8 \end{bmatrix}$$

$$= \begin{bmatrix} 20 & 5 \\ 0 & 15 \end{bmatrix} - \begin{bmatrix} 3 & -10 \\ 9 & 4 \end{bmatrix} = \begin{bmatrix} 17 & 15 \\ -9 & 11 \end{bmatrix}. \blacksquare$$

It is not difficult to show that if λ_1 and λ_2 are scalars, and if $\mathbf{A}$ and $\mathbf{B}$ are matrices of identical order, then

(S1) $\lambda_1 \mathbf{A} = \mathbf{A}\lambda_1$.
(S2) $\lambda_1 (\mathbf{A} + \mathbf{B}) = \lambda_1 \mathbf{A} + \lambda_1 \mathbf{B}$.
(S3) $(\lambda_1 + \lambda_2) \mathbf{A} = \lambda_1 \mathbf{A} + \lambda_2 \mathbf{A}$.
(S4) $\lambda_1 (\lambda_2 \mathbf{A}) = (\lambda_1 \lambda_2) \mathbf{A}$.

The reader is cautioned that there is *no* such operation as matrix division. We will, however, define a somewhat analogous operation, namely matrix inversion, in Chapter 3.

Problems 1.2

In Problems 1 through 26, let

$$\mathbf{A} = \begin{bmatrix} 1 & 2 \\ 3 & 4 \end{bmatrix}, \quad \mathbf{B} = \begin{bmatrix} 5 & 6 \\ 7 & 8 \end{bmatrix}, \quad \mathbf{C} = \begin{bmatrix} -1 & 0 \\ 3 & -3 \end{bmatrix},$$

$$\mathbf{D} = \begin{bmatrix} 3 & 1 \\ -1 & 2 \\ 3 & -2 \\ 2 & 6 \end{bmatrix}, \quad \mathbf{E} = \begin{bmatrix} -2 & 2 \\ 0 & -2 \\ 5 & -3 \\ 5 & 1 \end{bmatrix}, \quad \mathbf{F} = \begin{bmatrix} 0 & 1 \\ -1 & 0 \\ 0 & 0 \\ 2 & 2 \end{bmatrix}.$$

1. Find $2\mathbf{A}$.
2. Find $-5\mathbf{A}$.
3. Find $3\mathbf{D}$.
4. Find $10\mathbf{E}$.
5. Find $-\mathbf{F}$.
6. Find $\mathbf{A} + \mathbf{B}$.
7. Find $\mathbf{C} + \mathbf{A}$.

8. Find $\mathbf{D} + \mathbf{E}$.
9. Find $\mathbf{D} + \mathbf{F}$.
10. Find $\mathbf{A} + \mathbf{D}$.
11. Find $\mathbf{A} - \mathbf{B}$.
12. Find $\mathbf{C} - \mathbf{A}$.
13. Find $\mathbf{D} - \mathbf{E}$.
14. Find $\mathbf{D} - \mathbf{F}$.
15. Find $2\mathbf{A} + 3\mathbf{B}$.
16. Find $3\mathbf{A} - 2\mathbf{C}$.
17. Find $0.1\mathbf{A} + 0.2\mathbf{C}$.
18. Find $-2\mathbf{E} + \mathbf{F}$.
19. Find $\mathbf{X}$ if $\mathbf{A} + \mathbf{X} = \mathbf{B}$.
20. Find $\mathbf{Y}$ if $2\mathbf{B} + \mathbf{Y} = \mathbf{C}$.
21. Find $\mathbf{X}$ if $3\mathbf{D} - \mathbf{X} = \mathbf{E}$.
22. Find $\mathbf{Y}$ if $\mathbf{E} - 2\mathbf{Y} = \mathbf{F}$.
23. Find $\mathbf{R}$ if $4\mathbf{A} + 5\mathbf{R} = 10\mathbf{C}$.
24. Find $\mathbf{S}$ if $3\mathbf{F} - 2\mathbf{S} = \mathbf{D}$.
25. Verify directly that $(\mathbf{A} + \mathbf{B}) + \mathbf{C} = \mathbf{A} + (\mathbf{B} + \mathbf{C})$.
26. Verify directly that $\lambda(\mathbf{A} + \mathbf{B}) = \lambda\mathbf{A} + \lambda\mathbf{B}$.
27. Find $6\mathbf{A} - \theta\mathbf{B}$ if

$$\mathbf{A} = \begin{bmatrix} \theta^2 & 2\theta - 1 \\ 4 & 1/\theta \end{bmatrix} \quad \text{and} \quad \mathbf{B} = \begin{bmatrix} \theta^2 - 1 & 6 \\ 3/\theta & \theta^3 + 2\theta + 1 \end{bmatrix}.$$

28. Prove Property (A1).
29. Prove Property (A3).
30. Prove Property (S2).
31. Prove Property (S3).
32. **(a)** Mr. Jones owns 200 shares of IBM and 150 shares of AT&T. Determine a portfolio matrix that reflects Mr. Jones' holdings.
   (b) Over the next year, Mr. Jones triples his holdings in each company. What is his new portfolio matrix?
   (c) The following year Mr. Jones lists changes in his portfolio as $[-50 \ 100]$. What is his new portfolio matrix?
33. The inventory of an appliance store can be given by a 1×4 matrix in which the first entry represents the number of television sets, the second entry the number of air conditioners, the third entry the number of refrigerators, and the fourth entry the number of dishwashers.
   (a) Determine the inventory given on January 1 by $[15 \ 2 \ 8 \ 6]$.
   (b) January sales are given by $[4 \ 0 \ 2 \ 3]$. What is the inventory matrix on February 1?

(c) February sales are given by [5 0 3 3], and new stock added in February is given by [3 2 7 8]. What is the inventory matrix on March 1?

34. The daily gasoline supply of a local service station is given by a 1×3 matrix in which the first entry represents gallons of regular, the second entry gallons of premium, and the third entry gallons of super.

(a) Determine the supply of gasoline at the close of business on Monday given by [14,000 8000 6000].

(b) Tuesday's sales are given by [3500 2000 1500]. What is the inventory matrix at day's end?

(c) Wednesday's sales are given by [5000 1500 1200]. In addition, the station received a delivery of 30,000 gallons of regular, 10,000 gallons of premium, but no super. What is the inventory at day's end?

35. On a recent shopping trip, Mary purchased 6 oranges, a dozen grapefruits, 8 apples, and 3 lemons. John purchased 9 oranges, 2 grapefruits, and 6 apples. Express each of their purchases as 1×4 matrices. What is the physical significance of the sum of these matrices?

1.3 Matrix multiplication

Matrix multiplication is the first operation we encounter where our intuition fails. First, two matrices are *not* multiplied together elementwise. Second, it is not always possible to multiply matrices of the same order while it is possible to multiply certain matrices of different orders. Third, if **A** and **B** are two matrices for which multiplication is defined, it is generally not the case that **AB** = **BA**; that is, *matrix multiplication is not a commutative operation*. There are other properties of matrix multiplication, besides the three mentioned, that defy our intuition, and we shall illustrate them shortly. We begin by determining which matrices can be multiplied.

Rule 1. The product of two matrices **AB** is defined if the number of columns of **A** equals the number of rows of **B**.

Thus, if **A** and **B** are given by

$$A = \begin{bmatrix} 6 & 1 & 0 \\ -1 & 2 & 1 \end{bmatrix} \quad \text{and} \quad B = \begin{bmatrix} -1 & 0 & 1 & 0 \\ 3 & 2 & -2 & 1 \\ 4 & 1 & 1 & 0 \end{bmatrix}, \tag{6}$$

then, the product **AB** is defined since **A** has three columns and **B** has three rows. The product **BA**, however, is not defined since **B** has four columns while **A** has only two rows.

When the product is written **AB**, **A** is said to *premultiply* **B**, while **B** is said to *postmultiply* **A**.

Rule 2. If the product **AB** is defined, then the resultant matrix will have the same number of rows as **A** and the same number of columns as **B**.

Thus, the product **AB**, where **A** and **B** are given in (6), will have two rows and four columns since **A** has two rows and **B** has four columns.

An easy method of remembering these two rules is the following: write the orders of the matrices on paper in the sequence in which the multiplication is to be carried out; that is, if **AB** is to be found where **A** has order 2×3 and **B** has order 3×4, write

$$(2 \times 3)(3 \times 4). \tag{7}$$

If the two adjacent numbers (indicated in (7) by the curved arrow) are both equal (in the case they are both 3 the multiplication is defined. The order of the product matrix is obtained by cancelling the adjacent numbers and using the two remaining numbers. Thus, in (7), we cancel the adjacent 3s and are left with 2×4, which in this case is the order of **AB**.

As a further example, consider the case where **A** is a 4×3 matrix, while **B** is a 3×5 matrix. The product **AB** is defined since, in the notation $(4 \times 3)\,(3 \times 5)$, the adjacent numbers denoted by the curved arrow are equal. The product will be a 4×5 matrix. The product **BA**, however, is not defined since in the notation $(3 \times 5)\,(4 \times 3)$ the adjacent numbers are not equal. In general, one may schematically state the method as

$$(k \times n)(n \times p) = (k \times p).$$

Rule 3. If the product **AB** $=$ **C** is defined, where **C** is denoted by $[c_{ij}]$, then the element c_{ij} is obtained by multiplying the elements in the ith row of **A** by the corresponding elements in the jth column of **B** and adding. Thus, if **A** has order $k \times n$, and **B** has order $n \times p$ and

$$
\begin{bmatrix} a_{11} & a_{12} & \cdots & a_{1n} \\ a_{21} & a_{22} & \cdots & a_{2n} \\ \vdots & \vdots & & \vdots \\ a_{k1} & a_{k2} & \cdots & a_{kn} \end{bmatrix}
\begin{bmatrix} b_{11} & b_{12} & \cdots & b_{1p} \\ b_{21} & b_{22} & \cdots & b_{2p} \\ \vdots & \vdots & & \vdots \\ b_{n1} & b_{n2} & \cdots & b_{np} \end{bmatrix}
\begin{bmatrix} c_{11} & c_{12} & \cdots & c_{1p} \\ c_{21} & c_{22} & \cdots & c_{2p} \\ \vdots & \vdots & & \vdots \\ c_{k1} & c_{k2} & \cdots & c_{kp} \end{bmatrix},
$$

then c_{11} is obtained by multiplying the elements in the first row of **A** by the corresponding elements in the first column of **B** and adding; hence,

$$c_{11} = a_{11}b_{11} + a_{12}b_{21} + \cdots + a_{1n}b_{n1}.$$

The element c_{12} is found by multiplying the elements in the first row of **A** by the corresponding elements in the second column of **B** and adding; hence,

$$c_{12} = a_{11}b_{12} + a_{12}b_{22} + \cdots + a_{1n}b_{n2}.$$

The element c_{kp} is obtained by multiplying the elements in the kth row of **A** by the corresponding elements in the pth column of **B** and adding; hence,

$$c_{kp} = a_{k1}b_{1p} + a_{k2}b_{2p} + \cdots + a_{kn}b_{np}.$$

Example 1.   Find **AB** and **BA** if

$$A = \begin{bmatrix} 1 & 2 & 3 \\ 4 & 5 & 6 \end{bmatrix} \quad \text{and} \quad B = \begin{bmatrix} -7 & -8 \\ 9 & 10 \\ 0 & -11 \end{bmatrix}.$$

Solution

$$AB = \begin{bmatrix} 1 & 2 & 3 \\ 4 & 5 & 6 \end{bmatrix} \begin{bmatrix} -7 & -8 \\ 9 & 10 \\ 0 & -11 \end{bmatrix}$$

$$= \begin{bmatrix} 1(-7) + 2(9) + 3(0) & 1(-8) + 2(10) + 3(-11) \\ 4(-7) + 5(9) + 6(0) & 4(-8) + 5(10) + 6(-11) \end{bmatrix}$$

$$= \begin{bmatrix} -7 + 18 + 0 & -8 + 20 - 33 \\ -28 + 45 + 0 & -32 + 50 - 66 \end{bmatrix} = \begin{bmatrix} 11 & -21 \\ 17 & -48 \end{bmatrix},$$

$$BA = \begin{bmatrix} -7 & -8 \\ 9 & 10 \\ 0 & -11 \end{bmatrix} \begin{bmatrix} 1 & 2 & 3 \\ 4 & 5 & 6 \end{bmatrix}$$

$$= \begin{bmatrix} (-7)1 + (-8)4 & (-7)2 + (-8)5 & (-7)3 + (-8)6 \\ 9(1) + 10(4) & 9(2) + 10(5) & 9(3) + 10(6) \\ 0(1) + (-11)4 & 0(2) + (-11)5 & 0(3) + (-11)6 \end{bmatrix}$$

$$= \begin{bmatrix} -7 - 32 & -14 - 40 & -21 - 48 \\ 9 + 40 & 18 + 50 & 27 + 60 \\ 0 - 44 & 0 - 55 & 0 - 66 \end{bmatrix} = \begin{bmatrix} -39 & -54 & -69 \\ 49 & 68 & 87 \\ -44 & -55 & -66 \end{bmatrix}. \quad \blacksquare$$

The preceding three rules can be incorporated into the following formal definition:

Definition 1.   If $A = [a_{ij}]$ is a $k \times n$ matrix and $B = [b_{ij}]$ is an $n \times p$ matrix, then the product **AB** is defined to be a $k \times p$ matrix $C = [c_{ij}]$, where $c_{ij} = \sum_{l=1}^{n} a_{il}b_{lj} = a_{i1}b_{1j} + a_{i2}b_{2j} + \cdots + a_{in}b_{nj}$ ($i = 1, 2, \ldots, k; j = 1, 2, \ldots, p$).

Example 2.   Find **AB** if

$$A = \begin{bmatrix} 2 & 1 \\ -1 & 0 \\ 3 & 1 \end{bmatrix} \quad \text{and} \quad B = \begin{bmatrix} 3 & 1 & 5 & -1 \\ 4 & -2 & 1 & 0 \end{bmatrix}.$$

Solution

$$AB = \begin{bmatrix} 2 & 1 \\ -1 & 0 \\ 3 & 1 \end{bmatrix} \begin{bmatrix} 3 & 1 & 5 & -1 \\ 4 & -2 & 1 & 0 \end{bmatrix}$$

$$= \begin{bmatrix} 2(3)+1(4) & 2(1)+1(-2) & 2(5)+1(1) & 2(-1)+1(0) \\ -1(3)+0(4) & -1(1)+0(-2) & -1(5)+0(1) & -1(-1)+0(0) \\ 3(3)+1(4) & 3(1)+1(-2) & 3(5)+1(1) & 3(-1)+1(0) \end{bmatrix}$$

$$= \begin{bmatrix} 10 & 0 & 11 & -2 \\ -3 & -1 & -5 & 1 \\ 13 & 1 & 16 & -3 \end{bmatrix}.$$

Note that in this example the product **BA** is not defined. ■

Example 3.   Find **AB** and **BA** if

$$A = \begin{bmatrix} 2 & 1 \\ -1 & 3 \end{bmatrix} \quad \text{and} \quad B = \begin{bmatrix} 4 & 0 \\ 1 & 2 \end{bmatrix}.$$

Solution

$$AB = \begin{bmatrix} 2 & 1 \\ -1 & 3 \end{bmatrix} \begin{bmatrix} 4 & 0 \\ 1 & 2 \end{bmatrix} = \begin{bmatrix} 2(4)+1(1) & 2(0)+1(2) \\ -1(4)+3(1) & -1(0)+3(2) \end{bmatrix} = \begin{bmatrix} 9 & 2 \\ -1 & 6 \end{bmatrix};$$

$$BA = \begin{bmatrix} 4 & 0 \\ 1 & 2 \end{bmatrix} \begin{bmatrix} 2 & 1 \\ -1 & 3 \end{bmatrix} = \begin{bmatrix} 4(2)+0(-1) & 4(1)+0(3) \\ 1(2)+2(-1) & 1(1)+2(3) \end{bmatrix} = \begin{bmatrix} 8 & 4 \\ 0 & 7 \end{bmatrix}.$$

This, therefore, is an example where both products **AB** and **BA** are defined but unequal. ■

Example 4.   Find **AB** and **BA** if

$$A = \begin{bmatrix} 3 & 1 \\ 0 & 4 \end{bmatrix} \quad \text{and} \quad B = \begin{bmatrix} 1 & 1 \\ 0 & 2 \end{bmatrix}.$$

Solution

$$AB = \begin{bmatrix} 3 & 1 \\ 0 & 4 \end{bmatrix} \begin{bmatrix} 1 & 1 \\ 0 & 2 \end{bmatrix} = \begin{bmatrix} 3 & 5 \\ 0 & 8 \end{bmatrix},$$

$$BA = \begin{bmatrix} 1 & 1 \\ 0 & 2 \end{bmatrix} \begin{bmatrix} 3 & 1 \\ 0 & 4 \end{bmatrix} = \begin{bmatrix} 3 & 5 \\ 0 & 8 \end{bmatrix}.$$

This, therefore, is an example where both products **AB** and **BA** are defined and equal. ■

In general, it can be shown that matrix multiplication has the following properties:

(M1)	**A(BC) = (AB)C**	(Associative Law)
(M2)	**A(B + C) = AB + AC**	(Left Distributive Law)
(M3)	**(B + C)A = BA + CA**	(Right Distributive Law)

providing that the matrices **A, B, C** have the correct order so that the above multiplications and additions are defined. The one basic property that matrix multiplication does not possess is commutativity; that is, in general, **AB** does not equal **BA** (see Example 3). We hasten to add, however, that while matrices in general do not commute, it may very well be the case that, given two particular matrices, they do commute as can be seen from Example 4.

Commutativity is not the only property that matrix multiplication lacks. We know from our experiences with real numbers that if the product $xy = 0$, then either $x = 0$ or $y = 0$ or both are zero. Matrices do not possess this property as the following example shows:

Example 5. Find **AB** if

$$A = \begin{bmatrix} 4 & 2 \\ 2 & 1 \end{bmatrix} \quad \text{and} \quad B = \begin{bmatrix} 3 & -4 \\ -6 & 8 \end{bmatrix}.$$

Solution

$$AB = \begin{bmatrix} 4 & 2 \\ 2 & 1 \end{bmatrix} \begin{bmatrix} 3 & -4 \\ -6 & 8 \end{bmatrix} = \begin{bmatrix} 4(3) + 2(-6) & 4(-4) + 2(8) \\ 2(3) + 1(-6) & 2(-4) + 1(8) \end{bmatrix}$$
$$= \begin{bmatrix} 0 & 0 \\ 0 & 0 \end{bmatrix}.$$

Thus, even though neither **A** nor **B** is zero, their product is zero. ■

One final "unfortunate" property of matrix multiplication is that the equation **AB = AC** does not imply **B = C**.

Example 6. Find **AB** and **AC** if

$$A = \begin{bmatrix} 4 & 2 \\ 2 & 1 \end{bmatrix}, \quad B = \begin{bmatrix} 1 & 1 \\ 2 & 1 \end{bmatrix}, \quad C = \begin{bmatrix} 2 & 2 \\ 0 & -1 \end{bmatrix}.$$

Solution

$$AB = \begin{bmatrix} 4 & 2 \\ 2 & 1 \end{bmatrix} \begin{bmatrix} 1 & 1 \\ 2 & 1 \end{bmatrix} = \begin{bmatrix} 4(1) + 2(2) & 4(1) + 2(1) \\ 2(1) + 1(2) & 2(1) + 1(1) \end{bmatrix} = \begin{bmatrix} 8 & 6 \\ 4 & 3 \end{bmatrix};$$
$$AC = \begin{bmatrix} 4 & 2 \\ 2 & 1 \end{bmatrix} \begin{bmatrix} 2 & 2 \\ 0 & -1 \end{bmatrix} = \begin{bmatrix} 4(2) + 2(0) & 4(2) + 2(-1) \\ 2(2) + 1(0) & 2(2) + 1(-1) \end{bmatrix} = \begin{bmatrix} 8 & 6 \\ 4 & 3 \end{bmatrix}.$$

Thus, cancellation is not a valid operation in the matrix algebra. ■

The reader has no doubt wondered why this seemingly complicated procedure for matrix multiplication has been introduced when the more obvious methods of multiplying matrices termwise could be used. The answer lies in systems of simultaneous linear equations. Consider the set of simultaneous linear equations given by

$$5x - 3y + 2z = 14,$$
$$x + y - 4z = -7,$$
$$7x \quad - 3z = 1.$$

(8)

This system can easily be solved by the method of substitution. Matrix algebra, however, will give us an entirely new method for obtaining the solution.

Consider the matrix equation

$$\mathbf{Ax = b}$$

(9)

where

$$\mathbf{A} = \begin{bmatrix} 5 & -3 & 2 \\ 1 & 1 & -4 \\ 7 & 0 & -3 \end{bmatrix}, \quad \mathbf{x} = \begin{bmatrix} x \\ y \\ z \end{bmatrix}, \quad \text{and} \quad \mathbf{b} = \begin{bmatrix} 14 \\ -7 \\ 1 \end{bmatrix}.$$

Here $\mathbf{A}$, called the *coefficient matrix*, is simply the matrix whose elements are the coefficients of the unknowns x, y, z in (8). (Note that we have been very careful to put all the x coefficients in the first column, all the y coefficients in the second column, and all the z coefficients in the third column. The zero in the $(3, 2)$ entry appears because the y coefficient in the third equation of system (8) is zero.) $\mathbf{x}$ and $\mathbf{b}$ are obtained in the obvious manner. One note of warning: there is a basic difference between the unknown matrix $\mathbf{x}$ in (9) and the unknown variable x. The reader should be especially careful not to confuse their respective identities.

Now using our definition of matrix multiplication, we have that

$$\mathbf{Ax} = \begin{bmatrix} 5 & -3 & 2 \\ 1 & 1 & -4 \\ 7 & 0 & -3 \end{bmatrix} \begin{bmatrix} x \\ y \\ z \end{bmatrix} = \begin{bmatrix} (5)(x) + (-3)(y) + (2)(z) \\ (1)(x) + (1)(y) + (-4)(z) \\ (7)(x) + (0)(y) + (-3)(z) \end{bmatrix}$$

$$= \begin{bmatrix} 5x - 3y + 2z \\ x + y - 4z \\ 7x \quad - 3z \end{bmatrix} = \begin{bmatrix} 14 \\ -7 \\ 1 \end{bmatrix}.$$

(10)

Using the definition of matrix equality, we see that (10) is precisely system (8). Thus, (9) is an alternate way of representing the original system. It should come as no surprise, therefore, that by redefining the matrices $\mathbf{A}$, $\mathbf{x}$, $\mathbf{b}$, appropriately, we can represent any system of simultaneous linear equations by the matrix equation $\mathbf{Ax = b}$.

Example 7. Put the following system into matrix form:

$$x - y + z + w = 5,$$
$$2x + y - z = 4,$$
$$3x + 2y + 2w = 0,$$
$$x - 2y + 3z + 4w = -1.$$

Solution Define

$$A = \begin{bmatrix} 1 & -1 & 1 & 1 \\ 2 & 1 & -1 & 0 \\ 3 & 2 & 0 & 2 \\ 1 & -2 & 3 & 4 \end{bmatrix}, \quad x = \begin{bmatrix} x \\ y \\ z \\ w \end{bmatrix}, \quad b = \begin{bmatrix} 5 \\ 4 \\ 0 \\ -1 \end{bmatrix}.$$

The original system is then equivalent to the matrix system $Ax = b$. ∎

Unfortunately, we are not yet in a position to solve systems that are in matrix form $Ax = b$. One method of solution depends upon the operation of inversion, and we must postpone a discussion of it until the inverse has been defined. For the present, however, we hope that the reader will be content with the knowledge that matrix multiplication, as we have defined it, does serve some useful purpose.

Problems 1.3

1. The order of A is 2×4, the order of B is 4×2, the order of C is 4×1, the order of D is 1×2, and the order of E is 4×4. Find the orders of

   (a) AB.
   (b) BA.
   (c) AC.
   (d) CA.
   (e) CD.
   (f) AE.
   (g) EB.
   (h) EA.
   (i) ABC.
   (j) DAE.
   (k) EBA.
   (l) EECD.

In Problems 2 through 19, let

$$\mathbf{A} = \begin{bmatrix} 1 & 2 \\ 3 & 4 \end{bmatrix}, \quad \mathbf{B} = \begin{bmatrix} 5 & 6 \\ 7 & 8 \end{bmatrix}, \quad \mathbf{C} = \begin{bmatrix} -1 & 0 & 1 \\ 3 & -2 & 1 \end{bmatrix},$$

$$\mathbf{D} = \begin{bmatrix} 1 & 1 \\ -1 & 2 \\ 2 & -2 \end{bmatrix}, \quad \mathbf{E} = \begin{bmatrix} -2 & 2 & 1 \\ 0 & -2 & -1 \\ 1 & 0 & 1 \end{bmatrix}, \quad \mathbf{F} = \begin{bmatrix} 0 & 1 & 2 \\ -1 & -1 & 0 \\ 1 & 2 & 3 \end{bmatrix},$$

$$\mathbf{X} = \begin{bmatrix} 1 & -2 \end{bmatrix}, \quad \mathbf{Y} = \begin{bmatrix} 1 & 2 & 1 \end{bmatrix}.$$

2. Find **AB**.
3. Find **BA**.
4. Find **AC**.
5. Find **BC**.
6. Find **CB**.
7. Find **XA**.
8. Find **XB**.
9. Find **XC**.
10. Find **AX**.
11. Find **CD**.
12. Find **DC**.
13. Find **YD**.
14. Find **YC**.
15. Find **DX**.
16. Find **XD**.
17. Find **EF**.
18. Find **FE**.
19. Find **YF**.
20. Find **AB** if

$$\mathbf{A} = \begin{bmatrix} 2 & 6 \\ 3 & 9 \end{bmatrix} \quad \text{and} \quad \mathbf{B} = \begin{bmatrix} 3 & -6 \\ -1 & 2 \end{bmatrix}.$$

Note that $\mathbf{AB} = \mathbf{0}$ but neither $\mathbf{A}$ nor $\mathbf{B}$ equals the zero matrix.

21. Find **AB** and **CB** if

$$\mathbf{A} = \begin{bmatrix} 3 & 2 \\ 1 & 0 \end{bmatrix}, \quad \mathbf{B} = \begin{bmatrix} 2 & 4 \\ 1 & 2 \end{bmatrix}, \quad \mathbf{C} = \begin{bmatrix} 1 & 6 \\ 3 & -4 \end{bmatrix}.$$

Thus, show that $\mathbf{AB} = \mathbf{CB}$ but $\mathbf{A} \neq \mathbf{C}$.

22. Compute the product

$$\begin{bmatrix} 1 & 2 \\ 3 & 4 \end{bmatrix} \begin{bmatrix} x \\ y \end{bmatrix}.$$

23. Compute the product

$$\begin{bmatrix} 1 & 0 & -1 \\ 3 & 1 & 1 \\ 1 & 3 & 0 \end{bmatrix} \begin{bmatrix} x \\ y \\ z \end{bmatrix}.$$

24. Compute the product

$$\begin{bmatrix} a_{11} & a_{12} \\ a_{21} & a_{22} \end{bmatrix} \begin{bmatrix} x \\ y \end{bmatrix}.$$

25. Compute the product

$$\begin{bmatrix} b_{11} & b_{12} & b_{13} \\ b_{21} & b_{22} & b_{23} \end{bmatrix} \begin{bmatrix} 2 \\ -1 \\ 3 \end{bmatrix}.$$

26. Evaluate the expression $\mathbf{A}^2 - 4\mathbf{A} - 5\mathbf{I}$ for the matrix*

$$\mathbf{A} = \begin{bmatrix} 1 & 2 \\ 4 & 3 \end{bmatrix}.$$

27. Evaluate the expression $(\mathbf{A} - \mathbf{I})(\mathbf{A} + 2\mathbf{I})$ for the matrix*

$$\mathbf{A} = \begin{bmatrix} 3 & 5 \\ -2 & 4 \end{bmatrix}.$$

28. Evaluate the expression $(\mathbf{I} - \mathbf{A})(\mathbf{A}^2 - \mathbf{I})$ for the matrix*

$$\mathbf{A} = \begin{bmatrix} 2 & -1 & 1 \\ 3 & -2 & 1 \\ 0 & 0 & 1 \end{bmatrix}.$$

---

*I is defined in Section 1.4.

29. Verify Property (M1) for

$$\mathbf{A} = \begin{bmatrix} 2 & 1 \\ 1 & 3 \end{bmatrix}, \quad \mathbf{B} = \begin{bmatrix} 0 & 1 \\ -1 & 4 \end{bmatrix}, \quad \mathbf{C} = \begin{bmatrix} 5 & 1 \\ 2 & 1 \end{bmatrix}.$$

30. Prove Property (M2).

31. Prove Property (M3).

32. Put the following system of equations into matrix form:

$$2x + 3y = 10,$$
$$4x - 5y = 11.$$

33. Put the following system of equations into matrix form:

$$x + z + y = 2,$$
$$3z + 2x + y = 4,$$
$$y + x = 0.$$

34. Put the following system of equations into matrix form:

$$5x + 3y + 2z + 4w = 5,$$
$$x + y + w = 0,$$
$$3x + 2y + 2z = -3,$$
$$x + y + 2z + 3w = 4.$$

35. The price schedule for a Chicago to Los Angeles flight is given by $\mathbf{P} = [200\ 350\ 500]$, where the matrix elements pertain, respectively, to coach tickets, business-class tickets, and first-class tickets. The number of tickets purchased in each category for a particular flight is given by

$$\mathbf{N} = \begin{bmatrix} 130 \\ 20 \\ 10 \end{bmatrix}.$$

Compute the products (a) $\mathbf{PN}$ and (b) $\mathbf{NP}$ and determine their significance.

36. The closing prices of a person's portfolio during the past week are given by the matrix

$$\mathbf{P} = \begin{bmatrix} 40 & 40\frac{1}{2} & 40\frac{7}{8} & 41 & 41 \\ 3\frac{1}{4} & 3\frac{5}{8} & 3\frac{1}{2} & 4 & 3\frac{7}{8} \\ 10 & 9\frac{3}{4} & 10\frac{1}{8} & 10 & 9\frac{5}{8} \end{bmatrix},$$

where the columns pertain to the days of the week, Monday through Friday, and the rows pertain to the prices of Orchard Fruits, Lion Airways, and Arrow Oil. The person's holdings in each of these companies are given by the matrix $\mathbf{H} =$ [100 500 400]. Compute the products (a) $\mathbf{HP}$ and (b) $\mathbf{PH}$ and determine their significance.

37. The time requirements for a company to produce three products is given by the matrix

$$\mathbf{T} = \begin{bmatrix} 0.2 & 0.5 & 0.4 \\ 1.2 & 2.3 & 0.7 \\ 0.8 & 3.1 & 1.2 \end{bmatrix},$$

where the rows pertain to lamp bases, cabinets, and tables, respectively. The columns pertain to the hours of labor required for cutting the wood, assembling, and painting, respectively. The hourly wages of a carpenter to cut wood, of a craftsperson to assemble a product, and of a decorator to paint is given, respectively, by the elements of the matrix

$$\mathbf{W} = \begin{bmatrix} 10.50 \\ 14.00 \\ 12.25 \end{bmatrix}.$$

Compute the product $\mathbf{TW}$ and determine its significance.

38. Continuing with the data given in the previous problem, assume further that the number of items on order for lamp bases, cabinets, and tables, respectively, is given by the matrix $\mathbf{O} =$ [1000 100 200]. Compute the product $\mathbf{OTW}$ and determine its significance.

39. The results of a flu epidemic at a college campus are collected in the matrix

$$\mathbf{F} = \begin{bmatrix} 0.20 & 0.20 & 0.15 & 0.15 \\ 0.10 & 0.30 & 0.30 & 0.40 \\ 0.70 & 0.50 & 0.55 & 0.45 \end{bmatrix}.$$

The elements denote percents converted to decimals. The columns pertain to freshmen, sophomores, juniors, and seniors, respectively, while the rows represent bedridden students, infected but ambulatory students, and well students, respectively. The male−female composition of each class is given by the matrix

$$\mathbf{C} = \begin{bmatrix} 1050 & 950 \\ 1100 & 1050 \\ 360 & 500 \\ 860 & 1000 \end{bmatrix}.$$

Compute the product $\mathbf{FC}$ and determine its significance.

1.4 Special matrices

There are certain types of matrices that occur so frequently that it becomes advisable to discuss them separately. One such type is the *transpose*. Given a matrix $\mathbf{A}$, the transpose of $\mathbf{A}$, denoted by $\mathbf{A}^T$ and read A-transpose, is obtained by changing all the rows of $\mathbf{A}$ into columns of $\mathbf{A}^T$ while preserving the order; hence, the first row of $\mathbf{A}$ becomes the first column of $\mathbf{A}^T$, while the second row of $\mathbf{A}$ becomes the second column of $\mathbf{A}^T$, and the last row of $\mathbf{A}$ becomes the last column of $\mathbf{A}^T$. Thus, if

$$\mathbf{A} = \begin{bmatrix} 1 & 2 & 3 \\ 4 & 5 & 6 \\ 7 & 8 & 9 \end{bmatrix}, \quad \text{then } \mathbf{A}^T = \begin{bmatrix} 1 & 4 & 7 \\ 2 & 5 & 8 \\ 3 & 6 & 9 \end{bmatrix}$$

and if

$$\mathbf{A} = \begin{bmatrix} 1 & 2 & 3 & 4 \\ 5 & 6 & 7 & 8 \end{bmatrix}, \quad \text{then } \mathbf{A}^T = \begin{bmatrix} 1 & 5 \\ 2 & 6 \\ 3 & 7 \\ 4 & 8 \end{bmatrix}.$$

Definition 1. If $\mathbf{A}$, denoted by $[a_{ij}]$ is an $n \times p$ matrix, then the transpose of $\mathbf{A}$, denoted by $\mathbf{A}^T = \left[a_{ij}^T \right]$, is a $p \times n$ matrix, where $a_{ij}^T = a_{ji}$.

It can be shown that the transpose possesses the following properties:

(1) $(\mathbf{A}^T)^T = \mathbf{A}$,
(2) $(\lambda \mathbf{A})^T = \lambda \mathbf{A}^T$ where λ represents a scalar,
(3) $(\mathbf{A} + \mathbf{B})^T = \mathbf{A}^T + \mathbf{B}^T$,
(4) $(\mathbf{A} + \mathbf{B} + \mathbf{C})^T = \mathbf{A}^T + \mathbf{B}^T + \mathbf{C}^T$,
(5) $(\mathbf{AB})^T = \mathbf{B}^T \mathbf{A}^T$,
(6) $(\mathbf{ABC})^T = \mathbf{C}^T \mathbf{B}^T \mathbf{A}^T$.

Transposes of sums and products of more than three matrices are defined in the obvious manner. We caution the reader to be alert to the ordering of properties (5) and (6). In particular, one should be aware that the transpose of a product is not the product of the transposes but rather the *commuted* product of the transposes.

Example 1 Find $(\mathbf{AB})^T$ and $\mathbf{B}^T \mathbf{A}^T$ if

$$\mathbf{A} = \begin{bmatrix} 3 & 0 \\ 4 & 1 \end{bmatrix} \quad \text{and} \quad \mathbf{B} = \begin{bmatrix} -1 & 2 & 1 \\ 3 & -1 & 0 \end{bmatrix}.$$

Solution

$$\mathbf{AB} = \begin{bmatrix} -3 & 6 & 3 \\ -1 & 7 & 4 \end{bmatrix}, \quad (\mathbf{AB})^T = \begin{bmatrix} -3 & -1 \\ 6 & 7 \\ 3 & 4 \end{bmatrix};$$

$$\mathbf{B}^T\mathbf{A}^T = \begin{bmatrix} -1 & 3 \\ 2 & -1 \\ 1 & 0 \end{bmatrix} \begin{bmatrix} 3 & 4 \\ 0 & 1 \end{bmatrix} = \begin{bmatrix} -3 & -1 \\ 6 & 7 \\ 3 & 4 \end{bmatrix}.$$

Note that $(\mathbf{AB})^T = \mathbf{B}^T \mathbf{A}^T$ but $\mathbf{A}^T \mathbf{B}^T$ is not defined. ■

A zero row in a matrix is a row containing only zeros, while a nonzero row is one that contains at least one nonzero element. A matrix is in *row-reduced form*, if it satisfies four conditions:

(R1) All zero rows appear below nonzero rows when both types are present in the matrix.

(R2) The first nonzero element in any nonzero row is unity.

(R3) All elements directly below (that is, in the same column but in succeeding rows from) the first nonzero element of a nonzero row are zero.

(R4) The first nonzero element of any nonzero row appears in a later column (further to the right) than the first nonzero element in any preceding row.

Such matrices are invaluable for solving sets of simultaneous linear equations and developing efficient algorithms for performing important matrix operations. We shall have much more to say on these matters in later chapters. Here we are simply interested in recognizing when a given matrix is or is not in row-reduced form.

Example 2 Determine which of the following matrices are in row-reduced form:

$$\mathbf{A} = \begin{bmatrix} 1 & 1 & -2 & 4 & 7 \\ 0 & 0 & -6 & 5 & 7 \\ 0 & 0 & 0 & 0 & 0 \\ 0 & 0 & 0 & 0 & 0 \end{bmatrix}, \quad \mathbf{B} = \begin{bmatrix} 1 & 2 & 3 \\ 0 & 0 & 0 \\ 0 & 0 & 1 \end{bmatrix},$$

$$\mathbf{C} = \begin{bmatrix} 1 & 2 & 3 & 4 \\ 0 & 0 & 1 & 2 \\ 0 & 1 & 0 & 5 \end{bmatrix}, \quad \mathbf{D} = \begin{bmatrix} -1 & -2 & 3 & 3 \\ 0 & 0 & 1 & -3 \\ 0 & 0 & 1 & 0 \end{bmatrix}.$$

Solution Matrix A is not in row-reduced form because the first nonzero element of the second row is not unity. This violates (R2). If a_{23} had been unity instead of -6, then the matrix would be in row-reduced form. Matrix **B** is not in row-reduced form because the second row is a zero row and it appears before the third row, which is a nonzero row. This violates (R1). If the second and third rows had been interchanged, then the matrix would be in row-reduced form. Matrix **C** is not in row-reduced form because the first nonzero element in row two appears in a later column, column 3, than the first nonzero element of row three. This violates (R4). If the second and third rows had been interchanged, then the matrix would be in row-reduced form. Matrix **D** is not in row-reduced form because the first nonzero element in row two appears in the third column, and everything below d_{23} is not zero. This violates (R3). Had the 3−3 element been zero instead of unity, then the matrix would be in row-reduced form. ∎

For the remainder of this section, we concern ourselves with square matrices; that is, matrices having the same number of rows as columns. A *diagonal matrix* is a square matrix all of whose elements are zero except possibly those on the main diagonal. (Recall that the main diagonal consists of all the diagonal elements a_{11}, a_{22}, a_{33}, and so on.) Thus,

$$\begin{bmatrix} 5 & 0 \\ 0 & -1 \end{bmatrix} \quad \text{and} \quad \begin{bmatrix} 3 & 0 & 0 \\ 0 & 3 & 0 \\ 0 & 0 & 3 \end{bmatrix}$$

are both diagonal matrices of order 2×2 and 3×3, respectively. The zero matrix is the special diagonal matrix having all the elements on the main diagonal equal to zero.

An *identity* matrix is a diagonal matrix worthy of special consideration. Designated by **I**, an identity is defined to be a diagonal matrix having all diagonal elements equal to one. Thus,

$$\begin{bmatrix} 1 & 0 \\ 0 & 1 \end{bmatrix} \quad \text{and} \quad \begin{bmatrix} 1 & 0 & 0 & 0 \\ 0 & 1 & 0 & 0 \\ 0 & 0 & 1 & 0 \\ 0 & 0 & 0 & 1 \end{bmatrix}$$

are the 2×2 and 4×4 identities, respectively. The identity is perhaps the most important matrix of all. If the identity is of the appropriate order so that the following multiplication can be carried out, then for any arbitrary matrix **A**,

$$\mathbf{AI} = \mathbf{A} \quad \text{and} \quad \mathbf{IA} = \mathbf{A}.$$

A *symmetric* matrix is a matrix that is equal to its transpose, while a *skew-symmetric* matrix is a matrix that is equal to the negative of its transpose. Thus, a matrix **A** is symmetric if $\mathbf{A} = \mathbf{A}^{\mathrm{T}}$, while it is skew symmetric if $\mathbf{A} = -\mathbf{A}^{\mathrm{T}}$. Examples of each are, respectively,

$$\begin{bmatrix} 1 & 2 & 3 \\ 2 & 4 & 5 \\ 3 & 5 & 6 \end{bmatrix} \quad \text{and} \quad \begin{bmatrix} 0 & 2 & -3 \\ -2 & 0 & 1 \\ 3 & -1 & 0 \end{bmatrix}.$$

A matrix $\mathbf{A} = [a_{ij}]$ is called *lower triangular* if $a_{ij} = 0$ for $j > i$ (that is, if all the elements above the main diagonal are zero) and *upper triangular* if $a_{ij} = 0$ for $i > j$ (that is, if all the elements below the main diagonal are zero).

Examples of lower and upper triangular matrices are, respectively,

$$\begin{bmatrix} 5 & 0 & 0 & 0 \\ -1 & 2 & 0 & 0 \\ 0 & 1 & 3 & 0 \\ 2 & 1 & 4 & 1 \end{bmatrix} \quad \text{and} \quad \begin{bmatrix} -1 & 2 & 4 & 1 \\ 0 & 1 & 3 & -1 \\ 0 & 0 & 2 & 5 \\ 0 & 0 & 0 & 5 \end{bmatrix}.$$

Theorem 1. *The product of two lower (upper) triangular matrices is also lower (upper) triangular.*

Proof. Let $\mathbf{A}$ and $\mathbf{B}$ both be $n \times n$ lower triangular matrices. Set $\mathbf{C} = \mathbf{AB}$. We need to show that $\mathbf{C}$ is lower triangular, or equivalently, that $c_{ij} = 0$ when $i < j$. Now,

$$c_{ij} = \sum_{k=1}^{n} a_{ik}b_{kj} = \sum_{k=1}^{j-1} a_{ik}b_{kj} + \sum_{k=j}^{n} a_{ik}b_{kj}.$$

We are given that $a_{ik} = 0$ when $i < k$, and $b_{kj} = 0$, when $k < j$, because both $\mathbf{A}$ and $\mathbf{B}$ are lower triangular. Thus,

$$\sum_{k=1}^{j-1} a_{ik}b_{kj} = \sum_{k=1}^{j-1} a_{ik}(0) = 0$$

because k is always less than j. Furthermore, if we restrict $i < j$, then

$$\sum_{k=j}^{n} a_{ik}b_{kj} = \sum_{k=j}^{n} (0)b_{kj} = 0$$

because $k \geq j > i$. Therefore, $c_{ij} = 0$ when $i < j$.

Finally, we define positive integral powers of a matrix in the obvious manner: $\mathbf{A}^2 = \mathbf{AA}$, $\mathbf{A}^3 = \mathbf{AAA}$ and, in general, if n is a positive integer,

$$\mathbf{A}^n = \underbrace{\mathbf{AA}...\mathbf{A}}_{n \text{ times}}.$$

Thus, if

$$\mathbf{A} = \begin{bmatrix} 1 & -2 \\ 1 & 3 \end{bmatrix}, \quad \text{then } \mathbf{A}^2 = \begin{bmatrix} 1 & -2 \\ 1 & 3 \end{bmatrix} \begin{bmatrix} 1 & -2 \\ 1 & 3 \end{bmatrix} = \begin{bmatrix} -1 & -8 \\ 4 & 7 \end{bmatrix}.$$

It follows directly from Property 5 that

$$\left(\mathbf{A}^2\right)^{\mathrm{T}} = (\mathbf{AA})^{\mathrm{T}} = \mathbf{A}^{\mathrm{T}}\mathbf{A}^{\mathrm{T}} = \left(\mathbf{A}^{\mathrm{T}}\right)^2.$$

We can generalize this result to the following property for any integral positive power n:

(7) $(\mathbf{A}^n)^{\mathrm{T}} = (\mathbf{A}^{\mathrm{T}})^n$.

Problems 1.4

1. Verify that $(\mathbf{A} + \mathbf{B})^{\mathrm{T}} = \mathbf{A}^{\mathrm{T}} + \mathbf{B}^{\mathrm{T}}$, where

$$\mathbf{A} = \begin{bmatrix} 1 & 5 & -1 \\ 2 & 1 & 3 \\ 0 & 7 & -8 \end{bmatrix} \quad \text{and} \quad \mathbf{B} = \begin{bmatrix} 6 & 1 & 3 \\ 2 & 0 & -1 \\ -1 & -7 & 2 \end{bmatrix}.$$

2. Verify that $(\mathbf{AB})^{\mathrm{T}} = \mathbf{B}^{\mathrm{T}} \mathbf{A}^{\mathrm{T}}$, where

$$\mathbf{A} = \begin{bmatrix} t & t^2 \\ 1 & 2t \\ 1 & 0 \end{bmatrix} \quad \text{and} \quad \mathbf{B} = \begin{bmatrix} 3 & t & t+1 & 0 \\ t & 2t & t^2 & t^3 \end{bmatrix}.$$

3. Simplify the following expressions:
(a) $(\mathbf{AB}^{\mathrm{T}})^{\mathrm{T}}$,
(b) $\mathbf{A}^{\mathrm{T}} + (\mathbf{A} + \mathbf{B}^{\mathrm{T}})^{\mathrm{T}}$,
(c) $(\mathbf{A}^{\mathrm{T}} (\mathbf{B} + \mathbf{C}^{\mathrm{T}}))^{\mathrm{T}}$,
(d) $((\mathbf{AB})^{\mathrm{T}} + \mathbf{C})^{\mathrm{T}}$,
(e) $((\mathbf{A} + \mathbf{A}^{\mathrm{T}})(\mathbf{A} - \mathbf{A}^{\mathrm{T}}))^{\mathrm{T}}$.

4. Find $\mathbf{X}^{\mathrm{T}} \mathbf{X}$ and $\mathbf{XX}^{\mathrm{T}}$ when

$$\mathbf{X} = \begin{bmatrix} 2 \\ 3 \\ 4 \end{bmatrix}.$$

5. Find $\mathbf{X}^{\mathrm{T}} \mathbf{X}$ and $\mathbf{XX}^{\mathrm{T}}$ when $\mathbf{X} = [1 \ -2 \ 3 \ -4]$.

6. Find $\mathbf{X}^{\mathrm{T}} \mathbf{AX}$ when

$$\mathbf{A} = \begin{bmatrix} 2 & 3 \\ 3 & 4 \end{bmatrix} \quad \text{and} \quad \mathbf{X} = \begin{bmatrix} x \\ y \end{bmatrix}.$$

7. Determine which, if any, of the following matrices are in row-reduced form:

$$A = \begin{bmatrix} 0 & 1 & 0 & 4 & -7 \\ 0 & 0 & 0 & 1 & 2 \\ 0 & 0 & 0 & 0 & 1 \\ 0 & 0 & 0 & 0 & 0 \end{bmatrix}, \quad B = \begin{bmatrix} 1 & 1 & 0 & 4 & -7 \\ 0 & 1 & 0 & 1 & 2 \\ 0 & 0 & 1 & 0 & 1 \\ 0 & 0 & 0 & 1 & 5 \end{bmatrix},$$

$$C = \begin{bmatrix} 1 & 1 & 0 & 4 & -7 \\ 0 & 1 & 0 & 1 & 2 \\ 0 & 0 & 0 & 0 & 1 \\ 0 & 0 & 0 & 1 & -5 \end{bmatrix} \quad D = \begin{bmatrix} 0 & 1 & 0 & 4 & -7 \\ 0 & 0 & 0 & 0 & 0 \\ 0 & 0 & 0 & 0 & 1 \\ 0 & 0 & 0 & 0 & 0 \end{bmatrix},$$

$$E = \begin{bmatrix} 2 & 2 & 2 \\ 0 & 2 & 2 \\ 0 & 0 & 2 \end{bmatrix}, \quad F = \begin{bmatrix} 0 & 0 & 0 \\ 0 & 0 & 0 \\ 0 & 0 & 0 \end{bmatrix}, \quad G = \begin{bmatrix} 1 & 2 & 3 \\ 0 & 0 & 1 \\ 1 & 0 & 0 \end{bmatrix},$$

$$H = \begin{bmatrix} 0 & 0 & 0 \\ 0 & 1 & 0 \\ 0 & 0 & 0 \end{bmatrix}, \quad J = \begin{bmatrix} 0 & 1 & 1 \\ 1 & 0 & 2 \\ 0 & 0 & 0 \end{bmatrix}, \quad K = \begin{bmatrix} 1 & 0 & 2 \\ 0 & -1 & 1 \\ 0 & 0 & 0 \end{bmatrix},$$

$$L = \begin{bmatrix} 2 & 0 & 0 \\ 0 & 2 & 0 \\ 0 & 0 & 0 \end{bmatrix}, \quad M = \begin{bmatrix} 1 & \frac{1}{2} & \frac{1}{3} \\ 0 & 1 & \frac{1}{4} \\ 0 & 0 & 1 \end{bmatrix}, \quad N = \begin{bmatrix} 1 & 0 & 0 \\ 0 & 0 & 1 \\ 0 & 0 & 0 \end{bmatrix},$$

$$Q = \begin{bmatrix} 0 & 1 \\ 1 & 0 \end{bmatrix}, \quad R = \begin{bmatrix} 1 & 1 \\ 0 & 0 \end{bmatrix}, \quad S = \begin{bmatrix} 1 & 0 \\ 1 & 0 \end{bmatrix}, \quad T = \begin{bmatrix} 1 & 12 \\ 0 & 1 \end{bmatrix}.$$

8. Determine which, if any, of the matrices in Problem 7 are upper triangular.
9. Must a square matrix in row-reduced form necessarily be upper triangular?
10. Must an upper triangular matrix necessarily be in row-reduced form?
11. Can a matrix be both upper and lower triangular simultaneously?

12. Show that $\mathbf{AB} = \mathbf{BA}$, where

$$\mathbf{A} = \begin{bmatrix} -1 & 0 & 0 \\ 0 & 3 & 0 \\ 0 & 0 & 1 \end{bmatrix} \quad \text{and } \mathbf{B} = \begin{bmatrix} 5 & 0 & 0 \\ 0 & 3 & 0 \\ 0 & 0 & 2 \end{bmatrix}.$$

13. Prove that if $\mathbf{A}$ and $\mathbf{B}$ are diagonal matrices of the same order, then $\mathbf{AB} = \mathbf{BA}$.

14. Does a 2×2 diagonal matrix commute with every other 2×2 matrix?

15. Compute the products $\mathbf{AD}$ and $\mathbf{BD}$ for the matrices

$$\mathbf{A} = \begin{bmatrix} 1 & 1 & 1 \\ 1 & 1 & 1 \\ 1 & 1 & 1 \end{bmatrix}, \quad \mathbf{B} = \begin{bmatrix} 0 & 1 & 2 \\ 3 & 4 & 5 \\ 6 & 7 & 8 \end{bmatrix}, \quad \mathbf{D} = \begin{bmatrix} 2 & 0 & 0 \\ 0 & 3 & 0 \\ 0 & 0 & -5 \end{bmatrix}.$$

What conclusions can you make about postmultiplying a square matrix by a diagonal matrix?

16. Compute the products $\mathbf{DA}$ and $\mathbf{DB}$ for the matrices defined in Problem 15. What conclusions can you make about premultiplying a square matrix by a diagonal matrix?

17. Prove that if a 2×2 matrix $\mathbf{A}$ commutes with every 2×2 diagonal matrix, then $\mathbf{A}$ must also be diagonal. *Hint*: Consider, in particular, the diagonal matrix

$$\mathbf{D} = \begin{bmatrix} 1 & 0 \\ 0 & 0 \end{bmatrix}.$$

18. Prove that if an $n \times n$ matrix $\mathbf{A}$ commutes with every $n \times n$ diagonal matrix, then $\mathbf{A}$ must also be diagonal.

19. Compute $\mathbf{D}^2$ and $\mathbf{D}^3$ for the matrix $\mathbf{D}$ defined in Problem 15.

20. Find $\mathbf{A}^3$ if

$$\mathbf{A} = \begin{bmatrix} 1 & 0 & 0 \\ 0 & 2 & 0 \\ 0 & 0 & 3 \end{bmatrix}.$$

21. Using the results of Problems 19 and 20 as a guide, what can be said about $\mathbf{D}^n$ if $\mathbf{D}$ is a diagonal matrix and n is a positive integer?

22. Prove that if $\mathbf{D} = [d_{ij}]$ is a diagonal matrix, then $\mathbf{D}^2 = \left[d_{ij}^2 \right]$.

23. Calculate $\mathbf{D}^{50} - 5\mathbf{D}^{35} + 4\mathbf{I}$, where

$$\mathbf{D} = \begin{bmatrix} 0 & 0 & 0 \\ 0 & 1 & 0 \\ 0 & 0 & -1 \end{bmatrix}.$$

24. A square matrix $\mathbf{A}$ is *nilpotent* if $\mathbf{A}^n = \mathbf{0}$ for some positive integer n. If n is the smallest positive integer for which $\mathbf{A}^n = \mathbf{0}$ then $\mathbf{A}$ is nilpotent of *index n*. Show that

$$\mathbf{A} = \begin{bmatrix} -1 & -1 & -3 \\ -5 & -2 & -6 \\ 2 & 1 & 3 \end{bmatrix}$$

is nilpotent of index 3.

25. Show that

$$\mathbf{A} = \begin{bmatrix} 0 & 1 & 0 & 0 \\ 0 & 0 & 1 & 0 \\ 0 & 0 & 0 & 1 \\ 0 & 0 & 0 & 0 \end{bmatrix}$$

is nilpotent. What is its index?

26. Prove that if $\mathbf{A}$ is a square matrix, then $\mathbf{B} = (\mathbf{A} + \mathbf{A}^T)/2$ is a symmetric matrix.

27. Prove that if $\mathbf{A}$ is a square matrix, then $\mathbf{C} = (\mathbf{A} - \mathbf{A}^T)/2$ is a skew-symmetric matrix.

28. Using the results of the preceding two problems, prove that any square matrix can be written as the sum of a symmetric matrix and a skew-symmetric matrix.

29. Write the matrix $\mathbf{A}$ in Problem 1 as the sum of a symmetric matrix and skew-symmetric matrix.

30. Write the matrix $\mathbf{B}$ in Problem 1 as the sum of a symmetric matrix and a skew-symmetric matrix.

31. Prove that if $\mathbf{A}$ is any matrix, then $\mathbf{A}\mathbf{A}^T$ is symmetric.

32. Prove that the diagonal elements of a skew-symmetric matrix must be zero.

33. Prove that the transpose of an upper triangular matrix is lower triangular, and vice versa.

34. If $\mathbf{P} = [p_{ij}]$ is a transition matrix for a Markov chain (see Problem 16 of Section 1.1), then it can be shown with elementary probability theory that the $i - j$ element of $\mathbf{P}^2$ denotes the probability of an object moving from state i to state j over two time periods. More generally, the $i - j$ element of $\mathbf{P}^n$ for any positive integer n denotes the probability of an object moving from state i to state j over n time periods.

(a) Calculate $\mathbf{P}^2$ and $\mathbf{P}^3$ for the two-state transition matrix

$$\mathbf{P} = \begin{bmatrix} 0.1 & 0.9 \\ 0.4 & 0.6 \end{bmatrix}.$$

(b) Determine the probability of an object beginning in state 1 and ending in state 1 after two time periods.

(c) Determine the probability of an object beginning in state 1 and ending in state 2 after two time periods.

(d) Determine the probability of an object beginning in state 1 and ending in state 2 after three time periods.

(e) Determine the probability of an object beginning in state 2 and ending in state 2 after three time periods.

35. Consider a two-state Markov chain. List the number of ways an object in state 1 can end in state 1 after three time periods.

36. Consider the Markov chain described in Problem 16 of Section 1.1. Determine (a) the probability that a family living in the city will find themselves in the suburbs after 2 years, and (b) the probability that a family living in the suburbs will find themselves living in the city after 2 years.

37. Consider the Markov chain described in Problem 17 of Section 1.1. Determine (a) the probability that there will be a Republican mayor 8 years after a Republican mayor serves and (b) the probability that there will be a Republican mayor 12 years after a Republican mayor serves.

38. Consider the Markov chain described in Problem 18 of Section 1.1. It is known that this year the apple harvest was poor. Determine (a) the probability that next year's harvest will be poor and (b) the probability that the harvest in 2 years will be poor.

39. Consider the Markov chain described in Problem 19 of Section 1.1. Determine (a) the probability that a brand X customer will be a brand X customer after 4 years, (b) after 6 years, and (c) the probability that a brand X customer will be a brand Y customer after 4 years.

40. A graph consists of a set of nodes, which we shall designate by positive integers, and a set of arcs that connect various pairs of nodes. An *adjacency matrix* **M** associated with a particular graph is defined by

$$m_{ij} = \text{number of distinct arcs connecting node } i \text{ to node } j$$

(a) Construct an adjacency matrix for the graph shown in Fig. 1.1.

(b) Calculate $\mathbf{M}^2$ and note that the $i - j$ element of $\mathbf{M}^2$ is the number of paths consisting of two arcs that connect node i to node j.

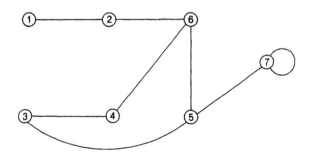

FIGURE 1.1

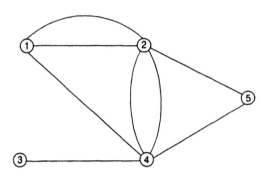

FIGURE 1.2

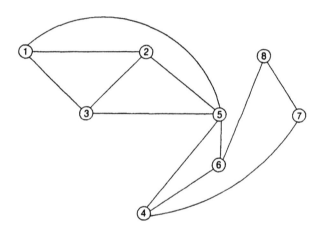

FIGURE 1.3

41. **(a)** Construct an adjacency matrix $\mathbf{M}$ for the graph shown in Fig. 1.2.
   (b) Calculate $\mathbf{M}^2$ and use that matrix to determine the number of paths consisting of two arcs that connect node 1 to node 5.
   (c) Calculate $\mathbf{M}^3$ and use that matrix to determine the number of paths consisting of three arcs that connect node 2 to node 4.
42. Fig. 1.3 depicts a road network linking various cities. A traveler in city 1 needs to drive to city 7 and would like to do so by passing through the least number of intermediate cities. Construct an adjacency matrix for this road network. Consider powers of this matrix to solve the traveler's problem.

1.5 **Submatrices and partitioning**

Given any matrix $\mathbf{A}$, a *submatrix* of $\mathbf{A}$ is a matrix obtained from $\mathbf{A}$ by the removal of any number of rows or columns. Thus, if

$$\mathbf{A} = \begin{bmatrix} 1 & 2 & 3 & 4 \\ 5 & 6 & 7 & 8 \\ 9 & 10 & 11 & 12 \\ 13 & 14 & 15 & 16 \end{bmatrix}, \quad \mathbf{B} = \begin{bmatrix} 10 & 12 \\ 14 & 16 \end{bmatrix}, \quad \text{and} \quad \mathbf{C} = \begin{bmatrix} 2 & 3 & 4 \end{bmatrix}, \quad (11)$$

then **B** and **C** are both submatrices of **A**. Here **B** was obtained by removing from **A** the first and second rows together with the first and third columns, while **C** was obtained by removing from **A** the second, third, and fourth rows together with the first column. By removing no rows and no columns from **A**, it follows that **A** is a submatrix of itself.

A matrix is said to be partitioned if it is divided into submatrices by horizontal and vertical lines between the rows and columns. By varying the choices of where to put the horizontal and vertical lines, one can partition a matrix in many different ways. Thus,

$$\begin{bmatrix} 1 & 2 & 3 & 4 \\ 5 & 6 & 7 & 8 \\ 9 & 10 & 11 & 12 \\ 13 & 14 & 15 & 16 \end{bmatrix} \quad \text{and} \quad \begin{bmatrix} 1 & 2 & 3 & 4 \\ 5 & 6 & 7 & 8 \\ 9 & 10 & 11 & 12 \\ 13 & 14 & 15 & 16 \end{bmatrix}$$

are examples of two different partitions of the matrix **A** given in (11).

If partitioning is carried out in a particularly judicious manner, it can be a great help in matrix multiplication. Consider the case where the two matrices **A** and **B** are to be multiplied together. If we partition both **A** and **B** into four submatrices, respectively, so that

$$\mathbf{A} = \begin{bmatrix} \mathbf{C} & \mathbf{D} \\ \mathbf{E} & \mathbf{F} \end{bmatrix} \quad \text{and} \quad \mathbf{B} = \begin{bmatrix} \mathbf{G} & \mathbf{H} \\ \mathbf{J} & \mathbf{K} \end{bmatrix}$$

where **C** through **K** represent submatrices, then the product **AB** may be obtained by simply carrying out the multiplication as if the submatrices were themselves elements. Thus,

$$\mathbf{AB} = \begin{bmatrix} \mathbf{CG} + \mathbf{DJ} & \mathbf{CH} + \mathbf{DK} \\ \mathbf{EG} + \mathbf{FJ} & \mathbf{EH} + \mathbf{FK} \end{bmatrix}, \quad (12)$$

providing the partitioning was such that the indicated multiplications are defined.

It is not unusual to need products of matrices having thousands of rows and thousands of columns. Problem 42 of Section 1.4 dealt with a road network connecting seven cities. A similar network for a state with connections between all cities in the state would have a very large adjacency matrix associated with it and its square is then the product of two such matrices. If we expand the network to include the entire United States, the associated matrix is huge, with one row and one column for each city and town in the country. Thus, it is not difficult to visualize large matrices that are too big to be stored in the internal memory of any modern day computer. And yet the product of such matrices must be computed.

The solution procedure is partitioning. Large matrices are stored in external memory on peripheral devices, such as disks, and then partitioned. Appropriate submatrices are fetched from the peripheral devices as needed, computed, and the results again stored on the peripheral devices. An example is the product given in (12). If **A** and **B** are too large for the internal memory of a particular computer but **C** through **K** are not, then the partitioned product can be computed. First, **C** and **G** are fetched from external memory and multiplied; the product is then stored in external memory. Next, **D** and **J** are fetched and multiplied. Then, the product **CG** is fetched and added to the product **DJ**. The result, which is the first partition of **AB**, is then stored in external memory and the process continues.

Example 1. Find **AB** if

$$A = \begin{bmatrix} 3 & 1 & 2 \\ 1 & 4 & -1 \\ 3 & 1 & 2 \end{bmatrix} \quad \text{and } B = \begin{bmatrix} 1 & 3 & 2 \\ -1 & 0 & 1 \\ 0 & 1 & 1 \end{bmatrix}.$$

Solution We first partition A and B in the following manner

$$A = \left[\begin{array}{cc|c} 3 & 1 & 2 \\ 1 & 4 & -1 \\ \hline 3 & 1 & 2 \end{array} \right] \quad \text{and } B = \left[\begin{array}{cc|c} 1 & 3 & 2 \\ -1 & 0 & 1 \\ \hline 0 & 1 & 2 \end{array} \right];$$

then,

$$AB = \left[\begin{array}{c|c} \begin{bmatrix} 3 & 1 \\ 1 & 4 \end{bmatrix} \begin{bmatrix} 1 & 3 \\ -1 & 0 \end{bmatrix} + \begin{bmatrix} 2 \\ -1 \end{bmatrix} [0 \ 1] & \begin{bmatrix} 3 & 1 \\ 1 & 4 \end{bmatrix} \begin{bmatrix} 2 \\ 1 \end{bmatrix} + \begin{bmatrix} 2 \\ -1 \end{bmatrix} [1] \\ \hline [3 \ 1] \begin{bmatrix} 1 & 3 \\ -1 & 0 \end{bmatrix} + [2] \ [0 \ 1] & [3 \ 1] \begin{bmatrix} 2 \\ 1 \end{bmatrix} + [2] \ [1] \end{array} \right]$$

$$= \left[\begin{array}{c|c} \begin{bmatrix} 2 & 9 \\ -3 & 3 \end{bmatrix} + \begin{bmatrix} 0 & 2 \\ 0 & -1 \end{bmatrix} & \begin{bmatrix} 7 \\ 6 \end{bmatrix} + \begin{bmatrix} 2 \\ -1 \end{bmatrix} \\ \hline [2 \ 9] + [0 \ 2] & [7] + [2] \end{array} \right]$$

$$= \left[\begin{array}{cc|c} 2 & 11 & 9 \\ -3 & 2 & 5 \\ \hline 2 & 11 & 9 \end{array} \right] = \begin{bmatrix} 2 & 11 & 9 \\ -3 & 2 & 5 \\ 2 & 11 & 9 \end{bmatrix}. \blacksquare$$

Example 2. Find **AB** if

$$
A = \begin{bmatrix} 3 & 1 & 0 \\ 2 & 0 & 0 \\ 0 & 0 & 3 \\ 0 & 0 & 1 \\ 0 & 0 & 0 \end{bmatrix} \quad \text{and} \quad B = \begin{bmatrix} 2 & 1 & 0 & 0 & 0 \\ -1 & 1 & 0 & 0 & 0 \\ 0 & 1 & 0 & 0 & 1 \end{bmatrix}.
$$

Solution From the indicated partitions, we find that

$$
AB = \begin{bmatrix} \begin{bmatrix} 3 & 1 \\ 2 & 0 \end{bmatrix}\begin{bmatrix} 2 & 1 \\ -1 & 1 \end{bmatrix} + \begin{bmatrix} 0 \\ 0 \end{bmatrix}[0\ 1] & \begin{bmatrix} 3 & 1 \\ 2 & 0 \end{bmatrix}\begin{bmatrix} 0 & 0 & 0 \\ 0 & 0 & 0 \end{bmatrix} + \begin{bmatrix} 0 \\ 0 \end{bmatrix}[0\ 0\ 1] \\ \begin{bmatrix} 0 & 0 \\ 0 & 0 \end{bmatrix}\begin{bmatrix} 2 & 1 \\ -1 & 1 \end{bmatrix} + \begin{bmatrix} 3 \\ 4 \end{bmatrix}[0\ 1] & \begin{bmatrix} 0 & 0 \\ 0 & 0 \end{bmatrix}\begin{bmatrix} 0 & 0 & 0 \\ 0 & 0 & 0 \end{bmatrix} + \begin{bmatrix} 3 \\ 1 \end{bmatrix}[0\ 0\ 1] \\ [0\ 0]\begin{bmatrix} 2 & 1 \\ -1 & 1 \end{bmatrix} + [0][0\ 1] & [0\ 0]\begin{bmatrix} 0 & 0 & 0 \\ 0 & 0 & 0 \end{bmatrix} + [0][0\ 0\ 1] \end{bmatrix}
$$

$$
AB = \begin{bmatrix} \begin{bmatrix} 5 & 4 \\ 4 & 2 \end{bmatrix}+\begin{bmatrix} 0 & 0 \\ 0 & 0 \end{bmatrix} & \begin{bmatrix} 0 & 0 & 0 \\ 0 & 0 & 0 \end{bmatrix}+\begin{bmatrix} 0 & 0 & 0 \\ 0 & 0 & 0 \end{bmatrix} \\ \begin{bmatrix} 0 & 0 \\ 0 & 0 \end{bmatrix}+\begin{bmatrix} 0 & 3 \\ 0 & 1 \end{bmatrix} & \begin{bmatrix} 0 & 0 & 0 \\ 0 & 0 & 0 \end{bmatrix}+\begin{bmatrix} 0 & 0 & 3 \\ 0 & 0 & 1 \end{bmatrix} \\ [0\ 0]+[0\ 0] & [0\ 0\ 0]+[0\ 0\ 0] \end{bmatrix}
$$

$$
= \begin{bmatrix} 5 & 4 & 0 & 0 & 0 \\ 4 & 2 & 0 & 0 & 0 \\ 0 & 3 & 0 & 0 & 3 \\ 0 & 1 & 0 & 0 & 1 \\ 0 & 0 & 0 & 0 & 0 \end{bmatrix} = \begin{bmatrix} 5 & 4 & 0 & 0 & 0 \\ 4 & 2 & 0 & 0 & 0 \\ 0 & 3 & 0 & 0 & 3 \\ 0 & 1 & 0 & 0 & 1 \\ 0 & 0 & 0 & 0 & 0 \end{bmatrix}
$$

Note that we partitioned in order to make maximum of the zero submatrices of both **A** and **B**. ■

A matrix **A** that can be partitioned into the form

$$
A = \begin{bmatrix} A_1 & & & & \\ & A_2 & & & \\ & & A_3 & & 0 \\ & & & \ddots & \\ & 0 & & & A_n \end{bmatrix}
$$

is called *block diagonal*. Such matrices are particularly easy to multiply because in partitioned form they act as diagonal matrices.

Problems 1.5

1. Which of the following are submatrices of the given **A** and why?

$$\mathbf{A} = \begin{bmatrix} 1 & 2 & 3 \\ 4 & 5 & 6 \\ 7 & 8 & 9 \end{bmatrix}$$

(a) $\begin{bmatrix} 1 & 3 \\ 7 & 9 \end{bmatrix}$ (b) $[1]$ (c) $\begin{bmatrix} 1 & 2 \\ 8 & 9 \end{bmatrix}$ (d) $\begin{bmatrix} 4 & 6 \\ 7 & 9 \end{bmatrix}$.

2. Determine all possible submatrices of

$$\mathbf{A} = \begin{bmatrix} a & b \\ c & d \end{bmatrix}.$$

3. Given the matrices **A** and **B** (as shown), find **AB** using the partitionings indicated:

$$\mathbf{A} = \left[\begin{array}{cc|c} 1 & -1 & 2 \\ 3 & 0 & 4 \\ \hline 0 & 1 & 2 \end{array} \right], \quad \mathbf{B} = \left[\begin{array}{ccc|c} 5 & 2 & 0 & 2 \\ \hline 1 & -1 & 3 & 1 \\ 0 & 1 & 1 & 4 \end{array} \right].$$

4. Partition the given matrices **A** and **B** and, using the results, find **AB**.

$$\mathbf{A} = \begin{bmatrix} 4 & 1 & 0 & 0 \\ 2 & 2 & 0 & 0 \\ 0 & 0 & 1 & 0 \\ 0 & 0 & 1 & 2 \end{bmatrix}, \quad \mathbf{B} = \begin{bmatrix} 3 & 2 & 0 & 0 \\ -1 & 1 & 0 & 0 \\ 0 & 0 & 2 & 1 \\ 0 & 0 & 1 & -1 \end{bmatrix}.$$

5. Compute $\mathbf{A}^2$ for the matrix **A** given in Problem 4 by partitioning **A** into block diagonal form.

6. Compute $\mathbf{B}^2$ for the matrix **B** given in Problem 4 by partitioning **B** into block diagonal form.

7. Use partitioning to compute $\mathbf{A}^2$ and $\mathbf{A}^3$ for

$$\mathbf{A} = \begin{bmatrix} 1 & 0 & 0 & 0 & 0 & 0 \\ 0 & 2 & 0 & 0 & 0 & 0 \\ 0 & 0 & 0 & 1 & 0 & 0 \\ 0 & 0 & 0 & 0 & 1 & 0 \\ 0 & 0 & 0 & 0 & 0 & 1 \\ 0 & 0 & 0 & 0 & 0 & 0 \end{bmatrix}.$$

What is $\mathbf{A}^n$ for any positive integral power of $n > 3$?

8. Use partitioning to compute $\mathbf{A}^2$ and $\mathbf{A}^3$ for

$$\mathbf{A} = \begin{bmatrix} 0 & -1 & 0 & 0 & 0 & 0 & 0 \\ -1 & 0 & 0 & 0 & 0 & 0 & 0 \\ 0 & 0 & 2 & -2 & -4 & 0 & 0 \\ 0 & 0 & -1 & 3 & 4 & 0 & 0 \\ 0 & 0 & 1 & -2 & -3 & 0 & 0 \\ 0 & 0 & 0 & 0 & 0 & -1 & 0 \\ 0 & 0 & 0 & 0 & 0 & 0 & -1 \end{bmatrix}.$$

What is $\mathbf{A}^n$ for any positive integral power of n?

1.6 Vectors

Definition 1. A *vector* is a $1 \times n$ or $n \times 1$ matrix.

A $1 \times n$ matrix is called a *row vector*, while an $n \times 1$ matrix is a *column vector*. The elements are called the *components* of the vector, while the number of components in the vector, in this case n, is its *dimension*. Thus,

$$\begin{bmatrix} 1 \\ 2 \\ 3 \end{bmatrix}$$

is an example of a three-dimensional column vector, while

$$\begin{bmatrix} t & 2t & -t & 0 \end{bmatrix}$$

is an example of a four-dimensional row vector.

The reader who is already familiar with vectors will notice that we have not defined vectors as directed line segments. We have done this intentionally, first because in more than three dimensions, this geometric interpretation loses its significance, and second, because in the general mathematical framework, vectors are not directed line segments. However, the idea of representing a finite dimensional vector by its components and hence as a matrix is one that is acceptable to the scientist, engineer, and mathematician. Also, as a bonus, since a vector is nothing more than a special matrix, we have already defined scalar multiplication, vector addition, and vector equality.

A vector $\mathbf{y}$ (vectors will be designated by boldface lowercase letters) has associated with it a nonnegative number called its *magnitude* or length designated by $\|\mathbf{y}\|$.

Definition 2. If $\mathbf{y} = [y_1 y_2 \ldots y_n]$ then $\|\mathbf{y}\| = \sqrt{(y_1)^2 + (y_2)^2 + \ldots + (y_n)^2}$.

Example 1. Find $\|\mathbf{y}\|$ if $\mathbf{y} = [1\ 2\ 3\ 4]$.

Solution $\|\mathbf{y}\| = \sqrt{(1)^2 + (2)^2 + (3)^2 + (4)^2} = \sqrt{30}.$ ∎
If $\mathbf{z}$ is a column vector, $\|\mathbf{z}\|$ is defined in a completely analogous manner.

Example 2. Find $\|\mathbf{z}\|$ if

$$\mathbf{z} = \begin{bmatrix} -1 \\ 2 \\ -3 \end{bmatrix}.$$

Solution $\|\mathbf{z}\| = \sqrt{(-1)^2 + (2)^2 + (-3)^2} = \sqrt{14}.$ ∎

A vector is called a *unit vector* if its magnitude is equal to one. A nonzero vector is said to be *normalized* if it is divided by its magnitude. Thus, a normalized vector is also a unit vector.

Example 3. Normalize the vector $[1\ 0\ -3\ 2\ -1]$.
Solution The magnitude of this vector is

$$\sqrt{(1)^2 + (0)^2 + (-3)^2 + (2)^2 + (-1)^2} = \sqrt{15}.$$

Hence, the normalized vector is

$$\left[\frac{1}{\sqrt{15}}\ \ 0\ \ \frac{-3}{\sqrt{15}}\ \ \frac{2}{\sqrt{15}}\ \ \frac{-1}{\sqrt{15}} \right]. \blacksquare$$

In passing, we note that when a general vector is written $\mathbf{y} = [y_1\ y_2\ ...\ y_n]$ one of the subscripts of each element of the matrix is deleted. This is done solely for the sake of convenience. Since a row vector has only one row (a column vector has only one column), it is redundant and unnecessary to exhibit the row subscript (the column subscript).

Problems 1.6

1. Find p if $5\mathbf{x} - 2\mathbf{y} = \mathbf{b}$, where

$$\mathbf{x} = \begin{bmatrix} 1 \\ 3 \\ 0 \end{bmatrix}, \quad \mathbf{y} = \begin{bmatrix} 2 \\ p \\ 1 \end{bmatrix}, \quad \text{and} \quad \mathbf{b} = \begin{bmatrix} 1 \\ 13 \\ -2 \end{bmatrix}.$$

2. Find **x** if $3\mathbf{x} + 2\mathbf{y} = \mathbf{b}$, where

$$\mathbf{y} = \begin{bmatrix} 3 \\ 1 \\ 6 \\ 0 \end{bmatrix} \quad \text{and} \quad \mathbf{b} = \begin{bmatrix} 2 \\ -2 \\ 4 \\ 1 \end{bmatrix}.$$

3. Find **y** if $2\mathbf{x} - 5\mathbf{y} = -\mathbf{b}$, where

$$\mathbf{x} = \begin{bmatrix} 2 & -1 & 3 \end{bmatrix} \quad \text{and} \quad \mathbf{b} = \begin{bmatrix} 1 & 0 & -1 \end{bmatrix}.$$

4. Using the vectors defined in Problem 2, calculate, if possible,
   (a) **yb**.
   (b) **yb**$^\mathrm{T}$.
   (c) **y**$^\mathrm{T}$ **b**.
   (d) **b**$^\mathrm{T}$ **y**.
5. Using the vectors defined in Problem 3, calculate, if possible,
   (a) **x** + 2**b**.
   (b) **xb**$^\mathrm{T}$.
   (c) **x**$^\mathrm{T}$ **b**.
   (d) **b**$^\mathrm{T}$ **b**.
6. Determine which of the following are unit vectors:

$$\text{(a) } \begin{bmatrix} 1 & 1 \end{bmatrix}, \quad \text{(b) } \begin{bmatrix} 1/2 & 1/2 \end{bmatrix}, \quad \text{(c) } \begin{bmatrix} 1/\sqrt{2} & -1/\sqrt{2} \end{bmatrix}$$

$$\text{(d) } \begin{bmatrix} 0 \\ 1 \\ 0 \end{bmatrix}, \quad \text{(e) } \begin{bmatrix} 1/2 \\ 1/3 \\ 1/6 \end{bmatrix}, \quad \text{(f) } \begin{bmatrix} 1/\sqrt{3} \\ 1/\sqrt{3} \\ 1/\sqrt{3} \end{bmatrix},$$

$$\text{(g) } \frac{1}{2} \begin{bmatrix} 1 \\ 1 \\ 1 \\ 1 \end{bmatrix}, \quad \text{(h) } \frac{1}{6} \begin{bmatrix} 1 \\ 5 \\ 3 \\ 1 \end{bmatrix}, \quad \text{(i) } \frac{1}{\sqrt{3}} \begin{bmatrix} -1 & 0 & 1 & -1 \end{bmatrix}.$$

7. Find $\|\mathbf{y}\|$ if

$$\text{(a) } \mathbf{y} = \begin{bmatrix} 1 & -1 \end{bmatrix}, \qquad \text{(b) } \mathbf{y} = \begin{bmatrix} 3 & 4 \end{bmatrix},$$

$$\text{(c) } \mathbf{y} = \begin{bmatrix} -1 & -1 & 1 \end{bmatrix}, \quad \text{(d) } \mathbf{y} = \begin{bmatrix} \frac{1}{2} & \frac{1}{2} & \frac{1}{2} \end{bmatrix},$$

$$\text{(e) } \mathbf{y} = \begin{bmatrix} 2 & 1 & -1 & 3 \end{bmatrix}, \quad \text{(f) } \mathbf{y} = \begin{bmatrix} 0 & -1 & 5 & 3 & 2 \end{bmatrix}.$$

8. Find $\|\mathbf{x}\|$ if

$$\text{(a) } \mathbf{x} = \begin{bmatrix} 1 \\ -1 \end{bmatrix},$$

$$\text{(b) } \mathbf{x} = \begin{bmatrix} 1 \\ 2 \end{bmatrix},$$

$$\text{(c) } \mathbf{x} = \begin{bmatrix} 1 \\ 1 \\ 1 \end{bmatrix},$$

$$\text{(d) } \mathbf{x} = \begin{bmatrix} 1 \\ -1 \\ 1 \\ -1 \end{bmatrix},$$

$$\text{(e) } \mathbf{x} = \begin{bmatrix} 1 \\ 2 \\ 3 \\ 4 \end{bmatrix},$$

$$\text{(f) } \mathbf{x} = \begin{bmatrix} 1 \\ 0 \\ 1 \\ 0 \end{bmatrix}.$$

9. Find $\|\mathbf{y}\|$ if

$$\text{(a) } \mathbf{y} = \begin{bmatrix} 2 & 1 & -1 & 3 \end{bmatrix},$$
$$\text{(b) } \mathbf{y} = \begin{bmatrix} 0 & -1 & 5 & 3 & 2 \end{bmatrix}.$$

10. Prove that a normalized vector must be a unit vector.

11. Show that the matrix equation

$$\begin{bmatrix} 1 & 1 & -2 \\ 2 & 5 & 3 \\ -1 & 3 & 1 \end{bmatrix} \begin{bmatrix} x \\ y \\ z \end{bmatrix} = \begin{bmatrix} -3 \\ 11 \\ 5 \end{bmatrix}$$

is equivalent to the vector equation

$$x\begin{bmatrix} 1 \\ 2 \\ -1 \end{bmatrix} + y\begin{bmatrix} 1 \\ 5 \\ 3 \end{bmatrix} + z\begin{bmatrix} -2 \\ 3 \\ 1 \end{bmatrix} = \begin{bmatrix} -3 \\ 11 \\ 5 \end{bmatrix}.$$

12. Convert the following system of equations into a vector equation:

$$2x + 3y = 10,$$
$$4x + 5y = 11.$$

13. Convert the following system of equations into a vector equation:

$$3x + 4y + 5z + 6w = 1,$$
$$y - 2z + 8w = 0,$$
$$-x + y + 2z - w = 0.$$

14. Using the definition of matrix multiplication, show that the jth column of $(\mathbf{AB}) = \mathbf{A} \times (j\text{th column of }\mathbf{B})$.

15. Verify the result of Problem 14 by showing that the first column of the product $\mathbf{AB}$ with

$$\mathbf{A} = \begin{bmatrix} 1 & 2 & 3 \\ 4 & 5 & 6 \end{bmatrix} \quad \text{and} \quad \mathbf{B} = \begin{bmatrix} 1 & 1 \\ -1 & 0 \\ 2 & -3 \end{bmatrix}$$

is

$$\mathbf{A}\begin{bmatrix} 1 \\ -1 \\ 2 \end{bmatrix},$$

while the second column of the product is

$$\mathbf{A}\begin{bmatrix} 1 \\ 0 \\ -3 \end{bmatrix}.$$

16. A *distribution row vector* **d** for an N-state Markov chain (see Problem 16 of Section 1.1 and Problem 34 of Section 1.4) is an N-dimensional row vector having as its components, one for each state, the probabilities that an object in the system is in each of the respective states. Determine a distribution vector for a three-state Markov chain if 50% of the objects are in state 1, 30% are in state 2, and 20% are in state 3.

17. Let $d^{(k)}$ denote the distribution vector for a Markov chain after k time periods. Thus, $d^{(0)}$ represents the initial distribution. It follows that

$$\mathbf{d}^{(k)} = \mathbf{d}^{(0)}\mathbf{P}^k = \mathbf{P}^{(k-1)}\mathbf{P},$$

where $\mathbf{P}$ is the transition matrix and $\mathbf{P}^k$ is its kth power.

Consider the Markov chain described in Problem 16 of Section 1.1.

(a) Explain the physical significance of saying $\mathbf{d}^{(0)} = [0.6\ 0.4]$.

(b) Find the distribution vectors $\mathbf{d}^{(1)}$ and $\mathbf{d}^{(2)}$.

18. Consider the Markov chain described in Problem 19 of Section 1.1.

(a) Explain the physical significance of saying $\mathbf{d}^{(0)} = [0.4\ 0.5\ 0.1]$.

(b) Find the distribution vectors $\mathbf{d}^{(1)}$ and $\mathbf{d}^{(2)}$.

19. Consider the Markov chain described in Problem 17 of Section 1.1. (a) Determine an initial distribution vector if the town currently has a Democratic mayor and (b) show that the components of $\mathbf{d}^{(1)}$ are the probabilities that the next mayor will be a Republican and a Democrat, respectively.

20. Consider the Markov chain described in Problem 18 of Section 1.1. (a) Determine an initial distribution vector if this year's crop is known to be poor. (b) Calculate $\mathbf{d}^{(2)}$ and use it to determine the probability that the harvest will be good in 2 years.

1.7 The geometry of vectors

Vector arithmetic can be described geometrically for two- and three-dimensional vectors. For simplicity, we consider two dimensions here; the extension to three-dimensional vectors is straightforward. For convenience, we restrict our examples to row vectors, but note that all constructions are equally valid for column vectors.

A two-dimensional vector $\mathbf{v} = [a\ b]$ is identified with the point (a, b) on the plane, measured from the origin a units along the horizontal axis and then b units parallel to the vertical axis. We can then draw an arrow beginning at the origin and ending at the point (a, b). This arrow or directed line segment, as shown in Fig. 1.4, represents the vector geometrically. It follows immediately from Pythagoras's theorem and Definition 2 of Section 1.6 that the length of the directed line segment is the magnitude of the vector. The angle associated with a vector, denoted by θ in Fig. 1.4, is the angle from the positive horizontal axis to the directed line segment measured in the counterclockwise direction.

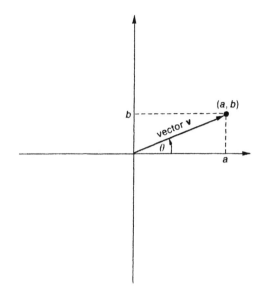

FIGURE 1.4

Example 1. Graph the vectors $\mathbf{v} = [2\ 4]$ and $\mathbf{u} = [-1\ 1]$ and determine the magnitude and angle of each.

Solution The vectors are drawn in Fig. 1.5. Using Pythagoras's theorem and elementary trigonometry, we have, for $\mathbf{v}$,

$$\|\mathbf{v}\| = \sqrt{(2)^2 + (4)^2} = 4.47, \quad \tan \theta = \frac{4}{2} = 2, \quad \text{and} \quad \theta = 63.4°.$$

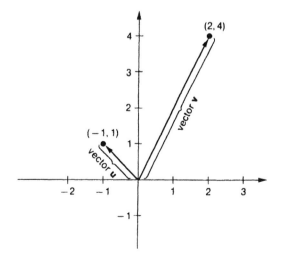

FIGURE 1.5

For **u**, similar computations yield

$$\|\mathbf{u}\| = \sqrt{(-1)^2 + (1)^2} = 1.14, \quad \tan\theta = \frac{1}{-1} = -1, \quad \text{and} \quad \theta = 135°.$$

To construct the sum of two vectors **u** + **v** geometrically, graph **u** normally, translate **v** so that its initial point coincides with the terminal point of **u**, *being careful to preserve both the magnitude and direction of* **v**, and then draw an arrow from the origin to the terminal point of **v** after translation. This arrow geometrically represents the sum **u** + **v**. The process is depicted in Fig. 1.6 for the two vectors defined in Example 1.

To construct the difference of two vectors **u** − **v** geometrically, graph both **u** and **v** normally and construct an arrow from the terminal point of **v** to the terminal point of **u**. This arrow geometrically represents the difference **u** − **v**. The process is depicted in Fig. 1.7 for the two vectors defined in Example 1. To measure the magnitude and direction of **u** − **v**, translate it so that its initial point is at the origin, *being careful to preserve both its magnitude and direction*, and then measure the translated vector.

Both geometrical sums and differences involve translations of vectors. This suggests that a vector is not altered by translating it to another position in the plane providing both its magnitude and direction are preserved.

Many physical phenomena such as velocity and force are completely described by their magnitudes and directions. For example, a velocity of 60 miles per hour in the northwest direction is a complete description of that velocity, and *it is independent of where that velocity occurs*. This independence is the rationale behind translating vectors geometrically. Geometrically, vectors having the same magnitude and direction are called *equivalent* and they are regarded as being equal even though they may be located at different positions in the plane.

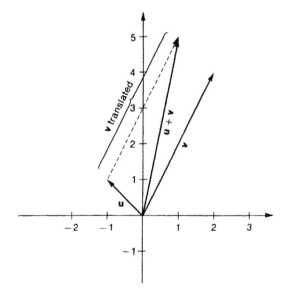

FIGURE 1.6

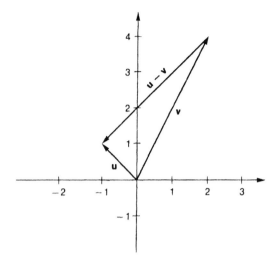

FIGURE 1.7

A scalar multiplication $k\mathbf{u}$ is defined geometrically to be a vector having length $\|k\|$ times the length of $\mathbf{u}$ with direction equal to $\mathbf{u}$ when k is positive, and opposite to $\mathbf{u}$ when k is negative. Effectively, $k\mathbf{u}$ is an elongation of $\mathbf{u}$ by a factor of $\|k\|$ when $\|k\|$ is greater than unity, or a contraction of $\mathbf{u}$ by a factor of $\|k\|$ when $\|k\|$ is less than unity, followed by no rotation when k is positive, or a rotation of 180 degrees when k is negative.

Example 2. Find $-2\mathbf{u}$ and $\frac{1}{2}\mathbf{v}$ geometrically for the vectors defined in Example 1.

Solution To construct $-2\mathbf{u}$, we double the length of $\mathbf{u}$ and then rotate the resulting vector by $180°$. To construct $\frac{1}{2}\mathbf{v}$, we halve the length of $\mathbf{v}$ and effect no rotation. These constructions are illustrated in Fig. 1.8. ∎

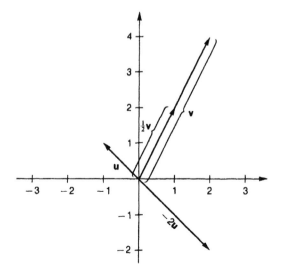

FIGURE 1.8

Problems 1.7

In Problems 1 through 16, geometrically construct the indicated vector operations for

$$\mathbf{u} = [3 \quad -1], \quad \mathbf{v} = [-2 \quad 5], \quad w = [-4 \quad -4],$$

$$\mathbf{x} = \begin{bmatrix} 3 \\ 5 \end{bmatrix}, \quad \text{and} \quad \mathbf{y} = \begin{bmatrix} 0 \\ -2 \end{bmatrix}.$$

1. $\mathbf{u} + \mathbf{v}$.
2. $\mathbf{u} + \mathbf{w}$.
3. $\mathbf{v} + \mathbf{w}$.
4. $\mathbf{x} + \mathbf{y}$.
5. $\mathbf{x} - \mathbf{y}$.
6. $\mathbf{y} - \mathbf{x}$.
7. $\mathbf{u} - \mathbf{v}$.
8. $\mathbf{w} - \mathbf{u}$.
9. $\mathbf{u} - \mathbf{w}$.
10. $2\mathbf{x}$.
11. $3\mathbf{x}$.
12. $-2\mathbf{x}$.
13. $\frac{1}{2}\mathbf{u}$.
14. $-\frac{1}{2}\mathbf{u}$.
15. $\frac{1}{3}\mathbf{v}$.
16. $-\frac{1}{4}\mathbf{w}$.
17. Determine the angle of $\mathbf{u}$.
18. Determine the angle of $\mathbf{v}$.
19. Determine the angle of $\mathbf{w}$.
20. Determine the angle of $\mathbf{x}$.
21. Determine the angle of $\mathbf{y}$.
22. For arbitrary two-dimensional row vectors constructed on the same graph $\mathbf{u} + \mathbf{v}$ and $\mathbf{v} + \mathbf{u}$.
    (a) Show that $\mathbf{u} + \mathbf{v} = \mathbf{v} + \mathbf{u}$.
    (b) Show that the sum is a diagonal of a parallelogram having $\mathbf{u}$ and $\mathbf{v}$ as two of its sides.

Simultaneous linear equations

Chapter outline

2.1 Linear systems

Systems of simultaneous equations appear frequently in engineering and scientific problems. Because of their importance and because they lend themselves to matrix analysis, we devote this entire chapter to their solutions.

We are interested in systems of the form

$$
\begin{aligned}
a_{11}x_1 + a_{12}x_2 + \cdots + a_{1n}x_n &= b_1, \\
a_{21}x_1 + a_{22}x_2 + \cdots + a_{2n}x_n &= b_2, \\
&\vdots \\
a_{m1}x_1 + a_{m2}x_2 + \cdots + a_{mn}x_n &= b_m.
\end{aligned}
\tag{1}
$$

We assume that the coefficients a_{ij} ($i = 1, 2, \ldots, m; j = 1, 2, \ldots, n$) and the quantities b_i ($i = 1, 2, \ldots, m$) are all known scalars. The quantities $x_1, x_2, \ldots, x_n$ represent unknowns.

Definition 1.   A *solution* to (1) is a set of n scalars $x_1, x_2, \ldots, x_n$ that when substituted into (1) satisfies the given equations (that is, the equalities are valid).

System (1) is a generalization of systems considered earlier in that m can differ from n. If $m > n$, the system has more equations than unknowns. If $m < n$, the system has more unknowns than equations. If $m = n$, the system has as many unknowns as equations. In any case, the methods of Section 1.3 may be used to convert (1) into the matrix form

$$\mathbf{Ax} = \mathbf{b}, \tag{2}$$

where

$$\mathbf{A} = \begin{bmatrix} a_{11} & a_{12} & \cdots & a_{1n} \\ a_{21} & a_{22} & \cdots & a_{2n} \\ \vdots & \vdots & & \vdots \\ a_{m1} & a_{m2} & \cdots & a_{mn} \end{bmatrix}, \quad \mathbf{x} = \begin{bmatrix} x_1 \\ x_2 \\ \vdots \\ x_n \end{bmatrix}, \quad \mathbf{b} = \begin{bmatrix} b_1 \\ b_2 \\ \vdots \\ b_m \end{bmatrix}.$$

Thus, if $m \neq n$, $\mathbf{A}$ will be rectangular and the dimensions of $\mathbf{x}$ and $\mathbf{b}$ will be different.

Example 1.   Convert the following system to matrix form:

$$x + 2y - z + w = 4,$$
$$x + 3y + 2z + 4w = 9.$$

Solution

$$\mathbf{A} = \begin{bmatrix} 1 & 2 & -1 & 1 \\ 1 & 3 & 2 & 4 \end{bmatrix}, \quad \mathbf{x} = \begin{bmatrix} x \\ y \\ z \\ w \end{bmatrix}, \quad \mathbf{b} = \begin{bmatrix} 4 \\ 9 \end{bmatrix}. \blacksquare$$

Example 2.   Convert the following system to matrix form:

$$x - 2y = -9,$$
$$4x + y = 9,$$
$$2x + y = 7,$$
$$x - y = -1.$$

Solution

$$\mathbf{A} = \begin{bmatrix} 1 & -2 \\ 4 & 1 \\ 2 & 1 \\ 1 & -1 \end{bmatrix}, \quad \mathbf{x} = \begin{bmatrix} x \\ y \end{bmatrix}, \quad \mathbf{b} = \begin{bmatrix} -9 \\ 9 \\ 7 \\ -1 \end{bmatrix}. \blacksquare$$

A system of equations given by (1) or (2) can possess no solutions, exactly one solution, or more than one solution (note that by a solution to (2) we mean a vector $\mathbf{x}$, which satisfies the matrix equality (2)). Examples of such systems are

$$\begin{aligned} x + y &= 1, \\ x + y &= 2, \end{aligned} \tag{3}$$

$$\begin{aligned} x + y &= 1, \\ x - y &= 0, \end{aligned} \tag{4}$$

$$\begin{aligned} x + y &= 0, \\ 2x + 2y &= 0. \end{aligned} \tag{5}$$

Eq. (3) has no solutions, (4) admits only the solution $x = y = \frac{1}{2}$, while (5) has solutions $x = -y$ for any value of y.

Definition 2. A system of simultaneous linear equations is *consistent* if it possesses at least one solution. If no solution exists, the system is *inconsistent*.

Eq. (3) is an example of an inconsistent system, while (4) and (5) represent examples of consistent systems.

Definition 3. A system given by (2) is *homogeneous* if $\mathbf{b} = \mathbf{0}$ (the zero vector). If $\mathbf{b} \neq \mathbf{0}$ (at least one component of $\mathbf{b}$ differs from zero) the system is *nonhomogeneous*.

Eq. (5) is an example of a homogeneous system.

Problems 2.1

In Problems 1 and 2, determine whether or not the proposed values of x, y, and z are solutions of the given systems.

1. $\begin{aligned} x + y + 2z &= 2, \\ x - y - 2z &= 0, \\ x + 2y + 2z &= 1. \end{aligned}$ (a) $x = 1, y = -3, z = 2.$
      (b) $x = 1, y = -1, z = 1.$

2. $\begin{aligned} x + 2y + 3z &= 6, \\ x - 3y + 2z &= 0, \\ 3x - 4y + 7z &= 6. \end{aligned}$ (a) $x = 1, y = 1, z = 1.$
      (b) $x = 2, y = 2, z = 0.$
      (c) $x = 14, y = 2, z = -4.$

3. Find a value for k such that $x = 1$, $y = 2$, and $z = k$ is a solution of the system

$$\begin{aligned} 2x + 2y + 4z &= 1, \\ 5x + y + 2z &= 5, \\ x - 3y - 2z &= -3. \end{aligned}$$

4. Find a value for k such that $x = 2$ and $y = k$ is a solution of the system

$$3x + 5y = 11,$$
$$2x - 7y = -3.$$

5. Find a value for k such that $x = 2k$, $y = -k$, and $z = 0$ is a solution of the system

$$x + 2y + z = 0,$$
$$-2x - 4y + 2z = 0,$$
$$3x - 6y - 4z = 1.$$

6. Find a value for k such that $x = 2k$, $y = -k$, and $z = 0$ is a solution of the system

$$x + 2y + 2z = 0,$$
$$2x - 4y + 2z = 0,$$
$$-3x - 6y - 4z = 0.$$

7. Find a value for k such that $x = 2k$, $y = -k$, and $z = 0$ is a solution of the system

$$x + 2y + 2z = 0,$$
$$2x + 4y + 2z = 0,$$
$$-3x - 6y - 4z = 1.$$

8. Put the system of equations given in Problem 4 into the matrix form $\mathbf{Ax = b}$.

9. Put the system of equations given in Problem 1 into the matrix form $\mathbf{Ax = b}$.

10. Put the system of equations given in Problem 2 into the matrix form $\mathbf{Ax = b}$.

11. Put the system of equations given in Problem 6 into the matrix form $\mathbf{Ax = b}$.

12. A manufacturer receives daily shipments of 70,000 springs and 45,000 pounds of stuffing for producing regular and support mattresses. Regular mattresses r require 50 springs and 30 pounds of stuffing; support mattresses s require 60 springs and 40 pounds of stuffing. The manufacturer wants to know how many mattresses of each type should be produced daily to utilize all available inventory. Show that this problem is equivalent to solving two equations in the two unknowns r and s.

13. A manufacturer produces desks and bookcases. Desks d require 5 hours of cutting time and 10 hours of assembling time. Bookcases b require 15 minutes of cutting time and 1 hour of assembling time. Each day, the manufacturer has available 200 hours for cutting and 500 hours for assembling. The manufacturer wants to know how many desks and bookcases should be scheduled for completion each day to utilize all available workpower. Show that this problem is equivalent to solving two equations in the two unknowns d and b.

14. A mining company has a contract to supply 70,000 tons of low-grade ore, 181,000 tons of medium-grade ore, and 41,000 tons of high-grade ore to a supplier. The company has three mines, which it can work. Mine A produces 8000 tons of low-grade ore, 5000 tons of medium-grade ore, and 1000 tons of high-grade ore during each day of operation. Mine B produces 3000 tons of low-grade ore, 12,000 tons of medium-grade ore, and 3000 tons of high-grade ore for each day it is in operation. The figures for mine C are 1000, 10,000, and 2000, respectively. Show that the problem of determining how many days each mine must be operated to meet contractual demands without surplus is equivalent to solving a set of three equations in A, B, and C, where the unknowns denote the number of days each mine will be in operation.

15. A pet store has determined that each rabbit in its care should receive 80 units of protein, 200 units of carbohydrates, and 50 units of fat daily. The store carries four different types of feed that are appropriate for rabbits with the following compositions:

Feed	Protein units/oz	Carbohydrates units/oz	Fat units/oz
A	5	20	3
B	4	30	3
C	8	15	10
D	12	5	7

The store wants to determine a blend of these four feeds that will meet the daily requirements of the rabbits. Show that this problem is equivalent to solving three equations in the four unknowns A, B, C, and D, where each unknown denotes the number of ounces of that feed in the blend.

16. A small company computes its end-of-the-year bonus b as 5% of the net profit after city and state taxes have been paid. The city tax c is 2% of taxable income, while the state tax s is 3% of taxable income with credit allowed for the city tax as a pretax deduction. This year, taxable income was $400,000. Show that b, c, and s are related by three simultaneous equations.

17. A gasoline producer has $800,000 in fixed annual costs and incurs an additional variable cost of $30 per barrel B of gasoline. The total cost C is the sum of the fixed and variable costs. The net sales S is computed on a wholesale price of $40 per barrel. (a) Show that C, B, and S are related by two simultaneous equations. (b) Show that the problem of determining how many barrels must be produced to break even, that is, for net sales to equal cost, is equivalent to solving a system of three equations.

18. (**Leontief Closed Models**) A closed economic model involves a society in which all the goods and services produced by members of the society are consumed by those members. No goods and services are imported from without and none are exported. Such a system involves N members, each of whom produces goods or services and charges for their use. The problem is to determine the prices each member should charge for his or her labor so that everyone breaks even after 1 year. For simplicity, it is assumed that each member produces one unit per year.

Consider a simple closed system consisting of a farmer, a carpenter, and a weaver. The farmer produces one unit of food each year, the carpenter produces one unit of finished wood products each year, and the weaver produces one unit of clothing each year. Let p_1 denote the farmer's annual income (that is, the price she charges for her unit of food), let p_2 denote the carpenter's annual income (that is, the price he charges for his unit of finished wood products), and let p_3 denote the weaver's annual income. Assume on an annual basis that the farmer and the carpenter consume 40% each of the available food, while the weaver eats the remaining 20%. Assume that the carpenter uses 25% of the wood products he makes, while the farmer uses 30% and the weaver uses 45%. Assume further that the farmer uses 50% of the weaver's clothing, while the carpenter uses 35% and the weaver consumes the remaining 15%. Show that a break-even equation for the farmer is

$$0.40p_1 + 0.30p_2 + 0.50p_3 = p_1,$$

while the break-even equation for the carpenter is

$$0.40p_1 + 0.25p_2 + 0.35p_3 = p_2.$$

What is the break-even equation for the weaver? Rewrite all three equations as a homogeneous system.

19. Paul, Jim, and Mary decide to help each other build houses. Paul will spend half his time on his own house and a quarter of his time on each of the houses of Jim and Mary. Jim will spend one third of his time on each of the three houses under construction. Mary will spend one sixth of her time on Paul's house, one third on Jim's house, and one half of her time on her own house. For tax purposes, each must place a price on his or her labor, but they want to do so in a way that each will break even. Show that the process of determining break-even wages is a Leontief closed model comprised of three homogeneous equations.

20. Four third world countries each grow a different fruit for export and each uses the income from that fruit to pay for imports of the fruits from the other countries. Country A exports 20% of its fruit to country B, 30% to country C, 35% to country D, and uses the rest of its fruit for internal consumption. Country B exports 10% of its fruit to country A, 15% to country C, 35% to country D, and retains the rest for its own citizens. Country C does not export to country A; it divides its crop equally between countries B and D and its own people. Country D does not consume its own fruit. All of its fruit is for export with 15% going to country A, 40% to country B, and 45% to country C. Show that the problem of determining prices on the annual harvests of fruit so that each country breaks even is equivalent to solving four homogeneous equations in four unknowns.

21. (**Leontief Input–Output Models**) Consider an economy consisting of N sectors, with each producing goods or services unique to that sector. Let x_i denote the amount produced by the ith sector, measured in dollars. Thus, x_i represents the dollar value of the supply of product i available in the economy. Assume that every sector in the economy has a demand for a proportion (which may be zero) of the output of every other sector. Thus, each sector j has a demand, measured in dollars, for the item produced in sector i. Let a_{ij} denote the proportion of item j's revenues that must be committed to the purchase of items from sector i in order for sector j to produce its goods or services. Assume also that there is an external demand, denoted by d_i and measured in dollars, for each item produced in the economy.

   The problem is to determine how much of each item should be produced to meet external demand without creating a surplus of any item. Show that for a two-sector economy, the solution to this problem is given by the supply/demand equations

$$\begin{array}{c}\underline{\text{supply}} \quad \underline{\text{demand}} \\ x_1 = a_{11}x_1 + a_{12}x_2 + d_1, \\ x_2 = a_{21}x_1 + a_{22}x_2 + d_2. \end{array}$$

   Show that this system is equivalent to the matrix equations

$$\mathbf{x} = \mathbf{A}\mathbf{x} + \mathbf{d} \quad \text{and} \quad (\mathbf{I} - \mathbf{A})\mathbf{x} = \mathbf{d}.$$

   In this formulation, $\mathbf{A}$ is called the *consumption matrix* and $\mathbf{d}$ the *demand vector*.

22. Determine $\mathbf{A}$ and $\mathbf{d}$ in the previous problem if sector 1 must expend half of its revenues purchasing goods from its own sector and one third of its revenues purchasing goods from the other sector, while sector 2 must expend one quarter of its revenues purchasing items from sector 1 and requires nothing from itself. In addition, the demand for items from these two sectors is $20,000 and $30,000, respectively.

23. A small town has three primary industries, coal mining (sector 1), transportation (sector 2), and electricity (sector 3). Production of one dollar of coal requires the purchase of 10 cents of electricity and 20 cents of transportation. Production of one dollar of transportation requires the purchase of 2 cents of coal and 35 cents of electricity. Production of one unit of electricity requires the purchase of 10 cents of electricity, 50 cents of coal, and 30 cents of transportation. The town has external contracts for $50,000 of coal, $80,000 of transportation, and $30,000 units of electricity. Show that the problem of determining how much coal, electricity, and transportation is required to supply the external demand without a surplus is equivalent to solving a Leontief input–output model. What are $\mathbf{A}$ and $\mathbf{d}$?

24. An economy consists of four sectors: energy, tourism, transportation, and construction. Each dollar of income from energy requires the expenditure of

20 cents on energy costs, 10 cents on transportation, and 30 cents on construction. Each dollar of income received by the tourism sector requires the expenditure of 20 cents on tourism (primarily in the form of complimentary facilities for favored customers), 15 cents on energy, 5 cents on transportation, and 30 cents on construction. Each dollar of income from transportation requires the expenditure of 40 cents on energy and 10 cents on construction; while each dollar of income from construction requires the expenditure of 5 cents on construction, 25 cents on energy, and 10 cents on transportation. The only external demand is for tourism, and this amounts to $5 million a year. Show that the problem of determining how much energy, tourism, transportation, and construction are required to supply the external demand without a surplus is equivalent to solving a Leontief input−output model. What are **A** and **d**?

25. A constraint is often imposed on each column of the consumption matrix of a Leontief input−output model that the sum of the elements in each column be less than unity. Show that this guarantees that each sector in the economy is profitable.

2.2 Solutions by substitution

Most readers have probably encountered simultaneous equations in high school algebra. At that time, matrices were not available; hence, other methods were developed to solve these systems, in particular, the method of substitution. We review this method in this section. In the next section, we develop its matrix equivalent, which is slightly more efficient and, more importantly, better suited for computer implementations.

Consider the system given by (1):

$$a_{11}x_1 + a_{12}x_2 + \cdots + a_{1n}x_n = b_1,$$
$$a_{21}x_1 + a_{22}x_2 + \cdots + a_{2n}x_n = b_2,$$
$$\vdots$$
$$a_{m1}x_1 + a_{m2}x_2 + \cdots + a_{mn}x_n = b_m.$$

The method of substitution is the following: take the first equation and solve for x_1 in terms of $x_2, x_3, \ldots, x_n$ and then substitute this value of x_1 into all the other equations, thus eliminating it from those equations. (If x_1 does not appear in the first equation, rearrange the equations so that it does. For example, one might have to interchange the order of the first and second equations.) This new set of equations is called the *first derived set*. Working with the first derived set, solve the second equation for x_2 in terms of $x_3, x_4, \ldots, x_n$ and then substitute this value of x_2 into the third, fourth, etc., equations, thus eliminating it. This new set is the second derived set. This process is kept up until the following set of equations is obtained:

$$x_1 = c_{12}x_2 + c_{13}x_3 + c_{14}x_4 + \cdots + c_{1n}x_n + d_1,$$
$$x_2 = \qquad c_{23}x_3 + c_{24}x_4 + \cdots + c_{2n}x_n + d_2,$$
$$x_3 = \qquad\qquad c_{34}x_4 + \cdots + c_{3n}x_n + d_3, \qquad\qquad (6)$$
$$\vdots$$
$$x_m = \qquad c_{m,m+1}x_{m+1} + \cdots + c_{mn}x_n + d_m,$$

where the c_{ij}'s and the d_i's are some combination of the original a_{ij}'s and b_i's. System (6) can be quickly solved by back-substitution.

Example 1.   Use the method of substitution to solve the system

$$r + 2s + t = 3,$$
$$2r + 3s - t = -6,$$
$$3r - 2s - 4t = -2.$$

Solution By solving the first equation for r and then substituting it into the second and third equations, we obtain the first derived set

$$r = 3 - 2s - t,$$
$$-s - 3t = -12,$$
$$-8s - 7t = -11.$$

By solving the second equation for s and then substituting it into the third equation, we obtain the second derived set

$$r = 3 - 2s - t,$$
$$s = 12 - 3t,$$
$$17t = 85.$$

By solving for t in the third equation and then substituting it into the remaining equations (of which there are none), we obtain the third derived set

$$r = 3 - 2s - t,$$
$$s = 12 - 3t,$$
$$t = 5.$$

Thus, the solution is $t = 5$, $s = -3$, $r = 4$. ∎

Example 2.   Use the method of substitution to solve the system

$$x + y + 3z = -1,$$
$$2x - 2y - z = 1,$$
$$5x + y + 8z = -2.$$

Solution The first derived set is

$$x = -1 - y - 3z,$$
$$-4y - 7z = 3,$$
$$-4y - 7z = 3.$$

The second derived set is

$$x = -1 - y - 3z,$$

$$y = -\frac{3}{4} - \frac{7}{4}z,$$

$$0 = 0.$$

Since the third equation cannot be solved for z, this is as far as we can go. Thus, since we cannot obtain a unique value for z, the first and second equations will not yield a unique value for x and y. *Caution*: The third equation does *not* imply that $z = 0$. On the contrary, this equation says nothing at all about z, consequently, z is completely arbitrary. The second equation gives y in terms of z. Substituting this value into the first equation, we obtain x in terms of z. The solution therefore is $x = -\frac{1}{4} - \frac{5}{4}z$ and $y = -\frac{3}{4} - \frac{7}{4}z$, z is arbitrary. Thus, there are infinitely many solutions to the above system. However, once z is chosen, x and y are determined. If z is chosen to be -1, then $x = y = 1$, while if z is chosen to be 3, then $x = -4$, $y = -6$. The solutions can be expressed in the vector form

$$\begin{bmatrix} x \\ y \\ z \end{bmatrix} = \begin{bmatrix} -\dfrac{1}{4} - \dfrac{5}{4}z \\ -\dfrac{3}{4} - \dfrac{7}{4}z \\ z \end{bmatrix} = \begin{bmatrix} -\dfrac{1}{4} \\ -\dfrac{3}{4} \\ 0 \end{bmatrix} + z \begin{bmatrix} -\dfrac{5}{4} \\ -\dfrac{7}{4} \\ 1 \end{bmatrix} . \blacksquare$$

Example 3.   Use the method of substitution to solve

$$a + 2b - 3c + d = 1,$$

$$2a + 6b + 4c + 2d = 8.$$

Solution The first derived set is

$$a = 1 - 2b + 3c - d,$$

$$2b + 10c = 6.$$

The second derived set is

$$a = 1 - 2b + 3c - d$$

$$b = 3 - 5c$$

Again, since there are no more equations, this is as far as we can go, and since there are no defining equations for c and d, these two unknowns must be arbitrary. Solving for a and b in terms of c and d, we obtain the solution $a = -5 + 13c - d$, $b = 3 - 5c$; c and d are arbitrary. The solutions can be expressed in the vector form

$$\begin{bmatrix} a \\ b \\ c \\ d \end{bmatrix} = \begin{bmatrix} -5 + 13c - d \\ 3 - 5c \\ c \\ d \end{bmatrix} = \begin{bmatrix} -5 \\ 3 \\ 0 \\ 0 \end{bmatrix} + c \begin{bmatrix} 13 \\ -5 \\ 1 \\ 0 \end{bmatrix} + d \begin{bmatrix} -1 \\ 0 \\ 0 \\ 1 \end{bmatrix} .$$

Note that while c and d are arbitrary, once they are given a particular value, a and b are automatically determined. For example, if c is chosen as -1 and d as 4, a solution is $a = -22$, $b = 8$, $c = -1$, $d = 4$, while if c is chosen as 0 and d as -3, a solution is $a = -2$, $b = 3$, $c = 0$, $d = -3$. $\blacksquare$

Example 4. Use the method of substitution to solve the following system:

$$x + 3y = 4,$$
$$2x - y = 1,$$
$$3x + 2y = 5,$$
$$5x + 15y = 20.$$

Solution The first derived set is

$$x = 4 - 3y,$$
$$-7y = -7,$$
$$-7y = -7,$$
$$0 = 0.$$

The second derived set is

$$x = 4 - 3y,$$
$$y = 1,$$
$$0 = 0,$$
$$0 = 0.$$

Thus, the solution is $y = 1$, $x = 1$, or in vector form

$$\begin{bmatrix} x \\ y \end{bmatrix} = \begin{bmatrix} 1 \\ 1 \end{bmatrix}. \blacksquare$$

Problems 2.2

Use the method of substitution to solve the following systems:

1. $x + 2y - 2z = -1,$
   $2x + y + z = 5,$
   $-x + y - z = -2.$

2. $x + y - z = 0,$
   $3x + 2y + 4z = 0.$

3. $x + 3y = 4,$
    $2x - y = 1,$
   $-2x - 6y = -8,$
    $4x - 9y = -5,$
   $-6x + 3y = -3.$

4. $4r - 3s + 2t = 1,$
    $r + s - 3t = 4,$
   $5r - 2s - t = 5.$

5. $2l - m + n - p = 1,$
$l + 2m - n + 2p = -1,$
$l - 3m + 2n - 3p = 2.$

6. $2x + y - z = 0,$
$x + 2y + z = 0,$
$3x - y + 2z = 0.$

7. $x + 2y - z = 5,$
$2x - y + 2z = 1,$
$2x + 2y - z = 7,$
$x + 2y + z = 3.$

8. $x + 2y + z - 2w = 1,$
$2x + 2y - z - w = 3,$
$2x - 2y + 2z + 3w = 3,$
$3x + y - 2z - 3w = 1.$

2.3 Gaussian elimination

Although the method of substitution is straightforward, it is not the most efficient way to solve simultaneous equations and it does not lend itself well to electronic computing. Computers have difficulty symbolically manipulating the unknowns in algebraic equations. A striking feature of the method of substitution, however, is that the unknowns remain unaltered throughout the process: x remains x, y remains y, z remains z. Only the coefficients of the unknowns and the numbers on the right side of the equations change from one derived set to the next. Thus, we can save a good deal of writing, and develop a useful representation for computer processing, if we direct our attention to just the numbers themselves.

Definition 1.   Given the system $\mathbf{Ax} = \mathbf{b}$, the *augmented matrix*, designated by $\mathbf{A^b}$, is a matrix obtained from $\mathbf{A}$ by adding to it one extra column, namely $\mathbf{b}$.
    Thus, if

$$\mathbf{A} = \begin{bmatrix} 1 & 2 & 3 \\ 4 & 5 & 6 \end{bmatrix} \quad \text{and} \quad \mathbf{b} = \begin{bmatrix} 7 \\ 8 \end{bmatrix},$$

then

$$\mathbf{A^b} = \begin{bmatrix} 1 & 2 & 3 & 7 \\ 4 & 5 & 6 & 8 \end{bmatrix},$$

while if

$$\mathbf{A} = \begin{bmatrix} 1 & 2 & 3 \\ 4 & 5 & 6 \\ 7 & 8 & 9 \end{bmatrix} \quad \text{and} \quad \mathbf{b} = \begin{bmatrix} -1 \\ -2 \\ -3 \end{bmatrix},$$

then

$$\mathbf{A^b} = \begin{bmatrix} 1 & 2 & 3 & -1 \\ 4 & 5 & 6 & -2 \\ 7 & 8 & 9 & -3 \end{bmatrix}.$$

In particular, the system

$$x + y - 2z = -3,$$
$$2x + 5y + 3z = 11,$$
$$-x + 3y + z = 5.$$

has the matrix representation

$$\begin{bmatrix} 1 & 1 & -2 \\ 2 & 5 & 3 \\ -1 & 3 & 1 \end{bmatrix} \begin{bmatrix} x \\ y \\ z \end{bmatrix} \begin{bmatrix} -3 \\ 11 \\ 5 \end{bmatrix},$$

with an augmented matrix of

$$\mathbf{A^b} = \begin{bmatrix} 1 & 1 & -2 & -3 \\ 2 & 5 & 3 & 11 \\ -1 & 3 & 1 & 5 \end{bmatrix}.$$

Example 1.   Write the set of equations in x, y, and z associated with the augmented matrix

$$\mathbf{A^b} = \begin{bmatrix} -2 & 1 & 3 & 8 \\ 0 & 4 & 5 & -3 \end{bmatrix}.$$

Solution

$$-2x + y + 3z = 8,$$
$$4y + 5z = -3. \quad \blacksquare$$

A second striking feature to the method of substitution is that every derived set is different from the system that preceded it. The method continues creating new derived sets until it has one that is particularly easy to solve by back-substitution. Of course, there is no purpose in solving any derived set, regardless how easy it is, unless we are assured beforehand that it has the same solution as the original system. Three elementary operations that alter equations but do not change their solutions are:

(i) Interchange the positions of any two equations.
(ii) Multiply an equation by a nonzero scalar.
(iii) Add to one equation a scalar times another equation.

If we restate these operations in words appropriate to an augmented matrix, we obtain the *elementary row operations*:

(E1) Interchange any two rows in a matrix.
(E2) Multiply any row of a matrix by a nonzero scalar.
(E3) Add to one row of a matrix a scalar times another row of that same matrix.

Gaussian elimination is a matrix method for solving simultaneous linear equations. The augmented matrix for the system is created and then it is transformed into a row-reduced matrix (see Section 1.4) using elementary row operations. This is most often accomplished by using operation (E3) with each diagonal element in a matrix to create zeros in all columns directly below it, beginning with the first column and moving successively through the matrix, column by column. The system of equations associated with a row-reduced matrix can be solved easily by back-substitution, if we solve each equation for the first unknown that appears in it. This is the unknown associated with the first nonzero element in each nonzero row of the final augmented matrix.

Example 2. Use Gaussian elimination to solve

$$x + 3y = 4,$$
$$2x - y = 1,$$
$$3x + 2y = 5,$$
$$5x + 15y = 20.$$

Solution The augmented matrix for this system is

$$\begin{bmatrix} 1 & 3 & 4 \\ 2 & -1 & 1 \\ 3 & 2 & 5 \\ 5 & 15 & 20 \end{bmatrix}.$$

Then,

$$
\begin{bmatrix} 1 & 3 & 4 \\ 2 & -1 & 1 \\ 3 & 2 & 5 \\ 5 & 15 & 20 \end{bmatrix} \rightarrow \begin{bmatrix} 1 & 3 & 4 \\ 0 & -7 & -7 \\ 3 & 2 & 5 \\ 5 & 15 & 20 \end{bmatrix}
$$
by adding to the second row (-2) times the first row

$$
\rightarrow \begin{bmatrix} 1 & 3 & 4 \\ 0 & -7 & -7 \\ 0 & -7 & -7 \\ 5 & 15 & 20 \end{bmatrix}
$$
by adding to the third row (-3) times the first row

$$
\rightarrow \begin{bmatrix} 1 & 3 & 4 \\ 0 & -7 & -7 \\ 0 & -7 & -7 \\ 0 & 0 & 0 \end{bmatrix}
$$
by adding to the fourth row (-5) times the first row

$$
\rightarrow \begin{bmatrix} 1 & 3 & 4 \\ 0 & 1 & 1 \\ 0 & -7 & -7 \\ 0 & 0 & 0 \end{bmatrix}
$$
by multiplying the second row by $\dfrac{-1}{7}$

$$
\rightarrow \begin{bmatrix} 1 & 3 & 4 \\ 0 & 1 & 1 \\ 0 & 0 & 0 \\ 0 & 0 & 0 \end{bmatrix} .
$$
by adding to the second row (7) times the first row

The system of equations associated with this last augmented matrix in row-reduced form is

$$
\begin{aligned}
x + 3y &= 4, \\
y &= 1, \\
0 &= 0, \\
0 &= 0.
\end{aligned}
$$

Solving the second equation for y and then the first equation for x, we obtain $x = 1$ and $y = 1$, which is also the solution to the original set of equations. Compare this solution with Example 4 of the previous section. ■

The notation ($\rightarrow$) should be read "is transformed into"; an equality sign is not correct because the transformed matrix is not equal to the original one.

Example 3. Use Gaussian elimination to solve

$$r + 2s + t = 3,$$
$$2r + 3s - t = -6,$$
$$3r - 2s - 4t = -2.$$

Solution The augmented matrix for this system is

$$\begin{bmatrix} 1 & 2 & 1 & 3 \\ 2 & 3 & -1 & -6 \\ 3 & -2 & -4 & -2 \end{bmatrix}.$$

Then,

$$\begin{bmatrix} 1 & 2 & 1 & 3 \\ 2 & 3 & -1 & -6 \\ 3 & -2 & -4 & -2 \end{bmatrix} \rightarrow \begin{bmatrix} 1 & 2 & 1 & 3 \\ 0 & -1 & -3 & -12 \\ 3 & -2 & -4 & -12 \end{bmatrix} \quad \left\{ \begin{array}{l} \text{by adding to the} \\ \text{second row } (-2) \text{ times} \\ \text{the first row} \end{array} \right.$$

$$\rightarrow \begin{bmatrix} 1 & 2 & 1 & 3 \\ 0 & -1 & -3 & -12 \\ 0 & -8 & -7 & -11 \end{bmatrix} \quad \left\{ \begin{array}{l} \text{by adding to the} \\ \text{third row } (-3) \text{ times} \\ \text{the first row} \end{array} \right.$$

$$\rightarrow \begin{bmatrix} 1 & 2 & 1 & 3 \\ 0 & 1 & 3 & 12 \\ 0 & -8 & -7 & -11 \end{bmatrix} \quad \left\{ \begin{array}{l} \text{by multiplying the} \\ \text{second row by } (-1) \end{array} \right.$$

$$\rightarrow \begin{bmatrix} 1 & 2 & 1 & 3 \\ 0 & 1 & 3 & 12 \\ 0 & 0 & 17 & 85 \end{bmatrix} \quad \left\{ \begin{array}{l} \text{by adding to the} \\ \text{third row } (8) \text{ times} \\ \text{the second row} \end{array} \right.$$

$$\rightarrow \begin{bmatrix} 1 & 2 & 1 & 3 \\ 0 & 1 & 3 & 12 \\ 0 & 0 & 1 & 5 \end{bmatrix}. \quad \left\{ \begin{array}{l} \text{by multiplying the} \\ \text{third row by } \left(\dfrac{1}{17}\right) \end{array} \right.$$

The system of equations associated with this last augmented matrix in row-reduced form is

$$r + 2s + t = 3,$$
$$s + 3t = 12,$$
$$t = 5.$$

Solving the third equation for t, then the second equation for s, and, lastly, the first equation for r, we obtain $r = 4$, $s = -3$, and $t = 5$, which is also the solution to the original set of equations. Compare this solution with Example 1 of the previous section. ∎

Whenever one element in a matrix is used to cancel another element to zero by elementary row operation (E3), the first element is called the *pivot*. In Example 3, we first used the element in the $1-1$ position to cancel the element in the $2-1$ position, and then to cancel the element in the $3-1$ position. In both of these operations, the unity element in the $1-1$ position was the pivot. Later, we used the unity element in the $2-2$ position to cancel the element -8 in the $3-2$ position; here, the $2-2$ element was the pivot.

While transforming a matrix into row-reduced form, it is advisable to adhere to three basic principles:

- Completely transform one column to the required form before considering another column.
- Work on columns in order, from left to right.
- Never use an operation if it will change a zero in a previously transformed column.

As a consequence of this last principle, one never involves the ith row of a matrix in an elementary row operation after the ith column has been transformed into its required form. That is, once the first column has the proper form, no pivot element should ever again come from the first row; once the second column has the proper form, no pivot element should ever again come from the second row; and so on.

When an element we want to use as a pivot is itself zero, we interchange rows using operation (E1).

Example 4.   Use Gaussian elimination to solve

$$2c + 3d = 4,$$
$$a + 3c + d = 2,$$
$$a + b + 2c = 0.$$

Solution The augmented matrix is

$$\begin{bmatrix} 0 & 0 & 2 & 3 & 4 \\ 1 & 0 & 3 & 1 & 2 \\ 1 & 1 & 2 & 0 & 0 \end{bmatrix}.$$

Normally, we would use the element in the $1-1$ position to cancel to zero the two elements directly below it, but we cannot because it is zero. To proceed with the reduction process, we must interchange the first row with either of the other two rows. The choice is arbitrary.

$$\begin{bmatrix} 0 & 0 & 2 & 3 & 4 \\ 1 & 0 & 3 & 1 & 2 \\ 1 & 1 & 2 & 0 & 0 \end{bmatrix} \rightarrow \begin{bmatrix} 1 & 0 & 3 & 1 & 2 \\ 0 & 0 & 2 & 3 & 4 \\ 1 & 1 & 2 & 0 & 0 \end{bmatrix} \quad \left\{ \begin{array}{l} \text{by interchanging the} \\ \text{first row with the} \\ \text{second row} \end{array} \right.$$

$$\rightarrow \begin{bmatrix} 1 & 0 & 3 & 1 & 2 \\ 0 & 0 & 2 & 3 & 4 \\ 0 & 1 & -1 & -1 & -2 \end{bmatrix}. \quad \left\{ \begin{array}{l} \text{by adding to the} \\ \text{third row } (-1) \text{ times} \\ \text{the first row} \end{array} \right.$$

Next, we would like to use the element in the 2–2 position to cancel to zero the element in the 3 –2 position, but we cannot because that prospective pivot is zero. We use elementary row operation (E1) once again. The transformation yields

$$\rightarrow \begin{bmatrix} 1 & 0 & 3 & 1 & 2 \\ 0 & 1 & -1 & -1 & -2 \\ 0 & 0 & 2 & 3 & 4 \end{bmatrix} \quad \left\{ \begin{array}{l} \text{by interchanging the} \\ \text{second row with the} \\ \text{third row} \end{array} \right.$$

$$\rightarrow \begin{bmatrix} 1 & 0 & 3 & 1 & 2 \\ 0 & 1 & -1 & -1 & -2 \\ 0 & 0 & 1 & 1.5 & 2 \end{bmatrix}. \quad \left\{ \begin{array}{l} \text{by multiplying the} \\ \text{third row by } (0.5) \end{array} \right.$$

The system of equations associated with this last augmented matrix in row-reduced form is

$$a + 3c + d = 2,$$
$$b - c - d = -2,$$
$$c + 1.5d = 2.$$

We use the third equation to solve for c, the second equation to solve for b, and the first equation to solve for a, because these are the unknowns associated with the first nonzero element of each nonzero row in the final augmented matrix. We have no defining equation for d, so this unknown remains arbitrary. The solution is $a = -4 + 3.5d$, $b = -0.5d$, $c = 2 - 1.5d$, and d arbitrary, or in vector form

$$\begin{bmatrix} a \\ b \\ c \\ d \end{bmatrix} = \begin{bmatrix} -4 + 3.5d \\ -0.5d \\ 2 - 1.5d \\ d \end{bmatrix} = \begin{bmatrix} -4 \\ 0 \\ 2 \\ 0 \end{bmatrix} + \frac{d}{2} \begin{bmatrix} 7 \\ -1 \\ -3 \\ 2 \end{bmatrix}.$$

This is also the solution to the original set of equations. ∎

The derived set of equations associated with a row-reduced, augmented matrix may contain an absurd equation, such as $0 = 1$. In such cases, we conclude that the derived set is inconsistent, because no values of the unknowns can simultaneously satisfy all the equations. In particular, it is impossible to choose values of the unknowns that will make the absurd equation true. Since the derived set has the same solutions as the original set, it follows that the original set of equations is also inconsistent.

Example 5.   Use Gaussian elimination to solve

$$2x + 4y + 3z = 8,$$
$$3x - 4y - 4z = 3,$$
$$5x - z = 12.$$

Solution The augmented matrix for this system is

$$\begin{bmatrix} 2 & 4 & 3 & 8 \\ 3 & -4 & -4 & 3 \\ 5 & 0 & -1 & 12 \end{bmatrix}.$$

Then,

$$
\begin{bmatrix} 2 & 4 & 3 & 8 \\ 3 & -4 & -4 & 3 \\ 5 & 0 & -1 & 12 \end{bmatrix} \rightarrow \begin{bmatrix} 1 & 2 & 1.5 & 4 \\ 3 & -4 & -4 & 3 \\ 5 & 0 & -1 & 12 \end{bmatrix}
$$

$$\left\{ \begin{array}{l} \text{by multiplying the} \\ \text{first row by } \left(\dfrac{1}{2}\right) \end{array} \right.$$

$$
\rightarrow \begin{bmatrix} 1 & 2 & 1.5 & 4 \\ 0 & -10 & -8.5 & -9 \\ 5 & 0 & -1 & -12 \end{bmatrix}
$$

$$\left\{ \begin{array}{l} \text{by adding to the} \\ \text{second row } (-3) \text{ times} \\ \text{the first row} \end{array} \right.$$

$$
\rightarrow \begin{bmatrix} 1 & 2 & 1.5 & 4 \\ 0 & -10 & -8.5 & -9 \\ 0 & -10 & -8.5 & -8 \end{bmatrix}
$$

$$\left\{ \begin{array}{l} \text{by adding to the} \\ \text{third row } (-5) \text{ times} \\ \text{the first row} \end{array} \right.$$

$$
\rightarrow \begin{bmatrix} 1 & 2 & 1.5 & 4 \\ 0 & 1 & 0.85 & 0.9 \\ 0 & -10 & -8.5 & -8 \end{bmatrix}
$$

$$\left\{ \begin{array}{l} \text{by multiplying the} \\ \text{second row by } \left(\dfrac{-1}{10}\right) \end{array} \right.$$

$$
\rightarrow \begin{bmatrix} 1 & 2 & 1.5 & 4 \\ 0 & 1 & 0.85 & 0.9 \\ 0 & 0 & 0 & 1 \end{bmatrix}.
$$

$$\left\{ \begin{array}{l} \text{by adding to the} \\ \text{third row } (10) \text{ times} \\ \text{the second row} \end{array} \right.$$

The system of equations associated with this last augmented matrix in row-reduced form is

$$x + 2y + 1.5z = 4,$$
$$y + 0.85z = 0.9,$$
$$0 = 1.$$

Since no values of x, y, and z can make this last equation true, this system, as well as the original one, has no solution. ■

Finally, we note that most matrices can be transformed into a variety of row-reduced forms. If a row-reduced matrix has two nonzero rows, then a different row-reduced matrix is easily constructed by adding to the first row any nonzero constant times the second row. The equations associated with both augmented matrices, however, will have identical solutions.

Problems 2.3

In Problems 1 through 5, construct augmented matrices for the given systems of equations:

1. $x + 2y = -3,$
   $3x + y = 1.$

2. $x + 2y - z = -1,$
   $2x - 3y + 2z = 4.$

3. $a + 2b = 5,$
   $-3a + b = 13,$
   $4a + 3b = 0.$

4. $2r + 4s = 2,$
   $3r + 2s + t = 8,$
   $5r - 3s + 7t = 15.$

5. $2r + 3s - 4t = 12,$
   $3r - 2s = -1,$
   $8r - s - 4t = 10.$

6. $\mathbf{A^b} = \begin{bmatrix} 1 & 2 & 5 \\ 0 & 1 & 7 \end{bmatrix}$ variables : x and y.

7. $\mathbf{A^b} = \begin{bmatrix} 1 & -2 & 3 & 10 \\ 0 & 1 & -5 & -3 \\ 0 & 0 & 1 & 4 \end{bmatrix}$ variables : x, y, and z.

8. $\mathbf{A^b} = \begin{bmatrix} 1 & -3 & 12 & 40 \\ 0 & 1 & -6 & -200 \\ 0 & 0 & 1 & 25 \end{bmatrix}$ variables : r, s, and t.

9. $\mathbf{A^b} = \begin{bmatrix} 1 & 3 & 0 & -8 \\ 0 & 1 & 4 & 2 \\ 0 & 0 & 0 & 0 \end{bmatrix}$ variables : x, y, and z.

10. $\mathbf{A^b} = \begin{bmatrix} 1 & -7 & 2 & 0 \\ 0 & 1 & -1 & 0 \\ 0 & 0 & 0 & 0 \end{bmatrix}$ variables : a, b, and c.

11. $\mathbf{A}^{\mathbf{b}} = \begin{bmatrix} 1 & -1 & 0 & 1 \\ 0 & 1 & -2 & 2 \\ 0 & 0 & 0 & 1 \end{bmatrix}$ variables : u, v, and w.

12. Solve the system of equations defined in Problem 6.
13. Solve the system of equations defined in Problem 7.
14. Solve the system of equations defined in Problem 8.
15. Solve the system of equations defined in Problem 9.
16. Solve the system of equations defined in Problem 10.
17. Solve the system of equations defined in Problem 11.

In Problems 18 through 24, use elementary row operations to transform the given matrices into row-reduced form:

18. $\begin{bmatrix} 1 & -2 & 5 \\ -3 & 7 & 8 \end{bmatrix}.$

19. $\begin{bmatrix} 4 & 24 & 20 \\ 2 & 11 & -8 \end{bmatrix}.$

20. $\begin{bmatrix} 0 & -1 & 6 \\ 2 & 7 & -5 \end{bmatrix}.$

21. $\begin{bmatrix} 1 & 2 & 3 & 4 \\ -1 & -1 & 2 & 3 \\ -2 & 3 & 0 & 0 \end{bmatrix}.$

22. $\begin{bmatrix} 0 & 1 & -2 & 4 \\ 1 & 3 & 2 & 1 \\ -2 & 3 & 1 & 2 \end{bmatrix}.$

23. $\begin{bmatrix} 1 & 3 & 2 & 0 \\ -1 & -4 & 3 & -1 \\ 2 & 0 & -1 & 3 \\ 2 & -1 & 4 & 2 \end{bmatrix}.$

24. $\begin{bmatrix} 2 & 3 & 4 & 6 & 0 & 10 \\ -5 & -8 & 15 & 1 & 3 & 40 \\ 3 & 3 & 5 & 4 & 4 & 20 \end{bmatrix}.$

25. Solve Problem 1.

26. Solve Problem 2.

27. Solve Problem 3.

28. Solve Problem 4.

29. Solve Problem 5.

30. Use Gaussian elimination to solve Problem 1 of Section 2.2.

31. Use Gaussian elimination to solve Problem 2 of Section 2.2.

32. Use Gaussian elimination to solve Problem 3 of Section 2.2.

33. Use Gaussian elimination to solve Problem 4 of Section 2.2.

34. Use Gaussian elimination to solve Problem 5 of Section 2.2.

35. Determine a production schedule that satisfies the requirements of the manufacturer described in Problem 12 of Section 2.1.

36. Determine a production schedule that satisfies the requirements of the manufacturer described in Problem 13 of Section 2.1.

37. Determine a production schedule that satisfies the requirements of the manufacturer described in Problem 14 of Section 2.1.

38. Determine feed blends that satisfy the nutritional requirements of the pet store described in Problem 15 of Section 2.1.

39. Determine the bonus for the company described in Problem 16 of Section 2.1.

40. Determine the number of barrels of gasoline that the producer described in Problem 17 of Section 2.1 must manufacture to break even.

41. Determine the annual incomes of each sector of the Leontief closed model described in Problem 18 of Section 2.1.

42. Determine the wages of each person in the Leontief closed model described in Problem 19 of Section 2.1.

43. Determine the total sales revenue for each country of the Leontief closed model described in Problem 20 of Section 2.1.

44. Determine the production quotas for each sector of the economy described in Problem 22 of Section 2.1.

45. An *elementary matrix* is a square matrix $\mathbf{E}$ having the property that the product $\mathbf{EA}$ is the result of applying a single elementary row operation on the matrix $\mathbf{A}$. Form a matrix $\mathbf{H}$ from the 4×4 identity matrix $\mathbf{I}$ by interchanging any two rows of $\mathbf{I},$ and then compute the product $\mathbf{HA}$ for any 4×4 matrix $\mathbf{A}$ of your choosing. Is $\mathbf{H}$ an elementary matrix? How would one construct elementary matrices corresponding to operation (E1)?

46. Form a matrix $\mathbf{G}$ from the 4×4 identity matrix $\mathbf{I}$ by multiplying any one row of $\mathbf{I}$ by the number 5 and then compute the product $\mathbf{GA}$ for any 4×4 matrix $\mathbf{A}$ of your choosing. Is $\mathbf{G}$ an elementary matrix? How would one construct elementary matrices corresponding to operation (E2)?

47. Form a matrix $\mathbf{F}$ from the 4×4 identity matrix $\mathbf{I}$ by adding to one row of $\mathbf{I}$ five times another row of $\mathbf{I}$. Use any two rows of your choosing. Compute the product $\mathbf{FA}$ for any 4×4 matrix $\mathbf{A}$ of your choosing. Is $\mathbf{F}$ an elementary matrix? How would one construct elementary matrices corresponding to operation (E3)?

48. A solution procedure uniquely suited to matrix equations of the form $\mathbf{x} = \mathbf{A}\mathbf{x} + \mathbf{d}$ is iteration. A trial solution $\mathbf{x}^{(0)}$ is proposed and then progressively better estimates $\mathbf{x}^{(1)}$, $\mathbf{x}^{(2)}$, $\mathbf{x}^{(3)}$,... for the solution are obtained iteratively from the formula

$$\mathbf{x}^{(i+1)} = \mathbf{A}\mathbf{x}^{(i)} + \mathbf{d}.$$

49. Use the iteration method described in the previous problem to solve the system defined in Problem 23 of Section 2.1. In particular, find the first two iterations by hand calculations, and then use a computer to complete the iteration process.

50. Use the iteration method described in Problem 48 to solve the system defined in Problem 24 of Section 2.1. In particular, find the first two iterations by hand calculations, and then use a computer to complete the iteration process.

2.4 **Pivoting strategies**

Gaussian elimination is often programmed for computer implementation. Since all computers round or truncate numbers to a finite number of digits (e.g., the fraction 1/3 might be stored as 0.33333, but never as the *infinite* decimal 0.333333 ...) round-off error can be significant. A number of strategies have been developed to minimize the effects of such errors.

The most popular strategy is *partial pivoting*, which requires that a pivot element is always larger in absolute value than any element below it in the same column. This is accomplished by interchanging rows whenever necessary.

Example 1.   Use partial pivoting with Gaussian elimination to solve the system

$$x + 2y + 4z = 18,$$
$$2x + 12y - 2z = 9,$$
$$5x + 26y + 5z = 14.$$

Solution The augmented matrix for this system is

$$\begin{bmatrix} 1 & 2 & 4 & 18 \\ 2 & 12 & -2 & 19 \\ 5 & 26 & 5 & 14 \end{bmatrix}.$$

Normally, the unity element in the 1−1 position would be the pivot. With partial pivoting, we compare this prospective pivot to all elements directly below it in the same column, and if any is larger in absolute value, as is the case here with the element 5 in the 3−1 position, we interchange rows to bring the largest element into the pivot position.

$$\begin{bmatrix} 1 & 2 & 4 & 18 \\ 2 & 12 & -2 & 9 \\ 5 & 26 & 5 & 14 \end{bmatrix} \rightarrow \begin{bmatrix} 5 & 26 & 5 & 14 \\ 2 & 12 & -2 & 9 \\ 1 & 2 & 4 & 18 \end{bmatrix} \quad \begin{cases} \text{by interchanging the} \\ \text{first and third rows} \end{cases}$$

Then,

$$\rightarrow \begin{bmatrix} 1 & 5.2 & 1 & 2.8 \\ 2 & 12 & -2 & 9 \\ 1 & 2 & 4 & 18 \end{bmatrix} \quad \left\{ \begin{array}{l} \text{by multiplying the} \\ \text{first row by } \dfrac{1}{5} \end{array} \right.$$

$$\rightarrow \begin{bmatrix} 1 & 5.2 & 1 & 2.8 \\ 0 & 1.6 & -4 & 3.4 \\ 1 & 2 & 4 & 18 \end{bmatrix} \quad \left\{ \begin{array}{l} \text{by adding to the} \\ \text{second row } (-2) \text{ times} \\ \text{the first row} \end{array} \right.$$

$$\rightarrow \begin{bmatrix} 1 & 5.2 & 1 & 2.8 \\ 0 & 1.6 & -4 & 3.4 \\ 0 & -3.2 & 3 & 15.2 \end{bmatrix}. \quad \left\{ \begin{array}{l} \text{by adding to the} \\ \text{third row } (-1) \text{ times} \\ \text{the first row} \end{array} \right.$$

The next pivot would normally be the element 1.6 in the 2−2 position. Before accepting it, however, we compare it to all elements directly below it in the same column. The largest element in absolute value is the element −3.2 in the 3−2 position. Therefore, we interchange rows to bring this larger element into the pivot position.

NOTE. We do not consider the element 5.2 in the 1−2 position, even though it is the largest element in its column. Comparisons are only made between a prospective pivot and all elements directly below it. Recall one of the three basic principles of row-reduction: never involve the first row of a matrix in a row operation after the first column has been transformed into its required form.

$$\rightarrow \begin{bmatrix} 1 & 5.2 & 1 & 2.8 \\ 0 & -3.2 & 3 & 15.2 \\ 0 & 1.6 & -4 & 3.4 \end{bmatrix} \quad \left\{ \begin{array}{l} \text{by interchanging the} \\ \text{second and third rows} \end{array} \right.$$

$$\rightarrow \begin{bmatrix} 1 & 5.2 & 1 & 2.8 \\ 0 & 1 & -0.9375 & -4.75 \\ 0 & 1.6 & -4 & 3.4 \end{bmatrix} \quad \left\{ \begin{array}{l} \text{by multiplying the} \\ \text{second row by } \dfrac{-1}{3.2} \end{array} \right.$$

$$\rightarrow \begin{bmatrix} 1 & 5.2 & 1 & 2.8 \\ -0 & 1 & -0.9375 & -4.75 \\ 0 & 0 & -2.5 & 11 \end{bmatrix} \quad \left\{ \begin{array}{l} \text{by adding to the} \\ \text{third row by } (-1.6) \text{ times} \\ \text{the second row} \end{array} \right.$$

$$\rightarrow \begin{bmatrix} 1 & 5.2 & 1 & 2.8 \\ 0 & 1 & -0.9375 & -4.75 \\ 0 & 0 & 1 & -4.4 \end{bmatrix} \quad \left\{ \begin{array}{l} \text{by multiplying the} \\ \text{third row by } \dfrac{-1}{2.5} \end{array} \right.$$

The new derived set of equations is

$$x + 5.2y + z = 2.8,$$
$$y - 0.9375z = -4.75,$$
$$z = -4.4,$$

which has as its solution $x = 53.35$, $y = -8.875$, and $z = -4.4$. ∎

Scaled pivoting involves ratios. A prospective pivot is divided by the largest element in absolute value in its row, ignoring the last column. The result is compared to the ratios formed by dividing every element directly below the pivot by the largest element in absolute value in its respective row, again ignoring the last column. Of these, the element that yields the largest ratio in absolute value is designated as the pivot, and if that element is not already in the pivot position, then row interchanges are performed to move it there.

Example 2.   Use scaled pivoting with Gaussian elimination to solve the system given in Example 1.
  Solution The augmented matrix for this system is

$$\begin{bmatrix} 1 & 2 & 4 & 18 \\ 2 & 12 & -2 & 9 \\ 5 & 26 & 5 & 14 \end{bmatrix}.$$

Normally, we would use the element in the $1-1$ position as the pivot. With scaled pivoting, however, we first compare ratios between elements in the first column to the largest elements in absolute value in each row, ignoring the last column. The ratios are

$$\frac{1}{4} = 0.25, \quad \frac{2}{12} = 0.167, \quad \text{and} \quad \frac{5}{26} = 0.1923.$$

The largest ratio in absolute value corresponds to the unity element in the $1-1$ position, so that element remains the pivot. Transforming the first column into reduced form, we obtain

$$\begin{bmatrix} 1 & 2 & 4 & 18 \\ 0 & 8 & -10 & -27 \\ 0 & 16 & -15 & -76 \end{bmatrix}.$$

Normally, the next pivot would be the element in the $2-2$ position. Instead, we consider the ratios

$$\frac{8}{10} = 0.8 \quad \text{and} \quad \frac{16}{16} = 1,$$

which are obtained by dividing the pivot element and every element directly below it by the largest element in absolute value appearing in their respective rows, ignoring elements in the last column. The largest ratio in absolute value corresponds to the element 16 appearing in the $3-2$ position. We move it into the pivot position by interchanging the second and third rows. The new matrix is

$$\begin{bmatrix} 1 & 2 & 4 & 18 \\ 0 & 16 & -15 & -76 \\ 0 & 8 & -10 & -27 \end{bmatrix}.$$

Completing the row-reduction transformation, we get

$$\begin{bmatrix} 1 & 2 & 4 & 18 \\ 0 & 1 & -0.9375 & -4.75 \\ 0 & 0 & 1 & -4.4 \end{bmatrix}.$$

The system of equations associated with this matrix is

$$x + 2y + 4z = 18,$$
$$y - 0.9375z = -4.75,$$
$$z = -4.4.$$

The solution is, as before, $x = 53.35$, $y = -8.875$, and $z = -4.4$. ∎

Complete pivoting compares prospective pivots with all elements in the largest submatrix for which the prospective pivot is in the upper left position, ignoring the last column. If any element in this submatrix is larger in absolute value than the prospective pivot, both row and column interchanges are made to move this larger element into the pivot position. Because column interchanges rearrange the order of the unknowns, a book-keeping method must be implemented to record all rearrangements. This is done by adding a new row, designated as row 0, to the matrix. The entries in the new row are initially the positive integers in ascending order, to denote that column 1 is associated with variable 1, column 2 with variable 2, and so on. This new top row is only affected by column interchanges; *none of the elementary row operations is applied to it.*

Example 3.   Use complete pivoting with Gaussian elimination to solve the system given in Example 1.
Solution The augmented matrix for this system is

$$\begin{bmatrix} 1 & 2 & 3 & \\ \hline 1 & 2 & 4 & 18 \\ 2 & 12 & -2 & 9 \\ 5 & 26 & 5 & 14 \end{bmatrix}.$$

Normally, we would use the element in the $1-1$ position of the coefficient matrix $\mathbf{A}$ as the pivot. With complete pivoting, however, we first compare this prospective pivot to all elements in the submatrix shaded below. In this case, the element 26 is the largest, so we interchange rows and columns to bring it into the pivot position.

$$\begin{bmatrix} 1 & 2 & 3 & \\ \hline 1 & 2 & 4 & 18 \\ 2 & 12 & -2 & 9 \\ 5 & 26 & 5 & 14 \end{bmatrix} \rightarrow \begin{bmatrix} 1 & 2 & 3 & \\ \hline 5 & 26 & 5 & 14 \\ 2 & 12 & -2 & 9 \\ 1 & 2 & 4 & 18 \end{bmatrix} \quad \left\{ \begin{array}{l} \text{by interchanging the} \\ \text{first and third rows} \end{array} \right.$$

$$\rightarrow \begin{bmatrix} 2 & 1 & 3 & \\ \hline 26 & 5 & 5 & 14 \\ 12 & 2 & -2 & 9 \\ 2 & 1 & 4 & 18 \end{bmatrix}. \quad \left\{ \begin{array}{l} \text{by interchanging the} \\ \text{first and second columns} \end{array} \right.$$

Applying Gaussian elimination to the first column, we obtain

$$\begin{bmatrix} 2 & 3 & 1 & \\ \hline 1 & 0.1923 & 0.1923 & 0.5385 \\ 0 & -0.3077 & -4.3077 & 2.5385 \\ 0 & 0.6154 & 3.6154 & 16.9231 \end{bmatrix}.$$

Normally, the next pivot would be -0.3077. Instead, we compare this number in absolute value to all the numbers in the submatrix shaded above. The largest such element in absolute value is -4.3077, which we move into the pivot position by interchanging the second and third column. The result is

$$\begin{bmatrix} 2 & 1 & 3 & \\ \hline 1 & 0.1923 & 0.1923 & 0.5385 \\ 0 & -4.3077 & -0.3077 & 2.5385 \\ 0 & 3.6154 & 0.6154 & 16.9231 \end{bmatrix}.$$

Continuing with Gaussian elimination, we obtain the row-reduced matrix

$$\begin{bmatrix} 2 & 3 & 1 & \\ \hline 1 & 0.1923 & 0.1923 & 0.5385 \\ 0 & 1 & 0.0714 & -0.5893 \\ 0 & 0 & 1 & 53.35 \end{bmatrix}.$$

The system associated with this matrix is

$$y + 0.1923z + 0.1923x = 0.5385,$$
$$z + 0.0714x = 0.5893,$$
$$x = 53.35.$$

Its solution is $x = 53.35$, $y = -8.8749$, and $z = -4.3985$, which is within the round-off error of the answers obtained previously. ■

Complete pivoting generally identifies a better pivot than scaled pivoting, which, in turn, identifies a better pivot than partial pivoting. Nonetheless, partial pivoting is most often the strategy of choice. Pivoting strategies are used to avoid round-off error. We do not need the best pivot; we only need to avoid bad pivots.

Problems 2.4

In Problems 1 through 6, determine the first pivot under (a) partial pivoting, (b) scaled pivoting, and (c) complete pivoting for given augmented matrices.

1. $\begin{bmatrix} 1 & 3 & 35 \\ 4 & 8 & 15 \end{bmatrix}.$

2. $\begin{bmatrix} 1 & -2 & -5 \\ 5 & 3 & 85 \end{bmatrix}.$

3. $\begin{bmatrix} 1 & 8 & 15 \\ 3 & -4 & 11 \end{bmatrix}.$

4. $\begin{bmatrix} -2 & 8 & -3 & 100 \\ 4 & 5 & 4 & 75 \\ -3 & -1 & 2 & 250 \end{bmatrix}.$

5. $\begin{bmatrix} 1 & 2 & 3 & 4 \\ 5 & 6 & 7 & 8 \\ 9 & 10 & 11 & 12 \end{bmatrix}.$

6.
$$\begin{bmatrix} 0 & 2 & 3 & 4 & 0 \\ 1 & 0.4 & 0.8 & 0.1 & 90 \\ 4 & 10 & 1 & 8 & 40 \end{bmatrix}.$$

7. Solve Problem 3 of Section 2.3 using Gaussian elimination with each of the three pivoting strategies.

8. Solve Problem 4 of Section 2.3 using Gaussian elimination with each of the three pivoting strategies.

9. Solve Problem 5 of Section 2.3 using Gaussian elimination with each of the three pivoting strategies.

10. Computers internally store numbers in formats similar to the scientific notation 0, −E−, representing the number 0. −multiplied by the power of 10 signified by the digits following E. Therefore, 0.1234E06 is 123,400, while 0.9935E02 is 99.35. The number of digits between the decimal point and E is finite and fixed; it is the number of significant figures. Arithmetic operations in computers are performed in registers, which have twice the number of significant figures as storage locations.

Consider the system

$$0.00001x + y = 1.00001,$$
$$x + y = 2.$$

Show that when Gaussian elimination is implemented on this system by a computer limited to four significant figures, the result is $x = 0$ and $y = 1$, which is incorrect. Show further that the difficulty is resolved when partial pivoting is employed.

2.5 **Linear independence**

We momentarily digress from our discussion of simultaneous equations to develop the concepts of linearly independent vectors and rank of a matrix, both of which will prove indispensable to us in the ensuing sections.

Definition 1. A vector $\mathbf{V}_1$ is a *linear combination* of the vectors $\mathbf{V}_2, \mathbf{V}_3, \ldots, \mathbf{V}_n$, if there exist scalars $d_2, d_3, \ldots, d_n$ such that

$$\mathbf{V}_1 = d_2\mathbf{V}_2 + d_3\mathbf{V}_3 + \cdots + d_n\mathbf{V}n.$$

Example 1. Show that [1 2 3] is a linear combination of [2 4 0] and [0 0 1].
  Solution $[1\ 2\ 3] = \frac{1}{2}[2\ 4\ 0] + 3[0\ 0\ 1].$ ∎

Referring to Example 1, we could say that the row vector [1 2 3] depends linearly on the other two vectors or, more generally, that the set of vectors {[1 2 3], [2 4 0],

[0 0 1]}is *linearly dependent*. Another way of expressing this dependence would be to say that there exist constants c_1, c_2, and c_3 not all zero such that c_1 [1 2 3] + c_2 [2 4 0] + c_3 [0 0 1] = [0 0 0]. Such a set would be $c_1 = -1$, $c_2 = \frac{1}{2}$, $c_3 = 3$. Note that the set $c_1 = c_2 = c_3 = 0$ is also a suitable set. The important fact about dependent sets, however, is that there exists a set of constants, *not all equal to zero*, that satisfies the equality.

Now consider the set given by $V_1 = [1\,0\,0]$ $V_2 = [0\,1\,0]$ $V_3 = [0\,0\,1]$. It is easy to verify that no vector in this set is a linear combination of the other two. Thus, each vector is linearly independent of the other two or, more generally, the set of vectors is *linearly independent*. Another way of expressing this independence would be to say the only scalars that satisfy the equation $c_1 [1\,0\,0] + c_2 [0\,1\,0] + c_3 [0\,0\,1] = [0\,0\,0]$ are $c_1 = c_2 = c_3 = 0$.

Definition 2. A set of vectors $\{V_1, V_2, \ldots, V_n\}$, of the same dimension, is *linearly dependent*, if there exist scalars $c_1, c_2, \ldots, c_n$, not all zero, such that

$$c_1 V_1 + c_2 V_2 + c_3 V_3 + \cdots + c_n V_n = 0 \qquad (7)$$

The vectors are *linearly independent*, if the only set of scalars that satisfies (7) is the set $c_1 = c_2 = \ldots = c_n = 0$.

Therefore, to test whether or not a given set of vectors is linearly independent, first form the vector Eq. (7) and ask "What values for the c's satisfy this equation?" Clearly $c_1 = c_2 = \ldots = c_n = 0$ is a suitable set. If this is the only set of values that satisfies (7), then the vectors are linearly independent. If there exists a set of values that is not all zero, then the vectors are linearly dependent.

Note that it is not necessary for all the c's to be different from zero for a set of vectors to be linearly dependent. Consider the vectors $V_1 = [1, 2]$, $V_2 = [1, 4]$, $V_3 = [2, 4]$. $c_1 = 2$, $c_2 = 0$, $c_3 = -1$ is a set of scalars, *not all zero*, such that $c_1 V_1 + c_2 V_2 + c_3 V_3 = 0$. Thus, this set is linearly dependent.

Example 2. Is the set $\{[1, 2], [3, 4]\}$ linearly independent?

Solution The vector equation is

$$c_1 [1\quad 2] + c_2 [3\quad 4] = [0\quad 0].$$

This equation can be rewritten as

$$[c_1\quad 2c_1] + [3c_2\quad 4c_2] = [0\quad 0]$$

or as

$$[c_1\ 3c_2\ 2c_1 + 4c_2] = [0\quad 0].$$

Equating components, we see that this vector equation is equivalent to the system

$$c_1 + 3c_2 = 0,$$
$$2c_1 + 4c_2 = 0.$$

Using Gaussian elimination, we find that the only solution to this system is $c_1 = c_2 = 0$, hence, the original set of vectors is linearly independent. ∎

Although we have worked exclusively with row vectors, the above definitions are equally applicable to column vectors.

Example 3. Is the set

$$
\left\{
\begin{bmatrix} 2 \\ 6 \\ -2 \end{bmatrix},
\begin{bmatrix} 3 \\ 1 \\ 2 \end{bmatrix},
\begin{bmatrix} 8 \\ 16 \\ -3 \end{bmatrix}
\right\}
$$

linearly independent?

Solution Consider the vector equation

$$
c_1 \begin{bmatrix} 2 \\ 6 \\ -2 \end{bmatrix}
+ c_2 \begin{bmatrix} 3 \\ 1 \\ 2 \end{bmatrix}
+ c_3 \begin{bmatrix} 8 \\ 16 \\ -3 \end{bmatrix}
= \begin{bmatrix} 0 \\ 0 \\ 0 \end{bmatrix}.
\tag{8}
$$

This equation can be rewritten as

$$
\begin{bmatrix} 2c_1 \\ 6c_1 \\ -2c_1 \end{bmatrix}
+ \begin{bmatrix} 3c_2 \\ c_2 \\ 2c_2 \end{bmatrix}
+ \begin{bmatrix} 8c_3 \\ 16c_3 \\ -3c_3 \end{bmatrix}
= \begin{bmatrix} 0 \\ 0 \\ 0 \end{bmatrix}
$$

or as

$$
\begin{bmatrix} 2c_1 + 3c_2 + 8c_3 \\ 6c_1 + c_2 + 16c_3 \\ -2c_1 + 2c_2 - 3c_3 \end{bmatrix}
= \begin{bmatrix} 0 \\ 0 \\ 0 \end{bmatrix}.
$$

By equating components, we see that this vector equation is equivalent to the system

$$
2c_1 + 3c_2 + 8c_3 = 0,
$$
$$
6c_1 + c_2 + 16c_3 = 0,
$$
$$
-2c_1 + 2c_2 - 3c_3 = 0.
$$

By using Gaussian elimination, we find that the solution to this system is $c_1 = \left(-\frac{5}{2}\right)c_3$, $c_2 = -c_3$, c_3 arbitrary. Thus, choosing $c_3 = 2$, we obtain $c_1 = -5$, $c_2 = -2$, $c_3 = 2$ as a particular nonzero set of constants that satisfies (8); hence, the original vectors are linearly dependent. ∎

Example 4. Is the set

$$
\left\{
\begin{bmatrix} 1 \\ 2 \end{bmatrix},
\begin{bmatrix} 5 \\ 7 \end{bmatrix},
\begin{bmatrix} -3 \\ 1 \end{bmatrix}
\right\}
$$

linearly independent?

Solution Consider the vector equation

$$
c_1 \begin{bmatrix} 1 \\ 2 \end{bmatrix}
+ c_2 \begin{bmatrix} 5 \\ 7 \end{bmatrix}
+ c_3 \begin{bmatrix} -3 \\ 1 \end{bmatrix}
= \begin{bmatrix} 0 \\ 0 \end{bmatrix}.
$$

This is equivalent to the system

$$
c_1 + 5c_2 - 3c_3 = 0,
$$
$$
2c_1 + 7c_2 + c_3 = 0.
$$

By using Gaussian elimination, we find that the solution to this system is $c_1 = (-26/3)c_3$, $c_2 = (7/3)c_3$, and c_3 arbitrary. Hence, a particular nonzero solution is found by choosing $c_3 = 3$; then $c_1 = -26$, $c_2 = 7$, and, therefore, the vectors are linearly dependent. ∎

We conclude this section with a few important theorems on linear independence and dependence.

Theorem 1. *A set of vectors is linearly dependent if and only if one of the vectors is a linear combination of the others.*

Proof. Let $\{V_1, V_2, ..., V_n\}$ be a linearly dependent set. Then there exist scalars $c_1, c_2, ..., c_n$, not all zero, such that (7) is satisfied. Assume $c_1 \neq 0$. (Since at least one of the c's must differ from zero, we lose no generality in assuming it is c_1.) Eq. (7) can be rewritten as

$$c_1 V_1 = -c_2 V_2 - c_3 V_3 - ... - c_n V_n,$$

or as

$$V_1 = -\frac{c_2}{c_1} V_2 - \frac{c_3}{c_1} V_3 - ... - \frac{c_n}{c_1} V_n.$$

Thus, V_1 is a linear combination of $V_2, V_3, ..., V_n$. To complete the proof, we must show that if one vector is a linear combination of the others, then the set is linearly dependent. We leave this as an exercise for the student (see Problem 36).

Observation 1. In order for a set of vectors to be linearly dependent, it is not necessary for *every* vector to be a linear combination of the others, only that there exists *one* vector that is a linear combination of the others. For example, consider the vectors [1 0], [2 0], [0 1]. Here, [0, 1] cannot be written as a linear combination of the other two vectors; however, [2 0] can be written as a linear combination of [1 0] and [0 1], namely, [2 0] = 2[1 0] + 0[0 1]; hence, the vectors are linearly dependent.

Theorem 2. *The set consisting of the single vector V_1 is a linearly independent set if and only if $V_1 \neq 0$.*

Proof. Consider the equation $c_1 V_1 = 0$. If $V_1 \neq 0$, then the only way this equation can be valid is if $c_1 = 0$; hence, the set is linearly independent. If $V_1 = 0$, then any $c_1 \neq 0$ will satisfy the equation; hence, the set is linearly dependent.

Theorem 3. *Any set of vectors that contains the zero vector is linearly dependent.*

Proof. Consider the set $\{V_1, V_2, ..., V_n, 0\}$. Pick $c_1 = c_2 = ... = c_n = 0$, $c_{n+1} = 5$ (any other number will do). Then this is a set of scalars, not all zero, such that

$$c_1 V_1 + c_2 V_2 + \cdots + c_n V_n + c_{n+1} 0 = 0;$$

hence, the set of vectors is linearly dependent.

Theorem 4. *If a set of vectors is linearly independent, any subset of these vectors is also linearly independent.*

Proof. See Problem 37.

Theorem 5. *If a set of vectors is linearly dependent, then any larger set, containing this set, is also linearly dependent.*

Proof. See Problem 38.

Problems 2.5

In Problems 1 through 19, determine whether or not the given set is linearly independent.

1. $\{[1 \quad 0], [0 \quad 1]\}$.

2. $\{[1 \quad 1], [1 \quad -1]\}$.

3. $\{[2 \quad -4], [-3 \quad 6]\}$.

4. $\{[1 \quad 3], [2 \quad -1], [1 \quad 1]\}$.

5. $\left\{ \begin{bmatrix} 1 \\ 2 \end{bmatrix}, \begin{bmatrix} 3 \\ 4 \end{bmatrix} \right\}$.

6. $\left\{ \begin{bmatrix} 1 \\ -1 \end{bmatrix}, \begin{bmatrix} 1 \\ 1 \end{bmatrix}, \begin{bmatrix} 1 \\ 2 \end{bmatrix} \right\}$.

7. $\left\{ \begin{bmatrix} 1 \\ 0 \\ 1 \end{bmatrix}, \begin{bmatrix} 1 \\ 1 \\ 0 \end{bmatrix}, \begin{bmatrix} 0 \\ 1 \\ 1 \end{bmatrix} \right\}$.

8. $\left\{ \begin{bmatrix} 1 \\ 0 \\ 1 \end{bmatrix}, \begin{bmatrix} 1 \\ 0 \\ 2 \end{bmatrix}, \begin{bmatrix} 2 \\ 0 \\ 1 \end{bmatrix} \right\}$.

9. $\left\{ \begin{bmatrix} 1 \\ 0 \\ 1 \end{bmatrix}, \begin{bmatrix} 1 \\ 1 \\ 1 \end{bmatrix}, \begin{bmatrix} 1 \\ -1 \\ 1 \end{bmatrix} \right\}$.

10. $\left\{ \begin{bmatrix} 0 \\ 0 \\ 0 \end{bmatrix}, \begin{bmatrix} 3 \\ 2 \\ 1 \end{bmatrix}, \begin{bmatrix} 2 \\ 1 \\ 3 \end{bmatrix} \right\}$.

11. $\left\{ \begin{bmatrix} 1 \\ 2 \\ 3 \end{bmatrix}, \begin{bmatrix} 3 \\ 2 \\ 1 \end{bmatrix}, \begin{bmatrix} 2 \\ 1 \\ 3 \end{bmatrix} \right\}$.

12. $\left\{ \begin{bmatrix} 1 \\ 2 \\ 3 \end{bmatrix}, \begin{bmatrix} 3 \\ 2 \\ 1 \end{bmatrix}, \begin{bmatrix} 2 \\ 1 \\ 3 \end{bmatrix}, \begin{bmatrix} -1 \\ 2 \\ 3 \end{bmatrix} \right\}.$

13. $\left\{ \begin{bmatrix} 4 \\ 5 \\ 1 \end{bmatrix}, \begin{bmatrix} 3 \\ 0 \\ 2 \end{bmatrix}, \begin{bmatrix} 1 \\ 1 \\ 1 \end{bmatrix} \right\}.$

14. $\{[1 \quad 1 \quad 0], [1 \quad -1 \quad 0]\}.$

15. $\{[1 \quad 2 \quad 3], [-3 \quad -6 \quad -9]\}.$

16. $\{[10 \quad 20 \quad 20], [10 \quad -10 \quad 10], [10 \quad 20 \quad 10]\}.$

17. $\{[10 \quad 20 \quad 20], [10 \quad -10 \quad 10], [10 \quad 20 \quad 10], [20 \quad 10 \quad 20]\}.$

18. $\{[2 \quad 1 \quad 1], [3 \quad -1 \quad 4], [1 \quad 3 \quad -2]\}.$

19. $\left\{ \begin{bmatrix} 2 \\ 1 \\ 1 \\ 3 \end{bmatrix}, \begin{bmatrix} 4 \\ -1 \\ 2 \\ -1 \end{bmatrix}, \begin{bmatrix} 8 \\ 1 \\ 4 \\ 5 \end{bmatrix} \right\}.$

20. Express the vector

$$\begin{bmatrix} 2 \\ 1 \\ 2 \end{bmatrix}$$

as a linear combination of

$$\left\{ \begin{bmatrix} 1 \\ 1 \\ 0 \end{bmatrix}, \begin{bmatrix} 1 \\ 0 \\ -1 \end{bmatrix}, \begin{bmatrix} 1 \\ 1 \\ 1 \end{bmatrix} \right\}.$$

21. Can the vector [2 3] be expressed as a linear combination of the vectors given in (a) Problem 1, (b) Problem 2, or (c) Problem 3?

22. Can the vector $[1 \ 1 \ 1]^T$ be expressed as a linear combination of the vectors given in (a) Problem 7, (b) Problem 8, or (c) Problem 9?

23. Can the vector $[2 \ 0 \ 3]^T$ be expressed as a linear combination of the vectors given in Problem 8?

24. A set of vectors S is a *spanning set* for another set of vectors R, if every vector in R can be expressed as a linear combination of the vectors in S. Show that the vectors given in Problem 1 are a spanning set for all two-dimensional row vectors. *Hint:* Show that for any arbitrary real numbers a and b, the vector $[a\ b]$ can be expressed as a linear combination of the vectors in Problem 1.

25. Show that the vectors given in Problem 2 are a spanning set for all two-dimensional row vectors.

26. Show that the vectors given in Problem 3 are not a spanning set for all two-dimensional row vectors.

27. Show that the vectors given in Problem 3 are a spanning set for all vectors of the form $[a\ -2a]$, where a designates any real number.

28. Show that the vectors given in Problem 4 are a spanning set for all two-dimensional row vectors.

29. Determine whether the vectors given in Problem 7 are a spanning set for all three-dimensional column vectors.

30. Determine whether the vectors given in Problem 8 are a spanning set for all three-dimensional column vectors.

31. Determine whether the vectors given in Problem 8 are a spanning set for vectors of the form $[a\ 0\ a]^{\mathrm{T}}$, where a denotes an arbitrary real number.

32. A set of vectors S is a *basis* for another set of vectors R if S is a spanning set for R and S is linearly independent. Determine which, if any, of the sets given in Problems 1 through 4 are a basis for the set of all two-dimensional row vectors.

33. Determine which, if any, of the sets given in Problems 7 through 12 are a basis for the set of all three-dimensional column vectors.

34. Prove that the columns of the 3×3 identity matrix form a basis for the set of all three-dimensional column vectors.

35. Prove that the rows of the 4×4 identity matrix form a basis for the set of all four-dimensional row vectors.

36. Finish the proof of Theorem 1. (*Hint:* Assume that $\mathbf{V}_1$ can be written as a linear combination of the other vectors.)

37. Prove Theorem 4.

38. Prove Theorem 5.

39. Prove that the set of vectors $\{\mathbf{x}, k\mathbf{x}\}$ is linearly dependent for any choice of the scalar k.

40. Prove that if $\mathbf{x}$ and $\mathbf{y}$ are linearly independent, then so too are $\mathbf{x} + \mathbf{y}$ and $\mathbf{x} - \mathbf{y}$.

41. Prove that if the set $\{\mathbf{x}_1, \mathbf{x}_2, \dots, \mathbf{x}_n\}$ is linearly independent then so too is the set $\{k_1\,\mathbf{x}_1, k_2\,\mathbf{x}_2, \dots, k_n\,\mathbf{x}_n\}$ for any choice of the *nonzero* scalars $k_1, k_2, \dots, k_n$.

42. Let $\mathbf{A}$ be an $n \times n$ matrix and let $\{\mathbf{x}_1, \mathbf{x}_2, \dots, \mathbf{x}_k\}$ and $\{\mathbf{y}_1, \mathbf{y}_2, \dots, \mathbf{y}_k\}$ be two sets of n-dimensional column vectors having the property that $\mathbf{A}\mathbf{x}_i = \mathbf{y}_i = 1, 2, \dots, k$. Show that the set $\{\mathbf{x}_1, \mathbf{x}_2, \dots, \mathbf{x}_k\}$ is linearly independent if the set $\{\mathbf{y}_1, \mathbf{y}_2, \dots, \mathbf{y}_k\}$ is.

2.6 Rank

If we interpret each row of a matrix as a row vector, the elementary row operations are precisely the operations used to form linear combinations, namely, multiplying vectors (rows) by scalars and adding vectors (rows) to other vectors (rows). This observation allows us to develop a straightforward matrix procedure for determining when a set of vectors is linearly independent. It rests on the concept of rank.

Definition 1. The *row rank* of a matrix is the maximum number of linearly independent vectors that can be formed from the rows of that matrix, considering each row as a separate vector. Analogically, the *column rank* of a matrix is the maximum number of linearly independent columns, considering each column as a separate vector.

Row rank is particularly easy to determine for matrices in row-reduced form.

Theorem 1. *The row rank of a row-reduced matrix is the number of nonzero rows in that matrix.*

Proof. We must prove two facts: First, that the nonzero rows, considered as vectors, form a linearly independent set, and second, that every larger set is linearly dependent. Consider the equation

$$c_1\mathbf{v}_1 + c_2\mathbf{v}_2 + \cdots + c_r\mathbf{v}_r = 0, \tag{9}$$

where $\mathbf{v}_1$ is the first nonzero row, $\mathbf{v}_2$ is the second nonzero row, ..., and $\mathbf{v}_r$ is the last nonzero row of a row-reduced matrix. The first nonzero element in the first nonzero row of a row-reduced matrix must be unity. Assume it appears in column j. Then, no other rows have a nonzero element in that column. Consequently, when the left side of Eq. (9) is computed, it will have c_1 as its jth component. Since the right side of Eq. (9) is the zero vector, it follows that $c_1 = 0$. A similar argument then shows iteratively that c_2, ..., c_r are all zero. Thus, the nonzero rows are linearly independent.

If all the rows of the matrix are nonzero, then they must comprise a maximum number of linearly independent vectors because the row rank cannot be greater than the number of rows in the matrix. If there are zero rows in the row-reduced matrix, then it follows from Theorem 3 of Section 2.5 that including them could not increase the number of linearly independent rows. Thus, the largest number of linearly independent rows comes from including just the nonzero rows.

Example 1. Determine the row rank of the matrix

$$\mathbf{A} = \begin{bmatrix} 1 & 0 & -2 & 5 & 3 \\ 0 & 0 & 1 & -4 & 1 \\ 0 & 0 & 0 & 1 & 0 \\ 0 & 0 & 0 & 0 & 0 \end{bmatrix}.$$

Solution $\mathbf{A}$ is in row-reduced form. Since it contains three nonzero rows, its row rank is three. ∎

The following two theorems, which are proved in the Final Comments to this chapter, are fundamental.

Theorem 2. *The row rank and column rank of a matrix are equal.*

For any matrix $\mathbf{A}$, we call this common number the *rank* of $\mathbf{A}$ and denote it by $r(\mathbf{A})$.

Theorem 3. *If $\mathbf{B}$ is obtained from $\mathbf{A}$ by an elementary row (or column)operation, then $r(\mathbf{B}) = r(\mathbf{A})$.*

Theorems 1 through 3 suggest a useful procedure for determining the rank of any matrix: Simply use elementary row operations to transform the given matrix to row-reduced form and then count the number of nonzero rows.

Example 2. Determine the rank of

$$\mathbf{A} = \begin{bmatrix} 1 & 3 & 4 \\ 2 & -1 & 1 \\ 3 & 2 & 5 \\ 5 & 15 & 20 \end{bmatrix}.$$

Solution In Example 2 of Section 2.3, we transferred this matrix into the row-reduced form

$$\begin{bmatrix} 1 & 3 & 4 \\ 0 & 1 & 1 \\ 0 & 0 & 0 \\ 0 & 0 & 0 \end{bmatrix}.$$

This matrix has two nonzero rows so its rank, as well as that of $\mathbf{A}$, is two. ∎

Example 3. Determine the rank of

$$\mathbf{B} = \begin{bmatrix} 1 & 2 & 1 & 3 \\ 2 & 3 & -1 & -6 \\ 3 & -2 & -4 & -2 \end{bmatrix}.$$

Solution In Example 3 of Section 2.3, we transferred this matrix into the row-reduced form

$$\begin{bmatrix} 1 & 2 & 1 & 3 \\ 0 & 1 & 3 & 12 \\ 0 & 0 & 1 & 5 \end{bmatrix}.$$

This matrix has three nonzero rows so its rank, as well as that of $\mathbf{B}$, is three. ∎

A similar procedure can be used for determining whether a set of vectors is linearly independent: Form a matrix in which each row is one of the vectors in the given set and then determine the rank of that matrix. If the rank equals the number of vectors, the set is linearly independent; if not, the set is linearly dependent. In either case, the rank is the maximal number of linearly independent vectors that can be formed from the given set.

Example 4. Determine whether the set

$$\left\{ \begin{bmatrix} 2 \\ 6 \\ -2 \end{bmatrix}, \begin{bmatrix} 3 \\ 1 \\ 2 \end{bmatrix}, \begin{bmatrix} 8 \\ 16 \\ -3 \end{bmatrix} \right\}$$

is linearly independent.

Solution We consider the matrix

$$\begin{bmatrix} 2 & 6 & -2 \\ 3 & 1 & 2 \\ 8 & 16 & -3 \end{bmatrix}.$$

Reducing this matrix to row-reduced form, we obtain

$$\begin{bmatrix} 1 & 3 & -1 \\ 0 & 1 & -\dfrac{5}{8} \\ 0 & 0 & 0 \end{bmatrix}.$$

This matrix has two nonzero rows, so its rank is two. Since this is less than the number of vectors in the given set, that set is linearly dependent.

We can say even more: The original set of vectors contains a subset of two linearly independent vectors, the same number as the rank. Also, since no row interchanges were involved in the transformation to row-reduced form, we can conclude that the third vector is a linear combination of the first two. ∎

Example 5. Determine whether the set

$$\{[0 \quad 1 \quad 2 \quad 3 \quad 0], [1 \quad 3 \quad -1 \quad 2 \quad 1],$$
$$[2 \quad 6 \quad -1 \quad -3 \quad 1], [4 \quad 0 \quad 1 \quad 0 \quad 2]\}$$

is linearly independent.

Solution We consider the matrix

$$\begin{bmatrix} 0 & 1 & 2 & 3 & 0 \\ 1 & 3 & -1 & 2 & 1 \\ 2 & 6 & -1 & -3 & 1 \\ 4 & 0 & 1 & 0 & 2 \end{bmatrix},$$

which can be reduced (after the first two rows are interchanged) to the row-reduced form

$$\begin{bmatrix} 1 & 3 & -1 & 2 & 1 \\ 0 & 1 & 2 & 3 & 0 \\ 0 & 0 & 1 & -7 & -1 \\ 0 & 0 & 0 & 1 & \dfrac{27}{175} \end{bmatrix}.$$

This matrix has four nonzero rows; hence, its rank is four, which is equal to the number of vectors in the given set. Therefore, the set is linearly independent. ∎

Example 6.   Can the vector

$$\begin{bmatrix} 1 \\ 1 \end{bmatrix}$$

be written as a linear combination of the vectors

$$\begin{bmatrix} 3 \\ 6 \end{bmatrix} \quad \text{and} \quad \begin{bmatrix} 2 \\ 4 \end{bmatrix} ?$$

Solution The matrix

$$\mathbf{A} = \begin{bmatrix} 3 & 6 \\ 2 & 4 \end{bmatrix}$$

can be transformed into the row-reduced form

$$\begin{bmatrix} 3 & 6 \\ 0 & 0 \end{bmatrix},$$

which has rank one; hence, **A** has just one linearly independent row vector. In contrast, the matrix

$$\mathbf{B} = \begin{bmatrix} 1 & 1 \\ 3 & 6 \\ 2 & 4 \end{bmatrix}$$

can be transformed into the row-reduced form

$$\begin{bmatrix} 1 & 1 \\ 0 & 1 \\ 0 & 0 \end{bmatrix},$$

which has rank two; hence, **B** has two linearly independent row vectors. Since **B** is precisely **A** with one additional row, it follows that the additional row $[1, 1]^T$ is independent of the other two and, therefore, cannot be written as a linear combination of the other two vectors. ∎

We did not have to transform **B** in Example 6 into row-reduced form to determine whether the three-vector set was linearly independent. There is a more direct approach. Since **B** has only two columns, its column rank must be less than or equal to two (why?). Thus, the column rank is less than three. It follows from Theorem 3 that the row rank of **B** is less than three, so the three vectors must be linearly dependent. Generalizing this reasoning, we deduce one of the more important results in linear algebra.

Theorem 4.   *In an n-dimensional vector space, every set of n + 1 vectors is linearly dependent.*

Problems 2.6

In Problems 1–5, find the rank of the given matrix.

1. $\begin{bmatrix} 1 & 2 & 0 \\ 3 & 1 & -5 \end{bmatrix}$.

2. $\begin{bmatrix} 4 & 1 \\ 2 & 3 \\ 2 & 2 \end{bmatrix}$.

3. $\begin{bmatrix} 1 & 4 & -2 \\ 2 & 8 & -4 \\ -1 & -4 & 2 \end{bmatrix}$.

4. $\begin{bmatrix} 1 & 2 & 4 & 2 \\ 1 & 1 & 3 & 2 \\ 1 & 2 & 4 & 2 \end{bmatrix}$.

5. $\begin{bmatrix} 1 & 7 & 0 \\ 0 & 1 & 1 \\ 1 & 1 & 0 \end{bmatrix}$.

6. $\{[1 \quad 0], [0 \quad 1]\}$.

7. $\{[1 \quad 1], [1 \quad -1]\}$.

8. $\{[2 \quad -4], [-3 \quad 6]\}$.

9. $\{[1 \quad 3], [2 \quad -1], [1 \quad 1]\}$.

10. $\left\{ \begin{bmatrix} 1 \\ 2 \end{bmatrix}, \begin{bmatrix} 3 \\ 4 \end{bmatrix} \right\}$.

11. $\left\{ \begin{bmatrix} 1 \\ -1 \end{bmatrix}, \begin{bmatrix} 1 \\ 1 \end{bmatrix}, \begin{bmatrix} 1 \\ 2 \end{bmatrix} \right\}$.

12. $\left\{ \begin{bmatrix} 1 \\ 0 \\ 1 \end{bmatrix}, \begin{bmatrix} 1 \\ 1 \\ 0 \end{bmatrix}, \begin{bmatrix} 0 \\ 1 \\ 1 \end{bmatrix} \right\}.$

13. $\left\{ \begin{bmatrix} 1 \\ 0 \\ 1 \end{bmatrix}, \begin{bmatrix} 1 \\ 0 \\ 2 \end{bmatrix}, \begin{bmatrix} 2 \\ 0 \\ 1 \end{bmatrix} \right\}.$

14. $\left\{ \begin{bmatrix} 1 \\ 0 \\ 1 \end{bmatrix}, \begin{bmatrix} 1 \\ 1 \\ 1 \end{bmatrix}, \begin{bmatrix} 1 \\ -1 \\ 1 \end{bmatrix} \right\}.$

15. $\left\{ \begin{bmatrix} 0 \\ 0 \\ 0 \end{bmatrix}, \begin{bmatrix} 3 \\ 2 \\ 1 \end{bmatrix}, \begin{bmatrix} 2 \\ 1 \\ 3 \end{bmatrix} \right\}.$

16. $\left\{ \begin{bmatrix} 1 \\ 2 \\ 3 \end{bmatrix}, \begin{bmatrix} 3 \\ 2 \\ 1 \end{bmatrix}, \begin{bmatrix} 2 \\ 1 \\ 3 \end{bmatrix} \right\}.$

17. $\left\{ \begin{bmatrix} 1 \\ 2 \\ 3 \end{bmatrix}, \begin{bmatrix} 3 \\ 2 \\ 1 \end{bmatrix}, \begin{bmatrix} 2 \\ 1 \\ 3 \end{bmatrix}, \begin{bmatrix} -1 \\ 2 \\ -3 \end{bmatrix} \right\}.$

18. $\{[1 \quad 1 \quad 0], [1 \quad -1 \quad 0]\}.$

19. $\{[1 \quad 2 \quad 3], [-3 \quad -6 \quad -9]\}.$

20. $\{[10 \quad 20 \quad 20], [10 \quad -10 \quad 10], [10 \quad 20 \quad 10]\}.$

21. $\{[10 \quad 20 \quad 20], [10 \quad -10 \quad 10], [10 \quad 20 \quad 10], [20 \quad 10 \quad 20]\}.$

22. $\{[2 \quad 1 \quad 1], [3 \quad -1 \quad 4], [1 \quad 3 \quad -2]\}.$

23. Can the vector [2 3] be expressed as a linear combination of the vectors given in
(a) Problem 6, (b) Problem 7, or (c) Problem 8?

24. Can the vector $[1\ 1\ 1]^T$ be expressed as a linear combination of the vectors given in (a) Problem 12, (b) Problem 13, or (c) Problem 14?

25. Can the vector $[2\ 0\ 3]^T$ be expressed as a linear combination of the vectors given in Problem 13?

26. Can $[3\ 7]$ be written as a linear combination of the vectors $[1\ 2]$ and $[3\ 2]$?

27. Can $[3\ 7]$ be written as a linear combination of the vectors $[1\ 2]$ and $[4\ 8]$?

28. Find a maximal linearly independent subset of the vectors given in Problem 9.

29. Find a maximal linearly independent subset of the vectors given in Problem 13.

30. Find a maximal linearly independent subset of the set

$$[1\ \ 2\ \ 4\ \ 0], [2\ \ 4\ \ 8\ \ 0], [1\ \ -1\ \ 0\ \ 1], [4\ \ 2\ \ 8\ \ 2], [4\ \ -1\ \ 4\ \ 3].$$

31. What is the rank of the zero matrix?

32. Show $r(\mathbf{A}^T) = r(\mathbf{A})$.

2.7 Theory of solutions

Consider once again the system $\mathbf{A}\mathbf{x} = \mathbf{b}$ of m equations and n unknowns given in Eq. (2). Designate the n columns of $\mathbf{A}$ by the vectors $\mathbf{V}_1, \mathbf{V}_2,..., \mathbf{V}_n$. Then Eq. (2) can be rewritten in the vector form

$$x_1\mathbf{V}_1 + x_2\mathbf{V}_2 + \cdots + x_n\mathbf{V}_n = \mathbf{b}. \tag{10}$$

Example 1.   Rewrite the following system in the vector form (10):

$$x - 2y + 3z = 7,$$
$$4x + 5y - 6z = 8.$$

Solution

$$x\begin{bmatrix}1\\4\end{bmatrix} + y\begin{bmatrix}-2\\5\end{bmatrix} + z\begin{bmatrix}3\\-6\end{bmatrix} = \begin{bmatrix}7\\8\end{bmatrix}. \blacksquare$$

Thus, finding solutions to (1) and (2) is equivalent to finding scalars $x_1, x_2, ..., x_n$ that satisfy (10). This, however, is asking precisely the question "Is the vector $\mathbf{b}$ a linear combination of $\mathbf{V}_1, \mathbf{V}_2, ..., \mathbf{V}_n$?" If $\mathbf{b}$ is a linear combination of $\mathbf{V}_1, \mathbf{V}_2, ..., \mathbf{V}_n$, then there will exist scalars $x_1, x_2, ..., x_n$ that satisfy (10) and the system is consistent. If $\mathbf{b}$ is not a linear combination of these vectors, that is, if $\mathbf{b}$ is linearly independent of the vectors $\mathbf{V}_1, \mathbf{V}_2, ..., \mathbf{V}_n$, then no scalars $x_1, x_2, ..., x_n$ will exist that satisfy (10) and the system is inconsistent.

Taking a hint from Example 6 of Section 2.6, we have the following theorem.

Theorem 1. *The system* $\mathbf{Ax} = \mathbf{b}$ *is consistent if and only if* $r(\mathbf{A}) = r(\mathbf{A^b})$.

Once a system is deemed consistent, the following theorem specifies the number of solutions.

Theorem 2. *If the system* $\mathbf{Ax} = \mathbf{b}$ *is consistent and* $r(\mathbf{A}) = k$ *then the solutions are expressible in terms of* $n - k$ *arbitrary unknowns (where n represents the number of unknowns in the system).*

Theorem 2 is almost obvious. To determine the rank of $\mathbf{A^b}$, we must reduce it to row-reduced form. The rank is the number of nonzero rows. With Gaussian elimination, we use each nonzero row to solve for the variable associated with the first nonzero entry in it. Thus, each nonzero row defines one variable, and all other variables remain arbitrary.

Example 2. Discuss the solutions of the system

$$x + y - z = 1,$$
$$x + y - z = 0.$$

Solution

$$\mathbf{A} = \begin{bmatrix} 1 & 1 & -1 \\ 1 & 1 & -1 \end{bmatrix}, \quad \mathbf{b} = \begin{bmatrix} 1 \\ 0 \end{bmatrix}, \quad \mathbf{A^b} = \begin{bmatrix} 1 & 1 & -1 & 1 \\ 1 & 1 & -1 & 0 \end{bmatrix}.$$

Here, $r(\mathbf{A}) = 1$, $r(\mathbf{A^b}) = 2$. Thus, $r(\mathbf{A}) \neq r(\mathbf{A^b})$ and no solution exists. ∎

Example 3. Discuss the solutions of the system

$$x + y + w = 3,$$
$$2x + 2y + 3w = 6,$$
$$-x - y - w = -3.$$

Solution

$$\mathbf{A} = \begin{bmatrix} 1 & 1 & 1 \\ 2 & 2 & 2 \\ -1 & -1 & -1 \end{bmatrix}, \quad \mathbf{b} = \begin{bmatrix} 3 \\ 6 \\ -3 \end{bmatrix}, \quad \mathbf{A^b} = \begin{bmatrix} 1 & 1 & 1 & 3 \\ 2 & 2 & 2 & 6 \\ -1 & -1 & -1 & -3 \end{bmatrix}.$$

Here $r(\mathbf{A}) = r(\mathbf{A^b}) = 1$; hence, the system is consistent. In this case, $n = 3$ and $k = 1$; thus, the solutions are expressible in terms of $3 - 1 = 2$ arbitrary unknowns. Using Gaussian elimination, we find that the solution is $x = 3 - y - w$, where y and w are both arbitrary. ∎

Example 4. Discuss the solutions of the system

$$2x - 3y + z = -1,$$
$$x - y + 2z = 2,$$
$$2x + y - 3z = 3.$$

Solution

$$A = \begin{bmatrix} 2 & -3 & 1 \\ 1 & -1 & 2 \\ 2 & 1 & -3 \end{bmatrix}, \quad b = \begin{bmatrix} -1 \\ 2 \\ 3 \end{bmatrix}, \quad A^b = \begin{bmatrix} 2 & -3 & 1 & -1 \\ 1 & -1 & 2 & 2 \\ 2 & 1 & -3 & 3 \end{bmatrix}.$$

Here $r(A) = r(A^b) = 3$; hence, the system is consistent. Since $n = 3$ and $k = 3$, the solution will be in $n - k = 0$ arbitrary unknowns. Thus, the solution is unique (none of the unknowns are arbitrary) and can be obtained by Gaussian elimination as $x = y = 2$, $z = 1$. ∎

Example 5.   Discuss the solutions of the system

$$x + y - 2z = 1,$$
$$2x + y + z = 2,$$
$$3x + 2y - z = 3,$$
$$4x + 2y + 2z = 4.$$

Solution

$$A = \begin{bmatrix} 1 & 1 & -2 \\ 2 & 1 & 1 \\ 3 & 2 & -1 \\ 4 & 2 & 2 \end{bmatrix}, \quad b = \begin{bmatrix} 1 \\ 2 \\ 3 \\ 4 \end{bmatrix}, \quad A^b = \begin{bmatrix} 1 & 1 & -2 & 1 \\ 2 & 1 & 1 & 2 \\ 3 & 2 & -1 & 3 \\ 4 & 2 & 2 & 4 \end{bmatrix}.$$

Here $r(A) = r(A^b) = 2$. Thus, the system is consistent and the solutions will be in terms of $3 - 2 = 1$ arbitrary unknowns. Using Gaussian elimination, we find that the solution is $x = 1 - 3z$, $y = 5z$, and z is arbitrary. ∎

In a consistent system, the solution is unique if $k = n$. If $k \neq n$, the solution will be in terms of arbitrary unknowns. Since these arbitrary unknowns can be chosen to be any constants whatsoever, it follows that there will be an infinite number of solutions. Thus, a consistent system will possess exactly one solution or an infinite number of solutions; there is no in between.

A homogeneous system of simultaneous linear equations has the form

$$a_{11}x_1 + a_{12}x_2 + \cdots + a_{1n}x_n = 0,$$
$$a_{21}x_1 + a_{22}x_2 + \cdots + a_{2n}x_n = 0,$$
$$\vdots$$
$$a_{m1}x_1 + a_{m2}x_2 + \cdots + a_{mn}x_n = 0,$$

(11)

or the matrix form

$$Ax = 0.$$

(12)

Since Eq. (12) is a special case of Eq. (2) with $b = 0$, all the theory developed for the system $Ax = b$ remains valid. Because of the simplified structure of a homogeneous system, one can draw conclusions about it that are not valid for a nonhomogeneous system. For instance, a homogeneous system is always consistent. To verify this statement, note that $x_1 = x_2 = \ldots = x_n = 0$ is always a solution to Eq. (12). Such a solution is called the *trivial solution*. It is, in general, the *nontrivial solutions* (solutions in which one or more of the unknowns is different from zero) that are of the greatest interest.

It follows from Theorem 2 that if the rank of **A** is less than n (n being the number of unknowns), then the solution will be in terms of arbitrary unknowns. Since these arbitrary unknowns can be assigned nonzero values, it follows that nontrivial solutions exist. On the other hand, if the rank of **A** equals n, then the solution will be unique, and, hence, must be the trivial solution (why?). Thus, it follows that:

Theorem 3.   *The homogeneous system* (12) *will admit nontrivial solutions if and only if $r(\mathbf{A}) \neq n$.*

Problems 2.7

In Problems 1–9, discuss the solutions of the given system in terms of consistency and number of solutions. Check your answers by solving the systems wherever possible.

1. $x - 2y = 0,$
    $x + y = 1,$
    $2x - y = 1.$

2. $x + y = 0,$
    $2x - 2y = 1,$
    $x - y = 0.$

3. $x + y + z = 1,$
    $x - y + z = 2,$
    $3x + y + 3z = 4.$

4. $x + 3y + 2z - w = 2,$
    $2x - y + z + w = 3.$

5. $2x - y + z = 0,$
    $x + 2y - z = 4,$
    $x + y + z = 1.$

6. $2x + 3y = 0,$
    $x - 4y = 0,$

7. $\begin{aligned} x - y + 2z &= 0, \\ 2x + 3y - z &= 0, \\ -2x + 7y - 7z &= 0. \end{aligned}$

8. $\begin{aligned} x - y + 2z &= 0, \\ 2x - 3y - z &= 0, \\ -2x + 7y - 9z &= 0. \end{aligned}$

9. $\begin{aligned} x - 2y + 3z + 3w &= 0, \\ y - 2z + 2w &= 0, \\ x + y - 3z + 9w &= 0. \end{aligned}$

2.8 **Final comments on Chapter 2**

We would like to show that the column rank of a matrix equals its row rank and that an elementary row operation of any kind does not alter the rank.

Lemma 1. *If* **B** *is obtained from* **A** *by interchanging two columns of* **A**, *then both* **A** *and* **B** *have the same column rank.*

Proof. The set of vectors formed from the columns of **A** is identical to the set formed from the columns of **B**, and, therefore, the two matrices must have the same column rank.

Lemma 2. *If* $\mathbf{Ax} = \mathbf{0}$ *and* $\mathbf{Bx} = \mathbf{0}$ *have the same set of solutions, then the column rank of* **A** *is less than or equal to the column rank of* **B**.

Proof. Let the order of **A** be $m \times n$. Then, the system $\mathbf{Ax} = \mathbf{0}$ is a set of m equations in the n unknowns $x_1, x_2, \ldots, x_n$, which has the vector form

$$x_1\mathbf{A}_1 + x_2\mathbf{A}_2 + \cdots + x_n\mathbf{A}_n = 0, \tag{13}$$

where $\mathbf{A}_1, \mathbf{A}_2, \ldots, \mathbf{A}_n$ denote the columns of **A**. Similarly, the system $\mathbf{Bx} = \mathbf{0}$ has the vector form

$$x_1\mathbf{B}_1 + x_2\mathbf{B}_2 + \cdots + x_n\mathbf{B}_n = 0. \tag{14}$$

We shall assume that the column rank of **A** is greater than the column rank of **B** and show that this assumption leads to a contradiction. It will then follow that the reverse must be true, which is precisely what we want to prove.

Denote the column rank of **A** as a and the column rank of **B** as b. We assume that $a > b$. Since the column rank of **A** is a, there must exist a columns of **A** that are linearly independent. If these columns are not the first a columns, rearrange the order of the columns so they are. Lemma 1 guarantees such reorderings do not alter the column rank. Thus, $\mathbf{A}_1, \mathbf{A}_2, \ldots, \mathbf{A}_a$ are linearly independent. Since a is assumed greater than b, we know that the first a columns of $\mathbf{B}$ are not linearly independent. Since they are linearly dependent, there must exist constants $c_1, c_2, \ldots, c_a$ — not all zero — such that

$$c_1\mathbf{B}_1 + c_2\mathbf{B}_2 + \cdots + c_a\mathbf{B}_a = \mathbf{0}.$$

It then follows that

$$c_1\mathbf{B}_1 + c_2\mathbf{B}_2 + \cdots + c_a\mathbf{B}_a + 0\mathbf{B}_{a+1} + \cdots + 0\mathbf{B}_n = \mathbf{0},$$

from which we conclude that

$$x_1 = c_1, \quad x_2 = c_2, \quad \cdots, x_a = c_a, \quad x_{a+1} = 0, \quad \cdots, x_n = 0$$

is a solution of Eq. (14). Since every solution to Eq. (14) is also a solution to Eq. (12), we have

$$c_1\mathbf{A}_1 + c_2\mathbf{A}_2 + \cdots + c_a\mathbf{A}_a + 0\mathbf{A}_{a+1} + \cdots + 0\mathbf{A}_n = \mathbf{0},$$

or more simply

$$c_1\mathbf{A}_1 + c_2\mathbf{A}_2 + \cdots + c_a\mathbf{A}_a = \mathbf{0},$$

where all the c's are not all zero. But this implies that the first a columns of $\mathbf{A}$ are linearly dependent, which is a contradiction of the assumption that they were linearly independent.

Lemma 3. *If $\mathbf{Ax} = \mathbf{0}$ and $\mathbf{Bx} = \mathbf{0}$ have the same set of solutions, then $\mathbf{A}$ and $\mathbf{B}$ have the same column rank.*

Proof. It follows from Lemma 2 that the column rank of $\mathbf{A}$ is less than or equal to the column rank of $\mathbf{B}$. By reversing the roles of $\mathbf{A}$ and $\mathbf{B}$, we can also conclude from Lemma 2 that the column rank of $\mathbf{B}$ is less than or equal to the column rank of $\mathbf{A}$. As a result, the two column ranks must be equal.

Theorem 1. *An elementary row operation does not alter the column rank of a matrix.*

Proof. Denote the original matrix as $\mathbf{A}$ and let $\mathbf{B}$ denote a matrix obtained by applying an elementary row operation to $\mathbf{A}$; consider the two homogeneous systems $\mathbf{Ax} = \mathbf{0}$ and $\mathbf{Bx} = \mathbf{0}$. Since elementary row operations do not alter solutions, both of these systems have the same solution set. Theorem 1 follows immediately from Lemma 3.

Lemma 4. *The column rank of a matrix is less than or equal to its row rank.*

Proof. Denote rows of $\mathbf{A}$ by $\mathbf{A}_1, \mathbf{A}_2, \ldots \mathbf{A}_m$, the column rank of matrix $\mathbf{A}$ by c and its row rank by r. There must exist r rows of $\mathbf{A}$, which are linearly independent. If these rows are not the first r rows, rearrange the order of the rows so they are. Theorem 1 guarantees such reorderings do not alter the column rank and they certainly do not alter the row rank. Thus, $\mathbf{A}_1, \mathbf{A}_2, \ldots, \mathbf{A}_r$ are linearly independent. Define partitioned matrices $\mathbf{R}$ and $\mathbf{S}$ by

$$\mathbf{R} = \begin{bmatrix} \mathbf{A}_1 \\ \mathbf{A}_2 \\ \vdots \\ \mathbf{A}_r \end{bmatrix} \quad \text{and} \quad \mathbf{S} = \begin{bmatrix} \mathbf{A}_{r+1} \\ \mathbf{A}_{r+2} \\ \vdots \\ \mathbf{A}_n \end{bmatrix}.$$

Then **A** has the partitioned form

$$\mathbf{A} = \begin{bmatrix} \mathbf{R} \\ \mathbf{S} \end{bmatrix}.$$

Every row of **S** is a linear combination of the rows of **R**. Therefore, there exist constants t_{ij} such that

$$\mathbf{A}_{r+1} = t_{r+1,1}\mathbf{A}_1 + t_{r+1,2}\mathbf{A}_2 + \cdots + t_{r+1,r}\mathbf{A}_r,$$
$$\mathbf{A}_{r+2} = t_{r+2,1}\mathbf{A}_1 + t_{r+2,2}\mathbf{A}_2 + \cdots + t_{r+2,r}\mathbf{A}_r,$$
$$\vdots$$
$$\mathbf{A}_n = t_{n,1}\mathbf{A}_1 + t_{n,2}\mathbf{A}_2 + \cdots + t_{n,r}\mathbf{A}_r,$$

which may be written in the matrix form

$$\mathbf{S} = \mathbf{TR},$$

where

$$\mathbf{T} = \begin{bmatrix} t_{r+1,1} & t_{r+1,2} & \cdots & t_{r+1,n} \\ t_{r+2,1} & t_{r+2,2} & \cdots & t_{r+2,n} \\ \vdots & \vdots & \vdots & \vdots \\ t_{n,1} & t_{n,2} & \cdots & t_{n,n} \end{bmatrix}.$$

Then, for any n-dimensional vector **x**, we have

$$\mathbf{Ax} = \begin{bmatrix} \mathbf{R} \\ \mathbf{S} \end{bmatrix}\mathbf{x} = \begin{bmatrix} \mathbf{Rx} \\ \mathbf{Sx} \end{bmatrix} = \begin{bmatrix} \mathbf{Rx} \\ \mathbf{TRx} \end{bmatrix}.$$

Thus, $\mathbf{Ax} = \mathbf{0}$ if and only if $\mathbf{Rx} = 0$, it follows from Lemma 3 that both **A** and **R** have the same column rank. But the columns of **R** are r-dimensional vectors, so its column rank cannot be larger than r. Thus,

$$c = \text{column rank of } \mathbf{A} = \text{column rank of } \mathbf{R} \leq r = \text{row rank of } \mathbf{A}$$

Lemma 5. *The row rank of a matrix is less than or equal to its column rank.*
Proof. By applying Lemma 4 to $\mathbf{A}^T$, we conclude that the column rank of $\mathbf{A}^T$ is less than or equal to the row rank of $\mathbf{A}^T$. But since the columns of $\mathbf{A}^T$ are the rows of **A** and vice versa, the result follows immediately.

Theorem 2. *The row rank of a matrix equals its column rank.*
Proof. The result is immediate from Lemmas 4 and 5.

Theorem 3. *An elementary row operation does not alter the row rank of a matrix.*
Proof. This theorem is an immediate consequence of both Theorems 1 and 2.

The inverse

3

Chapter outline

3.1 Introduction

Definition 1. An *inverse* of an $n \times n$ matrix $\mathbf{A}$ is an $n \times n$ matrix $\mathbf{B}$ having the property that

$$\mathbf{AB} = \mathbf{BA} = \mathbf{I}. \tag{1}$$

Here, $\mathbf{B}$ is called an inverse of $\mathbf{A}$ and is usually denoted by $\mathbf{A}^{-1}$. If a square matrix $\mathbf{A}$ has an inverse, it is said to be *invertible* or *nonsingular*. If $\mathbf{A}$ does not possess an inverse, it is said to be *singular*. Note that inverses are only defined for square matrices. In particular, the identity matrix is invertible and is its own inverse because

$$\mathbf{II} = \mathbf{I}.$$

Matrix Methods. https://doi.org/10.1016/B978-0-12-818419-6.00003-4

Example 1.   Determine whether

$$B = \begin{bmatrix} 1 & \frac{1}{2} \\ \frac{1}{3} & \frac{1}{4} \end{bmatrix} \text{ or } C = \begin{bmatrix} -2 & 1 \\ \frac{3}{2} & -\frac{1}{2} \end{bmatrix}$$

are inverses for

$$A = \begin{bmatrix} 1 & 2 \\ 3 & 4 \end{bmatrix}.$$

Solution B is an inverse if and only if $AB = BA = I$; C is an inverse if and only if $AC = CA = I$. Here,

$$AB = \begin{bmatrix} 1 & 2 \\ 3 & 4 \end{bmatrix} \begin{bmatrix} 1 & \frac{1}{2} \\ \frac{1}{3} & \frac{1}{4} \end{bmatrix} = \begin{bmatrix} \frac{5}{3} & 1 \\ \frac{13}{3} & \frac{5}{2} \end{bmatrix} \neq \begin{bmatrix} 1 & 0 \\ 0 & 1 \end{bmatrix},$$

while

$$AC = \begin{bmatrix} 1 & 2 \\ 3 & 4 \end{bmatrix} \begin{bmatrix} -2 & 1 \\ \frac{3}{2} & -\frac{1}{2} \end{bmatrix} = \begin{bmatrix} 1 & 0 \\ 0 & 1 \end{bmatrix} = \begin{bmatrix} -2 & 1 \\ \frac{3}{2} & -\frac{1}{2} \end{bmatrix} \begin{bmatrix} 1 & 2 \\ 3 & 4 \end{bmatrix} = CA.$$

Thus, B is not an inverse for A, but C is. We may write $A^{-1} = C$. ∎

Definition 1 is a test for checking whether a given matrix is an inverse of another given matrix. In the Final Comments to this chapter we prove that if $AB = I$ for two square matrices of the same order, then A and B commute and $BA = I$. Thus, we can reduce the checking procedure by half. A matrix B is an inverse for a square matrix A if either $AB = I$ or $BA = I$; each equality automatically guarantees the other for square matrices. We will show in Section 3.4 that an inverse is unique. If a square matrix has an inverse, it has only one.

Definition 1 does not provide a method for finding inverses. We develop such a procedure in the next section. Still, inverses for some matrices can be found directly.

The inverse for a diagonal matrix D having only nonzero elements on its main diagonal is also a diagonal matrix whose diagonal elements are the reciprocals of the corresponding diagonal elements of D. That is, if

$$D = \begin{bmatrix} \lambda_1 & & & & 0 \\ & \lambda_2 & & & \\ & & \lambda_3 & & \\ & & & \ddots & \\ 0 & & & & \lambda_n \end{bmatrix},$$

then,

$$\mathbf{D}^{-1} = \begin{bmatrix} \dfrac{1}{\lambda_1} & & & & & \mathbf{0} \\ & \dfrac{1}{\lambda_2} & & & & \\ & & \dfrac{1}{\lambda_3} & & & \\ & & & \ddots & & \\ \mathbf{0} & & & & & \dfrac{1}{\lambda_n} \end{bmatrix}.$$

It is easy to show that if any diagonal element in a diagonal matrix is zero, then that matrix is singular. (See Problem 57.)

An elementary matrix $\mathbf{E}$ is a square matrix that generates an elementary row operation on a matrix $\mathbf{A}$ (which need not be square) under the multiplication $\mathbf{EA}$. Elementary matrices are constructed by applying the desired elementary row operation to an identity matrix of appropriate order. The appropriate order for both $\mathbf{I}$ and $\mathbf{E}$ is a square matrix having as many columns as there are rows in $\mathbf{A}$; then, the multiplication $\mathbf{EA}$ is defined. Because identity matrices contain many zeros, the process for constructing elementary matrices can be simplified still further. After all, nothing is accomplished by interchanging the positions of zeros, multiplying zeros by nonzero constants, or adding zeros to zeros.

(i) To construct an elementary matrix that interchanges the ith row with the jth row, begin with an identity matrix of the appropriate order. First, interchange the unity element in the $i - i$ position with the zero in the $j - i$ position and then interchange the unity element in the $j - j$ position with the zero in the $i - j$ position.

(ii) To construct an elementary matrix that multiplies the ith row of a matrix by the nonzero scalar k, replace the unity element in the $i - i$ position of the identity matrix of appropriate order with the scalar k.

(iii) To construct an elementary matrix that adds to the jth row of a matrix k times the ith row, replace the zero element in the $j - i$ position of the identity matrix of appropriate order with the scalar k.

Example 2.   Find elementary matrices that when multiplied on the right by any 4×3 matrix $\mathbf{A}$ will (a) interchange the second and fourth rows of $\mathbf{A}$, (b) multiply the third row of $\mathbf{A}$ by 3, and (c) add to the fourth row of $\mathbf{A} - 5$ times its second row.

Solution

(a) $\begin{bmatrix} 1 & 0 & 0 & 0 \\ 0 & 0 & 0 & 1 \\ 0 & 0 & 1 & 0 \\ 0 & 1 & 0 & 0 \end{bmatrix}$, (b) $\begin{bmatrix} 1 & 0 & 0 & 0 \\ 0 & 1 & 0 & 0 \\ 0 & 0 & 3 & 0 \\ 0 & 0 & 0 & 1 \end{bmatrix}$, (c) $\begin{bmatrix} 1 & 0 & 0 & 0 \\ 0 & 1 & 0 & 0 \\ 0 & 0 & 1 & 0 \\ 0 & -5 & 0 & 1 \end{bmatrix}$. ∎

Example 3. Find elementary matrices that when multiplied on the right by any 3×5 matrix **A** will (a) interchange the first and second rows of **A**, (b) multiply the third row of **A** by -0.5, and (c) add to the third row of A -1 times its second row.

Solution

(a) $\begin{bmatrix} 0 & 1 & 0 \\ 1 & 0 & 0 \\ 0 & 0 & 1 \end{bmatrix}$, (b) $\begin{bmatrix} 1 & 0 & 0 \\ 0 & 1 & 0 \\ 0 & 0 & -0.5 \end{bmatrix}$, (c) $\begin{bmatrix} 1 & 0 & 0 \\ 0 & 1 & 0 \\ 0 & -1 & 1 \end{bmatrix}$. ∎

The inverse of an elementary matrix that interchanges two rows is the matrix itself, it is its own inverse. The inverse of an elementary matrix that multiplies one row by a nonzero scalar k is obtained by replacing k by $1/k$. The inverse of an elementary matrix that adds to one row a constant k times another row is obtained by replacing the scalar k by $-k$.

Example 4. Compute the inverses of the elementary matrices found in Example 2.

Solution

(a) $\begin{bmatrix} 1 & 0 & 0 & 0 \\ 0 & 0 & 0 & 1 \\ 0 & 0 & 1 & 0 \\ 0 & 1 & 0 & 0 \end{bmatrix}$, (b) $\begin{bmatrix} 1 & 0 & 0 & 0 \\ 0 & 1 & 0 & 0 \\ 0 & 0 & \frac{1}{3} & 0 \\ 0 & 0 & 0 & 1 \end{bmatrix}$, (c) $\begin{bmatrix} 1 & 0 & 0 & 0 \\ 0 & 1 & 0 & 0 \\ 0 & 0 & 1 & 0 \\ 0 & 5 & 0 & 1 \end{bmatrix}$. ∎

Example 5. Compute the inverses of the elementary matrices found in Example 3.

Solution

(a) $\begin{bmatrix} 0 & 1 & 0 \\ 1 & 0 & 0 \\ 0 & 0 & 1 \end{bmatrix}$, (b) $\begin{bmatrix} 1 & 0 & 0 \\ 0 & 1 & 0 \\ 0 & 0 & -2 \end{bmatrix}$, (c) $\begin{bmatrix} 1 & 0 & 0 \\ 0 & 1 & 0 \\ 0 & 1 & 1 \end{bmatrix}$.

Finally, if **A** can be partitioned into the block diagonal form,

$$\mathbf{A} = \begin{bmatrix} \mathbf{A}_1 & & & & \mathbf{0} \\ & \mathbf{A}_2 & & & \\ & & \mathbf{A}_3 & & \\ & & & \ddots & \\ \mathbf{0} & & & & \mathbf{A}_n \end{bmatrix},$$

then **A** is invertible if and only if each of the diagonal blocks $\mathbf{A}_1, \mathbf{A}_2, \ldots, \mathbf{A}_n$ is invertible and

$$\mathbf{A}^{-1} = \begin{bmatrix} \mathbf{A}_1^{-1} & & & & \mathbf{0} \\ & \mathbf{A}_2^{-1} & & & \\ & & \mathbf{A}_3^{-1} & & \\ & & & \ddots & \\ \mathbf{0} & & & & \mathbf{A}_n^{-1} \end{bmatrix}. \blacksquare$$

Example 6. Find the inverse of

$$\mathbf{A} = \begin{bmatrix} 2 & 0 & 0 & 0 & 0 & 0 & 0 \\ 0 & 5 & 0 & 0 & 0 & 0 & 0 \\ 0 & 0 & 1 & 0 & 0 & 0 & 0 \\ 0 & 0 & 4 & 1 & 0 & 0 & 0 \\ 0 & 0 & 0 & 0 & 1 & 0 & 0 \\ 0 & 0 & 0 & 0 & 0 & 0 & 1 \\ 0 & 0 & 0 & 0 & 0 & 1 & 0 \end{bmatrix}.$$

Solution Set

$$\mathbf{A}_1 = \begin{bmatrix} 2 & 0 \\ 0 & 5 \end{bmatrix}, \quad \mathbf{A}_2 = \begin{bmatrix} 1 & 0 \\ 4 & 1 \end{bmatrix}, \quad \text{and} \quad \mathbf{A}_3 = \begin{bmatrix} 1 & 0 & 0 \\ 0 & 0 & 1 \\ 0 & 1 & 0 \end{bmatrix};$$

then, **A** is in the block diagonal form

$$\mathbf{A} = \begin{bmatrix} \mathbf{A}_1 & & 0 \\ & \mathbf{A}_2 & \\ 0 & & \mathbf{A}_3 \end{bmatrix}.$$

Here $\mathbf{A}_1$ is a diagonal matrix with nonzero diagonal elements, $\mathbf{A}_2$ is an elementary matrix that adds to the second row four times the first row, and $\mathbf{A}_3$ is an elementary matrix that interchanges the second and third rows; thus,

$$\mathbf{A}_1^{-1} = \begin{bmatrix} \dfrac{1}{2} & 0 \\ 0 & \dfrac{1}{5} \end{bmatrix}, \quad \mathbf{A}_2^{-1} = \begin{bmatrix} 1 & 0 \\ -4 & 1 \end{bmatrix}, \quad \text{and} \quad \mathbf{A}_3^{-1} = \begin{bmatrix} 1 & 0 & 0 \\ 0 & 0 & 1 \\ 0 & 1 & 0 \end{bmatrix},$$

and

$$\mathbf{A}^{-1} = \begin{bmatrix} \dfrac{1}{2} & 0 & 0 & 0 & 0 & 0 & 0 \\ 0 & \dfrac{1}{5} & 0 & 0 & 0 & 0 & 0 \\ 0 & 0 & 1 & 0 & 0 & 0 & 0 \\ 0 & 0 & -4 & 1 & 0 & 0 & 0 \\ 0 & 0 & 0 & 0 & 1 & 0 & 0 \\ 0 & 0 & 0 & 0 & 0 & 0 & 1 \\ 0 & 0 & 0 & 0 & 0 & 1 & 0 \end{bmatrix}. \blacksquare$$

Problems 3.1

1. Determine if any of the following matrices are inverses for

$$A = \begin{bmatrix} 1 & 3 \\ 2 & 9 \end{bmatrix}:$$

(a) $\begin{bmatrix} 1 & \dfrac{1}{3} \\ \dfrac{1}{2} & \dfrac{1}{9} \end{bmatrix}$,   (b) $\begin{bmatrix} -1 & -3 \\ -2 & -9 \end{bmatrix}$,

(c) $\begin{bmatrix} 3 & -1 \\ -\dfrac{2}{3} & \dfrac{1}{3} \end{bmatrix}$,   (d) $\begin{bmatrix} 9 & -3 \\ -2 & 1 \end{bmatrix}$.

2. Determine if any of the following matrices are inverses for

$$B = \begin{bmatrix} 1 & 1 \\ 1 & 1 \end{bmatrix}:$$

(a) $\begin{bmatrix} 1 & 1 \\ 1 & 1 \end{bmatrix}$,   (b) $\begin{bmatrix} -1 & 1 \\ 1 & -1 \end{bmatrix}$,

(c) $\begin{bmatrix} 1 & 1 \\ -1 & -1 \end{bmatrix}$,   (d) $\begin{bmatrix} 2 & -1 \\ -1 & 2 \end{bmatrix}$.

3. Calculate directly the inverse of

$$A = \begin{bmatrix} 8 & 2 \\ 5 & 3 \end{bmatrix}.$$

Hint: Define

$$B = \begin{bmatrix} a & b \\ c & d \end{bmatrix}.$$

Calculate **AB**, set the product equal to **I**, and then solve for the elements of **B**.

4. Use the procedure described in Problem 3 to calculate the inverse of

$$C = \begin{bmatrix} 1 & 2 \\ 2 & 1 \end{bmatrix}.$$

5. Use the procedure described in Problem 3 to calculate the inverse of

$$\mathbf{D} = \begin{bmatrix} 1 & 1 \\ 1 & 1 \end{bmatrix}.$$

6. Show directly that the inverse of

$$\mathbf{A} = \begin{bmatrix} a & b \\ c & d \end{bmatrix}$$

when $ad - bc \neq 0$ is

$$\mathbf{A}^{-1} = \frac{1}{ad - bc} \begin{bmatrix} d & -b \\ -c & a \end{bmatrix}.$$

7. Use the results of Problem 6 to calculate the inverse of

$$\begin{bmatrix} 1 & 1 \\ 3 & 4 \end{bmatrix}.$$

8. Use the results of Problem 6 to calculate the inverse of

$$\begin{bmatrix} 2 & 1 \\ 4 & 3 \end{bmatrix}.$$

9. Use the results of Problem 6 to calculate the inverse of

$$\begin{bmatrix} 1 & \frac{1}{2} \\ \frac{1}{2} & \frac{1}{3} \end{bmatrix}.$$

10. Use the results of Problem 6 to calculate the inverse of

$$\begin{bmatrix} 10 & 20 \\ 30 & 40 \end{bmatrix}.$$

In Problems 11 through 26, find elementary matrices that when multiplied on the right by a matrix **A** will generate the specified result.

11. Interchange the order of the first and second row of the 2×2 matrix **A**.

12. Multiply the first row of a 2×2 matrix **A** by 3.

13. Multiply the second row of a 2×2 matrix **A** by -5.

14. Multiply the second row of a 3×3 matrix **A** by -5.

15. Add to the second row of a 2×2 matrix **A** 3 times its first row.

16. Add to the first row of a 2×2 matrix **A** 3 times its second row.
17. Add to the second row of a 3×3 matrix **A** 3 times its third row.
18. Add to the third row of a 3×4 matrix **A** 5 times its first row.
19. Add to the second row of a 4×4 matrix **A** 8 times its fourth row.
20. Add to the fourth row of a 5×7 matrix **A** -2 times its first row.
21. Interchange the second and fourth rows of a 4×6 matrix **A**.
22. Interchange the second and fourth rows of a 4×4 matrix **A**.
23. Interchange the second and fourth rows of a 6×6 matrix **A**.
24. Multiply the second row of a 2×5 matrix **A** by 7.
25. Multiply the third row of a 5×2 matrix **A** by 7.
26. Multiply the second row of a 3×5 matrix **A** by -0.2.
   In Problems 27 through 42, find the inverses of the given elementary matrices.

27. $\begin{bmatrix} 2 & 0 \\ 0 & 1 \end{bmatrix}$,

28. $\begin{bmatrix} 1 & 2 \\ 0 & 1 \end{bmatrix}$,

29. $\begin{bmatrix} 1 & 0 \\ -3 & 1 \end{bmatrix}$,

30. $\begin{bmatrix} 1 & 0 \\ 1 & 1 \end{bmatrix}$,

31. $\begin{bmatrix} 1 & 0 & 0 \\ 0 & 2 & 0 \\ 0 & 0 & 1 \end{bmatrix}$,

32. $\begin{bmatrix} 0 & 1 & 0 \\ 1 & 0 & 0 \\ 0 & 0 & 1 \end{bmatrix}$,

33. $\begin{bmatrix} 1 & 0 & 0 \\ 0 & 1 & 0 \\ 3 & 0 & 1 \end{bmatrix}$,

34. $\begin{bmatrix} 1 & 0 & 3 \\ 0 & 1 & 0 \\ 0 & 0 & 1 \end{bmatrix}$,

35. $\begin{bmatrix} 1 & 0 & 0 \\ 0 & 1 & -2 \\ 0 & 0 & 1 \end{bmatrix}$,

36. $\begin{bmatrix} 1 & 0 & 0 \\ 0 & 1 & 0 \\ 0 & 0 & -4 \end{bmatrix}$,

37. $\begin{bmatrix} 1 & 0 & 0 & 0 \\ 0 & 1 & 0 & 0 \\ 0 & 0 & 0 & 1 \\ 0 & 0 & 1 & 0 \end{bmatrix}$,

38. $\begin{bmatrix} 1 & 0 & 0 & 0 \\ 0 & 1 & 0 & 7 \\ 0 & 0 & 1 & 0 \\ 0 & 0 & 0 & 1 \end{bmatrix}$,

39. $\begin{bmatrix} 1 & 0 & 0 & 0 \\ 0 & 1 & 0 & 0 \\ -3 & 0 & 1 & 0 \\ 0 & 0 & 0 & 1 \end{bmatrix}$,

40. $\begin{bmatrix} 0 & 0 & 0 & 1 \\ 0 & 1 & 0 & 0 \\ 0 & 0 & 1 & 0 \\ 1 & 0 & 0 & 0 \end{bmatrix}$,

41. $\begin{bmatrix} 1 & 0 & 0 & 0 \\ 0 & 1 & 0 & 0 \\ 0 & 1 & 1 & 0 \\ 0 & 0 & 0 & 1 \end{bmatrix}$,

42. $\begin{bmatrix} 1 & 0 & 0 & 0 \\ 0 & 1 & 0 & 0 \\ 0 & 0 & -\dfrac{1}{2} & 0 \\ 0 & 0 & 0 & 1 \end{bmatrix}$,

In Problems 43 through 55, find the inverses, if they exist, of the given diagonal or block diagonal matrices.

43. $\begin{bmatrix} 2 & 0 \\ 0 & 3 \end{bmatrix}$,

44. $\begin{bmatrix} -1 & 0 \\ 0 & 0 \end{bmatrix}$,

45. $\begin{bmatrix} 3 & 0 \\ 0 & -3 \end{bmatrix}$,

46. $\begin{bmatrix} \dfrac{1}{2} & 0 \\ 0 & -\dfrac{2}{3} \end{bmatrix}$,

47. $\begin{bmatrix} 10 & 0 & 0 \\ 0 & 5 & 0 \\ 0 & 0 & 5 \end{bmatrix}$,

48. $\begin{bmatrix} 1 & 1 & 0 \\ 0 & 1 & 0 \\ 0 & 0 & -1 \end{bmatrix}$,

49. $\begin{bmatrix} -4 & 0 & 0 \\ 0 & -2 & 0 \\ 0 & 0 & \dfrac{3}{5} \end{bmatrix}$

50. $\begin{bmatrix} 1 & 2 & 0 & 0 \\ 0 & 1 & 0 & 0 \\ 0 & 0 & 1 & 0 \\ 0 & 0 & 2 & 1 \end{bmatrix}$,

51. $\begin{bmatrix} 2 & 0 & 0 & 0 \\ 0 & 3 & 0 & 0 \\ 0 & 0 & 1 & -3 \\ 0 & 0 & 0 & 1 \end{bmatrix}$,

52. $\begin{bmatrix} 4 & 0 & 0 & 0 \\ 0 & 5 & 0 & 0 \\ 0 & 0 & 6 & 0 \\ 0 & 0 & 0 & 1 \end{bmatrix}$,

53.
$$\begin{bmatrix} 0 & 1 & 0 & 0 \\ 1 & 0 & 0 & 0 \\ 0 & 0 & 0 & 1 \\ 0 & 0 & 1 & 0 \end{bmatrix},$$

54.
$$\begin{bmatrix} 0 & 0 & 1 & 0 \\ 0 & 1 & 0 & 0 \\ 1 & 0 & 0 & 0 \\ 0 & 0 & 0 & 7 \end{bmatrix},$$

55.
$$\begin{bmatrix} 4 & 0 & 0 & 0 \\ 0 & 5 & 0 & 0 \\ 0 & 0 & 1 & 6 \\ 0 & 0 & 0 & 1 \end{bmatrix},$$

56. Prove that a square zero matrix does not have an inverse.

57. Prove that if a diagonal matrix has at least one zero on its main diagonal, then that matrix cannot have an inverse.

58. Prove that if $\mathbf{A}^2 = \mathbf{I}$, then $\mathbf{A}^{-1} = \mathbf{A}$.

3.2 Calculating inverses

In Section 2.3, we developed a method for transforming any matrix into row-reduced form using elementary row operations. If we now restrict our attention to square matrices, we may say that the resulting row-reduced matrices are upper triangular matrices having either a unity or zero element in each entry on the main diagonal. This provides a simple test for determining which matrices have inverses.

Theorem 1. *A square matrix has an inverse if and only if reduction to row-reduced form by elementary row operations results in a matrix having all unity elements on the main diagonal.*

We shall prove this theorem in the Final Comments to this chapter as

Theorem 2. *An $n \times n$ matrix has an inverse if and only if it has rank n.*

Theorem 1 not only provides a test for determining when a matrix is invertible but it also suggests a technique for obtaining the inverse when it exists. Once a matrix has been transformed to a row-reduced matrix with unity elements on the main diagonal, it is a simple matter to reduce it still further to the identity matrix. This is done by applying elementary row operation (E3)—adding to one row of a matrix a scalar times another row of the same matrix—to each column of the matrix, *beginning with the last column and moving sequentially toward the first column*, placing zeros in all positions above the diagonal elements.

Example 1.   Use elementary row operations to transform the upper triangular matrix

$$A = \begin{bmatrix} 1 & 2 & 1 \\ 0 & 1 & 3 \\ 0 & 0 & 1 \end{bmatrix}$$

to the identity matrix.

Solution

$$\begin{bmatrix} 1 & 2 & 1 \\ 0 & 1 & 3 \\ 0 & 0 & 1 \end{bmatrix} \rightarrow \begin{bmatrix} 1 & 2 & 1 \\ 0 & 1 & 0 \\ 0 & 0 & 1 \end{bmatrix} \quad \begin{cases} \text{by adding to} \\ \text{the second row } (-3) \\ \text{times the third row} \end{cases}$$

$$\rightarrow \begin{bmatrix} 1 & 2 & 0 \\ 0 & 1 & 0 \\ 0 & 0 & 1 \end{bmatrix} \quad \begin{cases} \text{by adding to} \\ \text{the first row } (-1) \\ \text{times the third row} \end{cases}$$

$$\rightarrow \begin{bmatrix} 1 & 0 & 0 \\ 0 & 1 & 0 \\ 0 & 0 & 1 \end{bmatrix}. \quad \begin{cases} \text{by adding to} \\ \text{the first row } (-2) \quad \blacksquare \\ \text{times the second row} \end{cases}$$

To summarize, we now know that a square matrix **A** has an inverse if and only if it can be transformed into the identity matrix by elementary row operations. Moreover, it follows from the previous section that each elementary row operation is represented by an elementary matrix **E** that generates the row operation under the multiplication **EA**. Therefore, **A** has an inverse if and only if there exist a sequence of elementary matrices. E_1, E_2, …, E_k such that

$$E_k E_{k-1} \cdots E_3 E_2 E_1 A = I.$$

But, if we denote the product of these elementary matrices as **B**, we then have **BA** = **I**, which implies that $B = A^{-1}$. That is, the inverse of a square matrix **A** of full rank is the product of those elementary matrices that reduce **A** to the identity matrix! Thus, to calculate the inverse of **A**, we need only keep a record of the elementary row operations, or equivalently the elementary matrices, that were used to reduce **A** to **I**. This is accomplished by simultaneously applying the same elementary row operations to both **A** and an identity matrix of the same order, because if

$$E_k E_{k-1} \cdots E_3 E_2 E_1 A = I,$$

then,

$$(E_k E_{k-1} \cdots E_3 E_2 E_1) I = E_k E_{k-1} \cdots E_3 E_2 E_1 = A^{-1}.$$

We have, therefore, the following procedure for calculating inverses when they exist. Let **A** be the $n \times n$ matrix we wish to invert. Place next to it another $n \times n$ matrix **B**, which is initially the identity. Using elementary row operations on **A**, transform it into the identity. Each time an operation is performed on **A**, repeat the exact same operation on **B**. After **A** is transformed into the identity, the matrix obtained from transforming **B** will be $\mathbf{A}^{-1}$.

If **A** cannot be transformed into an identity matrix, which is equivalent to saying that its row-reduced form contains at least one zero row, then **A** does not have an inverse.

Example 2. Invert

$$\mathbf{A} = \begin{bmatrix} 1 & 2 \\ 3 & 4 \end{bmatrix}.$$

Solution

$$\begin{bmatrix} 1 & 2 & | & 1 & 0 \\ 3 & 4 & | & 0 & 1 \end{bmatrix} \rightarrow \begin{bmatrix} 1 & 2 & | & 1 & 0 \\ 0 & -2 & | & -3 & 1 \end{bmatrix} \quad \begin{cases} \text{by adding to} \\ \text{the second row } (-3) \\ \text{times the first row} \end{cases}$$

$$\rightarrow \begin{bmatrix} 1 & 2 & | & 1 & 0 \\ 0 & 1 & | & \frac{3}{2} & -\frac{1}{2} \end{bmatrix}. \quad \begin{cases} \text{by multiplying} \\ \text{the second row by } (-\frac{1}{2}) \end{cases}$$

A has been transformed into row-reduced form with a main diagonal of only unity elements; it has an inverse. Continuing with the transformation process, we get

$$\rightarrow \begin{bmatrix} 1 & 0 & | & -2 & 1 \\ 0 & 1 & | & \frac{3}{2} & -\frac{1}{2} \end{bmatrix}. \quad \begin{cases} \text{by adding to} \\ \text{the first row } (-2) \\ \text{times the second row} \end{cases}$$

Thus,

$$\mathbf{A}^{-1} = \begin{bmatrix} -2 & 1 \\ \frac{3}{2} & -\frac{1}{2} \end{bmatrix}. \ \blacksquare$$

Example 3. Find the inverse of

$$\mathbf{A} = \begin{bmatrix} 5 & 8 & 1 \\ 0 & 2 & 1 \\ 4 & 3 & -1 \end{bmatrix}.$$

Solution

$$
\begin{bmatrix} 5 & 8 & 1 & | & 1 & 0 & 0 \\ 0 & 2 & 1 & | & 0 & 1 & 0 \\ 4 & 3 & -1 & | & 0 & 0 & 1 \end{bmatrix} \rightarrow \begin{bmatrix} 1 & 1.6 & 0.2 & | & 0.2 & 0 & 0 \\ 0 & 2 & 1 & | & 0 & 1 & 0 \\ 4 & 3 & -1 & | & 0 & 0 & 1 \end{bmatrix}
\quad \begin{cases} \text{by multiplying the} \\ \text{first row by } (0.2) \end{cases}
$$

$$
\rightarrow \begin{bmatrix} 1 & 1.6 & 0.2 & | & 0.2 & 0 & 0 \\ 0 & 2 & 1 & | & 0 & 1 & 0 \\ 0 & -3.4 & -1.8 & | & -0.8 & 0 & 1 \end{bmatrix}
\quad \begin{cases} \text{by adding to the} \\ \text{third row } (-4) \\ \text{times the first row} \end{cases}
$$

$$
\rightarrow \begin{bmatrix} 1 & 1.6 & 0.2 & | & 0.2 & 0 & 0 \\ 0 & 1 & 0.5 & | & 0 & 0.5 & 0 \\ 0 & -3.4 & -1.8 & | & -0.8 & 0 & 1 \end{bmatrix}
\quad \begin{cases} \text{by multiplying the} \\ \text{second row by } (0.5) \end{cases}
$$

$$
\rightarrow \begin{bmatrix} 1 & 1.6 & 0.2 & | & 0.2 & 0 & 0 \\ 0 & 1 & 0.5 & | & 0 & 0.5 & 0 \\ 0 & 0 & -0.1 & | & -0.8 & 1.7 & 1 \end{bmatrix}
\quad \begin{cases} \text{by adding to the} \\ \text{third row } (3.4) \\ \text{times the second row} \end{cases}
$$

$$
\rightarrow \begin{bmatrix} 1 & 1.6 & 0.2 & | & 0.2 & 0 & 0 \\ 0 & 1 & 0.5 & | & 0 & 0.5 & 0 \\ 0 & 0 & 1 & | & 8 & -17 & -10 \end{bmatrix}.
\quad \begin{cases} \text{by multiplying the} \\ \text{third row by } (-0.1) \end{cases}
$$

A has been transformed into row-reduced form with a main diagonal of only unity elements; it has an inverse. Continuing with the transformation process, we get

$$
\rightarrow \begin{bmatrix} 1 & 1.6 & 0.2 & | & 0.2 & 0 & 0 \\ 0 & 1 & 0 & | & -4 & 9 & 5 \\ 0 & 0 & 1 & | & 8 & -17 & -10 \end{bmatrix}
\quad \begin{cases} \text{by adding to the} \\ \text{second row } (-0.5) \\ \text{times the third row} \end{cases}
$$

$$
\rightarrow \begin{bmatrix} 1 & 1.6 & 0 & | & -1.4 & 3.4 & 2 \\ 0 & 1 & 0 & | & -4 & 9 & 5 \\ 0 & 0 & 1 & | & 8 & -17 & -10 \end{bmatrix}
\quad \begin{cases} \text{by adding to the} \\ \text{first row } (-0.2) \\ \text{times the third row} \end{cases}
$$

$$
\rightarrow \begin{bmatrix} 1 & 0 & 0 & | & 5 & -11 & -6 \\ 0 & 1 & 0 & | & -4 & 9 & 5 \\ 0 & 0 & 1 & | & 8 & -17 & -10 \end{bmatrix}.
\quad \begin{cases} \text{by adding to the} \\ \text{first row } (-1.6) \\ \text{times the second row} \end{cases}
$$

Thus,

$$
\mathbf{A}^{-1} = \begin{bmatrix} 5 & -11 & -6 \\ -4 & 9 & 5 \\ 8 & -17 & -10 \end{bmatrix}. \blacksquare
$$

Example 4.   Find the inverse of

$$
\mathbf{A} = \begin{bmatrix} 0 & 1 & 1 \\ 1 & 1 & 1 \\ 1 & 1 & 3 \end{bmatrix}.
$$

Solution

$$\begin{bmatrix} 0 & 1 & 1 & | & 1 & 0 & 0 \\ 1 & 1 & 1 & | & 0 & 1 & 0 \\ 1 & 1 & 3 & | & 0 & 0 & 1 \end{bmatrix} \rightarrow \begin{bmatrix} 1 & 1 & 1 & | & 0 & 1 & 0 \\ 0 & 1 & 1 & | & 1 & 0 & 0 \\ 1 & 1 & 3 & | & 0 & 0 & 1 \end{bmatrix}$$
$\left\{\begin{array}{l}\text{by interchanging the}\\ \text{first and second rows}\end{array}\right.$

$$\rightarrow \begin{bmatrix} 1 & 1 & 1 & | & 0 & 1 & 0 \\ 0 & 1 & 1 & | & 1 & 0 & 0 \\ 0 & 0 & 2 & | & 0 & -1 & 1 \end{bmatrix}$$
$\left\{\begin{array}{l}\text{by adding to the}\\ \text{the third row } (-1)\\ \text{times the first row}\end{array}\right.$

$$\rightarrow \begin{bmatrix} 1 & 1 & 1 & | & 0 & 1 & 0 \\ 0 & 1 & 1 & | & 1 & 0 & 0 \\ 0 & 0 & 1 & | & 0 & -\frac{1}{2} & \frac{1}{2} \end{bmatrix}$$
$\left\{\begin{array}{l}\text{by multiplying the}\\ \text{third row by } (\frac{1}{2})\end{array}\right.$

$$\rightarrow \begin{bmatrix} 1 & 1 & 1 & | & 0 & 1 & 0 \\ 0 & 1 & 0 & | & 1 & \frac{1}{2} & -\frac{1}{2} \\ 0 & 0 & 1 & | & 0 & -\frac{1}{2} & \frac{1}{2} \end{bmatrix}$$
$\left\{\begin{array}{l}\text{by adding to the}\\ \text{second row } (-1)\\ \text{times the third row}\end{array}\right.$

$$\rightarrow \begin{bmatrix} 1 & 1 & 0 & | & 0 & \frac{3}{2} & -\frac{1}{2} \\ 0 & 1 & 0 & | & 1 & \frac{1}{2} & -\frac{1}{2} \\ 0 & 0 & 1 & | & 0 & -\frac{1}{2} & \frac{1}{2} \end{bmatrix}$$
$\left\{\begin{array}{l}\text{by adding to the}\\ \text{first row } (-1)\\ \text{times the third row}\end{array}\right.$

$$\rightarrow \begin{bmatrix} 1 & 0 & 0 & | & -1 & 1 & 0 \\ 0 & 1 & 0 & | & 1 & \frac{1}{2} & -\frac{1}{2} \\ 0 & 0 & 1 & | & 0 & -\frac{1}{2} & \frac{1}{2} \end{bmatrix}.$$
$\left\{\begin{array}{l}\text{by adding to the}\\ \text{first row } (-1)\\ \text{times the second row}\end{array}\right.$

Thus,

$$\mathbf{A}^{-1} = \begin{bmatrix} -1 & 1 & 0 \\ 1 & \frac{1}{2} & -\frac{1}{2} \\ 0 & -\frac{1}{2} & \frac{1}{2} \end{bmatrix}. \blacksquare$$

Example 5.   Invert

$$\mathbf{A} = \begin{bmatrix} 1 & 2 \\ 2 & 4 \end{bmatrix}.$$

Solution

$$\begin{bmatrix} 1 & 2 & 1 & 0 \\ 2 & 4 & 0 & 1 \end{bmatrix} \rightarrow \begin{bmatrix} 1 & 2 & 1 & 0 \\ 0 & 0 & -2 & 1 \end{bmatrix}. \quad \left\{ \begin{array}{l} \text{by adding to} \\ \text{the second row } (-2) \\ \text{times the first row} \end{array} \right.$$

A has been transformed into row-reduced form. Since the main diagonal contains a zero element, here in the 2–2 position, the matrix **A** does not have an inverse. It is singular. ■

Problems 3.2

In Problems 1–20, find the inverses of the given matrices, if they exist.

1. $\begin{bmatrix} 1 & 1 \\ 3 & 4 \end{bmatrix}$,

2. $\begin{bmatrix} 2 & 1 \\ 1 & 2 \end{bmatrix}$,

3. $\begin{bmatrix} 4 & 4 \\ 4 & 4 \end{bmatrix}$,

4. $\begin{bmatrix} 2 & -1 \\ 3 & 4 \end{bmatrix}$,

5. $\begin{bmatrix} 8 & 3 \\ 5 & 2 \end{bmatrix}$,

6. $\begin{bmatrix} 1 & \dfrac{1}{2} \\ \dfrac{1}{2} & \dfrac{1}{3} \end{bmatrix}$,

7. $\begin{bmatrix} 1 & 1 & 0 \\ 1 & 0 & 1 \\ 0 & 1 & 1 \end{bmatrix}$,

8. $\begin{bmatrix} 0 & 0 & 1 \\ 1 & 0 & 0 \\ 0 & 1 & 0 \end{bmatrix}$,

9. $\begin{bmatrix} 2 & 0 & -1 \\ 0 & 1 & 2 \\ 3 & 1 & 1 \end{bmatrix}$,

10. $\begin{bmatrix} 1 & 2 & 3 \\ 4 & 5 & 6 \\ 7 & 8 & 9 \end{bmatrix}$,

11. $\begin{bmatrix} 2 & 0 & 0 \\ 5 & 1 & 0 \\ 4 & 1 & 1 \end{bmatrix}$,

12. $\begin{bmatrix} 2 & 1 & 5 \\ 0 & 3 & -1 \\ 0 & 0 & 2 \end{bmatrix}$,

13. $\begin{bmatrix} 3 & 2 & 1 \\ 4 & 0 & 1 \\ 3 & 9 & 2 \end{bmatrix}$,

14. $\begin{bmatrix} 1 & 2 & -1 \\ 2 & 0 & 1 \\ -1 & 1 & 3 \end{bmatrix}$,

15. $\begin{bmatrix} 1 & 2 & 1 \\ 3 & -2 & -4 \\ 2 & 3 & -1 \end{bmatrix}$,

16. $\begin{bmatrix} 2 & 4 & 3 \\ 3 & -4 & -4 \\ 5 & 0 & -1 \end{bmatrix}$,

17. $\begin{bmatrix} 5 & 0 & -1 \\ 2 & -1 & 2 \\ 2 & 3 & -1 \end{bmatrix}$,

18. $\begin{bmatrix} 3 & 1 & 1 \\ 1 & 3 & -1 \\ 2 & 3 & -1 \end{bmatrix}$,

19. $\begin{bmatrix} 1 & 1 & 1 & 2 \\ 0 & 1 & -1 & 1 \\ 0 & 0 & 2 & 3 \\ 0 & 0 & 0 & -2 \end{bmatrix}$,

20.
$$\begin{bmatrix} 1 & 0 & 0 & 0 \\ 2 & -1 & 0 & 0 \\ 4 & 6 & 2 & 0 \\ 3 & 2 & 4 & -1 \end{bmatrix}.$$

21. Use the results of Problems 11−20 to deduce a theorem involving inverses of lower triangular matrices.

22. Use the results of Problems 12 and 19 to deduce a theorem involving the inverses of upper triangular matrices.

23. Matrix inversion can be used to encode and decode sensitive messages for transmission. Initially, each letter in the alphabet is assigned a unique positive integer, with the simplest correspondence being

A B C D E F G H I J K L M N O P Q R S T U V W X Y Z
↓ ↓
1 2 3 4 5 6 7 8 9 10 11 12 13 14 15 16 17 18 19 20 21 22 23 24 25 26

Zeros are used to separate words. Thus, the message

SHE IS A SEER

is encoded

19 8 5 0 9 19 0 1 0 19 5 5 18 0.

This scheme is too easy to decipher, however, so a scrambling effect is added prior to transmission. One scheme is to package the coded string as a set of 2-tuples, multiply each 2-tuple by a 2×2 invertible matrix, and then transmit the new string. For example, using the matrix

$$\mathbf{A} = \begin{bmatrix} 1 & 2 \\ 2 & 3 \end{bmatrix},$$

the coded message above would be scrambled into

$$\begin{bmatrix} 1 & 2 \\ 2 & 3 \end{bmatrix} \begin{bmatrix} 19 \\ 8 \end{bmatrix} = \begin{bmatrix} 35 \\ 62 \end{bmatrix},$$

$$\begin{bmatrix} 1 & 2 \\ 2 & 3 \end{bmatrix} \begin{bmatrix} 5 \\ 0 \end{bmatrix} = \begin{bmatrix} 5 \\ 10 \end{bmatrix},$$

$$\begin{bmatrix} 1 & 2 \\ 2 & 3 \end{bmatrix} \begin{bmatrix} 9 \\ 19 \end{bmatrix} = \begin{bmatrix} 47 \\ 75 \end{bmatrix}, \quad \text{etc.,}$$

and the scrambled message becomes

$$35 \quad 62 \quad 5 \quad 10 \quad 47 \quad 75 \ldots.$$

Note an immediate benefit from the scrambling: the letter S, which was originally always coded as 19 in each of its three occurrences, is now coded as a 35
the first time and as 75 the second time. Continue with the scrambling and
determine the final code for transmitting the above message.

24. Scramble the message SHE IS A SEER using matrix

$$\mathbf{A} = \begin{bmatrix} 2 & -3 \\ 4 & 5 \end{bmatrix}.$$

25. Scramble the message AARON IS A NAME using the matrix and steps
described in Problem 23.

26. Transmitted messages are unscrambled by again packaging the received message into 2-tuples and multiplying each vector by the inverse of **A**. To decode
the scrambled message

$$18 \quad 31 \quad 44 \quad 72$$

using the encoding scheme described in Problem 23, we first calculate

$$\mathbf{A}^{-1} = \begin{bmatrix} -3 & 2 \\ 2 & -1 \end{bmatrix},$$

and then,

$$\begin{bmatrix} -3 & 2 \\ 2 & -1 \end{bmatrix} \begin{bmatrix} 18 \\ 31 \end{bmatrix} = \begin{bmatrix} 8 \\ 5 \end{bmatrix},$$

$$\begin{bmatrix} -3 & 2 \\ 2 & -1 \end{bmatrix} \begin{bmatrix} 44 \\ 72 \end{bmatrix} = \begin{bmatrix} 12 \\ 16 \end{bmatrix}.$$

The unscrambled message is

$$8 \quad 5 \quad 12 \quad 16$$

which, according to the letter-integer correspondence given in Problem 23,
translates to HELP. Using the same procedure, decode the scrambled message

$$26 \quad 43 \quad 40 \quad 60 \quad 18 \quad 31 \quad 28 \quad 51.$$

27. Use the decoding procedure described in Problem 26, but with the matrix **A**
given in Problem 24, to decipher the transmitted message

$$16 \quad 120 \quad -39 \quad 131 \quad -27 \quad 45 \quad 38 \quad 76 \quad -51 \quad 129 \quad 28 \quad 56.$$

28. Scramble the message SHE IS A SEER by packaging the coded letters into 3-tuples and then multiplying by the 3×3 invertible matrix

$$\mathbf{A} = \begin{bmatrix} 1 & 0 & 1 \\ 0 & 1 & 1 \\ 1 & 1 & 0 \end{bmatrix}.$$

Add as many zeros as necessary to the end of the message to generate complete 3-tuples.

3.3 Simultaneous equations

One use of the inverse is in the solution of systems of simultaneous linear equations. Recall from Section 1.3 that any such system may be written in the form

$$\mathbf{Ax} = \mathbf{b}, \tag{2}$$

where $\mathbf{A}$ is the coefficient matrix, $\mathbf{b}$ is a known vector, and $\mathbf{x}$ is the unknown vector we wish to find. If $\mathbf{A}$ is invertible, then we can premultiply (2) by $\mathbf{A}^{-1}$ and obtain

$$\mathbf{A}^{-1}\mathbf{Ax} = \mathbf{A}^{-1}\mathbf{b}.$$

But $\mathbf{A}^{-1}\mathbf{A} = \mathbf{1}$; therefore,

$$\mathbf{Ix} = \mathbf{A}^{-1}\mathbf{b}$$

or

$$\mathbf{x} = \mathbf{A}^{-1}\mathbf{b}. \tag{3}$$

Hence, (3) shows that if $\mathbf{A}$ is invertible, then $\mathbf{x}$ can be obtained by premultiplying $\mathbf{b}$ by the inverse of $\mathbf{A}$.

Example 1.   Solve the following system for x and y:

$$x - 2y = -9,$$
$$-3x + y = 2.$$

Solution Define

$$\mathbf{A} = \begin{bmatrix} 1 & -2 \\ -3 & 1 \end{bmatrix}, \quad \mathbf{x} = \begin{bmatrix} x \\ y \end{bmatrix}, \quad \mathbf{b} = \begin{bmatrix} -9 \\ 2 \end{bmatrix};$$

then, the system can be written as $\mathbf{Ax} = \mathbf{b}$; hence, $\mathbf{x} = \mathbf{A}^{-1}\mathbf{b}$. Using the method given in Section 3.2 we find that

$$\mathbf{A}^{-1} = \left(-\frac{1}{5}\right)\begin{bmatrix} 1 & 2 \\ 3 & 1 \end{bmatrix}.$$

Thus,

$$\begin{bmatrix} x \\ y \end{bmatrix} = \mathbf{x} = \mathbf{A}^{-1}\mathbf{b} = \left(-\frac{1}{5}\right)\begin{bmatrix} 1 & 2 \\ 3 & 1 \end{bmatrix}\begin{bmatrix} -9 \\ 2 \end{bmatrix} = \left(-\frac{1}{5}\right)\begin{bmatrix} -5 \\ -25 \end{bmatrix} = \begin{bmatrix} 1 \\ 5 \end{bmatrix}.$$

Using the definition of matrix equality (two matrices are equal if and only if their corresponding elements are equal), we have that $x = 1$ and $y = 5$. ∎

Example 2.   Solve the following system for x, y, and z:

$$5x + 8y + z = 2,$$
$$2y + z = -1,$$
$$4x + 3y - z = 3.$$

Solution

$$\mathbf{A} = \begin{bmatrix} 5 & 8 & 1 \\ 0 & 2 & 1 \\ 4 & 3 & -1 \end{bmatrix}, \quad \mathbf{x} = \begin{bmatrix} x \\ y \\ z \end{bmatrix}, \quad \mathbf{b} = \begin{bmatrix} 2 \\ -1 \\ 3 \end{bmatrix}.$$

$\mathbf{A}^{-1}$ is found to be (see Example 3 of Section 3.2)

$$\begin{bmatrix} 5 & -11 & -6 \\ -4 & 9 & 5 \\ 8 & -17 & -10 \end{bmatrix}.$$

Thus,

$$\begin{bmatrix} x \\ y \\ z \end{bmatrix} = \mathbf{x} = \mathbf{A}^{-1}\mathbf{b} = \begin{bmatrix} 5 & -11 & -6 \\ -4 & 9 & 5 \\ 8 & -17 & -10 \end{bmatrix}\begin{bmatrix} 2 \\ -1 \\ 3 \end{bmatrix} = \begin{bmatrix} 3 \\ -2 \\ 3 \end{bmatrix};$$

hence $x = 3$, $y = -2$, and $z = 3$. ∎

Not only does the invertibility of $\mathbf{A}$ provide us with a solution of the system $\mathbf{Ax} = \mathbf{b}$ it also provides us with a means of showing that this solution is unique (that is, there is no other solution to the system).

Theorem 1.   *If $\mathbf{A}$ is invertible, then the system of simultaneous linear equations given by $\mathbf{Ax} = \mathbf{b}$ has one and only one solution.*

Proof. Define $\mathbf{w} = \mathbf{A}^{-1}\mathbf{b}$. Since we have already shown that $\mathbf{w}$ is a solution to $\mathbf{Ax} = \mathbf{b}$, it follows that

$$\mathbf{Aw} = \mathbf{b}. \tag{4}$$

Assume that there exists another solution $\mathbf{y}$. Since $\mathbf{y}$ is a solution, we have that

$$\mathbf{Ay} = \mathbf{b}. \tag{5}$$

Eqs. (4) and (5) imply that

$$\mathbf{Aw} = \mathbf{Ay}. \tag{6}$$

Premultiply both sides of (6) by $\mathbf{A}^{-1}$. Then,

$$\mathbf{A}^{-1}\mathbf{A}\mathbf{w} = \mathbf{A}^{-1}\mathbf{A}\mathbf{y},$$
$$\mathbf{I}\mathbf{w} = \mathbf{I}\mathbf{y},$$

or

$$\mathbf{w} = \mathbf{y}.$$

Thus, we see that if $\mathbf{y}$ is assumed to be a solution of $\mathbf{A}\mathbf{x} = \mathbf{b}$, it must, in fact, equal $\mathbf{w}$. Therefore, $\mathbf{w} = \mathbf{A}^{-1}\mathbf{b}$ is the only solution to the problem.

If $\mathbf{A}$ is singular, so that $\mathbf{A}^{-1}$ does not exist, then (3) is not valid and other methods, such as Gaussian elimination, must be used to solve the given system of simultaneous equations.

Problems 3.3

In Problems 1 through 12, use matrix inversion, if possible, to solve the given systems of equations:

1. $x + 2y = -3,$
   $3x + y = 1.$

2. $a + 2b = 5,$
   $-3a + b = 13.$

3. $4x + 2y = 6,$
   $2x - 3y = 7.$

4. $4l - p = 1,$
   $5l - 2p = -1.$

5. $2x + 3y = 8,$
   $6x + 9y = 24.$

6. $x + 2y - z = -1,$
   $2x + 3y + 2z = 5,$
   $y - z = 2.$

7. $2x + 3y - z = 4,$
   $-x - 2y + z = -2,$
   $3x - y = 2.$

8. $60l + 30m + 20n = 0,$
$30l + 20m + 15n = -10,$
$20l + 15m + 12n = -10.$

9. $2r + 4s = 2,$
$3r + 2s + t = 8,$
$5r - 3s + 7t = 15.$

10. $2r + 4s = 3,$
$3r + 2s + t = 8,$
$5r - 3s + 7t = 15.$

11. $2r + 3s - 4t = 12,$
$3r - 2s = -1,$
$8r - s - 4t = 10.$

12. $x + 2y - 2z = -1,$
$2x + y + z = 5,$
$-x + y - z = -2.$

13. Use matrix inversion to determine a production schedule that satisfies the requirements of the manufacturer described in Problem 12 of Section 2.1.

14. Use matrix inversion to determine a production schedule that satisfies the requirements of the manufacturer described in Problem 13 of Section 2.1.

15. Use matrix inversion to determine a production schedule that satisfies the requirements of the manufacturer described in Problem 14 of Section 2.1.

16. Use matrix inversion to determine the bonus for the company described in Problem 16 of Section 2.1.

17. Use matrix inversion to determine the number of barrels of gasoline that the producer described in Problem 17 of Section 2.1 must manufacture to break even.

18. Use matrix inversion to solve the Leontief input−output model described in Problem 22 of Section 2.1.

19. Use matrix inversion to solve the Leontief input−output model described in Problem 23 of Section 2.1.

3.4 **Properties of the inverse**

Theorem 1. *If* **A**, **B**, *and* **C** *are square matrices of the same order with* $\mathbf{AB} = \mathbf{I}$ *and* $\mathbf{CA} = \mathbf{I}$, *then,* $\mathbf{B} = \mathbf{C}$.

   Proof. $\mathbf{C} = \mathbf{CI} = \mathbf{C}(\mathbf{AB}) = (\mathbf{CA})\mathbf{B} = \mathbf{IB} = \mathbf{B}$.

Theorem 2. *The inverse of a matrix is unique.*

   Proof. Suppose that **B** and **C** are inverse of **A**. Then, by (1), we have that

$$\mathbf{AB} = \mathbf{I}, \quad \mathbf{BA} = \mathbf{I}, \quad \mathbf{AC} = \mathbf{I}, \quad \text{and} \quad \mathbf{CA} = \mathbf{I}.$$

   It follows from Theorem 1 that $\mathbf{B} = \mathbf{C}$. Thus, if **B** and **C** are both inverses of **A**, they must in fact be equal. Hence, the inverse is unique.

   Using Theorem 2, we can prove some useful properties of the inverse of a matrix **A** when **A** is nonsingular.

Property 1. $\left(\mathbf{A}^{-1}\right)^{-1} = \mathbf{A}$.

   Proof. See Problem 1.

Property 2. $(\mathbf{AB})^{-1} = \mathbf{B}^{-1}\mathbf{A}^{-1}$.

   Proof. $(\mathbf{AB})^{-1}$ denotes the inverse of **AB**. However, $(\mathbf{B}^{-1}\mathbf{A}^{-1})(\mathbf{AB}) = \mathbf{B}^{-1}(\mathbf{A}^{-1}\mathbf{A})\mathbf{B} = \mathbf{B}^{-1}\mathbf{IB} = \mathbf{B}^{-1}\mathbf{B} = \mathbf{I}$. Thus, $\mathbf{B}^{-1}\mathbf{A}^{-1}$ is also an inverse for **AB**, and by uniqueness of the inverse, $\mathbf{B}^{-1}\mathbf{A}^{-1} = (\mathbf{AB})^{-1}$.

Property 3. $(\mathbf{A}_1\mathbf{A}_2...\mathbf{A}_n)^{-1} = \mathbf{A}_n^{-1}\mathbf{A}_{n-1}^{-1}...\mathbf{A}_2^{-1}\mathbf{A}_1^{-1}$.

   Proof. This is an extension of Property 2 and, as such, is proved in a similar manner.

CAUTION. Note that Property 3 states that the inverse of a product is *not* the product of the inverses but rather the product of the inverses commuted.

Property 4. $(\mathbf{A}^{\mathrm{T}})^{-1} = \left(\mathbf{A}^{-1}\right)^{\mathrm{T}.}$

   Proof. $(\mathbf{A}^{\mathrm{T}})^{-1}$ denotes the inverse of $\mathbf{A}^{\mathrm{T}}$. However, using the property of the transpose that $(\mathbf{AB})^{\mathrm{T}} = \mathbf{B}^{\mathrm{T}}\mathbf{A}^{\mathrm{T}}$, we have that

$$\left(\mathbf{A}^{\mathrm{T}}\right)\left(\mathbf{A}^{-1}\right)^{\mathrm{T}} = \left(\mathbf{A}^{-1}\mathbf{A}\right)^{\mathrm{T}} = \mathbf{I}^{\mathrm{T}} = \mathbf{I}.$$

   Thus, $(\mathbf{A}^{-1})^{\mathrm{T}}$ is an inverse of $\mathbf{A}^{\mathrm{T}}$, and by uniqueness of the inverse, $(\mathbf{A}^{-1})^{\mathrm{T}} = (\mathbf{A}^{\mathrm{T}})^{-1}$.

Property 5. $(\lambda\mathbf{A})^{-1} = (1/\lambda)(\mathbf{A})^{-1}$ *if* λ *is a nonzero scalar.*

   Proof. $(\lambda\mathbf{A})^{-1}$ denotes the inverse of $\lambda\mathbf{A}$. However,

$$(\lambda\mathbf{A})(1/\lambda)\mathbf{A}^{-1} = \lambda(1/\lambda)\mathbf{AA}^{-1} = 1\cdot\mathbf{I} = \mathbf{I}.$$

   Thus, $(1/\lambda)\mathbf{A}^{-1}$ is an inverse of λA and by uniqueness of the inverse $(1/\lambda)\mathbf{A}^{-1} = (\lambda\mathbf{A})^{-1}$.

Property 6. *The inverse of a nonsingular symmetric matrix is symmetric.*
   Proof. See Problem 18.

Property 7. *The inverse of a nonsingular upper or lower triangular matrix is again an upper or lower triangular matrix respectively.*
   Proof. This is immediate from Theorem 2 and the constructive procedure described in Section 3.2 for calculating inverses.

Finally, the inverse provides us with a straightforward way of defining square matrices raised to negative integral powers. If $\mathbf{A}$ is nonsingular, then we define $\mathbf{A}^{-n} = (\mathbf{A}^{-1})^n$.

Example 1.   Find $\mathbf{A}^{-2}$ if

$$\mathbf{A} = \begin{bmatrix} \frac{1}{3} & \frac{1}{2} \\ \frac{1}{2} & 1 \end{bmatrix}.$$

Solution

$$\mathbf{A}^{-2} = (\mathbf{A}^{-1})^2$$

$$= \begin{bmatrix} 12 & -6 \\ -6 & 4 \end{bmatrix}^2 = \begin{bmatrix} 12 & -6 \\ -6 & 4 \end{bmatrix}\begin{bmatrix} 12 & -6 \\ -6 & 4 \end{bmatrix} = \begin{bmatrix} 180 & -96 \\ -96 & 52 \end{bmatrix}. \blacksquare$$

Problems 3.4

1. Prove Property 1.

2. Verify Property 2 for

$$\mathbf{A} = \begin{bmatrix} 1 & 1 \\ 2 & 3 \end{bmatrix} \quad \text{and} \quad \mathbf{B} = \begin{bmatrix} 2 & 5 \\ 1 & 2 \end{bmatrix}.$$

3. Verify Property 2 for

$$\mathbf{A} = \begin{bmatrix} 1 & 2 \\ 3 & 4 \end{bmatrix} \quad \text{and} \quad \mathbf{B} = \begin{bmatrix} 1 & -1 \\ 3 & 5 \end{bmatrix}.$$

4. Verify Property 2 for

$$\mathbf{A} = \begin{bmatrix} 1 & 1 & 1 \\ 0 & 1 & 1 \\ 0 & 0 & 1 \end{bmatrix} \quad \text{and} \quad \mathbf{B} = \begin{bmatrix} 1 & 2 & -1 \\ 0 & 1 & -1 \\ 0 & 0 & 1 \end{bmatrix}.$$

5. Prove that $(\mathbf{ABC})^{-1} = \mathbf{C}^{-1}\mathbf{B}^{-1}\mathbf{A}^{-1}$.

6. Verify the result of Problem 5 if

$$\mathbf{A} = \begin{bmatrix} 1 & 3 \\ 0 & 2 \end{bmatrix}, \quad \mathbf{B} = \begin{bmatrix} 4 & 0 \\ 0 & 2 \end{bmatrix}, \quad \text{and} \quad \mathbf{C} = \begin{bmatrix} -1 & 0 \\ 2 & 2 \end{bmatrix}.$$

7. Verify Property 4 for the matrix $\mathbf{A}$ defined in Problem 2.

8. Verify Property 4 for the matrix $\mathbf{A}$ defined in Problem 3.

9. Verify Property 4 for the matrix $\mathbf{A}$ defined in Problem 4.

10. Verify Property 5 for $\lambda = 2$ and

$$\mathbf{A} = \begin{bmatrix} 1 & 0 & 2 \\ 2 & 3 & -1 \\ -1 & 0 & 3 \end{bmatrix}.$$

11. Find $\mathbf{A}^{-2}$ and $\mathbf{B}^{-2}$ for the matrices defined in Problem 2.

12. Find $\mathbf{A}^{-3}$ and $\mathbf{B}^{-3}$ for the matrices defined in Problem 2.

13. Find $\mathbf{A}^{-2}$ and $\mathbf{B}^{-4}$ for the matrices defined in Problem 3.

14. Find $\mathbf{A}^{-2}$ and $\mathbf{B}^{-2}$ for the matrices defined in Problem 4.

15. Find $\mathbf{A}^{-3}$ and $\mathbf{B}^{-3}$ for the matrices defined in Problem 4.

16. Find $\mathbf{A}^{-3}$ if

$$\mathbf{A} = \begin{bmatrix} 1 & -2 \\ 2 & 1 \end{bmatrix}.$$

17. If $\mathbf{A}$ is symmetric, prove the identity

$$\left(\mathbf{BA}^{-1}\right)^{\mathrm{T}}\left(\mathbf{A}^{-1}\mathbf{B}^{\mathrm{T}}\right)^{-1} = \mathbf{I}.$$

18. Prove Property 6.

3.5 LU decomposition

Matrix inversion of elementary matrices (see Section 3.1) can be combined with the third elementary row operation (see Section 2.3) to generate a good numerical technique for solving simultaneous equations. It rests on being able to decompose a *nonsingular* square matrix $\mathbf{A}$ into the product of lower triangular matrix $\mathbf{L}$ with an upper triangular matrix $\mathbf{U}$. Generally, there are many such factorizations. If, however, we add the additional condition that all diagonal elements of $\mathbf{L}$ be unity, then the decomposition, when it exists, is unique, and we may write

$$\mathbf{A} = \mathbf{LU} \tag{7}$$

with

$$
L = \begin{bmatrix}
1 & 0 & 0 & \cdots & 0 \\
l_{21} & 1 & 0 & \cdots & 0 \\
l_{31} & l_{32} & 1 & \cdots & 0 \\
\vdots & \vdots & \vdots & \ddots & \vdots \\
l_{n1} & l_{n2} & l_{n3} & \cdots & 1
\end{bmatrix}
$$

and

$$
U = \begin{bmatrix}
u_{11} & u_{12} & u_{13} & \cdots & u_{1n} \\
0 & u_{22} & u_{23} & \cdots & u_{2n} \\
0 & 0 & u_{33} & \cdots & u_{3n} \\
\vdots & \vdots & \vdots & \ddots & \vdots \\
0 & 0 & 0 & \cdots & u_{nn}
\end{bmatrix}.
$$

To decompose A into from (7), we first reduce A to upper triangular form using just the third elementary row operation: namely, add to one row of a matrix a scalar times another row of that same matrix. This is completely analogous to transforming a matrix to row-reduced form, except that we no longer use the first two elementary row operations. We do not interchange rows and we do not multiply a row by a nonzero constant. Consequently, we no longer require the first nonzero element of each nonzero row to be unity, and if any of the pivots are zero—which in the row-reduction scheme would require a row interchange operation—then the decomposition scheme we seek cannot be done.

Example 1.   Use the third elementary row operation to transform the matrix

$$
A = \begin{bmatrix}
2 & -1 & 3 \\
4 & 2 & 1 \\
-6 & -1 & 2
\end{bmatrix}
$$

into upper triangular form.

Solution

$$
A = \begin{bmatrix}
2 & -1 & 3 \\
4 & 2 & 1 \\
-6 & -1 & 2
\end{bmatrix}
\rightarrow
\begin{bmatrix}
2 & -1 & 3 \\
0 & 4 & -5 \\
-6 & -1 & 2
\end{bmatrix}
$$
$\left\{ \begin{array}{l} \text{by adding to the} \\ \text{second row } (-2) \text{ times} \\ \text{the first row} \end{array} \right.$

$$
\rightarrow
\begin{bmatrix}
2 & -1 & 3 \\
0 & 4 & -5 \\
0 & -4 & 11
\end{bmatrix}
$$
$\left\{ \begin{array}{l} \text{by adding to the} \\ \text{third row } (3) \text{ times} \\ \text{the first row} \end{array} \right.$

$$
\rightarrow
\begin{bmatrix}
2 & -1 & 3 \\
0 & 4 & -5 \\
0 & 0 & 6
\end{bmatrix}.
$$
$\left\{ \begin{array}{l} \text{by adding to the} \\ \text{third row } (1) \text{ times } \blacksquare \\ \text{the second row} \end{array} \right.$

If a square matrix A can be reduced to upper triangular form U by a sequence of elementary row operations of the third type, then there exists a sequence of elementary matrices $E_{21}, E_{31}, E_{41}, \ldots, E_{n, n-1}$ such that

$$\left(\mathbf{E}_{n-1,n}\cdots\mathbf{E}_{41}\mathbf{E}_{31}\mathbf{E}_{21}\right)\mathbf{A} = \mathbf{U}, \tag{8}$$

where $\mathbf{E}_{21}$ denotes the elementary matrix that places a zero in the 2−1 position, $\mathbf{E}_{31}$ denotes the elementary matrix that places a zero in the 3−1 position, $\mathbf{E}_{41}$ denotes the elementary matrix that places a zero in the 4−1 position, and so on. Since elementary matrices have inverses, we can write (8) as

$$\mathbf{A} = \left(\mathbf{E}_{21}^{-1}\mathbf{E}_{31}^{-1}\mathbf{E}_{41}^{-1}\cdots\mathbf{E}_{n,n-1}^{-1}\right)\mathbf{U}. \tag{9}$$

Each elementary matrix in (8) is lower triangular. It follows from Property 7 of Section 3.4 that each of the inverses in (9) are lower triangular and then from Theorem 1 of Section 1.4 that the product of these lower triangular matrices is itself lower triangular. Setting

$$\mathbf{L} = \mathbf{E}_{21}^{-1}\mathbf{E}_{31}^{-1}\mathbf{E}_{41}^{-1}\cdots\mathbf{E}_{n,n-1}^{-1},$$

we see that (9) is identical to (7), and we have the decomposition we seek.

Example 2.   Construct an **LU** decomposition for the matrix given in Example 1.

   Solution The elementary matrices associated with the elementary row operations described in Example 1 are

$$\mathbf{E}_{21} = \begin{bmatrix} 1 & 0 & 0 \\ -2 & 1 & 0 \\ 0 & 0 & 1 \end{bmatrix}, \quad \mathbf{E}_{31} = \begin{bmatrix} 1 & 0 & 0 \\ 0 & 1 & 0 \\ 3 & 0 & 1 \end{bmatrix}, \quad \text{and} \quad \mathbf{E}_{42} = \begin{bmatrix} 1 & 0 & 0 \\ 0 & 1 & 0 \\ 0 & 1 & 1 \end{bmatrix},$$

with inverses given respectively by

$$\mathbf{E}_{21}^{-1} = \begin{bmatrix} 1 & 0 & 0 \\ 2 & 1 & 0 \\ 0 & 0 & 1 \end{bmatrix}, \quad \mathbf{E}_{31}^{-1} = \begin{bmatrix} 1 & 0 & 0 \\ 0 & 1 & 0 \\ -3 & 0 & 1 \end{bmatrix}, \quad \text{and} \quad \mathbf{E}_{42}^{-1} = \begin{bmatrix} 1 & 0 & 0 \\ 0 & 1 & 0 \\ 0 & -1 & 1 \end{bmatrix}.$$

Then,

$$\begin{bmatrix} 2 & -1 & 3 \\ 4 & 2 & 1 \\ -6 & -1 & 2 \end{bmatrix} = \begin{bmatrix} 1 & 0 & 0 \\ 2 & 1 & 0 \\ 0 & 0 & 1 \end{bmatrix}\begin{bmatrix} 1 & 0 & 0 \\ 0 & 1 & 0 \\ -3 & 0 & 1 \end{bmatrix}\begin{bmatrix} 1 & 0 & 0 \\ 0 & 1 & 0 \\ 0 & -1 & 1 \end{bmatrix}\begin{bmatrix} 2 & -1 & 3 \\ 0 & 4 & -5 \\ 0 & 0 & 6 \end{bmatrix}$$

or, upon multiplying together the inverses of the elementary matrices,

$$\begin{bmatrix} 2 & -1 & 3 \\ 4 & 2 & 1 \\ -6 & -1 & 2 \end{bmatrix} = \begin{bmatrix} 1 & 0 & 0 \\ 2 & 1 & 0 \\ -3 & -1 & 1 \end{bmatrix}\begin{bmatrix} 2 & -1 & 3 \\ 0 & 4 & -5 \\ 0 & 0 & 6 \end{bmatrix}. \blacksquare$$

Example 2 suggests an important simplification of the decomposition process. Note the elements in **L** below the main diagonal are *the negatives of the scalars* used in the elementary row operations to reduce the original matrix to upper triangular form! This is no coincidence. In general

Observation 1. If an elementary row operation is used to put a zero in the $i-j$ position of $\mathbf{A}(i > j)$ by adding to row i a scalar k times row j, then the $i-j$ element of $\mathbf{L}$ in the **LU** decomposition of $\mathbf{A}$ is $-k$.

We summarize the decomposition process as follows: Use only the third elementary row operation to transform a given square matrix $\mathbf{A}$ to upper triangular form. If this is not possible, because of a zero pivot, then stop; otherwise, the **LU** decomposition is found by defining the resulting upper triangular matrix as $\mathbf{U}$ and constructing the lower triangular matrix $\mathbf{L}$ utilizing Observation 1.

Example 3. Construct an LU decomposition for the matrix

$$\mathbf{A} = \begin{bmatrix} 2 & 1 & 2 & 3 \\ 6 & 2 & 4 & 8 \\ 1 & -1 & 0 & 4 \\ 0 & 1 & -3 & -4 \end{bmatrix}.$$

Solution Transforming $\mathbf{A}$ to upper triangular form, we get

$$\begin{bmatrix} 2 & 1 & 2 & 3 \\ 6 & 2 & 4 & 8 \\ 1 & -1 & 0 & 4 \\ 0 & 1 & -3 & -4 \end{bmatrix} \rightarrow \begin{bmatrix} 2 & 1 & 2 & 3 \\ 0 & -1 & -2 & -1 \\ 1 & -1 & 0 & 4 \\ 0 & 1 & -3 & -4 \end{bmatrix}$$

by adding to the second row (-3) times the first row

$$\rightarrow \begin{bmatrix} 2 & 1 & 2 & 3 \\ 0 & -1 & -2 & -1 \\ 0 & -\dfrac{3}{2} & -1 & \dfrac{5}{2} \\ 0 & 1 & -3 & -4 \end{bmatrix}$$

by adding to the third row $\left(-\dfrac{1}{2}\right)$ times the first row

$$\rightarrow \begin{bmatrix} 2 & 1 & 2 & 3 \\ 0 & -1 & -2 & -1 \\ 0 & 0 & 2 & 4 \\ 0 & 1 & -3 & -4 \end{bmatrix}$$

by adding to the third row $\left(-\dfrac{3}{2}\right)$ times the second row

$$\rightarrow \begin{bmatrix} 2 & 1 & 2 & 3 \\ 0 & -1 & -2 & -1 \\ 0 & 0 & 2 & 4 \\ 0 & 0 & -5 & -5 \end{bmatrix}$$

by adding to the fourth row (1) times the second row

$$\rightarrow \begin{bmatrix} 2 & 1 & 2 & 3 \\ 0 & -1 & -2 & -1 \\ 0 & 0 & 2 & 4 \\ 0 & 0 & 0 & 5 \end{bmatrix}.$$

by adding to the fourth row $\left(\dfrac{5}{2}\right)$ times the third row

We now have an upper triangular matrix $\mathbf{U}$. To get the lower triangular matrix $\mathbf{L}$ in the decomposition, we note that we used the scalar -3 to place a zero in the $2-1$ position, so its negative $-(-3) = 3$ goes into the $2-1$ position of $\mathbf{L}$. We used the scalar $-\frac{1}{2}$ to place a zero in the $3-1$ position in the second step of the above triangularization process, so its negative, $\frac{1}{2}$, becomes the $3-1$ element in $\mathbf{L}$; we used the scalar $\frac{5}{2}$ to place a zero in the $4-3$ position during the last step of the triangularization process, so its negative, $-\frac{5}{2}$, becomes the $4-3$ element in $\mathbf{L}$. Continuing in this manner, we generate the decomposition

$$
\begin{bmatrix} 2 & 1 & 2 & 3 \\ 6 & 2 & 4 & 8 \\ 1 & -1 & 0 & 4 \\ 0 & 1 & -3 & -4 \end{bmatrix} = \begin{bmatrix} 1 & 0 & 0 & 0 \\ 3 & 1 & 0 & 0 \\ \frac{1}{2} & \frac{3}{2} & 1 & 0 \\ 0 & -1 & -\frac{5}{2} & 1 \end{bmatrix} \begin{bmatrix} 2 & 1 & 2 & 3 \\ 0 & -1 & -2 & -1 \\ 0 & 0 & 2 & 4 \\ 0 & 0 & 0 & 5 \end{bmatrix}. \blacksquare
$$

LU decompositions, when they exist, can be used to solve systems of simultaneous linear equations. If a square matrix $\mathbf{A}$ can be factored into $\mathbf{A} = \mathbf{LU}$, then the system of equations $\mathbf{Ax} = \mathbf{b}$ can be written as $\mathbf{L(Ux)} = \mathbf{b}$. To find $\mathbf{x}$, we first solve the system

$$\mathbf{Ly} = \mathbf{b} \tag{10}$$

for $\mathbf{y}$, and then, once $\mathbf{y}$ is determined, we solve the system

$$\mathbf{Ux} = \mathbf{y} \tag{11}$$

for $\mathbf{x}$. Both systems (10) and (11) are easy to solve, the first by forward substitution and the second by backward substitution.

Example 4. Solve the system of equations

$$
\begin{aligned}
2x - y + 3z &= 9, \\
4x + 2y + z &= 9, \\
-6x - y + 2z &= 12.
\end{aligned}
$$

Solution This system has the matrix form

$$
\begin{bmatrix} 2 & -1 & 3 \\ 4 & 2 & 1 \\ -6 & -1 & 2 \end{bmatrix} \begin{bmatrix} x \\ y \\ z \end{bmatrix} = \begin{bmatrix} 9 \\ 9 \\ 12 \end{bmatrix}.
$$

The **LU** decomposition for the coefficient matrix $\mathbf{A}$ is given in Example 2. If we define the components of $\mathbf{y}$ by α, β, and γ, respectively, the matrix system $\mathbf{Ly} = \mathbf{b}$ is

$$
\begin{bmatrix} 1 & 0 & 0 \\ 2 & 1 & 0 \\ -3 & -1 & 1 \end{bmatrix} \begin{bmatrix} \alpha \\ \beta \\ \gamma \end{bmatrix} = \begin{bmatrix} 9 \\ 9 \\ 12 \end{bmatrix},
$$

which is equivalent to the system of equations

$$\alpha = 9,$$
$$2\alpha + \beta = 9,$$
$$-3\alpha - \beta + \gamma = 12.$$

Solving this system from top to bottom, we get $\alpha = 9$, $\beta = -9$, and $\gamma = 30$. Consequently, the matrix system $\mathbf{Ux} = \mathbf{y}$ is

$$\begin{bmatrix} 2 & -1 & 3 \\ 0 & 4 & -5 \\ 0 & 0 & 6 \end{bmatrix} \begin{bmatrix} x \\ y \\ z \end{bmatrix} = \begin{bmatrix} 9 \\ -9 \\ 30 \end{bmatrix}.$$

which is equivalent to the system of equations

$$2x - y + 3z = 9,$$
$$4y - 5z = -9,$$
$$6z = 30.$$

Solving this system from bottom to top, we obtain the final solution $x = -1$, $y = 4$, and $z = 5$. ∎

Example 5.   Solve the system

$$2a + b + 2c + 3d = 5,$$
$$6a + 2b + 4c + 8d = 8,$$
$$a - b + 4d = -4,$$
$$b - 3c - 4d = -3.$$

Solution The matrix representation for this system has as its coefficient matrix the matrix $\mathbf{A}$ of Example 3. Define

$$\mathbf{y} = [\alpha, \ \beta, \ \gamma, \ \delta]^{\mathrm{T}}.$$

Then, using the decomposition determined in Example 3, we can write the matrix system $\mathbf{Ly} = \mathbf{b}$ as the system of equations

$$\alpha = 5,$$
$$3\alpha + \beta = 8,$$
$$\frac{1}{2}\alpha + \frac{3}{2}\beta + \gamma = -4,$$
$$-\beta - \frac{5}{2}\gamma + \delta = -3,$$

which has as its solution $\alpha = 5$, $\beta = -7$, $\gamma = 4$, and $\delta = 0$. Thus, the matrix system $\mathbf{Ux} = \mathbf{y}$ is equivalent to the system of equations

$$2a + b + 2c + 3d = 5,$$
$$-b - 2c - d = -7,$$
$$2c + 4d = 4,$$
$$5d = 0.$$

Solving this set from bottom to top, we calculate the final solution $a = -1$, $b = 3$, $c = 2$, and $d = 0$. ∎

LU decomposition and Gaussian elimination are equally efficient for solving $\mathbf{Ax} = \mathbf{b}$, when the decomposition exists. LU decomposition is superior when $\mathbf{Ax} = \mathbf{b}$ must be solved repeatedly for different values of $\mathbf{b}$ but the same $\mathbf{A}$, because once the factorization of $\mathbf{A}$ is determined it can be used with all $\mathbf{b}$. (See Problems 17 and 18.) A disadvantage of LU decomposition is that it does not exist for all nonsingular matrices, in particular whenever a pivot is zero. Fortunately, this occurs rarely and when it does the difficulty usually overcome by simply rearranging the order of the equations. (See Problems 19 and 20.)

Problems 3.5

In Problems 1 through 14, $\mathbf{A}$ and $\mathbf{b}$ are given. Construct an LU decomposition for the matrix $\mathbf{A}$ and then use it to solve the system $\mathbf{Ax} = \mathbf{b}$ for $\mathbf{x}$.

1. $\mathbf{A} = \begin{bmatrix} 1 & 1 \\ 3 & 4 \end{bmatrix}$, $\mathbf{b} = \begin{bmatrix} 1 \\ -6 \end{bmatrix}$.

2. $\mathbf{A} = \begin{bmatrix} 2 & 1 \\ 1 & 2 \end{bmatrix}$, $\mathbf{b} = \begin{bmatrix} 11 \\ -2 \end{bmatrix}$.

3. $\mathbf{A} = \begin{bmatrix} 8 & 3 \\ 5 & 2 \end{bmatrix}$, $\mathbf{b} = \begin{bmatrix} 625 \\ 550 \end{bmatrix}$.

4. $\mathbf{A} = \begin{bmatrix} 1 & 1 & 0 \\ 1 & 0 & 1 \\ 0 & 1 & 1 \end{bmatrix}$, $\mathbf{b} = \begin{bmatrix} 4 \\ 1 \\ -1 \end{bmatrix}$.

5. $\mathbf{A} = \begin{bmatrix} -1 & 2 & 0 \\ 1 & -3 & 1 \\ 2 & -2 & 3 \end{bmatrix}$, $\mathbf{b} = \begin{bmatrix} -1 \\ -2 \\ 3 \end{bmatrix}$.

6. $\mathbf{A} = \begin{bmatrix} 2 & 1 & 3 \\ 4 & 1 & 0 \\ -2 & -1 & 2 \end{bmatrix}$, $\mathbf{b} = \begin{bmatrix} 10 \\ -40 \\ 0 \end{bmatrix}$.

7. $\mathbf{A} = \begin{bmatrix} 3 & 2 & 1 \\ 4 & 0 & 1 \\ 3 & 9 & 2 \end{bmatrix}$, $\mathbf{b} = \begin{bmatrix} 50 \\ 80 \\ 20 \end{bmatrix}$.

8. $\mathbf{A} = \begin{bmatrix} 1 & 2 & -1 \\ 2 & 0 & 1 \\ -1 & 1 & 3 \end{bmatrix}$, $\quad \mathbf{b} = \begin{bmatrix} 80 \\ 159 \\ -75 \end{bmatrix}$,

9. $\mathbf{A} = \begin{bmatrix} 1 & 2 & -1 \\ 0 & 2 & 1 \\ 0 & 0 & 1 \end{bmatrix}$, $\quad \mathbf{b} = \begin{bmatrix} 8 \\ -1 \\ 5 \end{bmatrix}$,

10. $\mathbf{A} = \begin{bmatrix} 1 & 0 & 0 \\ 3 & 2 & 0 \\ 1 & 1 & 2 \end{bmatrix}$, $\quad \mathbf{b} = \begin{bmatrix} 2 \\ 4 \\ 2 \end{bmatrix}$,

11. $\mathbf{A} = \begin{bmatrix} 1 & 0 & 1 & 1 \\ 1 & 1 & 0 & 1 \\ 1 & 1 & 1 & 0 \\ 0 & 1 & 1 & 1 \end{bmatrix}$, $\quad \mathbf{b} = \begin{bmatrix} 4 \\ -3 \\ -2 \\ -2 \end{bmatrix}$.

12. $\mathbf{A} = \begin{bmatrix} 2 & 1 & -1 & 3 \\ 1 & 4 & 2 & 1 \\ 0 & 0 & -1 & 1 \\ 0 & 1 & 0 & 1 \end{bmatrix}$, $\quad \mathbf{b} = \begin{bmatrix} 1000 \\ 200 \\ 100 \\ 100 \end{bmatrix}$.

13. $\mathbf{A} = \begin{bmatrix} 1 & 2 & 1 & 1 \\ 1 & 1 & 2 & 1 \\ 1 & 1 & 1 & 2 \\ 0 & 1 & 1 & 1 \end{bmatrix}$, $\quad \mathbf{b} = \begin{bmatrix} 30 \\ 30 \\ 10 \\ 10 \end{bmatrix}$.

14. $\mathbf{A} = \begin{bmatrix} 2 & 0 & 2 & 0 \\ 2 & 2 & 0 & 6 \\ -4 & 3 & 1 & 1 \\ 1 & 0 & 3 & 1 \end{bmatrix}$, $\quad \mathbf{b} = \begin{bmatrix} -2 \\ 4 \\ 9 \\ 4 \end{bmatrix}$.

15. (a) Use **LU** decomposition to solve the system

$$-x + 2y = 9,$$
$$2x + 3y = 4.$$

(b) Resolve when the right sides of each equation are replaced by 1 and -1, respectively.

16. (a) Use **LU** decomposition to solve the system

$$x + 3y - z = -1,$$
$$2x + 5y + z = 4,$$
$$2x + 7y - 4z = -6.$$

(b) Resolve when the right sides of each equation are replaced by 10, 10, and 10, respectively.

17. Solve the system $\mathbf{Ax} = \mathbf{b}$ for the following vectors $\mathbf{b}$ when $\mathbf{A}$ is given as in Problem 4:

$$\text{(a)} \begin{bmatrix} 5 \\ 7 \\ -4 \end{bmatrix}, \quad \text{(b)} \begin{bmatrix} 2 \\ 2 \\ 0 \end{bmatrix}, \quad \text{(c)} \begin{bmatrix} 40 \\ 50 \\ 20 \end{bmatrix}, \quad \text{(d)} \begin{bmatrix} 1 \\ 1 \\ 3 \end{bmatrix}.$$

18. Solve the system $\mathbf{Ax} = \mathbf{b}$ for the following vectors $\mathbf{x}$ when $\mathbf{A}$ is given as in Problem 13:

$$\text{(a)} \begin{bmatrix} -1 \\ 1 \\ 1 \\ 1 \end{bmatrix}, \quad \text{(b)} \begin{bmatrix} 0 \\ 0 \\ 0 \\ 0 \end{bmatrix}, \quad \text{(c)} \begin{bmatrix} 190 \\ 130 \\ 160 \\ 60 \end{bmatrix}, \quad \text{(d)} \begin{bmatrix} 1 \\ 1 \\ 1 \\ 1 \end{bmatrix}.$$

19. Show that **LU** decomposition cannot be used to solve the system

$$2y + z = -1,$$
$$x + y + 3z = 8,$$
$$2x - y - z = 1,$$

but that the decomposition can be used if the first two equations are interchanged.

20. Show that **LU** decomposition cannot be used to solve the system

$$x + 2y + z = 2,$$
$$2x + 4y - z = 7,$$
$$x + y + 2z = 2,$$

but that the decomposition can be used if the first and third equations are interchanged.

21. **(a)** Show that the **LU** decomposition procedure given in this chapter cannot be applied to

$$A = \begin{bmatrix} 0 & 2 \\ 0 & 9 \end{bmatrix}.$$

(b) Verify that $A = LU$, when

$$L = \begin{bmatrix} 1 & 0 \\ 1 & 1 \end{bmatrix} \quad \text{and} \quad U = \begin{bmatrix} 0 & 2 \\ 0 & 7 \end{bmatrix}.$$

(c) Verify that $A = LU$, when

$$L = \begin{bmatrix} 1 & 0 \\ 3 & 1 \end{bmatrix} \quad \text{and} \quad U = \begin{bmatrix} 0 & 2 \\ 0 & 3 \end{bmatrix}.$$

(d) Why do you think the **LU** decomposition procedure fails for this **A**? What might explain the fact that **A** has more than one **LU** decomposition?

3.6 Final comments on Chapter 3

We now prove the answers to two questions raised earlier. First, what matrices have inverses? Second, if $AB = I$, is it necessarily true that $AB = I$ too?

Lemma 1.   *Let* **A** *and* **B** *be* $n \times n$ *matrices. If* $AB = I$, *then the system of equations* $Ax = y$ *has a solution for every choice of the vector* **y**.

   Proof. Once **y** is specified, set $x = By$. Then,

$$Ax = A(By) = (AB)y = Iy = y,$$

so $x = By$ is a solution of $Ax = y$.

Lemma 2.   *If* **A** *and* **B** *are* $n \times n$ *matrices with* $AB = I$, *then* **A** *has rank* n.

   Proof. Designate the rows of **A** by $A_1, A_2, \ldots, A_n$. We want to show that these n rows constitute a linearly independent set of vectors, in which case the rank of **A** is n. Designate the columns of **I** as the vectors $e_1, e_2, \ldots, e_n$, respectively. It follows from Lemma 1 that the set of equations $Ax = e_j$ ($j = 1, 2, \ldots, n$) has a solution for each j. Denote these solutions by $x_1, x_2, \ldots x_n$, respectively. Therefore,

$$Ax_j = e_j.$$

   Since e_j ($j = 1, 2, \ldots, n$) is an n-dimensional column vector having a unity element in row j and zeros everywhere else, it follows from the last equation that

$$\mathbf{A}_i \mathbf{x}_j = \begin{cases} 1 & \text{when } i = j, \\ 0 & \text{when } i \neq j. \end{cases}$$

This equation can be notationally simplified if we make use of the *Kronecker delta* δ_{ij} defined by

$$\delta_{ij} = \begin{cases} 1 & \text{when } i = j, \\ 0 & \text{when } i \neq j. \end{cases}$$

Then,

$$\mathbf{A}_i \mathbf{x}_j = \delta_{ij}.$$

Now consider the equation

$$\sum_{i=0}^{n} c_i \mathbf{A}_i = \mathbf{0}.$$

We wish to show that each constant c_i must be zero. Multiplying both sides of this last equation on the right by the vector $\mathbf{x}_j$, we have

$$\left(\sum_{i=0}^{n} c_i \mathbf{A}_i \right) \mathbf{x}_j = \mathbf{0}\mathbf{x}_j,$$

$$\sum_{i=0}^{n} (c_i \mathbf{A}_i) \mathbf{x}_j = 0,$$

$$\sum_{i=0}^{n} c_i (\mathbf{A}_i \mathbf{x}_j) = 0,$$

$$\sum_{i=0}^{n} c_i \delta_{ij} = 0,$$

$$c_j = 0.$$

Thus for each $\mathbf{x}_j$ ($j = 1, 2, \ldots, n$), we have $c_j = 0$, which implies that $c_1 = c_2 = \cdots = c_n = 0$ and that the rows $\mathbf{A}_1, \mathbf{A}_2, \ldots, \mathbf{A}_n$ are linearly independent.

It follows directly from Lemma 2 and the definition of an inverse that if an $n \times n$ matrix $\mathbf{A}$ has an inverse, then $\mathbf{A}$ must have rank n. This in turn implies directly that if $\mathbf{A}$ does not have rank n, then it does not have an inverse. We now want to show the converse: that is, if $\mathbf{A}$ has rank n, then $\mathbf{A}$ has an inverse.

We already have part of the result. If an $n \times n$ matrix $\mathbf{A}$ has rank n, then the procedure described in Section 3.2 is a constructive method for obtaining a matrix $\mathbf{C}$ having the property that $\mathbf{CA} = \mathbf{I}$. The procedure transforms $\mathbf{A}$ to an identity matrix by a sequence of elementary row operations $\mathbf{E}_1, \mathbf{E}_2, \ldots, \mathbf{E}_{k-1}, \mathbf{E}_k$. That is,

$$\mathbf{E}_k \mathbf{E}_{k-1} \ldots \mathbf{E}_2 \mathbf{E}_1 \mathbf{A} = \mathbf{I}.$$

Setting

$$C = E_k E_{k-1} \ldots E_2 E_1, \tag{12}$$

we have

$$CA = I. \tag{13}$$

We need only show that $AC = I$, too.

Theorem 1. *If A and B are $n \times n$ matrices such that $AB = I$, then $BA = I$.*

Proof. If $AB = I$, then from Lemma 1 A has rank n, and from (12) and (13) there exists a matrix C such that $CA = I$. It follows from Theorem 1 of Section 3.4 that $B = C$.

The major implication of Theorem 1 is that if B is a right inverse of A, then B is also a left inverse of A; and also if A is a left inverse of B, then A is also a right inverse of B. Thus, one needs only check whether a matrix is a right or left inverse; once one is verified for square matrices, the other is guaranteed. In particular, if an $n \times n$ matrix A has rank n, then (13) is valid. Thus, C is a left inverse of A. As a result of Theorem 1, however, C is also a right inverse of A—just replace A with C and B with A in Theorem 1—so C is both a left and right inverse of A, which means that C is the inverse of A. We have now proven:

Theorem 2. *An $n \times n$ matrix A has an inverse if and only if A has rank n.*

An introduction to optimization

Chapter outline

4.1 Graphing inequalities

Many times in real life, solving simple *equations* can give us solutions to everyday problems.

Example 1. Suppose we enter a supermarket and are informed that a certain brand of coffee is sold in 3-lb bags for $6.81. If we wanted to determine the cost per unit pound, we could *model* this problem as follows:

Let x be the cost per unit pound of coffee; then, the following equation represents the total cost of the coffee:

$$x + x + x = 3x = 6.81. \tag{1}$$

Dividing both sides of (1) by 3 gives the cost of $2.27 per pound of coffee. ∎

Example 2. Let's suppose that we are going to rent a car. If the daily fixed cost is $100.00, with the added price of $1.25 per mile driven, then

$$C = 100 + 1.25m \tag{2}$$

represents the total daily cost, C, where m is the number of miles traveled on a particular day.

What if we had a daily budget of $1000.00? We would then use (2) to determine the number of miles we could travel given this budget. Using elementary algebra, we see that we would be able to drive 720 miles. ∎

Matrix Methods. https://doi.org/10.1016/B978-0-12-818419-6.00004-6

These two simple examples illustrate how equations can assist us in our daily lives. But sometimes things can be a bit more complicated.

Example 3.   Suppose we are employed in a factory that produces two types of bicycles: a standard model (S) and a deluxe model (D). Let us assume that the revenue (R) on the former is $250 per bicycle and the revenue on the latter is $300 per bicycle. Then the total revenue can be expressed by the following equation:

$$R = 250S + 300D. \tag{3}$$

Now suppose manufacturing costs are $10,000; so to make a profit, R has to be *greater than* $10,000. Hence the following *inequality* is used to relate the bicycles and revenue with respect to showing a profit:

$$250S + 300D > 10,000. \tag{4}$$

Relationship (4) illustrates the occurrence of inequalities. However, before we can solve problems related to this example, it is important to "visualize" inequalities because the graphing of such relationships will assist us in many ways. ∎

For the rest of this section, we will sketch inequalities in two dimensions.

Example 4.   Sketch the inequality $x + y \leq 2$. The *equation* $x + y = 2$ is a straight line passing through the points (2, 0)—the x-intercept—and (0, 2)—the y-intercept. The *inequality* $x + y \leq 2$ merely includes the region "under" the line. ∎

Remark 1.   Notice that the lower left-hand part of the graph is *shaded.* An easy way to check is to pick a point, say $(-50, -50)$; clearly $-50 + -50 \leq 2$; therefore, the "half-region" containing this point must be the shaded portion.

Remark 2.   The graph of the *strict* inequality $x + y < 2$ yields the same picture with the line *dashed* (instead of solid) to indicate that points on the line $x + y = 2$ are *not* included.

Example 5.   Sketch $2x + 3y \geq 450$. ∎

Remark 3.   Notice that we have restricted this graph to the first quadrant. Many times the variables involved will have nonnegative values, such as volume, area, etc. Notice, too, that the region is *infinite,* as is the region in Example 4.

Example 6.   Sketch $4x + y \leq 12$ and $2x + 5y \leq 24$, where $x \geq 0$ and $y \geq 0$. ∎

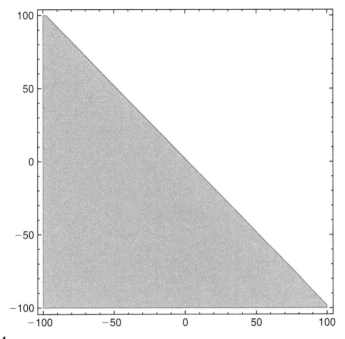

FIGURE 4.1

Remark 4. Note that the "upper-right" corner point is (2, 4). This point is the *intersection* of the straight lines given by equations $4x + y = 12$ and $2x + 5y = 24$; in Chapter 2, we covered techniques used in solving simultaneous equations. Here the added constraints of $x \geq 0$ and $y \geq 0$ render a *bounded* or *finite* region.

We will see regions like Fig. 4.3 again both in Section 4.2, as we shall see below in Fig 4.2 and Fig 4.3, below, (with regard to modeling) and Section 4.3 (using the technique of *linear programming*).

Problems 4.1

Sketch the following inequalities:

1. $y \leq 0$
2. $x \geq 0$
3. $y \geq \pi$
4. $x + 4y \leq 12$
5. $x + 4y < 12$
6. $x + 4y \geq 12$
7. $x + 4y > 12$

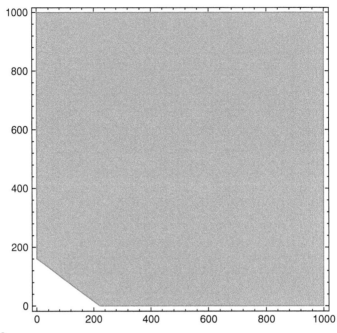

FIGURE 4.2

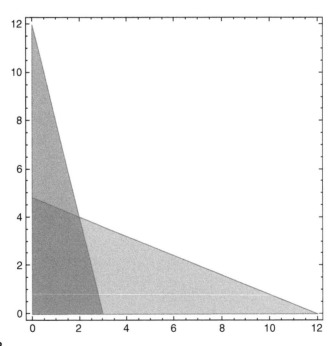

FIGURE 4.3

Sketch the inequalities on the same set of axes:

8. $x + 4y \leq 12, x \geq 0, y \geq 0$
9. $x + 4y \leq 12, 5x + 2y \leq 24$
10. $x + 4y \geq 12, 5x + 2y \geq 24$
11. $x + 2y \leq 12, 2x + y \leq 16, x + 2y \leq 20$
12. $x - y \geq 100$
13. $x + y \geq 100, 3x + 3y \leq 60$
14. $x + y \leq 10, -x + y \leq 10, x - y \leq 10, -x-y \leq 10$

4.2 Modeling with inequalities

Consider the following situation. Suppose a toy company makes two types of wagons, X and Y. Let us further assume that during any work period, each X takes 3 hours to construct and 2 hours to paint, while each Y takes 1 hour to construct and 2 hours to paint. Finally, the maximum number of hours allotted for construction is 1500 and the limit on hours available for painting is 1200 hours. If the profit on each X is $50 and the profit on each Y is $60, how many of each type of wagon should be produced to maximize the profit?

We can model the above with a number of inequalities. First, we must define our variables. Let X represent the number of X wagons produced and Y represent the number of Y wagons produced. This leads to the following four relationships:

$$3X = Y \leq 1500 \tag{5}$$

$$2X + 2Y \leq 1200 \tag{6}$$

$$X \geq 0 \tag{7}$$

$$Y \geq 0. \tag{8}$$

Note that (5) represents the constraint due to *construction* (in hours), while (6) represents the constraint due to *painting* (also in hours). The inequalities (7) and (8) merely state that the number of each type of wagon cannot be negative.

These four inequalities can be graphed as shown in Fig. 4.4:

Let us make a few observations. We will call the shaded region that satisfies all four inequalities the *region of feasibility*. Next, the shaded region has four "corner points" called *vertices*. The coordinates of these points are given by (0, 0), (0, 600), (450, 150), and (500, 0). Lastly, this region has the property that, given any two points in the interior of the region, the straight line segment connecting these two points lies entirely within the region. We call regions with this property *convex*.

The following equation gives the profit (in dollars):

$$P(X, Y) = 50X + 60Y. \tag{9}$$

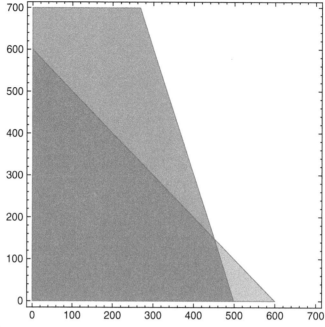

FIGURE 4.4

Note that Eq. (9) is called the *objective function.* The notation $P(X, Y)$ is read "P of X and Y" and is evaluated by simply substituting the respective values into the expression. For example, $P(0{,}600) = 50(0) + 60(600) = 0 + 36{,}000 = 36{,}000$ dollars, while $P(450{,}150) = 50(450) + 60(150) = 22{,}500 + 9000 = 31{,}500$ dollars.

Eq. (9), the inequalities (5)–(8), and Fig. 4.4 model the situation above, which is an example of an *optimization problem.* In this particular example, our goal was to *maximize* a quantity (profit). Our next example deals with *minimization.*

Suppose a specific diet calls for the following minimum daily requirements: 186 units of Vitamin A and 120 units of Vitamin B. Pill X contains 6 units of Vitamin A and 3 units of Vitamin B, while pill Y contains 2 units of Vitamin A and 2 units of Vitamin B. What is the least number of pills needed to satisfy both vitamin requirements?

Let us allow X to represent the number of X pills ingested and let Y represent the number of Y pills taken. Then the following inequalities hold:

$$6X + 2Y \geq 186 \tag{10}$$

$$3X + 2Y \geq 120 \tag{11}$$

$$X \geq 0 \tag{12}$$

$$Y \geq 0. \tag{13}$$

Note that (10) models the minimum daily requirement of units of Vitamin A, while (11) refers to the minimum daily requirement of units of Vitamin B. The quantity to be minimized, the total number of pills, is given by the objective function

$$N(X, Y) = X + Y. \tag{14}$$

We note that while this region of feasibility is convex, it is also *unbounded.* Our vertices are (40, 0), (0, 93), and (22, 27).

In the next section, we will solve problems such as these by applying a very simple, yet extremely powerful, theorem of *linear programming.*

Problems 4.2

Model the following situations by defining all variables and giving all inequalities, the objective function, and the region of feasibility.

1. Farmer John gets $5000 for every truck of wheat sold and $6000 for every truck of corn sold. He has two fields: field A has 23 acres and field B has 17 acres. For every 2 acres of field A, Farmer John produces a truck of wheat, while 3 acres are required of field B for the same amount of wheat. Regarding the corn, 3 acres of field A are required for a truck, while only 1 acre of field B is needed. How many trucks of each commodity should be produced to maximize Farmer John's profit?

2. Redo Problem (1) if Farmer John gets $8000 for every truck of wheat and $5000 for every truck of corn.

3. Dr. Lori Pesciotta, a research scientist, is experimenting with two forms of a special compound, *H-Turebab.* She needs at least 180 units of one form of the compound (α) and at least 240 units of the second form of the compound (β). Two mixtures are used: X and Y. Every unit of X contains two units of α and three units of β, while each unit of Y has the opposite concentration. What combination of X and Y will minimize Dr. Pesciotta's costs, if each unit of X costs $500 and each unit of Y costs $750?

4. Redo Problem (3) if X costs $750 per unit and Y costs $500 per unit.

5. Redo Problem (3) if, in addition, Dr. Pesciotta needs at least 210 units of a third form (γ) of *H-Turebab*, and it is known that every unit of both X and Y contains 10 units of γ.

6. Cereal X costs $.05 per ounce, while Cereal Y costs $.04 per ounce. Every ounce of X contains 2 mg (mg) of zinc and 1 mg of calcium, while every ounce of Y contains 1 mg of zinc and 4 mg of calcium. The minimum daily requirement (MDR) is 10 mg of zinc and 15 mg of calcium. Find the least expensive combination of the cereals that would satisfy the MDR.

7. Redo Problem (6) with the added constraint of at least 12 mg of sodium if each ounce of X contains 3 mg of sodium and every ounce of Y has 2 mg of sodium.

8. Redo Problem (7) if Cereal X costs $.07 an ounce and Cereal Y costs $.08 an ounce.

9. Consider the following group of inequalities along with a corresponding objective function. For each one, sketch the region of feasibility (except for 9 g) and construct a scenario that might model each set of inequalities:

   (a) $x \geq 0$, $y \geq 0$, $2x + 5y \leq 10$, $3x + 4y \leq 12$, $F(x, y) = 100x + 55y$
   (b) $x \geq 0$, $y \leq 0$, $x + y \leq 40$, $x + 2y \leq 60$, $G(x, y) = 7x + 6y$
   (c) $x \geq 2$, $y \geq 3$, $x + y \leq 40$, $x + 2y \leq 60$, $H(x, y) = x + 3y$
   (d) $x \geq 0$, $y \geq 0$, $x + y \leq 600$, $3x + y \leq 900$, $x + 2y \leq 1000$, $J(x, y) = 10x + 4y$
   (e) $2x + 9y \geq 1800$, $3x + y \geq 750$, $K(x, y) = 4x + 11y$
   (f) $x + y \geq 100$, $x + 3y \geq 270$, $3x + y \geq 240$, $L(x, y) = 600x + 375y$
   (g) $x \geq 0$, $y \geq 0$, $z \geq 0$, $x + y + 2z \leq 12$, $2x + y + z \leq 14$, $x + 3y + z \leq 15$, $M(x, y, z) = 2x + 3y + 4z$ (Do *not* sketch the region of feasibility for this problem.)

4.3 Solving problems using linear programming

We are now ready to solve a fairly large class of optimization problems using a special form of the *Fundamental Theorem of Linear Programming*. We will not prove this theorem, but many references to the proof of a more general result are available (for example, see Luenberger, D. G., *Linear and Nonlinear Programming*, 2nd Ed., Springer 2003).

The Fundamental Theorem of Linear Programming Let Γ be a convex region of feasibility in the xy-plane. Then the objective function $F(x, y) = ax + by$, where a and b are real numbers, takes on both maximum and minimum values—if they exist—on one or more vertices of Γ.

Remark 1. The theorem holds only *if* maximum and/or minimum values exist.

Remark 2. It is possible to have infinitely many values where an optimal (maximum or minimum) value exists. In this case, they would lie on one of the line segments that form the boundary of the region of feasibility. See Example 3 below.

Remark 3. The word *programming* has nothing to do with computer programming, but rather the systematic order followed by the procedure, which can also be termed an *algorithm*.

Some examples are in order.

Example 1. (Wagons): Consider the inequalities (5)–(8), along with Eq. (9), from Section 4.2. We again give the region of feasibility in Fig. 4.6 (same as Fig. 4.4):

Evaluation of our objective function,

$$P(X, Y) = 50X + 60Y, \tag{15}$$

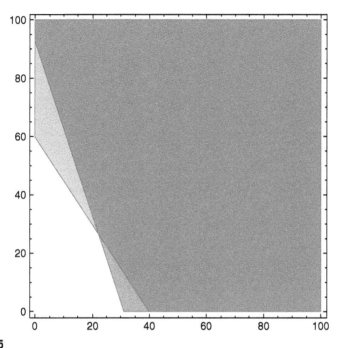

FIGURE 4.5

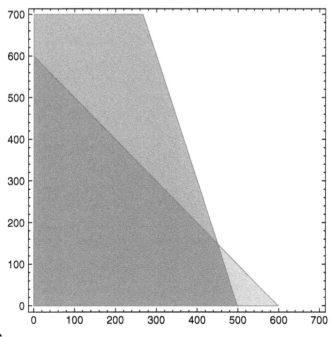

FIGURE 4.6

at each of the four vertices yields the following results:

$$\begin{cases} P(0,0) = 0 \\ P(0,600) = 36,000 \\ P(450,150) = 31,500 \\ P(500,0) = 25,000. \end{cases}$$

By the fundamental theorem of linear programming, we see that the maximum profit of $36,000 occurs if no X wagons are produced and 600 Y wagons are made. ∎

Example 2.   (Wagons): Suppose the profit function in the previous example is given by

$$R(X,Y) = 80X = 50Y. \tag{16}$$

Then

$$\begin{cases} R(0,0) = 0 \\ R(0,600) = 30,000 \\ R(450,150) = 43,500 \\ R(500,0) = 40,000. \end{cases}$$

We see, in this situation, that the maximum profit of $43,500 occurs if 450 X wagons are produced, along with 150 Y wagons. ∎

Example 3.   (Wagons): Consider the examples above with the profit function given by

$$L(X,Y) = 75X + 75Y. \tag{17}$$

Then

$$\begin{cases} L(0,0) = 0 \\ L(0,600) = 45,000 \\ L(450, 150) = 45,000 \\ L(500,0) = 37,500. \end{cases}$$

Note that we have *two* situations in which the profit is maximized at $45,000; in fact, there are *many* points where this occurs. For example,

$$L(300, 300) = 45,000. \tag{18}$$

This occurs *at any point* along the constraint given by inequality (2). The reason lies in the fact that *coefficients* of X and Y in (2) and in Eq. (7) have the same ratio. ∎

Example 4.   (Vitamins): Consider constraints (10)–(13) above in Section 4.2; minimize the objective function given by Eq. (14):

$$N(X,Y) = X + Y. ∎ \tag{19}$$

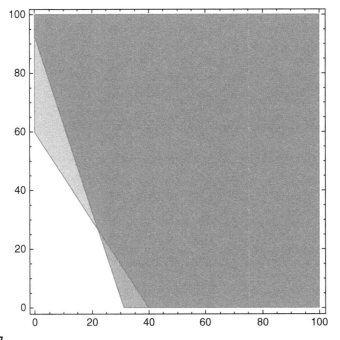

FIGURE 4.7

The region of feasibility (same as Fig. 4.5) is given in Fig. 4.7: Evaluating our objective function (19) at the three vertices, we find that

$$\begin{cases} N(40,0) = 40 \\ N(0,93) = 93 \\ N(22, 27) = 49, \end{cases}$$

so the minimum number of pills needed to satisfy the minimum daily requirement is 40.

Sometimes a constraint is *redundant*; that is, the other constraints "include" the redundant constraint.

For example, suppose we want to maximize the objective function

$$Z(X, Y) = 4X + 3Y, \tag{20}$$

given the constraints

$$4X + 2Y \leq 40 \tag{21}$$

$$3X + 4Y \leq 60 \tag{22}$$

$$X \geq 0 \tag{23}$$

$$Y \geq 0. \tag{24}$$

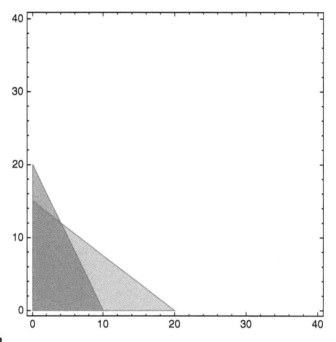

FIGURE 4.8

The vertices of the region of feasibility are *(0, 0)*, *(0, 15)*, *(4, 12)*, and *(10, 0)*, as seen in Fig. 4.8.

Note that (11) is maximized at $Z(4, 12) = 52$.

Suppose we now add a third constraint:

$$X + Y \leq 30. \tag{25}$$

Fig. 4.9 reflects this added condition. Note, however, that the region of feasibility is *not* changed and the four vertices are unaffected by this *redundant* constraint. It follows, therefore, that our objective function $Z(X, Y) = 4X + 3Y$ is still maximized at $Z(4, 12) = 52$.

Remark 4. Sometimes a vertex does not have whole number coordinates (see Problem (15) below). If the physical model does not make sense to have a fractional or decimal answer—for example, 2.5 bicycles or 1/3 cars—then we should check the closest points with whole number coordinates, *provided these points lie in the region of feasibility.* For example, if (2.3, 7.8) is the vertex that gives the optimal value for an objective function, then the following points should be checked: (2, 7), (2, 8), (3, 7), and (3, 8).

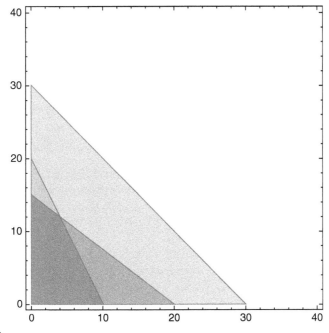

FIGURE 4.9

Problems 4.3

Using linear programming techniques, solve the following problems:

1. Section 4.2, Problem (1).
2. Section 4.2, Problem (2).
3. Section 4.2, Problem (3).
4. Section 4.2, Problem (4).
5. Section 4.2, Problem (5).
6. Section 4.2, Problem (6).
7. Section 4.2, Problem (7).
8. Section 4.2, Problem (8).
9. Section 4.2, Problem (9a); maximize $F(x, y)$.
10. Section 4.2, Problem (9b); maximize $G(x, y)$.
11. Section 4.2, Problem (9c); maximize $H(x, y)$.
12. Section 4.2, Problem (9d); maximize $J(x, y)$.
13. Section 4.2, Problem (9e); minimize $K(x, y)$.
14. Section 4.2, Problem (9f); minimize $L(x, y)$.
15. Maximize $P(x, y) = 7x + 6y$ subject to the constraints $x \geq 0$, $y \geq 0$, $2x + 3y \leq 1200$, and $6x + y \leq 1500$.

4.4 **An introduction to the simplex method**

In most of the problems considered in the previous section, we had but two variables (usually X and Y) and two constraints, not counting the usual conditions of the non-negativity of X and Y. Once a third constraint is imposed, the region of feasibility becomes more complicated; and, with a fourth constraint, even more so.

Also, if a third variable, say Z, is brought into the discussion, then the region of feasibility becomes *three dimensional*! This certainly makes the technique employed in the previous section much more difficult to apply, although theoretically it can be used.

We are fortunate that an alternate method exists, which is valid for any number of variables and any number of constraints. It is known as the *simplex method*. This is a *classic* method that has been in use for *many* years. The reader may wish to consult G. Hadley's *Linear Programming* published by Addison-Wesley in 1963 for the theoretical underpinnings of this algorithm.

Before we illustrate this technique with a number of examples, describing and defining terms as we go along, we point out that this section will deal exclusively with *maximization* problems. We will address minimization in the next, and final, section of this chapter.

Example 1.   Suppose we want to maximize the following function of two variables:

$$z = 7x_1 + 22x_2. \tag{26}$$

Note that we are using x_i instead of the usual x and y, because, in later examples, we will have more than two independent variables.

Let us assume that the following constraints are imposed:

$$3x_1 + 10x_2 \leq 33,000 \tag{27}$$

$$5x_1 + 8x_2 \leq 42,000 \tag{28}$$

$$x_1 \geq 0 \tag{29}$$

$$x_2 \geq 0. \tag{30}$$

We now introduce the concept of *slack variables*, which we denote by s_i. These variables, (which can never be negative), will "pick up the slack" in the relationships (27) and (28) and convert these inequalities into equations. That is, (27) and (28) can now be written, respectively, as

$$3x_1 + 10x_2 + s_1 = 33,000 \tag{31}$$

and

$$5x_1 + 8x_2 + s_2 = 42,000. \tag{32}$$

We also incorporate these slack variables into our objective function (26), rewriting it as

$$-7x_1 - 22x_2 + 0s_1 + 0s_2 + z = 0. \tag{33}$$

Finally, we rewrite (27) and (28) as

$$3x_1 + 10x_2 + s_1 + 0s_2 + 0z = 33,000 \tag{34}$$

$$5x_1 + 8x_2 + 0s_1 + 1s_2 + 0z = 42,000. \blacksquare \tag{35}$$

Remark 1. Admittedly, Eqs. (33)–(35) seem somewhat strange. However, the reader will soon see why we have recast these equations as they now appear.

We are now ready to put these last three equations into a table known as the *initial tableau*. This is nothing more than a kind of augmented matrix. To do this, we merely "detach" the coefficients of the five unknowns (x_1, x_2, s_1, s_2, and z) and form the following table:

$$
\begin{array}{ccccc}
x_1 & x_2 & s_1 & s_2 & z \\
\end{array}
$$
$$
\left[
\begin{array}{ccccc|c}
3 & 10 & 1 & 0 & 0 & 33,000 \\
5 & 8 & 0 & 1 & 0 & 42,000 \\
\hline
-7 & -22 & 0 & 0 & 1 & 0
\end{array}
\right].
\tag{36}
$$

Remark 2. Note that the objective function equation—here, Eq. (33)—is in the bottom row. Also, unless otherwise stipulated, we shall always assume that the decision variables—that is, x_1 and x_2—are nonnegative. Notice, too, the *vertical bar* that appears to the *left* of the rightmost column and the *horizontal bar* placed *above* the bottom row. Finally, we point out that the entry in the last row and last column is always zero for this initial tableau. These conventions will assist us in interpreting the end state of the simplex method.

Before continuing with the simplex method, let us consider another example.

Example 2. Put the following maximization problem into the initial tableau form: $z = 4x_1 + 7x_2 + 9x_3$, where $x_1 + x_2 + 6x_3 \leq 50$, $2x_1 + 3x_2 \leq 40$, and $4x_1 + 9x_2 + 3x_3 \leq 10$.

Note that we have three independent (decision) variables (the x_i) and that the three constraints will give us three slack variables (the s_i). These lead us to the following four equations:

$$-4x_1 - 7x_2 - 9x_3 + 0s_1 + 0s_2 + 0s_3 + z = 0 \tag{37}$$

$$x_1 + x_2 - 6x_3 + s_1 + 0s_2 + 0s_3 + 0z = 50 \tag{38}$$

$$2x_1 + 0x_2 + 3x_3 + 0s_1 + s_2 + 0s_3 + 0z = 40 \tag{39}$$

$$4x_1 + 9x_2 + 3x_3 + 0s_1 + 0s_2 + s_3 + 0z = 10. \tag{40}$$

The initial tableau for this example is given below:

$$
\begin{array}{ccccccc}
x_1 & x_2 & x_3 & s_1 & s_2 & s_3 & z \\
\end{array}
$$
$$
\left[
\begin{array}{ccccccc|c}
1 & 1 & 6 & 1 & 0 & 0 & 0 & 50 \\
2 & 0 & 3 & 0 & 1 & 0 & 0 & 40 \\
4 & 9 & 3 & 0 & 0 & 1 & 0 & 10 \\
\hline
-4 & -7 & -9 & 0 & 0 & 0 & 1 & 0
\end{array}
\right]. \; \blacksquare
\tag{41}
$$

We will now outline the steps in the simplex method:

- Change all inequalities into equations via the use of slack variables.
- Rewrite the objective function, z, in terms of slack variables, setting one side of the equation equal to zero and keeping the coefficient of z equal to $+1$.
- The number of equations should equal the *sum* of the constraints $+1$ (the equation given by the objective function).
- Form the initial tableau, listing the constraints above the objective function, labeling the columns, beginning with the decision variables, followed by the slack variables, with z represented by the last column before the vertical bar. The last column should have all the "constants".
- Locate the *most negative number* in the last row. If more than one equally negative number is present, arbitrarily choose any one of them. Call this number k. This column will be called the *work column*.
- Consider each *positive* element in the work column. Divide each of these elements into the corresponding row entry element in the last column. The ratio that is the *smallest* will be used as the work column's *pivot*. If there is more than one smallest ratio, arbitrarily choose any one of them.
- Use elementary row operations (see Chapter 2) to change the pivot element to 1, unless it is already 1.
- Use elementary row operations to transform all the other elements in the work column to 0.
- A column is *reduced* when all the elements are 0, with the exception of the pivot, which is 1.
- Repeat the process until there are *no negative elements in the last row.*
- We are then able to *determine the answers* from this final tableau.

Let us illustrate this by returning to Example 1, where the initial tableau is given by

$$
\begin{array}{ccccc}
x_1 & x_2 & s_1 & s_2 & z \\
\end{array}
$$
$$
\left[
\begin{array}{ccccc|c}
3 & 10 & 1 & 0 & 0 & 33,000 \\
5 & 8 & 0 & 1 & 0 & 42,000 \\
\hline
-7 & -22 & 0 & 0 & 1 & 0 \\
\end{array}
\right]. \tag{42}
$$

We first note that -22 is the *most negative number* in the last row of (42). So the "x_2" column is our *work column.*

We next divide 33,000 by $10 = 3300$ and 42,000 by $8 = 5250$; since 3300 is the *lesser positive number*, we will use 10 as the *pivot*. Note that we have put a carat ($\hat{\ }$) over the 10 to signify it is the pivot element:

$$
\begin{array}{ccccc}
x_1 & x_2 & s_1 & s_2 & z \\
\end{array}
$$
$$
\left[
\begin{array}{ccccc|c}
3 & \hat{10} & 1 & 0 & 0 & 33,000 \\
5 & 8 & 0 & 1 & 0 & 42,000 \\
\hline
-7 & -22 & 0 & 0 & 1 & 0 \\
\end{array}
\right]. \tag{43}
$$

We now divide every element in the row containing the pivot by 10:

$$
\begin{array}{ccccc}
x_1 & x_2 & s_1 & s_2 & z \\
\end{array}
$$

$$
\left[\begin{array}{ccccc|c}
0.3 & \hat{1} & 0.1 & 0 & 0 & 3300 \\
5 & 8 & 0 & 1 & 0 & 42,000 \\
\hline
-7 & -22 & 0 & 0 & 1 & 0
\end{array}\right]. \tag{44}
$$

Next, we use elementary row operations; we multiply the first row by -8 and add it to the second row and multiply the first row by 22 and add it to the third row. This will give us a 0 for every element (other than the pivot) in the work column:

$$
\begin{array}{ccccc}
x_1 & x_2 & s_1 & s_2 & z \\
\end{array}
$$

$$
\left[\begin{array}{ccccc|c}
0.3 & \hat{1} & 0.1 & 0 & 0 & 3300 \\
2.6 & 0 & -0.8 & 1 & 0 & 15,600 \\
\hline
-0.4 & 0 & 2.2 & 0 & 1 & 72,600
\end{array}\right]. \tag{45}
$$

And now we repeat the process because we still have a negative entry in the last row, that is, -0.4 is in the "x_1" column. Hence, this becomes our new work column.

Dividing 3300 by 0.3 yields 11,000; dividing 15,600 by 2.6 gives us 6000; since 6000 is the lesser of the two positive ratios, we will use the 2.6 entry as the pivot (again denoting it with a carat, and removing the carat from our first pivot):

$$
\begin{array}{ccccc}
x_1 & x_2 & s_1 & s_2 & z \\
\end{array}
$$

$$
\left[\begin{array}{ccccc|c}
0.3 & 1 & 0.1 & 0 & 0 & 3300 \\
2.6 & 0 & -0.8 & 1 & 0 & 15,600 \\
\hline
-0.4 & 0 & 2.2 & 0 & 1 & 72,600
\end{array}\right]. \tag{46}
$$

Dividing each element in this row by 2.6 gives us the following tableau:

$$
\begin{array}{ccccc}
x_1 & x_2 & s_1 & s_2 & z \\
\end{array}
$$

$$
\left[\begin{array}{ccccc|c}
0.3 & 1 & 0.1 & 0 & 0 & 3300 \\
\hat{1} & 0 & -.31 & .38 & 0 & 6000 \\
\hline
-0.4 & 0 & 2.2 & 0 & 1 & 72,600
\end{array}\right]. \tag{47}
$$

Using our pivot and elementary row operations, we transform every other element in this work column to 0. That is, we multiply each element in the second row by -0.3 and add the row to the first row and we multiply every element in the second row by 0.4 and add the row to the last row. This gives us the following tableau:

$$
\begin{array}{ccccc}
x_1 & x_2 & s_1 & s_2 & z \\
\end{array}
$$

$$
\left[\begin{array}{ccccc|c}
0 & 1 & 0.19 & -0.12 & 0 & 1500 \\
1 & 0 & -.31 & .38 & 0 & 6000 \\
\hline
0 & 0 & 2.08 & 0.15 & 1 & 75,000
\end{array}\right]. \tag{48}
$$

We are now *finished* with the process because there are no negative elements in the last row. We *interpret* this final tableau as follows:

- $x_1 = 6000$ (note the "1" in the x_1 column and the "0" in the x_2 column).
- $x_2 = 1500$ (note the "0" in the x_1 column and the "1" in the x_2 column).
- Both slack variables equal 0. To verify this, please see Eqs. (31) and (32) and substitute our values for x_1 and x_2 into these equations.
- The maximum value of z is 75,000 (found in the lower right-hand corner box).

We now give another example.

Example 3.   Maximize $z = x_1 + 2x_2$,   subject   to   the   constraints $4x_1 + 2x_2 \leq 40$ and $3x_1 + 4x_2 \leq 60$.

Following the practice discussed in this section and introducing the slack variables, we have

$$4x_1 + 2x_2 + s_1 = 40 \tag{49}$$

$$3x_1 + 4x_2 + s_2 = 60 \tag{50}$$

and

$$-x_1 - 2x_2 + z = 0. \tag{51}$$

We form the initial tableau, using coefficients of 0 where needed, as follows:

$$\begin{array}{ccccc} x_1 & x_2 & s_1 & s_2 & z \\ \left[\begin{array}{ccccc|c} 4 & 2 & 1 & 0 & 0 & 40 \\ 3 & 4 & 0 & 1 & 0 & 60 \\ \hline -1 & -2 & 0 & 0 & 1 & 0 \end{array}\right] \end{array}. \tag{52}$$

The second column will be our work column, since -2 is the most negative entry. Dividing 40 by 2 gives 20; dividing 60 by 4 yields 15. Since 15 is a lesser positive ratio than 20, we will use the 4 as the pivot:

$$\begin{array}{ccccc} x_1 & x_2 & s_1 & s_2 & z \\ \left[\begin{array}{ccccc|c} 4 & 2 & 1 & 0 & 0 & 40 \\ 3 & \widehat{4} & 0 & 1 & 0 & 60 \\ -1 & -2 & 0 & 0 & 1 & 0 \end{array}\right] \end{array}. \tag{53}$$

Dividing every element of the second row will make our pivoting element 1:

$$\begin{array}{ccccc} x_1 & x_2 & s_1 & s_2 & z \\ \left[\begin{array}{ccccc|c} 4 & 2 & 1 & 0 & 0 & 40 \\ 0.75 & \widehat{1} & 0 & 0.25 & 0 & 15 \\ -1 & -2 & 0 & 0 & 1 & 0 \end{array}\right] \end{array}. \tag{54}$$

We now use our pivot, along with the proper elementary row operations, to make every other element in the column zero. This leads to the following tableau:

$$\begin{array}{ccccc} x_1 & x_2 & s_1 & s_2 & z \\ \left[\begin{array}{ccccc|c} 2.5 & 0 & 1 & -0.5 & 0 & 10 \\ 0.75 & \widehat{1} & 0 & 0.25 & 0 & 15 \\ 0.5 & 0 & 0 & 0.5 & 1 & 30 \end{array}\right] \end{array}. \tag{55}$$

Since the last row has no negative entries, we are finished and have the final tableau:

$$
\begin{array}{ccccc}
x_1 & x_2 & s_1 & s_2 & z \\
\end{array}
$$

$$
\left[
\begin{array}{ccccc|c}
2.5 & 0 & 1 & -0.5 & 0 & 10 \\
0.75 & 1 & 0 & 0.25 & 0 & 15 \\
\hline
0.5 & 0 & 0 & 0.5 & 1 & 30 \\
\end{array}
\right] . \tag{56}
$$

This final tableau is a little more complicated to interpret than (48).

Notice the "1" in the second row; this implies that $x_2 = 15$. The corresponding equation represented by this second row thereby reduces to

$$
0.75x_1 + 15 + 0.25s_2 = 15. \tag{57}
$$

Which forces both x_1 and s_2 to be zero, since neither can be negative. This forces $s_1 = 10$, as we can infer from the equation represented by the first row:

$$
0.25x_1 + s_1 - 0.5s_2 = 10. \tag{58}
$$

In practice, we are not concerned with the values of the slack variables, so we summarize by simply saying that our answers are $x_1 = 0$ and $x_2 = 15$ with a maximum value of $z = 30$. ∎

As we have pointed out, this is a *classic* technique. However, as the number of variables (decision and/or slack) increases, the calculations can be somewhat burdensome. Thankfully, there are many software packages to assist in this matter. Please refer to the Final Comments at the end of this chapter.

One final remark: As is the case with linear programming, if there are an infinite number of optimal solutions, the simplex method does not give *all* solutions.

Problems 4.4

Using the simplex method, solve the following problems:

1. Section 4.2, Problem (1).
2. Section 4.2, Problem (2).
3. Maximize $z = 3x_1 + 5x_2$, subject to $x_1 + x_2 \le 6$ and $2x_1 + x_2 \le 8$.
4. Maximize $z = 8x_1 + x_2$, subject to the same constraints in (3).
5. Maximize $z = x_1 + 12x_2$, subject to the same constraints in (3).
6. Maximize $z = 3x_1 + 6x_2$, subject to the constraints $x_1 + 3x_2 \le 30$, $2x_1 + 2x_2 \le 40$, and $3x_1 + x_2 \le 30$.
7. Consider Problem (9) at the end of Section 4.2. Set up the initial tableaus for Problems (9a) through (9d).

4.5 **Final comments on Chapter 4**

In this chapter, we covered two approaches to optimization, the linear programming method and the simplex method. Both of these techniques are classical and their geometrics and algebraic simplicity reflect both the beauty and power of mathematics.

Our goal was to introduce the reader to the basics of these "simple" methods. However, he or she should be cautioned with regard to the underlying theory. That is, many times in mathematics we have elegant results (theorems), which are proved using very *deep* and *subtle* mathematical concepts with respect to the proofs of these theorems.

As we mentioned in the last section, the calculations, while not difficult, can be a burden. Calculators and software packages can be of great assistance here.

We close with two observations. Please note that we have considered very special cases where the constraints of the "$\leq$" variety had *positive* quantities on the right-hand side. If this is not the case for all the constraints, then we must use an *enhanced* version of the simplex method (see, for example, *Finite Mathematics: A Modeling Approach* by R. Bronson and G. Bronson, published by West in 1996).

Similarly, regarding the solving of *minimization* problems via the simplex method, we essentially consider the "negation" of the objective function, and then apply a modified version of the simplex method. For example, suppose we wanted to minimize $z = 3x_1 + 2x_2$, subject to the same constraints. In this case we would *maximize $Z = -z = -3x_1 - 2x_2$*, while recasting our constraints, and then proceed with the simplex method.

Determinants

Chapter outline

5.1 Introduction

Every square matrix has associated with it a scalar called its *determinant*. To be extremely rigorous, we would have to define this scalar in terms of permutations on positive integers. However, since in practice it is difficult to apply a definition of this sort, other procedures have been developed, which yield the determinant in a more straightforward manner. In this chapter, therefore, we concern ourselves solely with those methods that can be applied easily. We note here for reference that determinants are only defined for square matrices.

Given a square matrix $\mathbf{A}$, we use $\det(\mathbf{A})$ or $|\mathbf{A}|$ to designate its determinant. If the matrix can actually be exhibited, we then designate the determinant of $\mathbf{A}$ by replacing the brackets by vertical straight lines. For example, if

$$\mathbf{A} = \begin{bmatrix} 1 & 2 & 3 \\ 4 & 5 & 6 \\ 7 & 8 & 9 \end{bmatrix} \tag{1}$$

then,

Matrix Methods. https://doi.org/10.1016/B978-0-12-818419-6.00005-8

$$\det(\mathbf{A}) = \begin{vmatrix} 1 & 2 & 3 \\ 4 & 5 & 6 \\ 7 & 8 & 9 \end{vmatrix}. \tag{2}$$

We cannot overemphasize the fact that (1) and (2) represent entirely different animals. (1) represents a matrix, a rectangular array, an entity unto itself, while (2) represents a scalar, a number associated with the matrix in (1). There is absolutely no similarity between the two other than form!

We are now ready to calculate determinants.

Definition 1. The determinant of a 1×1 matrix $[a]$ is the scalar a.

Thus, the determinant of the matrix $[5]$ is 5 and the determinant of the matrix $[-3]$ is -3.

Definition 2. The determinant of a 2×2 matrix

$$\begin{bmatrix} a & b \\ c & d \end{bmatrix}$$

is the scalar $ad - bc$.

Example 1. Find $\det(\mathbf{A})$ if

$$\mathbf{A} = \begin{bmatrix} 1 & 2 \\ 4 & 3 \end{bmatrix}.$$

Solution

$$\det(\mathbf{A}) = \begin{vmatrix} 1 & 2 \\ 4 & 3 \end{vmatrix} = (1)(3) - (2)(4) = 3 - 8 = -5. \ \blacksquare$$

Example 2. Find $|\mathbf{A}|$ if

$$\mathbf{A} = \begin{bmatrix} 2 & -1 \\ 4 & 3 \end{bmatrix}.$$

Solution

$$|\mathbf{A}| = \begin{vmatrix} 2 & -1 \\ 4 & 3 \end{vmatrix} = (2)(3) - (-1)(4) = 6 + 4 = 10. \ \blacksquare$$

We now could proceed to give separate rules, which would enable one to compute determinants of 3×3, 4×4, and higher order matrices. This is unnecessary. In the next section, we will give a method that enables us to reduce all determinants of order $n(n > 2)$ (if $\mathbf{A}$ has order $n \times n$ then $\det(\mathbf{A})$ is said to have order n) to a sum of determinants of order 2.

Problems 5.1

In Problems 1 through 18, find the determinants of the given matrices.

1. $\begin{bmatrix} 3 & 4 \\ 5 & 6 \end{bmatrix}$,

2. $\begin{bmatrix} 3 & -4 \\ 5 & 6 \end{bmatrix}$,

3. $\begin{bmatrix} 3 & 4 \\ -5 & 6 \end{bmatrix}$,

4. $\begin{bmatrix} 5 & 6 \\ 7 & 8 \end{bmatrix}$,

5. $\begin{bmatrix} 5 & 6 \\ -7 & 8 \end{bmatrix}$,

6. $\begin{bmatrix} 5 & 6 \\ 7 & -8 \end{bmatrix}$,

7. $\begin{bmatrix} 1 & -1 \\ 2 & 7 \end{bmatrix}$,

8. $\begin{bmatrix} -2 & -3 \\ -4 & 4 \end{bmatrix}$,

9. $\begin{bmatrix} 3 & -1 \\ -3 & 8 \end{bmatrix}$,

10. $\begin{bmatrix} 0 & 1 \\ -2 & 6 \end{bmatrix}$,

11. $\begin{bmatrix} -2 & 3 \\ -4 & -4 \end{bmatrix}$,

12. $\begin{bmatrix} 9 & 0 \\ 2 & 0 \end{bmatrix}$,

13. $\begin{bmatrix} 12 & 20 \\ -3 & -5 \end{bmatrix}$,

14. $\begin{bmatrix} -36 & -3 \\ -12 & -1 \end{bmatrix}$,

15. $\begin{bmatrix} -8 & -3 \\ -7 & 9 \end{bmatrix}$,

16. $\begin{bmatrix} t & 2 \\ 3 & 4 \end{bmatrix}$,

17. $\begin{bmatrix} 2t & 3 \\ -2 & t \end{bmatrix}$,

18. $\begin{bmatrix} 3t & -t^2 \\ 2 & t \end{bmatrix}$.

19. Find t so that

$$\begin{vmatrix} t & 2t \\ 1 & t \end{vmatrix} = 0.$$

20. Find t so that

$$\begin{vmatrix} t-2 & t \\ 3 & t+2 \end{vmatrix} = 0.$$

21. Find λ so that

$$\begin{vmatrix} 4-\lambda & 2 \\ -1 & 1-\lambda \end{vmatrix} = 0.$$

22. Find λ so that

$$\begin{vmatrix} 1-\lambda & 5 \\ 1 & -1-\lambda \end{vmatrix} = 0.$$

23. Find $\det(\mathbf{A} - \lambda\mathbf{I})$ if $\mathbf{A}$ is the matrix defined in Problem 1.
24. Find $\det(\mathbf{A} - \lambda\mathbf{I})$ if $\mathbf{A}$ is the matrix defined in Problem 2.
25. Find $\det(\mathbf{A} - \lambda\mathbf{I})$ if $\mathbf{A}$ is the matrix defined in Problem 4.
26. Find $\det(\mathbf{A} - \lambda\mathbf{I})$ if $\mathbf{A}$ is the matrix defined in Problem 7.
27. Find $|\mathbf{A}|$, $|\mathbf{B}|$, and $|\mathbf{AB}|$ if

$$\mathbf{A} = \begin{bmatrix} 1 & 3 \\ 2 & 1 \end{bmatrix} \quad \text{and} \quad \mathbf{B} = \begin{bmatrix} 4 & 2 \\ -1 & 2 \end{bmatrix}.$$

What is the relationship between these three determinants?

28. Interchange the rows for each of the matrices given in Problems 1 through 15 and calculate the new determinants. How do they compare with the determinants of the original matrices?

29. The second elementary row operation is to multiply any row of a matrix by a nonzero constant. Apply this operation to the matrices given in Problems 1 through 15 for any constants of your choice and calculate the new determinants. How do they compare with the determinants of the original matrix?

30. Redo Problem 29 for the third elementary row operation.

31. What is the determinant of a 2×2 matrix if one row or one column contains only zero entries?

32. What is the relationship between the determinant of a 2×2 matrix and its transpose?

33. What is the determinant of a 2×2 matrix if one row is a linear combination of the other row?

5.2 Expansion by cofactors

Definition 1. Given a matrix $\mathbf{A}$, a *minor* is the determinant of any square submatrix of $\mathbf{A}$.

That is, given a square matrix $\mathbf{A}$, a minor is the determinant of any matrix formed from $\mathbf{A}$ by the removal of an equal number of rows and columns. As an example, if

$$\mathbf{A} = \begin{bmatrix} 1 & 2 & 3 \\ 4 & 5 & 6 \\ 7 & 8 & 9 \end{bmatrix},$$

then,

$$\begin{vmatrix} 1 & 2 \\ 7 & 8 \end{vmatrix} \quad \text{and} \quad \begin{vmatrix} 5 & 6 \\ 8 & 9 \end{vmatrix}$$

are both minors because

$$\begin{bmatrix} 1 & 2 \\ 7 & 8 \end{bmatrix} \quad \text{and} \quad \begin{bmatrix} 5 & 6 \\ 8 & 9 \end{bmatrix}$$

are both submatrices of $\mathbf{A}$, while

$$\begin{bmatrix} 1 & 2 \\ 8 & 9 \end{bmatrix} \quad \text{and} \quad \begin{vmatrix} 1 & 2 \end{vmatrix}$$

are not minors because

$$\begin{bmatrix} 1 & 2 \\ 8 & 9 \end{bmatrix}$$

is not a submatrix of $\mathbf{A}$ and [1 2], although a submatrix of $\mathbf{A}$, is not square.

A more useful concept for our immediate purposes, since it will enable us to calculate determinants, is that of the cofactor of an element of a matrix.

Definition 2. Given a matrix $\mathbf{A} = [a_{ij}]$, the *cofactor of the element* a_{ij} is a scalar obtained by multiplying together the term $(-1)^{i+j}$ and the minor obtained from $\mathbf{A}$ by removing the ith row and jth column.

In other words, to compute the cofactor of the element a_{ij}, we first form a submatrix of $\mathbf{A}$ by crossing out both the row and column in which the element a_{ij} appears. We then find the determinant of the submatrix and finally multiply it by the number $(-1)^{i+j}$.

Example 1.   Find the cofactor of the element 4 in the matrix

$$\mathbf{A} = \begin{bmatrix} 1 & 2 & 3 \\ 4 & 5 & 6 \\ 7 & 8 & 9 \end{bmatrix}.$$

Solution We first note that 4 appears in the (2, 1) position. The submatrix obtained by crossing out the second row and first column is

$$\begin{bmatrix} 1 & 2 & 3 \\ 4 & 5 & 6 \\ 7 & 8 & 9 \end{bmatrix} = \begin{bmatrix} 2 & 3 \\ 8 & 9 \end{bmatrix},$$

which has a determinant equal to $(2)(9) - (3)(8) = -6$. Since 4 appears in the (2, 1) position, $i = 2$ and $j = 1$. Thus, $(-1)^{i+j} = (-1)^{2+1} = (-1)^3 = (-1)$. The cofactor of 4 is $(-1)(-6) = 6$. ∎

Example 2.   Using the same $\mathbf{A}$ as in Example 1, find the cofactor of the element 9.

Solution The element 9 appears in the (3, 3) position. Thus, crossing out the third row and third column, we obtain the submatrix

$$\begin{bmatrix} 1 & 2 & 3 \\ 4 & 5 & 6 \\ 7 & 8 & 9 \end{bmatrix} = \begin{bmatrix} 1 & 2 \\ 4 & 5 \end{bmatrix},$$

which has a determinant equal to $(1)(5) - (2)(4) = -3$. Since, in this case, $i = j = 3$, the cofactor of 9 is $(-1)^{3+3}(-3) = (-1)^6(-3) = -3$. ∎

We now have enough tools at hand to find the determinant of any matrix.

EXPANSION by COFACTORS. To find the determinant of a matrix $\mathbf{A}$ of arbitrary order, (a) pick any one row or any one column of the matrix (dealer's choice), (b) for each element in the row or column chosen, find its cofactor, (c) multiply each element in the row or column chosen by its cofactor and sum the results. This sum is the determinant of the matrix.

Example 3.   Find det(A) if

$$A = \begin{bmatrix} 3 & 5 & 0 \\ -1 & 2 & 1 \\ 3 & -6 & 4 \end{bmatrix}.$$

Solution In this example, we expand by the second column.

$$|A| = (5)(\text{cofactor of } 5) + (2)(\text{cofactor of } 2) + (-6)(\text{cofactor of } -6)$$

$$= (5)(-1)^{1+2}\begin{vmatrix} -1 & 1 \\ 3 & 4 \end{vmatrix} + (2)(-1)^{2+2}\begin{vmatrix} 3 & 0 \\ 3 & 4 \end{vmatrix} + (-6)(-1)^{3+2}\begin{vmatrix} 3 & 0 \\ -1 & 1 \end{vmatrix}$$

$$= 5(-1)(-4-3) + (2)(1)(12-0) + (-6)(-1)(3-0)$$

$$= (-5)(-7) + (2)(12) + (6)(3) = 35 + 24 + 18 = 77. \ \blacksquare$$

Example 4.   Using the A of Example 3 and expanding by the first row, find det(A).
Solution

$$|A| = 3(\text{cofactor of } 3) + 5(\text{cofactor of } 5) + 0(\text{cofactor of } 10)$$

$$= (3)(-1)^{1+1}\begin{vmatrix} 2 & 1 \\ -6 & 4 \end{vmatrix} + 5(-1)^{1+2}\begin{vmatrix} -1 & 1 \\ 3 & 4 \end{vmatrix} + 0$$

$$= (3)(1)(8+6) + (5)(-1)(-4-3)$$

$$= (3)(14) + (-5)(-7) = 42 + 35 = 77. \ \blacksquare$$

The previous examples illustrate two important properties of the method. First, the value of the determinant is the same regardless of which row or column we choose to expand by and second, expanding by a row or column that contains zeros significantly reduces the number of computations involved.

Example 5.   Find det(A) if

$$A = \begin{bmatrix} 1 & 0 & 5 & 2 \\ -1 & 4 & 1 & 0 \\ 3 & 0 & 4 & 1 \\ -2 & 1 & 1 & 3 \end{bmatrix}.$$

Solution We first check to see which row or column contains the most zeros and expand by it. Thus, expanding by the second column gives

$$|\mathbf{A}| = 0(\text{cofactor of } 0) + 4(\text{cofactor of } 4) + 0(\text{cofactor of } 0) + 1(\text{cofactor of } 1)$$

$$= 0 + 4(-1)^{2+2} \begin{vmatrix} 1 & 5 & 2 \\ 3 & 4 & 1 \\ -2 & 1 & 3 \end{vmatrix} + 0 + 1(-1)^{4+2} \begin{vmatrix} 1 & 5 & 2 \\ -1 & 1 & 0 \\ 3 & 4 & 1 \end{vmatrix}$$

$$= 4 \begin{vmatrix} 1 & 5 & 2 \\ 3 & 4 & 1 \\ -2 & 1 & 3 \end{vmatrix} + \begin{vmatrix} 1 & 5 & 2 \\ -1 & 1 & 0 \\ 3 & 4 & 1 \end{vmatrix}.$$

Using expansion by cofactors on each of the determinants of order 3 yields

$$\begin{vmatrix} 1 & 5 & 2 \\ 3 & 4 & 1 \\ -2 & 1 & 3 \end{vmatrix} = 1(-1)^{1+1} \begin{vmatrix} 4 & 1 \\ 1 & 3 \end{vmatrix} + 5(-1)^{1+2} \begin{vmatrix} 3 & 1 \\ -2 & 3 \end{vmatrix} + 2(-1)^{1+3} \begin{vmatrix} 3 & 4 \\ -2 & 1 \end{vmatrix}$$

$$= -22 \quad \text{(expanding by the first row)}$$

and

$$\begin{vmatrix} 1 & 5 & 2 \\ -1 & 1 & 0 \\ 3 & 4 & 1 \end{vmatrix} = 2(-1)^{1+3} \begin{vmatrix} -1 & 1 \\ 3 & 4 \end{vmatrix} + 0 + 1(-1)^{3+3} \begin{vmatrix} 1 & 5 \\ -1 & 1 \end{vmatrix}$$

$$= -8 \quad \text{(expanding by the third column)}.$$

Hence,

$$|\mathbf{A}| = 4(-22) - 8 = -88 - 8 = -96. \ \blacksquare$$

For $n \times n$ matrices with $n > 3$, expansion by cofactors is an inefficient procedure for calculating determinants. It simply takes too long. A more elegant method, based on elementary row operations, is given in Section 5.4 for matrices whose elements are all numbers.

Problems 5.2

In Problems 1 through 22, use expansion by cofactors to evaluate the determinants of the given matrices.

1. $\begin{bmatrix} 1 & 2 & -2 \\ 0 & 2 & 3 \\ 0 & 0 & -3 \end{bmatrix}$,

2. $\begin{bmatrix} 3 & 2 & -2 \\ 1 & 0 & 4 \\ 2 & 0 & -3 \end{bmatrix}$,

3. $\begin{bmatrix} 1 & -2 & -2 \\ 7 & 3 & -3 \\ 0 & 0 & 0 \end{bmatrix}$,

4. $\begin{bmatrix} 2 & 0 & -1 \\ 1 & 1 & 1 \\ 3 & 2 & -3 \end{bmatrix}$,

5. $\begin{bmatrix} 3 & 5 & 2 \\ -1 & 0 & 4 \\ -2 & 2 & 7 \end{bmatrix}$,

6. $\begin{bmatrix} 1 & -3 & -3 \\ 2 & 8 & 3 \\ 4 & 5 & 0 \end{bmatrix}$,

7. $\begin{bmatrix} 2 & 1 & -9 \\ 3 & -1 & 1 \\ 3 & -1 & 2 \end{bmatrix}$,

8. $\begin{bmatrix} -1 & 3 & 3 \\ 1 & 1 & 4 \\ -1 & 1 & 2 \end{bmatrix}$,

9. $\begin{bmatrix} 1 & -3 & -3 \\ 2 & 8 & 4 \\ 3 & 5 & 1 \end{bmatrix}$,

10. $\begin{bmatrix} 2 & 1 & 3 \\ 3 & -1 & 2 \\ 2 & 3 & 5 \end{bmatrix}$,

11. $\begin{bmatrix} -1 & 3 & 3 \\ 4 & 5 & 6 \\ -1 & 3 & 3 \end{bmatrix}$,

12. $\begin{bmatrix} 1 & 2 & -3 \\ 5 & 5 & 1 \\ 2 & -5 & -1 \end{bmatrix}$,

13. $\begin{bmatrix} -4 & 0 & 0 \\ 2 & -1 & 0 \\ 3 & 1 & -2 \end{bmatrix}$,

14. $\begin{bmatrix} 1 & 3 & 2 \\ -1 & 4 & 1 \\ 5 & 3 & 8 \end{bmatrix}$,

15. $\begin{bmatrix} 3 & -2 & 0 \\ 1 & 1 & 2 \\ -3 & 4 & 1 \end{bmatrix}$,

16. $\begin{bmatrix} -4 & 0 & 0 & 0 \\ 1 & -5 & 0 & 0 \\ 2 & 1 & -2 & 0 \\ 3 & 1 & -2 & 1 \end{bmatrix}$,

17. $\begin{bmatrix} -1 & 2 & 1 & 2 \\ 1 & 0 & 3 & -1 \\ 2 & 2 & -1 & 1 \\ 2 & 0 & -3 & 2 \end{bmatrix}$,

18. $\begin{bmatrix} 1 & 1 & 2 & -2 \\ 1 & 5 & 2 & -1 \\ -2 & -2 & 1 & 3 \\ -3 & 4 & -1 & 8 \end{bmatrix}$,

19. $\begin{bmatrix} -1 & 3 & 2 & -2 \\ 1 & -5 & -4 & 6 \\ 3 & -6 & 1 & 1 \\ 3 & -4 & 3 & -3 \end{bmatrix}$,

20. $\begin{bmatrix} 1 & 1 & 0 & -2 \\ 1 & 5 & 0 & -1 \\ -2 & -2 & 0 & 3 \\ -3 & 4 & 0 & 8 \end{bmatrix}$,

21. $\begin{bmatrix} 1 & 2 & 1 & -1 \\ 4 & 0 & 3 & 0 \\ 1 & 1 & 0 & 5 \\ 2 & -2 & 1 & 1 \end{bmatrix}$,

22. $\begin{bmatrix} 11 & 1 & 0 & 9 & 0 \\ 2 & 1 & 1 & 0 & 0 \\ 4 & -1 & 1 & 0 & 0 \\ 3 & 2 & 2 & 1 & 0 \\ 0 & 0 & 1 & 2 & 0 \end{bmatrix}$.

23. Use the results of Problems 1, 13, and 16 to develop a theorem about the determinants of triangular matrices.

24. Use the results of Problems 3, 20, and 22 to develop a theorem regarding determinants of matrices containing a zero row or column.

25. Find $\det(\mathbf{A} - \lambda\mathbf{I})$ if $\mathbf{A}$ is the matrix given in Problem 2.

26. Find $\det(\mathbf{A} - \lambda\mathbf{I})$ if $\mathbf{A}$ is the matrix given in Problem 3.

27. Find $\det(\mathbf{A} - \lambda\mathbf{I})$ if $\mathbf{A}$ is the matrix given in Problem 4.

28. Find $\det(\mathbf{A} - \lambda\mathbf{I})$ if $\mathbf{A}$ is the matrix given in Problem 5.

5.3 **Properties of determinants**

In this section, we list some useful properties of determinants. For the sake of expediency, we only give proofs for determinants of order three, keeping in mind that these proofs may be extended in a straightforward manner to determinants of higher order.

Property 1. *If one row of a matrix consists entirely of zeros, then the determinant is zero.*

  Proof. Expanding by the zero row, we immediately obtain the desired result.

Property 2. *If two rows of a matrix are interchanged, the determinant changes sign.*

  Proof. Consider

$$\mathbf{A} = \begin{bmatrix} a_{11} & a_{12} & a_{13} \\ a_{21} & a_{22} & a_{23} \\ a_{31} & a_{32} & a_{33} \end{bmatrix}.$$

Expanding by the third row, we obtain

$$|\mathbf{A}| = a_{31}(a_{12}a_{23} - a_{13}a_{22}) - a_{32}(a_{11}a_{23} - a_{13}a_{21})$$
$$+a_{33}(a_{11}a_{22} - a_{12}a_{21}).$$

Now consider the matrix $\mathbf{B}$ obtained from $\mathbf{A}$ by interchanging the second and third rows:

$$\mathbf{B} = \begin{bmatrix} a_{11} & a_{12} & a_{13} \\ a_{31} & a_{32} & a_{33} \\ a_{21} & a_{22} & a_{23} \end{bmatrix}.$$

Expanding by the second row, we find that

$$|\mathbf{B}| = -a_{31}(a_{12}a_{23} - a_{13}a_{22}) + a_{32}(a_{11}a_{23} - a_{13}a_{21})$$
$$-a_{33}(a_{11}a_{22} - a_{12}a_{21}).$$

Thus, $|\mathbf{B}| = -|\mathbf{A}|$. Through similar reasoning, one can demonstrate that the result is valid regardless of which two rows are interchanged.

Property 3. *If two rows of a determinant are identical, the determinant is zero.*

Proof. If we interchange the two identical rows of the matrix, the matrix remains unaltered; hence, the determinant of the matrix remains constant. From Property 2, however, by interchanging two rows of a matrix, we change the sign of the determinant. Thus, the determinant must on the one hand remain the same, while on the other hand change the sign. The only way both of these conditions can be met simultaneously is for the determinant to be zero.

Property 4. *If the matrix $\mathbf{B}$ is obtained from the matrix $\mathbf{A}$ by multiplying every element in one row of $\mathbf{A}$ by the scalar λ, then $|\mathbf{B}| = \lambda| \mathbf{A} |$.*

Proof.

$$\begin{vmatrix} \lambda a_{11} & \lambda a_{12} & \lambda a_{13} \\ a_{21} & a_{22} & a_{23} \\ a_{31} & a_{32} & a_{33} \end{vmatrix} = \lambda a_{11}\begin{vmatrix} a_{22} & a_{23} \\ a_{32} & a_{33} \end{vmatrix} - \lambda a_{12}\begin{vmatrix} a_{21} & a_{23} \\ a_{31} & a_{33} \end{vmatrix} + \lambda a_{13}\begin{vmatrix} a_{12} & a_{22} \\ a_{31} & a_{32} \end{vmatrix}$$

$$= \lambda \left(a_{11}\begin{vmatrix} a_{22} & a_{23} \\ a_{32} & a_{33} \end{vmatrix} - a_{12}\begin{vmatrix} a_{21} & a_{23} \\ a_{31} & a_{33} \end{vmatrix} + a_{13}\begin{vmatrix} a_{21} & a_{22} \\ a_{31} & a_{32} \end{vmatrix} \right)$$

$$= \lambda \begin{vmatrix} a_{11} & a_{12} & a_{13} \\ a_{21} & a_{22} & a_{23} \\ a_{31} & a_{32} & a_{33} \end{vmatrix}.$$

In essence, Property 4 shows us how to multiply a scalar times a determinant. We know from Chapter 1 that multiplying a scalar times a matrix simply multiplies every element of the matrix by that scalar. Property 4, however, implies that multiplying a scalar times a determinant simply multiplies *one* row of the determinant by the scalar. Thus, while in matrices

$$8 \begin{bmatrix} 1 & 2 \\ 3 & 4 \end{bmatrix} = \begin{bmatrix} 8 & 16 \\ 24 & 32 \end{bmatrix},$$

in determinants we have

$$8 \begin{vmatrix} 1 & 2 \\ 3 & 4 \end{vmatrix} = \begin{vmatrix} 1 & 2 \\ 24 & 32 \end{vmatrix},$$

or alternatively

$$8 \begin{vmatrix} 1 & 2 \\ 3 & 4 \end{vmatrix} = 4(2) \begin{vmatrix} 1 & 2 \\ 3 & 4 \end{vmatrix} = 4 \begin{vmatrix} 2 & 4 \\ 3 & 4 \end{vmatrix} = \begin{vmatrix} 2 & 4 \\ 12 & 16 \end{vmatrix}.$$

Property 5. *For an $n \times n$ matrix $\mathbf{A}$ and any scalar λ, $\det(\lambda \mathbf{A}) = \lambda^n \det(\mathbf{A})$.*
  Proof. This proof makes continued use of Property 4.

$$\det(\lambda \mathbf{A}) = \det \left\{ \lambda \begin{bmatrix} a_{11} & a_{12} & a_{13} \\ a_{21} & a_{22} & a_{23} \\ a_{31} & a_{32} & a_{33} \end{bmatrix} \right\} = \det \left\{ \begin{bmatrix} \lambda a_{11} & \lambda a_{12} & \lambda a_{13} \\ \lambda a_{21} & \lambda a_{22} & \lambda a_{23} \\ \lambda a_{31} & \lambda a_{32} & \lambda a_{33} \end{bmatrix} \right\}$$

$$= \begin{vmatrix} \lambda a_{11} & \lambda a_{12} & \lambda a_{13} \\ \lambda a_{21} & \lambda a_{22} & \lambda a_{23} \\ \lambda a_{31} & \lambda a_{32} & \lambda a_{33} \end{vmatrix} = \lambda \begin{vmatrix} a_{11} & a_{12} & a_{13} \\ \lambda a_{21} & \lambda a_{22} & \lambda a_{23} \\ \lambda a_{31} & \lambda a_{32} & \lambda a_{33} \end{vmatrix}$$

$$= (\lambda)(\lambda) \begin{vmatrix} a_{11} & a_{12} & a_{13} \\ a_{21} & a_{22} & a_{23} \\ \lambda a_{31} & \lambda a_{32} & \lambda a_{33} \end{vmatrix} = \lambda(\lambda)(\lambda) \begin{vmatrix} a_{11} & a_{12} & a_{13} \\ a_{21} & a_{22} & a_{23} \\ a_{31} & a_{32} & a_{33} \end{vmatrix}$$

$$= \lambda^3 \det(\mathbf{A}).$$

  Note that for a *3 × 3* matrix, $n = 3$.

Property 6. *If a matrix $\mathbf{B}$ is obtained from a matrix $\mathbf{A}$ by adding to one row of $\mathbf{A}$, a scalar times another row of $\mathbf{A}$, then $|\mathbf{A}| = |\mathbf{B}|$.*
  Proof. Let

$$\mathbf{A} = \begin{bmatrix} a_{11} & a_{12} & a_{13} \\ a_{21} & a_{22} & a_{23} \\ a_{31} & a_{32} & a_{33} \end{bmatrix}$$

and

$$\mathbf{B} = \begin{bmatrix} a_{11} & a_{12} & a_{13} \\ a_{21} & a_{22} & a_{23} \\ a_{31} + \lambda a_{11} & a_{32} + \lambda a_{12} & a_{33} + \lambda a_{13} \end{bmatrix},$$

where $\mathbf{B}$ has been obtained from $\mathbf{A}$ by adding λ times the first row of $\mathbf{A}$ to the third row of $\mathbf{A}$. Expanding $|\mathbf{B}|$ by its third row, we obtain

$$|\mathbf{B}| = (a_{31} + \lambda a_{11}) \begin{vmatrix} a_{12} & a_{13} \\ a_{22} & a_{23} \end{vmatrix} - (a_{32} + \lambda a_{12}) \begin{vmatrix} a_{11} & a_{13} \\ a_{21} & a_{23} \end{vmatrix}$$

$$+ (a_{33} + \lambda a_{13}) \begin{vmatrix} a_{11} & a_{12} \\ a_{21} & a_{22} \end{vmatrix}$$

$$= a_{31} \begin{vmatrix} a_{12} & a_{13} \\ a_{22} & a_{23} \end{vmatrix} - a_{32} \begin{vmatrix} a_{11} & a_{13} \\ a_{21} & a_{23} \end{vmatrix} + a_{33} \begin{vmatrix} a_{11} & a_{12} \\ a_{21} & a_{22} \end{vmatrix}$$

$$+ \lambda \left\{ a_{11} \begin{vmatrix} a_{12} & a_{13} \\ a_{22} & a_{23} \end{vmatrix} - a_{12} \begin{vmatrix} a_{11} & a_{13} \\ a_{21} & a_{23} \end{vmatrix} + a_{13} \begin{vmatrix} a_{11} & a_{12} \\ a_{21} & a_{22} \end{vmatrix} \right\}.$$

The first three terms of this sum are exactly $|\mathbf{A}|$ (expand $|\mathbf{A}|$ by its third row), while the last three terms of the sum are

$$\lambda \begin{vmatrix} a_{11} & a_{12} & a_{13} \\ a_{21} & a_{22} & a_{23} \\ a_{11} & a_{12} & a_{13} \end{vmatrix}$$

(expand this determinant by its third row). Thus, it follows that

$$|\mathbf{B}| = |\mathbf{A}| + \lambda \begin{vmatrix} a_{11} & a_{12} & a_{13} \\ a_{21} & a_{22} & a_{23} \\ a_{11} & a_{12} & a_{13} \end{vmatrix}.$$

From Property 3, however, this second determinant is zero since its first and third rows are identical, hence $|\mathbf{B}| = |\mathbf{A}|$.

The same type of argument will quickly show that this result is valid regardless of the two rows chosen.

Example 1. Without expanding, show that

$$\begin{vmatrix} a & b & c \\ r & s & t \\ x & y & z \end{vmatrix} = \begin{vmatrix} a-r & b-s & c-t \\ r+2x & s+2y & t+2z \\ x & y & z \end{vmatrix}.$$

Solution Using Property 6, we have that

$$\begin{vmatrix} a & b & c \\ r & s & t \\ x & y & z \end{vmatrix} = \begin{vmatrix} a-r & b-s & c-t \\ r & s & t \\ x & y & z \end{vmatrix}, \quad \left\{ \begin{array}{l} \text{by adding to the first} \\ \text{row } (-1) \text{ times the} \\ \text{second row} \end{array} \right.$$

$$= \begin{vmatrix} a-r & b-s & c-t \\ r+2x & s+2y & t+2z \\ x & y & z \end{vmatrix}. \quad \left\{ \begin{array}{l} \text{by adding to the} \\ \text{second row } (2) \text{ times the} \\ \text{third row} \end{array} \right. \quad ■$$

Property 7. $\det(\mathbf{A}) = \det(\mathbf{A}^T)$.
   Proof. If

$$\mathbf{A} = \begin{bmatrix} a_{11} & a_{12} & a_{13} \\ a_{21} & a_{22} & a_{23} \\ a_{31} & a_{32} & a_{33} \end{bmatrix}, \quad \text{then} \quad \mathbf{A}^T = \begin{bmatrix} a_{11} & a_{21} & a_{31} \\ a_{12} & a_{22} & a_{32} \\ a_{13} & a_{23} & a_{33} \end{bmatrix}.$$

Expanding $\det(\mathbf{A}^T)$ by the first column, it follows that

$$\left| \mathbf{A}^T \right| = a_{11} \begin{vmatrix} a_{22} & a_{32} \\ a_{23} & a_{33} \end{vmatrix} - a_{12} \begin{vmatrix} a_{21} & a_{31} \\ a_{23} & a_{33} \end{vmatrix} + a_{13} \begin{vmatrix} a_{21} & a_{31} \\ a_{22} & a_{32} \end{vmatrix}$$

$$= a_{11}(a_{22}a_{33} - a_{32}a_{23}) - a_{12}(a_{21}a_{33} - a_{31}a_{23}) + a_{13}(a_{21}a_{32} - a_{31}a_{22}).$$

This, however, is exactly the expression we would obtain if we expand $\det(\mathbf{A})$ by the first row. Thus, $\left| \mathbf{A}^T \right| = \left| \mathbf{A} \right|$.

It follows from Property 7 that any property about determinants dealing with row operations is equally true for column operations (the analogous elementary row operation applied to columns) because a row operation on $\mathbf{A}^T$ is the same as a column operation on $\mathbf{A}$. Thus, if one column of a matrix consists entirely of zeros, then its determinant is zero; if two columns of a matrix are interchanged, the determinant changes the sign; if two columns of a matrix are identical, its determinant is zero; multiplying a determinant by a scalar is equivalent to multiplying one column of the matrix by that scalar and then calculating the new determinant; and the third elementary column operation when applied to a matrix does not change its determinant.

Property 8. *The determinant of a triangular matrix, either upper or lower, is the product of the elements on the main diagonal.*

Proof. See Problem 2.

Property 9.   *If* **A** *and* **B** *are of the same order, then* $\det(\mathbf{A})\,\det(\mathbf{B}) = \det(\mathbf{AB})$.
Because of its difficulty, the proof of Property 9 is omitted here.

Example 2.   Show that Property 9 is valid for

$$\mathbf{A} = \begin{bmatrix} 2 & 3 \\ 1 & 4 \end{bmatrix} \quad \text{and} \quad \mathbf{B} = \begin{bmatrix} 6 & -1 \\ 7 & 4 \end{bmatrix}.$$

Solution $|\mathbf{A}| = 5$, $|\mathbf{B}| = 31$.

$$\mathbf{AB} = \begin{bmatrix} 33 & 10 \\ 34 & 15 \end{bmatrix} \quad \text{thus} \quad |\mathbf{AB}| = 155 = |\mathbf{A}||\mathbf{B}|. \;\blacksquare$$

Problems 5.3

1. Prove that the determinant of a diagonal matrix is the product of the elements on the main diagonal.
2. Prove that the determinant of an upper or lower triangular matrix is the product of the elements on the main diagonal.
3. Without expanding, show that

$$\begin{vmatrix} a+x & r-x & x \\ b+y & s-y & y \\ c+z & t-z & z \end{vmatrix} = \begin{vmatrix} a & r & x \\ b & s & y \\ c & t & z \end{vmatrix}.$$

4. Verify Property 5 for $\lambda = -3$ and

$$\mathbf{A} = \begin{bmatrix} 2 & 1 & 0 \\ 5 & -1 & 3 \\ 2 & 1 & 1 \end{bmatrix}.$$

5. Verify Property 9 for

$$\mathbf{A} = \begin{vmatrix} 6 & 1 \\ 1 & 2 \end{vmatrix} \quad \text{and} \quad \mathbf{B} = \begin{vmatrix} 3 & -1 \\ 2 & 1 \end{vmatrix}.$$

6. Without expanding, show that

$$\begin{vmatrix} 2a & 3r & x \\ 4b & 6s & 2y \\ -2c & -3t & -z \end{vmatrix} = -12 \begin{vmatrix} a & r & x \\ b & s & y \\ c & t & z \end{vmatrix}.$$

7. Without expanding, show that

$$\begin{vmatrix} a - 3b & r - 3s & x - 3y \\ b - 2c & s - 2t & y - 2z \\ 5c & 5t & 5z \end{vmatrix} = 5 \begin{vmatrix} a & r & x \\ b & s & y \\ c & t & z \end{vmatrix}.$$

8. Without expanding, show that

$$\begin{vmatrix} a & b & c \\ r & s & t \\ x & y & z \end{vmatrix} = - \begin{vmatrix} a & x & r \\ b & y & s \\ c & z & t \end{vmatrix}.$$

9. Without expanding, show that

$$\begin{vmatrix} a & b & c \\ r & s & t \\ x & y & z \end{vmatrix} = -\frac{1}{4} \begin{vmatrix} 2a & 4b & 2c \\ -r & -2c & -t \\ x & 2y & z \end{vmatrix}.$$

10. Without expanding, show that

$$\begin{vmatrix} a - 3x & b - 3y & c - 3z \\ a + 5x & b + 5y & c + 5z \\ x & y & z \end{vmatrix} = 0.$$

11. Without expanding, show that

$$\begin{vmatrix} 2a & 3a & c \\ 2r & 3r & t \\ 2x & 3x & z \end{vmatrix} = 0.$$

12. Prove that if one column of a square matrix is a linear combination of another column, then the determinant of that matrix is zero.

13. Prove that if $\mathbf{A}$ is invertible, then $\det(\mathbf{A}^{-1}) = 1/\det(\mathbf{A})$.

5.4 Pivotal condensation

Properties 2, 4, and 6 of the previous section describe the effects on the determinant of a matrix of applying elementary row operations to the matrix itself. They comprise part of an efficient algorithm for calculating determinants of matrices whose elements are numbers. The technique is known as *pivotal condensation*: A given matrix is transformed into row-reduced form using elementary row operations. A record is kept of the changes to the determinant as a result of Properties 2, 4, and 6. Once the transformation is complete, the row-reduced matrix is in upper triangular form and its determinant is found easily by Property 8. In fact, since a row-reduced matrix has either unity elements or zeros on its main diagonal, its determinant will be unity if all its diagonal elements are unity, or zero if any one diagonal element is zero.

Example 1.   Use pivotal condensation to evaluate

$$\begin{vmatrix} 1 & 2 & 3 \\ -2 & 3 & 2 \\ 3 & -1 & 1 \end{vmatrix}.$$

Solution

$$\begin{vmatrix} 1 & 2 & 3 \\ -2 & 3 & 2 \\ 3 & -1 & 1 \end{vmatrix} = \begin{vmatrix} 1 & 2 & 3 \\ 0 & 7 & 8 \\ 3 & -1 & 1 \end{vmatrix}$$

$\left\{\begin{array}{l} \text{Property 6: adding to} \\ \text{the second row } (2) \\ \text{times the first row} \end{array}\right.$

$$= \begin{vmatrix} 1 & 2 & 3 \\ 0 & 7 & 8 \\ 0 & -7 & -8 \end{vmatrix}$$

$\left\{\begin{array}{l} \text{Property 6: adding to} \\ \text{the third row } (-3) \\ \text{times the first row} \end{array}\right.$

$$= 7 \begin{vmatrix} 1 & 2 & 3 \\ 0 & 1 & \dfrac{8}{7} \\ 0 & -7 & -8 \end{vmatrix}$$

$\left\{\begin{array}{l} \text{Property 4: applied} \\ \text{to the second row} \end{array}\right.$

$$= 7 \begin{vmatrix} 1 & 2 & 3 \\ 0 & 1 & \dfrac{8}{7} \\ 0 & 0 & 0 \end{vmatrix}$$

$\left\{\begin{array}{l} \text{Property 6: adding} \\ \text{to the third row } (7) \\ \text{times the second row} \end{array}\right.$

$$= 7(0) = 0.   \qquad \{\text{Property 8} \quad \blacksquare$$

Example 2.   Use pivotal condensation to evaluate

$$
\begin{vmatrix}
0 & -1 & 4 \\
1 & -5 & 1 \\
-6 & 2 & -3
\end{vmatrix}.
$$

Solution

$$
\begin{vmatrix}
0 & -1 & 4 \\
1 & -5 & 1 \\
-6 & 2 & -3
\end{vmatrix}
= (-1)
\begin{vmatrix}
1 & -5 & 1 \\
0 & -1 & 4 \\
-6 & 2 & -3
\end{vmatrix}
$$

$\left\{\begin{array}{l}\text{Property 2: interchanging} \\ \text{the first and second rows}\end{array}\right.$

$$
= (-1)
\begin{vmatrix}
1 & -5 & 1 \\
0 & 1 & 4 \\
0 & -28 & 3
\end{vmatrix}
$$

$\left\{\begin{array}{l}\text{Property 6: adding} \\ \text{to the third row } (6) \\ \text{times the first row}\end{array}\right.$

$$
= (-1)(-1)
\begin{vmatrix}
1 & -5 & 1 \\
0 & 1 & -4 \\
0 & -28 & 3
\end{vmatrix}
$$

$\left\{\begin{array}{l}\text{Property 4: applied} \\ \text{to the second row}\end{array}\right.$

$$
=
\begin{vmatrix}
1 & -5 & 1 \\
0 & 1 & -4 \\
0 & 0 & -109
\end{vmatrix}
$$

$\left\{\begin{array}{l}\text{Property 6: adding} \\ \text{to the third row } (28) \\ \text{times the second row}\end{array}\right.$

$$
= (-109)
\begin{vmatrix}
1 & -5 & 1 \\
0 & 1 & -4 \\
0 & 0 & 1
\end{vmatrix}
$$

$\left\{\begin{array}{l}\text{Property 4: applied} \\ \text{to the third row}\end{array}\right.$

$$
= (-109)(1) = -109.
$$

{Property 8 ■

Pivotal condensation is easily coded for implementation on a computer. Although shortcuts can be had by creative individuals evaluating determinants by hand, this rarely happens. The orders of most matrices that occur in practice are too large and, therefore, too time consuming to consider hand calculations in the evaluation of their determinants. In fact, such determinants can bring computer algorithms to their knees. As a result, calculating determinants is avoided whenever possible.

Still, when determinants are evaluated by hand, appropriate shortcuts are taken, as illustrated in the next two examples. The general approach involves operating on a matrix so that one row or one column is transformed into a new row or column containing at most one nonzero element. Expansion by cofactors is then applied to that row or column.

Example 3.   Evaluate

$$\begin{vmatrix} 10 & -6 & -9 \\ 6 & -5 & -7 \\ -10 & 9 & 12 \end{vmatrix}.$$

Solution

$$\begin{vmatrix} 10 & -6 & -9 \\ 6 & -5 & -7 \\ -10 & 9 & 12 \end{vmatrix} = \begin{vmatrix} 10 & -6 & -9 \\ 6 & -5 & -7 \\ 0 & 3 & 3 \end{vmatrix} \qquad \left\{ \begin{array}{l} \text{by adding (1) times the} \\ \text{first row to the third row} \\ \text{(Property 6)} \end{array} \right.$$

$$= \begin{vmatrix} 10 & -6 & -3 \\ 6 & -5 & -2 \\ 0 & 3 & 0 \end{vmatrix} \qquad \left\{ \begin{array}{l} \text{by adding } (-1) \text{ times the} \\ \text{second column to the} \\ \text{third column (Property 6)} \end{array} \right.$$

$$= -3 \begin{vmatrix} 10 & -3 \\ 6 & -2 \end{vmatrix} \qquad \{ \text{by expansion by cofactors}$$

$$= -3(-20 + 18) = 6. \quad \blacksquare$$

Example 4.   Evaluate

$$\begin{vmatrix} 3 & -1 & 0 & 2 \\ 0 & 1 & 4 & 1 \\ 3 & -2 & 3 & 5 \\ 9 & 7 & 0 & 2 \end{vmatrix}.$$

Solution Since the third column already contains two zeros, it would seem advisable to work on that one.

$$\begin{vmatrix} 3 & -1 & 0 & 2 \\ 0 & 1 & 4 & 1 \\ 3 & -2 & 3 & 5 \\ 9 & 7 & 0 & 2 \end{vmatrix} = \begin{vmatrix} 3 & -1 & 0 & 2 \\ 0 & 1 & 4 & 1 \\ 3 & -\dfrac{11}{4} & 0 & \dfrac{17}{4} \\ 9 & 7 & 0 & 2 \end{vmatrix} \qquad \left\{ \begin{array}{l} \text{by adding } \left(-\dfrac{3}{4}\right) \text{ times} \\ \text{the second row to} \\ \text{the third row} \end{array} \right.$$

$$= -4 \begin{vmatrix} 3 & -1 & 2 \\ 3 & -\dfrac{11}{4} & \dfrac{17}{4} \\ 9 & 7 & 2 \end{vmatrix} \qquad \left\{ \begin{array}{l} \text{by expansion} \\ \text{by cofactors} \end{array} \right.$$

$$= -4 \left(\dfrac{1}{4}\right) \begin{vmatrix} 3 & -1 & 2 \\ 12 & -11 & 17 \\ 9 & 7 & 2 \end{vmatrix} \qquad \{ \text{by Property 4}$$

$$= (-1) \begin{vmatrix} 3 & -1 & 2 \\ 0 & -7 & 9 \\ 9 & 7 & 2 \end{vmatrix} \qquad \begin{cases} \text{by adding } (-4) \text{ times} \\ \text{the first row to the} \\ \text{second row} \end{cases}$$

$$= (-1) \begin{vmatrix} 3 & -1 & 2 \\ 0 & -7 & 9 \\ 0 & 10 & -4 \end{vmatrix} \qquad \begin{cases} \text{by adding } (-3) \text{ times} \\ \text{the first row to the} \\ \text{third row} \end{cases}$$

$$= (-1)(3) \begin{vmatrix} -7 & 9 \\ 10 & -4 \end{vmatrix} \qquad \begin{cases} \text{by expansion by} \\ \text{cofactors} \end{cases}$$

$$= (-3)(28 - 90) = 186. \quad \blacksquare$$

Problems 5.4

In Problems 1 through 18, evaluate the determinants of the given matrices.

1. $\begin{bmatrix} 1 & 2 & -2 \\ 1 & 3 & 3 \\ 2 & 5 & 0 \end{bmatrix}$,

2. $\begin{bmatrix} 1 & 2 & 3 \\ 4 & 5 & 6 \\ 7 & 8 & 9 \end{bmatrix}$,

3. $\begin{bmatrix} 3 & -4 & 2 \\ -1 & 5 & 7 \\ 1 & 9 & -6 \end{bmatrix}$,

4. $\begin{bmatrix} -1 & 3 & 3 \\ 1 & 1 & 4 \\ -1 & 1 & 2 \end{bmatrix}$,

5. $\begin{bmatrix} 1 & -3 & -3 \\ 2 & 8 & 4 \\ 3 & 5 & 1 \end{bmatrix}$,

6. $\begin{bmatrix} 2 & 1 & -9 \\ 3 & -1 & 1 \\ 3 & -1 & 2 \end{bmatrix}$,

7. $\begin{bmatrix} 2 & 1 & 3 \\ 3 & -1 & 2 \\ 2 & 3 & 5 \end{bmatrix}$,

8. $\begin{bmatrix} -1 & 3 & 3 \\ 4 & 5 & 6 \\ -1 & 3 & 3 \end{bmatrix}$,

9. $\begin{bmatrix} 1 & 2 & -3 \\ 5 & 5 & 1 \\ 2 & -5 & -1 \end{bmatrix}$,

10. $\begin{bmatrix} 2 & 0 & -1 \\ 1 & 1 & 1 \\ 3 & 2 & -3 \end{bmatrix}$,

11. $\begin{bmatrix} 3 & 5 & 2 \\ -1 & 0 & 4 \\ -2 & 2 & 7 \end{bmatrix}$,

12. $\begin{bmatrix} 1 & -3 & -3 \\ 2 & 8 & 3 \\ 4 & 5 & 0 \end{bmatrix}$,

13. $\begin{bmatrix} 3 & 5 & 4 & 6 \\ -2 & 1 & 0 & 7 \\ -5 & 4 & 7 & 2 \\ 8 & -3 & 1 & 1 \end{bmatrix}$,

14. $\begin{bmatrix} -1 & 2 & 1 & 2 \\ 1 & 0 & 3 & -1 \\ 2 & 2 & -1 & 1 \\ 2 & 0 & -3 & 2 \end{bmatrix}$,

15. $\begin{bmatrix} 1 & 1 & 2 & -2 \\ 1 & 5 & 2 & -1 \\ -2 & -2 & 1 & 3 \\ -3 & 4 & -1 & 8 \end{bmatrix}$,

16. $\begin{bmatrix} -1 & 3 & 2 & -2 \\ 1 & -5 & -4 & 6 \\ 3 & -6 & 1 & 1 \\ 3 & -4 & 3 & -3 \end{bmatrix}$,

17. $\begin{bmatrix} 1 & 1 & 0 & -2 \\ 1 & 5 & 0 & -1 \\ -2 & -2 & 0 & 3 \\ -3 & 4 & 0 & 8 \end{bmatrix}$,

18. $\begin{bmatrix} -2 & 0 & 1 & 3 \\ 4 & 0 & 2 & -2 \\ -3 & 1 & 0 & 1 \\ 5 & 4 & 1 & 7 \end{bmatrix}$.

19. What can you say about the determinant of an $n \times n$ matrix that has rank less than n?

20. What can you say about the determinant of a singular matrix?

5.5 Inversion

As an immediate consequence of Theorem 1 of Section 3.2 and the method of pivotal condensation, we have:

Theorem 1. *A square matrix has an inverse if and only if its determinant is not zero.*

In this section, we develop a method to calculate inverses of nonsingular matrices using determinants. For matrices with order greater than 3×3, this method is less efficient than the one described in Section 3.2, and is generally avoided.

Definition 1. The *cofactor matrix* associated with an $n \times n$ matrix $\mathbf{A}$ is an $n \times n$ matrix $\mathbf{A}^c$ obtained from $\mathbf{A}$ by replacing each element of $\mathbf{A}$ by its cofactor.

Example 1. Find A^c if

$$A = \begin{bmatrix} 3 & 1 & 2 \\ -2 & 4 & 4 \\ 1 & 3 & 6 \end{bmatrix}.$$

Solution

$$A^c = \begin{bmatrix} (-1)^{1+1}\begin{vmatrix} 5 & 4 \\ 3 & 6 \end{vmatrix} & (-1)^{1+2}\begin{vmatrix} -2 & 4 \\ 1 & 6 \end{vmatrix} & (-1)^{1+3}\begin{vmatrix} -2 & 5 \\ 1 & 3 \end{vmatrix} \\ (-1)^{2+1}\begin{vmatrix} 1 & 2 \\ 3 & 6 \end{vmatrix} & (-1)^{2+2}\begin{vmatrix} 3 & 2 \\ 1 & 6 \end{vmatrix} & (-1)^{2+3}\begin{vmatrix} 3 & 1 \\ 1 & 3 \end{vmatrix} \\ (-1)^{3+1}\begin{vmatrix} 1 & 2 \\ 5 & 4 \end{vmatrix} & (-1)^{3+2}\begin{vmatrix} 3 & 2 \\ -2 & 4 \end{vmatrix} & (-1)^{3+3}\begin{vmatrix} 3 & 1 \\ -2 & 5 \end{vmatrix} \end{bmatrix},$$

$$A^c = \begin{bmatrix} 18 & 16 & -11 \\ 0 & 16 & -8 \\ -6 & -16 & 17 \end{bmatrix}. \quad \blacksquare$$

If $A = [a_{ij}]$, we will use the notation $A^c = [a^c_{ij}]$ to represent the cofactor matrix. Thus a^c_{ij} represents the cofactor of a_{ij}.

Definition 2. The *adjugate* of an $n \times n$ matrix A is the transpose of the cofactor matrix of A.

Thus, if we designate the adjugate of A by A^a, we have that $A^a = (A^c)^T$.

Example 2. Find A^a for the A given in Example 1.

Solution

$$A^a = \begin{bmatrix} 18 & 0 & -6 \\ 16 & 16 & -6 \\ -11 & -8 & 17 \end{bmatrix}. \quad \blacksquare$$

The importance of the adjugate is given in the following theorem, which is proved in the Final Comments to this chapter.

Theorem 2. $AA^a = A^aA = |A|I$.

If $|A| \neq 0$, we may divide by it in Theorem 2 and obtain

$$A\left(\frac{A^a}{|A|}\right) = \left(\frac{A^a}{|A|}\right)A = I.$$

Thus, using the definition of the inverse, we have

$$\mathbf{A}^{-1} = \frac{1}{|\mathbf{A}|}\mathbf{A}^a \quad if\,|\mathbf{A}| \neq 0.$$

That is, if $|\mathbf{A}| \neq 0$, then $\mathbf{A}^{-1}$ may be obtained by dividing the adjugate of $\mathbf{A}$ by the determinant of $\mathbf{A}$.

Example 3.   Find $\mathbf{A}^{-1}$ for the $\mathbf{A}$ given in Example 1.

    Solution The determinant of $\mathbf{A}$ is found to be 48. Using the solution to Example 2, we have

$$\mathbf{A}^{-1} = \left(\frac{\mathbf{A}^a}{|\mathbf{A}|}\right) = 1/48 \begin{bmatrix} 18 & 0 & -6 \\ 16 & 16 & -16 \\ -11 & -8 & 17 \end{bmatrix} = \begin{bmatrix} 3/8 & 0 & -1/8 \\ 1/3 & 1/3 & -1/3 \\ -11/48 & -1/6 & 17/48 \end{bmatrix}.\ \blacksquare$$

Example 4.   Find $\mathbf{A}^{-1}$ if

$$\mathbf{A} = \begin{bmatrix} 5 & 8 & 1 \\ 0 & 2 & 1 \\ 4 & 3 & -1 \end{bmatrix}.$$

Solution $\det(\mathbf{A}) = -1 \neq 0$, therefore $\mathbf{A}^{-1}$ exists.

$$\mathbf{A}^c = \begin{bmatrix} -5 & 4 & -8 \\ 11 & -9 & 17 \\ 6 & -5 & 10 \end{bmatrix}, \quad \mathbf{A}^a = (\mathbf{A}^c)^T = \begin{bmatrix} -5 & 11 & 6 \\ 4 & -9 & -5 \\ -8 & 17 & 10 \end{bmatrix}$$

$$\mathbf{A}^{-1} = \frac{\mathbf{A}^a}{|\mathbf{A}|} = \begin{bmatrix} 5 & -11 & -6 \\ -4 & 9 & 5 \\ 8 & -17 & -10 \end{bmatrix}.\ \blacksquare$$

Example 5.   Find $\mathbf{A}^{-1}$ if

$$\mathbf{A} = \begin{bmatrix} 1 & 2 \\ 3 & 4 \end{bmatrix}.$$

Solution $|\mathbf{A}| = -2$, therefore $\mathbf{A}^{-1}$ exists.

$$\mathbf{A}^c = \begin{bmatrix} 4 & -3 \\ -2 & 1 \end{bmatrix}, \quad \mathbf{A}^a = (\mathbf{A}^c)^T = \begin{bmatrix} 4 & -2 \\ -3 & 1 \end{bmatrix},$$

$$\mathbf{A}^{-1} = \frac{\mathbf{A}^a}{|\mathbf{A}|} = \left(-\frac{1}{2}\right)\begin{bmatrix} 4 & -2 \\ -3 & 1 \end{bmatrix} = \begin{bmatrix} -2 & 1 \\ \frac{3}{2} & -\frac{1}{2} \end{bmatrix}.\ \blacksquare$$

Problems 5.5

In Problems 1 through 15, find the inverses of the given matrices, if they exist.

1. $\begin{bmatrix} 4 & 4 \\ 4 & 4 \end{bmatrix}$,

2. $\begin{bmatrix} 1 & 1 \\ 3 & 4 \end{bmatrix}$,

3. $\begin{bmatrix} 1 & \dfrac{1}{2} \\ \dfrac{1}{2} & \dfrac{1}{3} \end{bmatrix}$,

4. $\begin{bmatrix} 2 & -1 \\ 3 & 4 \end{bmatrix}$,

5. $\begin{bmatrix} 8 & 3 \\ 5 & 2 \end{bmatrix}$,

6. $\begin{bmatrix} 2 & -1 \\ 4 & -2 \end{bmatrix}$,

7. $\begin{bmatrix} 1 & 1 & 0 \\ 1 & 0 & 1 \\ 0 & 1 & 1 \end{bmatrix}$,

8. $\begin{bmatrix} 0 & 0 & 1 \\ 1 & 0 & 0 \\ 0 & 1 & 0 \end{bmatrix}$,

9. $\begin{bmatrix} 2 & 0 & -1 \\ 0 & 1 & 2 \\ 3 & 1 & 1 \end{bmatrix}$,

10. $\begin{bmatrix} 1 & 2 & 3 \\ 4 & 5 & 6 \\ 7 & 8 & 9 \end{bmatrix}$,

11. $\begin{bmatrix} 2 & 0 & 0 \\ 5 & 1 & 0 \\ 4 & 1 & 1 \end{bmatrix}$,

12. $\begin{bmatrix} 1 & 2 & 1 \\ 3 & -2 & -4 \\ 2 & 3 & -1 \end{bmatrix}$,

13.
$$\begin{bmatrix} 2 & 4 & 3 \\ 3 & -4 & -4 \\ 5 & 0 & -1 \end{bmatrix},$$

14.
$$\begin{bmatrix} 5 & 0 & -1 \\ 2 & -1 & 2 \\ 2 & 3 & -1 \end{bmatrix},$$

15.
$$\begin{bmatrix} 3 & 1 & 1 \\ 1 & 3 & -1 \\ 2 & 3 & -1 \end{bmatrix}.$$

16. Find a formula for the inverse of

$$\mathbf{A} = \begin{bmatrix} a & b \\ c & d \end{bmatrix}.$$

17. Prove that if **A** and **B** are square matrices of the same order, then the product **AB** is nonsingular if and only if both **A** and **B** are.

18. Prove Theorem 1.

19. What can be said about the rank of a square matrix having a nonzero determinant?

5.6 Cramer's rule

Cramer's rule is a method, based on determinants, for solving systems of simultaneous linear equations. In this section, we first state the rule, then illustrate its usage by an example, and finally prove its validity using the properties derived in Section 5.3. We also discuss many limitations of the method.

Cramer's rule states that given a system of simultaneous linear equations in the matrix form $\mathbf{Ax} = \mathbf{b}$ (see Section 1.3), the ith component of $\mathbf{x}$ (or equivalently the ith unknown) is the quotient of two determinants. The determinant in the numerator is the determinant of a matrix obtained from **A** by replacing the ith column of **A** by the vector **b**, while the determinant in the denominator is just $|\mathbf{A}|$. Thus, if we are considering the system

$$a_{11}x_1 + a_{12}x_2 + a_{13}x_3 = b_1,$$
$$a_{21}x_1 + a_{22}x_2 + a_{23}x_3 = b_2,$$
$$a_{31}x_1 + a_{32}x_2 + a_{33}x_3 = b_3,$$

where x_1, x_2, and x_3 represent the unknowns, then Cramer's rule states that

$$x_1 = \frac{\begin{vmatrix} b_1 & a_{12} & a_{13} \\ b_2 & a_{22} & a_{23} \\ b_3 & a_{32} & a_{33} \end{vmatrix}}{|\mathbf{A}|}, \qquad x_2 = \frac{\begin{vmatrix} a_{11} & b_1 & a_{13} \\ a_{21} & b_2 & a_{23} \\ a_{31} & b_3 & a_{33} \end{vmatrix}}{|\mathbf{A}|},$$

$$x_3 = \frac{\begin{vmatrix} a_{11} & a_{12} & b_1 \\ a_{21} & a_{22} & b_2 \\ a_{31} & a_{32} & b_3 \end{vmatrix}}{|\mathbf{A}|}, \qquad \text{where} \qquad |\mathbf{A}| = \begin{vmatrix} a_{11} & a_{12} & a_{13} \\ a_{21} & a_{22} & a_{23} \\ a_{31} & a_{32} & a_{33} \end{vmatrix}.$$

Two restrictions on the application of Cramer's rule are immediate. First, the systems under consideration must have exactly the same number of equations as unknowns to insure that all matrices involved are square and hence have determinants. Second, the determinant of the coefficient matrix must not be zero since it appears in the denominator. If $|\mathbf{A}| = 0$, then Cramer's rule cannot be applied.

Example 1.   Solve the system

$$x + 2y - 3z + w = -5,$$
$$y + 3z + w = 6,$$
$$2x + 3y + z + w = 4,$$
$$x + z + w = 1.$$

Solution

$$\mathbf{A} = \begin{bmatrix} 1 & 2 & -3 & 1 \\ 0 & 1 & 3 & 1 \\ 2 & 3 & 1 & 1 \\ 1 & 0 & 1 & 1 \end{bmatrix}, \qquad \mathbf{x} = \begin{bmatrix} x \\ y \\ z \\ w \end{bmatrix}, \qquad \mathbf{b} = \begin{bmatrix} -5 \\ 6 \\ 4 \\ 1 \end{bmatrix}.$$

Since $|\mathbf{A}| = 20$, Cramer's rule can be applied, and

$$x = \frac{\begin{vmatrix} -5 & -2 & -3 & 1 \\ 6 & 1 & 3 & 1 \\ 4 & 3 & 1 & 1 \\ 1 & 0 & 1 & 1 \end{vmatrix}}{20} = \frac{0}{20} = 0, \quad y = \frac{\begin{vmatrix} 1 & -5 & -3 & 1 \\ 0 & 6 & 3 & 1 \\ 2 & 4 & 1 & 1 \\ 1 & 1 & 1 & 1 \end{vmatrix}}{20} = \frac{20}{20} = 1,$$

$$z = \frac{\begin{vmatrix} 1 & 2 & -5 & 1 \\ 0 & 1 & 6 & 1 \\ 2 & 3 & 4 & 1 \\ 1 & 0 & 1 & 1 \end{vmatrix}}{20} = \frac{40}{20} = 2, \quad w = \frac{\begin{vmatrix} 1 & 2 & -3 & -5 \\ 0 & 1 & 3 & 6 \\ 2 & 3 & 1 & 4 \\ 1 & 0 & 1 & 1 \end{vmatrix}}{20} = \frac{-20}{20} = -1. \blacksquare$$

We now derive Cramer's rule using only those properties of determinants given in Section 5.3. We consider the general system $\mathbf{Ax} = \mathbf{b}$, where

$$\mathbf{A} = \begin{bmatrix} a_{11} & a_{12} & a_{13} & \cdots & a_{1n} \\ a_{21} & a_{22} & a_{23} & \cdots & a_{2n} \\ a_{31} & a_{32} & a_{33} & \cdots & a_{3n} \\ \vdots & \vdots & \vdots & & \vdots \\ a_{n1} & a_{n2} & a_{n3} & \cdots & a_{mn} \end{bmatrix}, \quad \mathbf{x} = \begin{bmatrix} x_1 \\ x_2 \\ x_3 \\ \vdots \\ x_n \end{bmatrix}, \quad \text{and } \mathbf{b} = \begin{bmatrix} b_1 \\ b_2 \\ b_3 \\ \vdots \\ b_n \end{bmatrix}.$$

Then,

$$x_1 |\mathbf{A}| = \begin{vmatrix} a_{11}x_1 & a_{12} & a_{13} & \cdots & a_{1n} \\ a_{21}x_1 & a_{22} & a_{23} & \cdots & a_{3n} \\ a_{31}x_1 & a_{32} & a_{33} & \cdots & a_{3n} \\ \vdots & \vdots & \vdots & & \vdots \\ a_{n1}x_1 & a_{n2} & a_{n3} & \cdots & a_{nn} \end{vmatrix} \quad \{\text{by Property 4 modified to columns}$$

$$= \begin{vmatrix} a_{11}x_1 + a_{12}x_2 & a_{12} & a_{13} & \cdots & a_{1n} \\ a_{21}x_1 + a_{22}x_2 & a_{22} & a_{23} & \cdots & a_{2n} \\ a_{31}x_1 + a_{32}x_2 & a_{32} & a_{33} & \cdots & a_{3n} \\ \vdots & & & & \vdots \\ a_{n1}x_1 + a_{n2}x_2 & a_{n2} & a_{n3} & \cdots & a_{nn} \end{vmatrix} \quad \left\{ \begin{array}{l} \text{by adding } (x_2) \text{ times} \\ \text{the second column to} \\ \text{the first column} \end{array} \right.$$

$$= \begin{vmatrix} a_{11}x_1 + a_{12}x_2 + a_{13}x_3 & a_{12} & a_{13} & \cdots & a_{1n} \\ a_{21}x_1 + a_{22}x_2 + a_{23}x_3 & a_{22} & a_{23} & \cdots & a_{2n} \\ a_{31}x_1 + a_{32}x_2 + a_{33}x_3 & a_{32} & a_{33} & \cdots & a_{3n} \\ \vdots & & & & \vdots \\ a_{n1}x_1 + a_{n2}x_2 + a_{n3}x_3 & a_{n2} & a_{n3} & \cdots & a_{nn} \end{vmatrix} \quad \left\{ \begin{array}{l} \text{by adding } (x_3) \\ \text{times the third} \\ \text{column to the} \\ \text{first column} \end{array} \right.$$

$$= \begin{vmatrix} a_{11}x_1 + a_{12}x_2 + a_{13}x_3 + \cdots + a_{1n}x_n & a_{12} & a_{13} & \cdots & a_{1n} \\ a_{21}x_1 + a_{22}x_2 + a_{23}x_3 + \cdots + a_{2n}x_n & a_{22} & a_{23} & \cdots & a_{2n} \\ a_{31}x_1 + a_{32}x_2 + a_{33}x_3 + \cdots + a_{3n}x_n & a_{32} & a_{33} & \cdots & a_{3n} \\ \vdots & & \vdots & \vdots & \vdots \\ a_{n1}x_1 + a_{n2}x_2 + a_{n3}x_3 + \cdots + a_{nn}x_n & a_{n2} & a_{n3} & \cdots & a_{nn} \end{vmatrix}$$

by making continued use of Property 6 in the obvious manner. We now note that the first column of the new determinant is nothing more than $\mathbf{Ax}$, and since $\mathbf{Ax} = \mathbf{b}$, the first column reduces to $\mathbf{b}$.

Thus,

$$
x_1 |\mathbf{A}| = \begin{vmatrix} b_1 & a_{12} & a_{13} & \cdots & a_{1n} \\ b_2 & a_{22} & a_{23} & \cdots & a_{2n} \\ b_3 & a_{32} & a_{33} & \cdots & a_{3n} \\ \vdots & \vdots & \vdots & & \vdots \\ b_n & a_{n2} & a_{n3} & \cdots & a_{nn} \end{vmatrix}
$$

or

$$
x_1 = \frac{\begin{vmatrix} b_1 & a_{12} & \cdots & a_{1n} \\ b_2 & a_{22} & \cdots & a_{2n} \\ \vdots & \vdots & & \vdots \\ b_n & a_{n2} & \cdots & a_{nn} \end{vmatrix}}{|\mathbf{A}|}
$$

providing $|\mathbf{A}| \neq 0$. This expression is Cramer's rule for obtaining x_1. A similar argument applied to the jth column, instead of the first column, quickly shows that Cramer's rule is valid for every x_j, $j = 1, 2, \ldots,$ n.

Although Cramer's rule gives a systematic method for the solution of simultaneous linear equations, the number of computations involved can become awesome if the order of the determinant is large. Thus, for large systems, Cramer's rule is never used. The recommended algorithms include Gaussian elimination (Section 2.3) and **LU** decomposition (Section 3.5).

Problems 5.6

Solve the following systems of equations by Cramer's rule.

1. $x + 2y = -3$,
   $3x + y = 1$.

2. $2x + y = 3$,
   $x - y = 6$.

3. $4a + 2b = 0$,
   $5a - 3b = 10$.

4. $3s - 4t = 30$,
   $-2s + 3t = -10$.

5. $2x - 8y = 200,$
   $-x + 4y = 150.$

6. $x + y - 2z = 3,$
   $2x - y + 3z = 2.$

7. $x + y = 15,$
   $x + z = 15,$
   $y + z = 10.$

8. $3x + y + z = 4,$
   $x - y + 2z = 15,$
   $2x - 2y - z = 5.$

9. $x + 2y - 2z = -1,$
   $2x + y + z = 5,$
   $-x + y - z = -2.$

10. $2a + 3b - c = 4,$
   $-a - 2b + c = -2,$
   $3a - b = 2.$

11. $2x + 3y + 2z = 3,$
   $3x + y + 5z = 2,$
   $7y - 4z = 5.$

12. $5r + 8s + t = 2,$
   $2s + t = -1,$
   $4r + 3s - t = 3.$

13. $x + 2y + z + w = 7,$
   $3x + 4y - 2z - 4w = 13,$
   $2x + y - z + w = -4,$
   $x - 3y + 4z + 5w = 0.$

5.7 **Final comments on Chapter 5**

We shall now prove Theorem 2 of Section 5.5 dealing with the product of a matrix with its adjugate. For this proof we will need the following lemma:

Lemma 1. *If each element of one row of a matrix is multiplied by the cofactor of the corresponding element of a different row, the sum is zero.*

Proof. We prove this lemma only for an arbitrary 3×3 matrix $\mathbf{A}$, where

$$\mathbf{A} = \begin{bmatrix} a_{11} & a_{12} & a_{13} \\ a_{21} & a_{22} & a_{23} \\ a_{31} & a_{32} & a_{33} \end{bmatrix}.$$

Consider the case in which we multiply every element of the third row by the cofactor of the corresponding element in the second row and then sum the results. Thus,

$$a_{31}(\text{cofactor of } a_{21}) + a_{32}(\text{cofactor of } a_{22}) + a_{33}(\text{cofactor of } a_{23})$$

$$= a_{31}(-1)^3 \begin{vmatrix} a_{12} & a_{13} \\ a_{32} & a_{33} \end{vmatrix} + a_{32}(-1)^4 \begin{vmatrix} a_{11} & a_{13} \\ a_{31} & a_{33} \end{vmatrix} + a_{33}(-1)^5 \begin{vmatrix} a_{11} & a_{12} \\ a_{31} & a_{32} \end{vmatrix}$$

$$= \begin{vmatrix} a_{11} & a_{12} & a_{13} \\ a_{31} & a_{32} & a_{33} \\ a_{31} & a_{32} & a_{33} \end{vmatrix} = 0 \ \{\text{from Property 3, Section 5.3.}$$

Note that this property is equally valid if we replace the word row by the word column.

Theorem 1. $\mathbf{AA}^a = |\mathbf{A}|\mathbf{I}$.

Proof. We prove this theorem only for matrices of order 3×3. The proof may be easily extended to cover matrices of any arbitrary order. This extension is left as an exercise for the student:

$$\mathbf{AA}^a = \begin{bmatrix} a_{11} & a_{12} & a_{13} \\ a_{21} & a_{22} & a_{23} \\ a_{31} & a_{32} & a_{33} \end{bmatrix} \begin{bmatrix} a_{11}^c & a_{21}^c & a_{31}^c \\ a_{12}^c & a_{22}^c & a_{32}^c \\ a_{13}^c & a_{23}^c & a_{33}^c \end{bmatrix}.$$

If we denote this product matrix by $[b_{ij}]$, then

$$b_{11} = a_{11}a_{11}^c + a_{12}a_{12}^c + a_{13}a_{13}^c,$$
$$b_{12} = a_{11}a_{21}^c + a_{12}a_{22}^c + a_{13}a_{23}^c,$$
$$b_{23} = a_{21}a_{31}^c + a_{22}a_{32}^c + a_{23}a_{33}^c,$$
$$b_{22} = a_{21}a_{21}^c + a_{22}a_{22}^c + a_{23}a_{23}^c.$$

We now note that $b_{11} = |\mathbf{A}|$, since it is precisely the term obtained when one computes $\det(\mathbf{A})$ by cofactors, expanding by the first row. Similarly, $b_{22} = |\mathbf{A}|$, since it is precisely the term obtained by computing $\det(\mathbf{A})$ by cofactors after expanding by the second row. It follows from the above lemma that $b_{12} = 0$ and $b_{23} = 0$ since b_{12} is the term obtained by multiplying each element in the first row of $\mathbf{A}$ by the cofactor of the corresponding element in the second row and adding, while b_{23} is the term obtained by multiplying each element in the second row of $\mathbf{A}$ by the cofactor of the corresponding element in the third row and adding. Continuing this analysis for each b_{ij}, we find that

$$\mathbf{A}\mathbf{A}^a = \begin{bmatrix} |\mathbf{A}| & 0 & 0 \\ 0 & |\mathbf{A}| & 0 \\ 0 & 0 & |\mathbf{A}| \end{bmatrix} = |\mathbf{A}| \begin{bmatrix} 1 & 0 & 0 \\ 0 & 1 & 0 \\ 0 & 0 & 1 \end{bmatrix},$$

$$\mathbf{A}\mathbf{A}^a = |\mathbf{A}|\mathbf{I}.$$

Theorem 2. $\mathbf{A}^a\mathbf{A} = |\mathbf{A}|\mathbf{I}.$

Proof. This proof is completely analogous to the previous one and is left as an exercise for the student. $\square$

Eigenvalues and eigenvectors

6

Chapter outline

6.1 Definitions

Consider the matrix $\mathbf{A}$ and the vectors $\mathbf{x}_1$, $\mathbf{x}_2$, $\mathbf{x}_3$ given by

$$\mathbf{A} = \begin{bmatrix} 1 & 4 & -1 \\ 0 & 2 & 1 \\ 0 & 0 & 3 \end{bmatrix}, \quad \mathbf{x}_1 = \begin{bmatrix} 4 \\ 1 \\ 0 \end{bmatrix}, \quad \mathbf{x}_2 = \begin{bmatrix} 3 \\ 2 \\ 2 \end{bmatrix}, \quad \mathbf{x}_3 = \begin{bmatrix} 3 \\ 0 \\ 0 \end{bmatrix}.$$

Forming the products $\mathbf{A}\mathbf{x}_1$, $\mathbf{A}\mathbf{x}_2$, and $\mathbf{A}\mathbf{x}_3$, we obtain

$$\mathbf{A}\mathbf{x}_1 = \begin{bmatrix} 8 \\ 2 \\ 0 \end{bmatrix}, \quad \mathbf{A}\mathbf{x}_2 = \begin{bmatrix} 9 \\ 6 \\ 6 \end{bmatrix}, \quad \mathbf{A}\mathbf{x}_3 = \begin{bmatrix} 3 \\ 0 \\ 0 \end{bmatrix}.$$

But

$$\begin{bmatrix} 8 \\ 2 \\ 0 \end{bmatrix} = 2\mathbf{x}_1, \quad \begin{bmatrix} 9 \\ 6 \\ 6 \end{bmatrix} = 3\mathbf{x}_2, \quad \text{and} \quad \begin{bmatrix} 3 \\ 0 \\ 0 \end{bmatrix} = 1\mathbf{x}_3;$$

Matrix Methods. https://doi.org/10.1016/B978-0-12-818419-6.00006-X

hence,

$$\mathbf{A}\mathbf{x}_1 = 2\mathbf{x}_1,$$

$$\mathbf{A}\mathbf{x}_2 = 3\mathbf{x}_2,$$

$$\mathbf{A}\mathbf{x}_3 = 1\mathbf{x}_3.$$

That is, multiplying $\mathbf{A}$ by any one of the vectors $\mathbf{x}_1$, $\mathbf{x}_2$, or $\mathbf{x}_3$ is equivalent to simply multiplying the vector by a suitable scalar.

Definition 1. A nonzero vector $\mathbf{x}$ is an *eigenvector* (or characteristic vector) of a square matrix $\mathbf{A}$ if there exists a scalar λ such that $\mathbf{A}\mathbf{x} = \lambda\mathbf{x}$. Then, λ is an *eigenvalue* (or characteristic value) of $\mathbf{A}$.

Thus, in the above example, $\mathbf{x}_1$, $\mathbf{x}_2$, and $\mathbf{x}_3$ are eigenvectors of $\mathbf{A}$ and 2, 3, 1 are eigenvalues of $\mathbf{A}$.

Note that eigenvectors and eigenvalues are only defined for square matrices. Furthermore, note that the zero vector *cannot* be an eigenvector even though $\mathbf{A}\cdot\mathbf{0} = \lambda\cdot\mathbf{0}$ for every scalar λ. An eigenvalue, however, can be zero.

Example 1.   Show that

$$\mathbf{x} = \begin{bmatrix} 5 \\ 0 \\ 0 \end{bmatrix}$$

is an eigenvector of

$$\mathbf{A} = \begin{bmatrix} 0 & 5 & 7 \\ 0 & -1 & 2 \\ 0 & 3 & 1 \end{bmatrix}.$$

Solution

$$\mathbf{A}\mathbf{x} = \begin{bmatrix} 0 & 5 & 7 \\ 0 & -1 & 2 \\ 0 & 3 & 1 \end{bmatrix} \begin{bmatrix} 5 \\ 0 \\ 0 \end{bmatrix} = \begin{bmatrix} 0 \\ 0 \\ 0 \end{bmatrix} = 0 \begin{bmatrix} 5 \\ 0 \\ 0 \end{bmatrix}.$$

Thus, x is an eigenvector of $\mathbf{A}$ and $\lambda = 0$ is an eigenvalue. ∎

Example 2.   Is

$$\mathbf{x} = \begin{bmatrix} 1 \\ 1 \end{bmatrix}$$

an eigenvector of

$$\mathbf{A} = \begin{bmatrix} 1 & 2 \\ 3 & 4 \end{bmatrix}?$$

Solution

$$\mathbf{Ax} = \begin{bmatrix} 1 & 2 \\ 3 & 4 \end{bmatrix} \begin{bmatrix} 1 \\ 1 \end{bmatrix} = \begin{bmatrix} 3 \\ 7 \end{bmatrix}.$$

Thus, if $\mathbf{x}$ is to be an eigenvector of $\mathbf{A}$, there must exist a scalar λ such that $\mathbf{Ax} = \lambda\mathbf{x}$ or such that

$$\begin{bmatrix} 3 \\ 7 \end{bmatrix} = \lambda \begin{bmatrix} 1 \\ 1 \end{bmatrix} = \begin{bmatrix} \lambda \\ \lambda \end{bmatrix}.$$

It is quickly verified that no such λ exists; hence, $\mathbf{x}$ is not an eigenvector of $\mathbf{A}$. ∎

Problems 6.1

1. Determine which of the following vectors are eigenvectors for

$$\mathbf{A} = \begin{bmatrix} 1 & 2 \\ -4 & 7 \end{bmatrix}.$$

(a) $\begin{bmatrix} 1 \\ 1 \end{bmatrix}$,

(b) $\begin{bmatrix} 1 \\ -1 \end{bmatrix}$,

(c) $\begin{bmatrix} 2 \\ 1 \end{bmatrix}$,

(d) $\begin{bmatrix} 1 \\ 2 \end{bmatrix}$,

(e) $\begin{bmatrix} 2 \\ 2 \end{bmatrix}$,

(f) $\begin{bmatrix} -4 \\ -4 \end{bmatrix}$,

(g) $\begin{bmatrix} 4 \\ -4 \end{bmatrix}$,

(h) $\begin{bmatrix} 2 \\ 4 \end{bmatrix}$.

2. What are the eigenvalues that correspond to the eigenvectors found in Problem 1?

3. Determine which of the following vectors are eigenvectors for

$$\mathbf{B} = \begin{bmatrix} 2 & -4 \\ 3 & -6 \end{bmatrix}.$$

(a) $\begin{bmatrix} 1 \\ 1 \end{bmatrix}$,

(b) $\begin{bmatrix} 1 \\ -1 \end{bmatrix}$,

(c) $\begin{bmatrix} 2 \\ 1 \end{bmatrix}$,

(d) $\begin{bmatrix} 0 \\ 0 \end{bmatrix}$,

(e) $\begin{bmatrix} 6 \\ 3 \end{bmatrix}$,

(f) $\begin{bmatrix} 2 \\ 3 \end{bmatrix}$,

(g) $\begin{bmatrix} -4 \\ -6 \end{bmatrix}$,

(h) $\begin{bmatrix} 1 \\ 0 \end{bmatrix}$.

4. What are the eigenvalues that correspond to the eigenvectors found in Problem 3?

5. Determine which of the following vectors are eigenvectors for

$$ \mathbf{A} = \begin{bmatrix} 2 & 0 & -1 \\ 1 & 2 & 1 \\ -1 & 0 & 2 \end{bmatrix}. $$

(a) $\begin{bmatrix} 1 \\ 0 \\ 0 \end{bmatrix}$,

(b) $\begin{bmatrix} 0 \\ 1 \\ 0 \end{bmatrix}$,

(c) $\begin{bmatrix} 1 \\ -2 \\ 1 \end{bmatrix}$,

(d) $\begin{bmatrix} -3 \\ 6 \\ -3 \end{bmatrix}$,

(e) $\begin{bmatrix} -1 \\ 0 \\ 1 \end{bmatrix}$,

(f) $\begin{bmatrix} 1 \\ 0 \\ 1 \end{bmatrix}$,

(g) $\begin{bmatrix} 2 \\ 0 \\ -2 \end{bmatrix}$,

(h) $\begin{bmatrix} 1 \\ 1 \\ 1 \end{bmatrix}$.

6. What are the eigenvalues that correspond to the eigenvectors found in Problem 5?

7. Determine which of the following vectors are eigenvectors for

$$\mathbf{A} = \begin{bmatrix} 1 & 3 & 0 & 0 \\ 1 & -1 & 0 & 0 \\ 0 & 0 & 1 & 2 \\ 0 & 0 & 4 & 3 \end{bmatrix}.$$

(a) $\begin{bmatrix} 1 \\ -1 \\ 0 \\ 0 \end{bmatrix}$,

(b) $\begin{bmatrix} 0 \\ 0 \\ 1 \\ -1 \end{bmatrix}$,

(c) $\begin{bmatrix} 1 \\ 0 \\ 0 \\ -1 \end{bmatrix}$,

(d) $\begin{bmatrix} 3 \\ 1 \\ 0 \\ 0 \end{bmatrix}$,

(e) $\begin{bmatrix} 0 \\ 0 \\ 0 \\ 0 \end{bmatrix}$,

(f) $\begin{bmatrix} 1 \\ 1 \\ 0 \\ 0 \end{bmatrix}$.

8. What are the eigenvalues that correspond to the eigenvectors found in Problem 7?

6.2 Eigenvalues

Let $\mathbf{x}$ be an eigenvector of the matrix $\mathbf{A}$. Then, there must exist an eigenvalue λ such that

$$\mathbf{A}\mathbf{x} = \lambda\mathbf{x} \tag{1}$$

or, equivalently,

$$\mathbf{A}\mathbf{x} - \lambda\mathbf{x} = \mathbf{0}$$

or

$$(\mathbf{A} - \lambda\mathbf{I})\mathbf{x} = \mathbf{0}. \tag{2}$$

Caution. We could not have written (2) as $(\mathbf{A} - \lambda)\mathbf{x} = \mathbf{0}$ since the term $\mathbf{A} - \lambda$ would require subtracting a scalar from a matrix, an operation that is not defined. The quantity $\mathbf{A} - \lambda\mathbf{I}$, however, is defined since we are now subtracting one matrix from another.

Define a new matrix

$$\mathbf{B} = \mathbf{A} - \lambda\mathbf{I}. \tag{3}$$

Then (2) may be rewritten as

$$\mathbf{B}\mathbf{x} = \mathbf{0}, \tag{4}$$

which is a linear homogeneous system of equations for the unknown $\mathbf{x}$. If $\mathbf{B}$ has an inverse, then we can solve Eq. (4) for $\mathbf{x}$, obtaining $\mathbf{x} = \mathbf{B}^{-1}\mathbf{0}$, or $\mathbf{x} = \mathbf{0}$. This result, however, is absurd since $\mathbf{x}$ is an eigenvector and cannot be zero. Thus, it follows that $\mathbf{x}$ will be an eigenvector of $\mathbf{A}$ if and only if $\mathbf{B}$ does not have an inverse. But if a square matrix does not have an inverse, then its determinant must be zero (Theorem 1 of Section 5.5). Therefore, $\mathbf{x}$ will be an eigenvector of $\mathbf{A}$ if and only if

$$\det(\mathbf{A} - \lambda\mathbf{I}) = 0. \tag{5}$$

Eq. (5) is called the *characteristic equation of* **A**. The roots of (5) determine the eigenvalues of **A**.

Example 1.   Find the eigenvalues of

$$A = \begin{bmatrix} 1 & 2 \\ 4 & 3 \end{bmatrix}.$$

Solution

$$A - \lambda I = \begin{bmatrix} 1 & 2 \\ 4 & 3 \end{bmatrix} - \lambda \begin{bmatrix} 1 & 0 \\ 0 & 1 \end{bmatrix} = \begin{bmatrix} 1 & 2 \\ 4 & 3 \end{bmatrix} - \begin{bmatrix} \lambda & 0 \\ 0 & \lambda \end{bmatrix}$$

$$= \begin{bmatrix} 1 - \lambda & 2 \\ 4 & 3 - \lambda \end{bmatrix}$$

$\det(A - \lambda I) = (1 - \lambda)(3 - \lambda) - 8 = \lambda^2 - 4\lambda - 5$. The characteristic equation of **A** is det $(A - \lambda I) = 0$, or $\lambda^2 - 4\lambda - 5 = 0$. Solving for λ, we have that $\lambda = -1, 5$; hence, the eigenvalues of **A** are $\lambda_1 = -1, \lambda_2 = 5$. ■

Example 2.   Find the eigenvalues of

$$A = \begin{bmatrix} 1 & -2 \\ 1 & 1 \end{bmatrix}.$$

Solution

$$A - \lambda I = \begin{bmatrix} 1 & -2 \\ 1 & 1 \end{bmatrix} - \lambda \begin{bmatrix} 1 & 0 \\ 0 & 1 \end{bmatrix} = \begin{bmatrix} 1 - \lambda & -2 \\ 1 & 1 - \lambda \end{bmatrix}$$

$\det(A - \lambda I) = (1 - \lambda)(1 - \lambda) + 2 = \lambda^2 - 2\lambda + 3$.

The characteristic equation is $\lambda^2 - 2\lambda + 3 = 0$; hence, solving for λ by the quadratic formula, we have that $\lambda_1 = 1 + \sqrt{2}i, \lambda_2 = 1 - \sqrt{2}i$, which are eigenvalues of **A**. ■

NOTE: Even if the elements of a matrix are real, the eigenvalues may be complex.

Example 3.   Find the eigenvalues of

$$A = \begin{bmatrix} t & 2t \\ 2t & -t \end{bmatrix}.$$

Solution

$$A - \lambda I = \begin{bmatrix} t & 2t \\ 2t & -t \end{bmatrix} - \lambda \begin{bmatrix} 1 & 0 \\ 0 & 1 \end{bmatrix} = \begin{bmatrix} t - \lambda & 2t \\ 2t & -t - \lambda \end{bmatrix}$$

$\det(A - \lambda I) = (t - \lambda)(-t - \lambda) - 4t^2 = \lambda^2 - 5t^2$.

The characteristic equation is $\lambda^2 - 5t^2 = 0$; hence, the eigenvalues are $\lambda_1 = \sqrt{5}t, \lambda_2 = \sqrt{5}t$.

NOTE: If the matrix **A** depends on a parameter (in this case the parameter is t), then the eigenvalues may also depend on the parameter. ■

Example 4.   Find the eigenvalues for

$$A = \begin{bmatrix} 2 & -1 & 1 \\ 3 & -2 & 1 \\ 0 & 0 & 1 \end{bmatrix}.$$

Solution

$$A - \lambda I = \begin{bmatrix} 2 & -1 & 1 \\ 3 & -2 & 1 \\ 0 & 0 & 1 \end{bmatrix} - \lambda \begin{bmatrix} 1 & 0 & 0 \\ 0 & 1 & 0 \\ 0 & 0 & 1 \end{bmatrix} = \begin{bmatrix} 2-\lambda & -1 & 1 \\ 3 & -2-\lambda & 1 \\ 0 & 0 & 1-\lambda \end{bmatrix}$$

$\det(A - \lambda I) = (1 - \lambda)[(2 - \lambda)(-2-\lambda) + 3] = (1 - \lambda)(\lambda^2 - 1).$

The characteristic equation is $(1 - \lambda)(\lambda^2 - 1) = 0$; hence, the eigenvalues are $\lambda_1 = \lambda_2 = 1, \lambda_3 = -1.$ ■

NOTE: The roots of the characteristic equation can be repeated. That is, $\lambda_1 = \lambda_2 = \lambda_3 = \cdots = \lambda_k$. When this happens, the eigenvalue is said to be of *multiplicity k*. Thus, in Example 4, $\lambda = 1$ is an eigenvalue of multiplicity 2, while $\lambda = -1$ is an eigenvalue of multiplicity 1.

From the definition of the characteristic Eq. (5), it can be shown that if A is an $n \times n$ matrix then the characteristic equation of A is an nth degree polynomial in λ. It follows from the fundamental theorem of algebra that the characteristic equation has n roots, counting multiplicity. Hence, A has exactly n eigenvalues, counting multiplicity (see Examples 1 and 4).

In general, it is very difficult to find the eigenvalues of a matrix. First, the characteristic equation must be obtained, and for matrices of high order this is a lengthy task. Then, the characteristic equation must be solved for its roots. If the equation is of high order, this can be an impossibility in practice. For example, the reader is invited to find the eigenvalues of

$$A = \begin{bmatrix} 10 & 7 & 8 & 7 \\ 7 & 5 & 6 & 5 \\ 8 & 6 & 10 & 9 \\ 7 & 5 & 9 & 10 \end{bmatrix}.$$

For these reasons, eigenvalues are rarely found by the method just given and numerical techniques are used to obtain approximate values (see Sections 6.6 and 10.4).

Problems 6.2

In Problems 1 through 35, find the eigenvalues of the given matrices.

1. $\begin{bmatrix} 1 & 2 \\ -1 & 4 \end{bmatrix},$

2. $\begin{bmatrix} 2 & 1 \\ 2 & 3 \end{bmatrix}$,

3. $\begin{bmatrix} 2 & 3 \\ 4 & 6 \end{bmatrix}$,

4. $\begin{bmatrix} 3 & 6 \\ 9 & 6 \end{bmatrix}$,

5. $\begin{bmatrix} 2 & -1 \\ 1 & 4 \end{bmatrix}$,

6. $\begin{bmatrix} 1 & 2 \\ 4 & -1 \end{bmatrix}$,

7. $\begin{bmatrix} 3 & 5 \\ 5 & -3 \end{bmatrix}$,

8. $\begin{bmatrix} 3 & 5 \\ -5 & -3 \end{bmatrix}$,

9. $\begin{bmatrix} 2 & 5 \\ -1 & -2 \end{bmatrix}$,

10 $\begin{bmatrix} 1 & 0 \\ 0 & 1 \end{bmatrix}$,

11. $\begin{bmatrix} 0 & 1 \\ 0 & 0 \end{bmatrix}$,

12. $\begin{bmatrix} 0 & 0 \\ 0 & 0 \end{bmatrix}$,

13. $\begin{bmatrix} 2 & 2 \\ -1 & -2 \end{bmatrix}$,

14. $\begin{bmatrix} 4 & 10 \\ 9 & -5 \end{bmatrix}$,

15. $\begin{bmatrix} 5 & 10 \\ 9 & -4 \end{bmatrix}$,

16. $\begin{bmatrix} 0 & t \\ 2t & -t \end{bmatrix}$,

17. $\begin{bmatrix} 0 & 2t \\ -2t & 4t \end{bmatrix}$,

18. $\begin{bmatrix} 4\theta & 2\theta \\ -\theta & \theta \end{bmatrix}$,

19. $\begin{bmatrix} 1 & 0 & 3 \\ 1 & 2 & 1 \\ 3 & 0 & 1 \end{bmatrix}$,

20. $\begin{bmatrix} 2 & 0 & -1 \\ 2 & 2 & 2 \\ -1 & 0 & 2 \end{bmatrix}$,

21. $\begin{bmatrix} 2 & 0 & -1 \\ 2 & 1 & 2 \\ -1 & 0 & 2 \end{bmatrix}$,

22. $\begin{bmatrix} 1 & 1 & -1 \\ 0 & 0 & 0 \\ 1 & 2 & 3 \end{bmatrix}$,

23. $\begin{bmatrix} 3 & 0 & 0 \\ 2 & 6 & 4 \\ 2 & 3 & 5 \end{bmatrix}$,

24. $\begin{bmatrix} 5 & -7 & 7 \\ 4 & -3 & 4 \\ 4 & -1 & 2 \end{bmatrix}$,

25. $\begin{bmatrix} 3 & 1 & -1 \\ 1 & 3 & -1 \\ -1 & -1 & 5 \end{bmatrix}$,

26. $\begin{bmatrix} 1 & 2 & 3 \\ 2 & 4 & 6 \\ 3 & 6 & 9 \end{bmatrix}$,

27. $\begin{bmatrix} 10 & 2 & 0 \\ 2 & 4 & 6 \\ 0 & 6 & 10 \end{bmatrix}$,

28. $\begin{bmatrix} 3 & -1 & 1 \\ -1 & 3 & -1 \\ 1 & -1 & 3 \end{bmatrix}$,

29. $\begin{bmatrix} 1 & 2 & 1 \\ 2 & 4 & 2 \\ 1 & 2 & 1 \end{bmatrix}$,

30. $\begin{bmatrix} 4 & 2 & 1 \\ 2 & 7 & 2 \\ 1 & 2 & 4 \end{bmatrix}$,

31. $\begin{bmatrix} 1 & 5 & 1 \\ -1 & -1 & 1 \\ 0 & 0 & 3 \end{bmatrix}$,

32. $\begin{bmatrix} 0 & 1 & 0 \\ 0 & 0 & 1 \\ 0 & -1 & 0 \end{bmatrix}$,

33. $\begin{bmatrix} 0 & 1 & 0 \\ 0 & 0 & 1 \\ 27 & -27 & 9 \end{bmatrix}$,

34. $\begin{bmatrix} 1 & -1 & 0 & 0 \\ 3 & 5 & 0 & 0 \\ 0 & 0 & 1 & 5 \\ 0 & 0 & -1 & 1 \end{bmatrix}$,

35. $\begin{bmatrix} 0 & 1 & 0 & 0 \\ 0 & 0 & 1 & 0 \\ 0 & 0 & 0 & 1 \\ -4 & 12 & -13 & 6 \end{bmatrix}$.

36. Consider the matrix

$$\mathbf{C} = \begin{bmatrix} 0 & 1 & 0 & \cdots & 0 \\ 0 & 0 & 1 & \cdots & 0 \\ \bullet & \bullet & \bullet & \cdots & \bullet \\ 0 & 0 & 0 & \cdots & 1 \\ -a_0 & -a_1 & -a_2 & \cdots & -a_{n-1} \end{bmatrix}.$$

Use mathematical induction to prove that

$$\det(\mathbf{C} - \lambda \mathbf{I}) = (-1)^n \left(\lambda^2 + a_{n-1}\lambda^{n-1} + \cdots + a_2\lambda^2 + a_1\lambda^2 + a_1\lambda + a_0 \right).$$

Deduce that the characteristic equation for this matrix is

$$\lambda^n + a_{n-1}\lambda^{n-1} + \cdots + a_2\lambda^2 + a_1\lambda + a_0 = 0.$$

The matrix $\mathbf{C}$ is called the *companion matrix* for this characteristic equation.

37. Show that if λ is an eigenvalue of $\mathbf{A}$, then $k\lambda$ is an eigenvalue of k$\mathbf{A}$, where k denotes an arbitrary scalar.

38. Show that if $\lambda \neq 0$ is an eigenvalue of $\mathbf{A}$, then $1/\lambda$ is an eigenvalue of $\mathbf{A}^{-1}$, providing the inverse exists.

39. Show that if λ is an eigenvalue of $\mathbf{A}$, then it is also an eigenvalue of $\mathbf{A}^{\mathrm{T}}$.

6.3 Eigenvectors

To each distinct eigenvalue of a matrix $\mathbf{A}$, there will correspond at least one eigenvector, which can be found by solving the appropriate set of homogeneous equations. If an eigenvalue λ_i is substituted into (2), the corresponding eigenvector $\mathbf{x}_i$ is the solution of

$$(\mathbf{A} - \lambda_i\mathbf{I})\mathbf{X}_i = \mathbf{0}. \tag{6}$$

Example 1. Find the eigenvectors of

$$\mathbf{A} = \begin{bmatrix} 1 & 2 \\ 4 & 3 \end{bmatrix}.$$

Solution The eigenvalues of $\mathbf{A}$ have already been found to be $\lambda_1 = -1$, $\lambda_2 = 5$ (see Example 1 of Section 6.2). We first calculate the eigenvectors corresponding to λ_1. From (6),

$$(\mathbf{A} - (-1)\mathbf{I})\mathbf{x_1} = 0. \tag{7}$$

If we designate the unknown vector $\mathbf{x_1}$ by

$$\begin{bmatrix} x_1 \\ y_1 \end{bmatrix},$$

Eq. (7) becomes

$$\left\{ \begin{bmatrix} 1 & 2 \\ 4 & 3 \end{bmatrix} + \begin{bmatrix} 1 & 0 \\ 0 & 1 \end{bmatrix} \right\} \begin{bmatrix} x_1 \\ y_1 \end{bmatrix} = \begin{bmatrix} 0 \\ 0 \end{bmatrix}$$

or

$$\begin{bmatrix} 2 & 2 \\ 4 & 4 \end{bmatrix} \begin{bmatrix} x_1 \\ y_1 \end{bmatrix} = \begin{bmatrix} 0 \\ 0 \end{bmatrix},$$

or, equivalently,

$$2x_1 + 2y_1 = 0,$$
$$4x_1 + 4y_1 = 0.$$

A nontrivial solution to this set of equations is $x_1 = -y_1$, y_1 arbitrary; hence, the eigenvector is

$$\mathbf{x_1} = \begin{bmatrix} x_1 \\ y_1 \end{bmatrix} = \begin{bmatrix} -y_1 \\ y_1 \end{bmatrix} = y_1 \begin{bmatrix} -1 \\ 1 \end{bmatrix}, \quad y_1 \text{ arbitrary.}$$

By choosing different values of y_1, different eigenvectors for $\lambda_1 = -1$ can be obtained. Note, however, that any two such eigenvectors would be scalar multiples of each other, hence linearly dependent. Thus, there is only one linearly independent eigenvector corresponding to $\lambda_1 = -1$. For convenience, we choose $y_1 = 1$, which gives us the eigenvector

$$\mathbf{x}_1 = \begin{bmatrix} -1 \\ 1 \end{bmatrix}.$$

Many times, however, the scalar y_1 is chosen in such a manner that the resulting eigenvector becomes a unit vector. If we wished to achieve this result for the above vector, we would have to choose $y_1 = 1/\sqrt{2}$.

Having found an eigenvector corresponding to $\lambda_1 = -1$, we proceed to find an eigenvector x_2 corresponding to $\lambda_2 = 5$. Designating the unknown vector x_2 by

$$\begin{bmatrix} x_2 \\ y_2 \end{bmatrix}$$

and substituting it with λ_2 into (6), we obtain

$$\left\{ \begin{bmatrix} 1 & 2 \\ 4 & 2 \end{bmatrix} - 5 \begin{bmatrix} 1 & 0 \\ 0 & 1 \end{bmatrix} \right\} \begin{bmatrix} x_2 \\ y_2 \end{bmatrix} = \begin{bmatrix} 0 \\ 0 \end{bmatrix},$$

or

$$\begin{bmatrix} -4 & 2 \\ 4 & -2 \end{bmatrix} \begin{bmatrix} x_2 \\ y_2 \end{bmatrix} = \begin{bmatrix} 0 \\ 0 \end{bmatrix},$$

or, equivalently,

$$-4x_2 + 2y_2 = 0,$$
$$4x_2 - 2y_2 = 0.$$

A nontrivial solution to this set of equations is $x_2 = \frac{1}{2}y_2$, where y_2 is arbitrary; hence,

$$x_2 = \begin{bmatrix} x_2 \\ x_2 \end{bmatrix} = \begin{bmatrix} x_2/2 \\ x_2 \end{bmatrix} = y_2 \begin{bmatrix} 1 \\ 2 \\ 1 \end{bmatrix}.$$

For convenience, we choose $y_2 = 2$, thus

$$\mathbf{x}_2 = \begin{bmatrix} 1 \\ 2 \end{bmatrix}.$$

In order to check whether or not x_2 is an eigenvector corresponding to $\lambda_2 = 5$, we need only check if $\mathbf{A}\mathbf{x}_2 = \lambda_2 \, x_2$:

$$\mathbf{A}\mathbf{x}_2 = \begin{bmatrix} 1 & 2 \\ 4 & 3 \end{bmatrix} \begin{bmatrix} 1 \\ 2 \end{bmatrix} = \begin{bmatrix} 5 \\ 10 \end{bmatrix} = 5 \begin{bmatrix} 1 \\ 2 \end{bmatrix} = \lambda_2 \mathbf{x}_2.$$

Again, note that x_2 is *not* unique! Any scalar multiple of x_2 is also an eigenvector corresponding to λ_2. However, in this case, there is just one *linearly independent* eigenvector corresponding to λ_2. ■

Example 2.   Find the eigenvectors of

$$\mathbf{A} = \begin{bmatrix} 2 & 0 & 0 \\ 0 & 2 & 5 \\ 0 & -1 & -2 \end{bmatrix}.$$

Solution By using the method of the previous section, we find the eigenvalues to be $\lambda_1 = 2$, $\lambda_2 = i$, $\lambda_3 = -i$. We first calculate the eigenvectors corresponding to $\lambda_1 = 2$. Designate x_1 by

$$\begin{bmatrix} x_1 \\ y_1 \\ z_1 \end{bmatrix}.$$

Then, (6) becomes

$$\left\{ \begin{bmatrix} 2 & 0 & 0 \\ 0 & 2 & 5 \\ 0 & -1 & -2 \end{bmatrix} - 2 \begin{bmatrix} 1 & 0 & 0 \\ 0 & 1 & 0 \\ 0 & 0 & 1 \end{bmatrix} \right\} \begin{bmatrix} x_1 \\ y_1 \\ z_1 \end{bmatrix} = \begin{bmatrix} 0 \\ 0 \\ 0 \end{bmatrix},$$

$$\begin{bmatrix} 0 & 0 & 0 \\ 0 & 0 & 5 \\ 0 & -1 & -4 \end{bmatrix} \begin{bmatrix} x_1 \\ y_1 \\ z_1 \end{bmatrix} = \begin{bmatrix} 0 \\ 0 \\ 0 \end{bmatrix},$$

or, equivalently,

$$0 = 0,$$
$$5z_1 = 0,$$
$$-y_1 - 4z_1 = 0.$$

or
A nontrivial solution to this set of equations is $y_1 = z_1 = 0$, x_1 arbitrary; hence,

$$\mathbf{x}_1 = \begin{bmatrix} x_1 \\ y_1 \\ z_1 \end{bmatrix} = \begin{bmatrix} x_1 \\ 0 \\ 0 \end{bmatrix} = x_1 \begin{bmatrix} 1 \\ 0 \\ 0 \end{bmatrix}.$$

We now find the eigenvectors corresponding to $\lambda_2 = i$. If we designate $\mathbf{x}_2$ by

$$\begin{bmatrix} x_2 \\ y_2 \\ z_2 \end{bmatrix},$$

Eq. (6) becomes

$$\begin{bmatrix} 2-i & 0 & 0 \\ 0 & 2-i & 5 \\ 0 & -1 & -2-i \end{bmatrix} \begin{bmatrix} x_2 \\ y_2 \\ z_2 \end{bmatrix} = \begin{bmatrix} 0 \\ 0 \\ 0 \end{bmatrix}$$

or

$$(2-i)x_2 = 0,$$
$$(2-i)y_2 + 5z_2 = 0,$$
$$-y_2 + (-2-i)z_2 = 0.$$

A nontrivial solution to this set of equations is $x_2 = 0$, $y_2 = (-2-i)z_2$, z_2 arbitrary; hence,

$$\mathbf{x}_2 = \begin{bmatrix} x_2 \\ y_2 \\ z_2 \end{bmatrix} = \begin{bmatrix} 0 \\ (-2-i)z_2 \\ z_2 \end{bmatrix} = z_2 \begin{bmatrix} 0 \\ -2-i \\ 1 \end{bmatrix}.$$

The eigenvectors corresponding to $\lambda_3 = -i$ are found in a similar manner to be

$$\mathbf{x}_3 = z_3 \begin{bmatrix} 0 \\ -2 - i \\ 1 \end{bmatrix} \quad z_3 \text{ arbitrary.} \blacksquare$$

It should be noted that even if a mistake is made in finding the eigenvalues of a matrix, the error will become apparent when the eigenvectors corresponding to the incorrect eigenvalues are found. For instance, suppose that λ_1 in Example 2 was calculated erroneously to be 3. If we now try to find $\mathbf{x}_1$, we obtain the equations

$$-x_1 = 0,$$
$$-y_1 + 5z_1 = 0,$$
$$-y_1 - 5z_1 = 0.$$

The only solution to this set of equations is $x_1 = y_1 = z_1 = 0$; hence,

$$\mathbf{x}_1 = \begin{bmatrix} 0 \\ 0 \\ 0 \end{bmatrix}.$$

However, by definition, an eigenvector cannot be the zero vector. Since every eigenvalue must have a corresponding eigenvector, there is a mistake. A quick check shows that all the calculations above are valid; hence, the error must lie in the value of the eigenvalue.

Problems 6.3

In Problems 1 through 23, find an eigenvector corresponding to each eigenvalue of the given matrix.

1. $\begin{bmatrix} 1 & 2 \\ -1 & 4 \end{bmatrix}$,

2. $\begin{bmatrix} 2 & 1 \\ 2 & 3 \end{bmatrix}$,

3. $\begin{bmatrix} 2 & 3 \\ 4 & 6 \end{bmatrix}$,

4. $\begin{bmatrix} 3 & 6 \\ 9 & 6 \end{bmatrix}$,

5. $\begin{bmatrix} 1 & 2 \\ 4 & -4 \end{bmatrix}$,

6. $\begin{bmatrix} 3 & 5 \\ 5 & -3 \end{bmatrix}$,

7. $\begin{bmatrix} 3 & 5 \\ -5 & -3 \end{bmatrix}$,

8. $\begin{bmatrix} 2 & 5 \\ -1 & -2 \end{bmatrix}$,

9. $\begin{bmatrix} 2 & 2 \\ -1 & -2 \end{bmatrix}$

10. $\begin{bmatrix} 4 & 10 \\ 9 & -5 \end{bmatrix}$,

11. $\begin{bmatrix} 0 & t \\ 2t & -t \end{bmatrix}$,

12. $\begin{bmatrix} 4\theta & 2\theta \\ -\theta & \theta \end{bmatrix}$,

13. $\begin{bmatrix} 1 & 0 & 3 \\ 1 & 2 & 1 \\ 3 & 0 & 1 \end{bmatrix}$,

14. $\begin{bmatrix} 2 & 0 & -1 \\ 2 & 2 & 2 \\ -1 & 0 & 2 \end{bmatrix}$,

15. $\begin{bmatrix} 3 & 0 & -1 \\ 2 & 3 & 2 \\ -1 & 0 & 3 \end{bmatrix}$,

16. $\begin{bmatrix} 3 & 0 & 0 \\ 2 & 6 & 4 \\ 2 & 3 & 5 \end{bmatrix}$,

17. $\begin{bmatrix} 5 & -7 & 7 \\ 4 & -3 & 4 \\ 4 & -1 & 2 \end{bmatrix}$,

18. $\begin{bmatrix} 3 & 1 & -1 \\ 1 & 3 & -1 \\ -1 & -1 & 5 \end{bmatrix}$,

19. $\begin{bmatrix} 1 & 5 & 1 \\ -1 & -1 & 1 \\ 0 & 0 & 3 \end{bmatrix}$,

20. $\begin{bmatrix} 0 & 1 & 0 \\ 0 & 0 & 1 \\ 0 & -1 & 0 \end{bmatrix}$,

21. $\begin{bmatrix} 3 & 2 & 1 \\ 0 & 4 & 0 \\ 0 & 1 & 5 \end{bmatrix}$,

22. $\begin{bmatrix} 1 & -1 & 0 & 0 \\ 3 & 5 & 0 & 0 \\ 0 & 0 & 1 & 4 \\ 0 & 0 & 1 & 1 \end{bmatrix}$,

23. $\begin{bmatrix} 2 & 4 & 2 & -2 \\ 0 & 1 & 0 & 0 \\ 0 & 3 & 3 & -1 \\ 0 & 2 & 0 & 4 \end{bmatrix}$.

24. Find unit eigenvectors (i.e., eigenvectors whose magnitudes equal unity) for the matrix in Problem 1.

25. Find unit eigenvectors for the matrix in Problem 2.

26. Find unit eigenvectors for the matrix in Problem 3.

27. Find unit eigenvectors for the matrix in Problem 13.

28. Find unit eigenvectors for the matrix in Problem 14.

29. Find unit eigenvectors for the matrix in Problem 16.

30. A nonzero vector $\mathbf{x}$ is a left eigenvector for a matrix $\mathbf{A}$ if there exists a scalar λ such that $\mathbf{xA} = \lambda\mathbf{x}$. Find a set of left eigenvectors for the matrix in Problem 1.

31. Find a set of left eigenvectors for the matrix in Problem 2.

32. Find a set of left eigenvectors for the matrix in Problem 3.

33. Find a set of left eigenvectors for the matrix in Problem 4.

34. Find a set of left eigenvectors for the matrix in Problem 13.

35. Find a set of left eigenvectors for the matrix in Problem 14.

36. Find a set of left eigenvectors for the matrix in Problem 16.

37. Find a set of left eigenvectors for the matrix in Problem 18.

38. Prove that if $\mathbf{x}$ is a right eigenvector of a symmetric matrix $\mathbf{A}$, then $\mathbf{x}^T$ is a left eigenvector of $\mathbf{A}$.

39. A left eigenvector for a given matrix is known to be [1 1]. Find another left eigenvector for the same matrix satisfying the property that the sum of the vector components must equal unity.

40. A left eigenvector for a given matrix is known to be [2 3]. Find another left eigenvector for the same matrix satisfying the property that the sum of the vector components must equal unity.

41. A left eigenvector for a given matrix is known to be [1 2 5]. Find another left eigenvector for the same matrix satisfying the property that the sum of the vector components must equal unity.

42. A Markov chain (see Problem 16 of Section 1.1 and Problem 16 of Section 1.6) is *regular* if some power of the transition matrix contains only positive elements. If the matrix itself contains only positive elements, then the power is one and the matrix is automatically regular. Transition matrices that are regular always have an eigenvalue of unity. They also have limiting distribution vectors denoted by $\mathbf{x}^{(\infty)}$, where the ith component of $\mathbf{x}^{(\infty)}$ represents the probability of an object being in state i after a large number of time periods have elapsed. The limiting distribution $\mathbf{x}^{(\infty)}$ is a left eigenvector of the transition matrix corresponding to the eigenvalue of unity and having the sum of its components equal to one.

(a) Find the limiting distribution vector for the Markov chain described in Problem 16 of Section 1.1.

(b) Ultimately, what is the probability that a family will reside in the city?

43. Find the limiting distribution vector for the Markov chain described in Problem 17 of Section 1.1. What is the probability of having a Republican mayor over the long run?

44. Find the limiting distribution vector for the Markov chain described in Problem 18 of Section 1.1. What is the probability of having a good harvest over the long run?

45. Find the limiting distribution vector for the Markov chain described in Problem 19 of Section 1.1. Ultimately, what is the probability that a person will use Brand Y?

6.4 Properties of eigenvalues and eigenvectors

Definition 1. The *trace* of a matrix $\mathbf{A}$, designated by tr($\mathbf{A}$), is the sum of the elements on the main diagonal.

Example 1.   Find the tr($\mathbf{A}$) if

$$\mathbf{A} = \begin{bmatrix} 3 & -1 & 2 \\ 0 & 4 & 1 \\ 1 & -1 & -5 \end{bmatrix}.$$

Solution tr($\mathbf{A}$) $= 3 + 4 + (-5) = 2$. ∎

Property 1. *The sum of the eigenvalues of a matrix equals the trace of the matrix.*

    Proof. See Problem 20.

    Property 1 provides us with a quick and useful procedure for checking eigenvalues.

Example 2. Verify Property 1 for

$$A = \begin{bmatrix} 11 & 3 \\ -5 & -5 \end{bmatrix}.$$

    Solution The eigenvalues of **A** are $\lambda_1 = 10$, $\lambda_2 = -4$.

$$\mathrm{tr}(\mathbf{A}) = 11 + (-5) = 6 = \lambda_1 + \lambda_2. \quad \blacksquare$$

Property 2. *A matrix is singular if and only if it has a zero eigenvalue.*

    Proof. A matrix **A** has a zero eigenvalue if and only if $\det(\mathbf{A} - 0\mathbf{I}) = 0$, or (since $0\mathbf{I} = 0$) if and only if $\det(\mathbf{A}) = 0$. But $\det(\mathbf{A}) = 0$ if and only if **A** is singular; thus, the result is immediate.

Property 3. *The eigenvalues of an upper (or lower) triangular matrix are the elements on the main diagonal.*

    Proof. See Problem 15.

Example 3. Find the eigenvalues of

$$\mathbf{A} = \begin{bmatrix} 1 & 0 & 0 \\ 2 & 1 & 0 \\ 3 & 4 & -1 \end{bmatrix}.$$

    Solution Since **A** is lower triangular, the eigenvalues must be $\lambda_1 = \lambda_2 = 1$, $\lambda_3 = -1$. $\blacksquare$

Property 4. *If λ is an eigenvalue of **A** and if **A** is invertible, then $1/\lambda$ is an eigenvalue of $\mathbf{A}^{-1}$.*

    Proof. Since **A** is invertible, Property 2 implies that $\lambda \neq 0$; hence, $1/\lambda$ exists. Since λ is an eigenvalue of **A**, there must exist an eigenvector **x** such that $\mathbf{Ax} = \lambda\mathbf{x}$. Premultiplying both sides of this equation by $\mathbf{A}^{-1}$, we obtain

$$\mathbf{x} = \lambda\mathbf{A}^{-1}\mathbf{x}$$

or, equivalently, $\mathbf{A}^{-1}\mathbf{x} = (1/\lambda)\mathbf{x}$. Thus, $1/\lambda$ is an eigenvalue of $\mathbf{A}^{-1}$.

Observation 1. If **x** is an eigenvector of **A** corresponding to the eigenvalue λ and if **A** is invertible, then **x** is an eigenvector of $\mathbf{A}^{-1}$ corresponding to the eigenvalue $1/\lambda$.

Property 5. *If λ is an eigenvalue of **A**, then $\alpha\lambda$ is an eigenvalue of $\alpha\mathbf{A}$, where α is any arbitrary scalar.*

Proof. If λ is an eigenvalue of $\mathbf{A}$, then there must exist an eigenvector $\mathbf{x}$ such that $\mathbf{Ax} = \lambda\mathbf{x}$. Multiplying both sides of this equation by α, we obtain $(\alpha\mathbf{A})\mathbf{x} = (\alpha\lambda)\mathbf{x}$, which implies Property 5.

Observation 2. If $\mathbf{x}$ is an eigenvector of $\mathbf{A}$ corresponding to the eigenvalue λ, then $\mathbf{x}$ is an eigenvector of $\alpha\mathbf{A}$ corresponding to eigenvalue $\alpha\lambda$.

Property 6. *If λ is an eigenvalue of $\mathbf{A}$, then λ^k is an eigenvalue of $\mathbf{A}^k$, for any positive integer k.*

Proof. We prove the result for the special cases $k = 2$ and $k = 3$. Other cases are handled by mathematical induction. (See Problem 16.) If λ is an eigenvalue of $\mathbf{A}$, there must exist an eigenvector $\mathbf{x}$ such that $\mathbf{Ax} = \lambda\mathbf{x}$. Then,

$$\mathbf{A}^2\mathbf{x} = \mathbf{A}(\mathbf{Ax}) = \mathbf{A}(\lambda\mathbf{x}) = \lambda(\mathbf{Ax}) = \lambda(\lambda\mathbf{x}) = \lambda^2\mathbf{x},$$

which implies that λ^2 is an eigenvalue of $\mathbf{A}^2$. As a result, we also have that

$$\mathbf{A}^3\mathbf{x} = \mathbf{A}(\mathbf{A}^2\mathbf{x}) = \mathbf{A}(\lambda^2\mathbf{x}) = \lambda^2(\mathbf{Ax}) = \lambda^2(\lambda\mathbf{x}) = \lambda^2\mathbf{x},$$

which implies that λ^3 is an eigenvalue of $\mathbf{A}^3$. ■

Observation 3. If $\mathbf{x}$ is an eigenvector of $\mathbf{A}$ corresponding to the eigenvalue λ, then $\mathbf{x}$ is an eigenvector $\mathbf{A}^k$ corresponding to the eigenvalue λ^k, for any positive integer k.

Property 7. *If λ is an eigenvalue of $\mathbf{A}$, then for any scalar c, $\lambda - c$ is an eigenvalue of $\mathbf{A} - c\mathbf{I}$.*

Proof. If λ is an eigenvalue of $\mathbf{A}$, then there exists an eigenvector $\mathbf{x}$ such that $\mathbf{A} = \lambda\mathbf{x}$. Consequently,

$$\mathbf{Ax} - c\mathbf{x} = \lambda\mathbf{x} - c\mathbf{x},$$

or

$$(\mathbf{A} - c\mathbf{I})\mathbf{x} = (\lambda - c)\mathbf{x}.$$

Thus, $\lambda - c$ is an eigenvalue of $\mathbf{A} - c\mathbf{I}$. ■

Observation 4. If $\mathbf{x}$ is an eigenvector of $\mathbf{A}$ corresponding to the eigenvalue λ, then $\mathbf{x}$ is an eigenvector $\mathbf{A} - c\mathbf{I}$ corresponding to the eigenvalue $\lambda - c$.

Property 8. *If λ is an eigenvalue of $\mathbf{A}$, then λ is an eigenvalue of $\mathbf{A}^T$.*

Proof. Since λ is an eigenvalue of $\mathbf{A}$, $\det(\mathbf{A} - \lambda\mathbf{I}) = 0$. Hence,

$$
\begin{aligned}
0 \quad &= \quad |\mathbf{A} - \lambda\mathbf{I}| = \left|(\mathbf{A}^T)^T - \lambda\mathbf{I}^T\right| \quad &&\{\text{Property 1 (Section 1.4)}\\
&= \quad \left|(\mathbf{A}^T - \lambda\mathbf{I})^T\right| \quad &&\{\text{Property 3 (Section 1.4)}\\
&= \quad |\mathbf{A}^T - \lambda\mathbf{I}| \quad &&\{\text{Property 7 (Section 5.3)}
\end{aligned}
$$

Thus, $\det(\mathbf{A}^T - \lambda\mathbf{I}) = 0$, which implies that λ is an eigenvalue of $\mathbf{A}^T$. ■

Property 9. *The product of the eigenvalues (counting multiplicity) of a matrix equals the determinant of the matrix.*

Proof. See Problem 21. ■

Example 4.   Verify Property 9 for the matrix **A** given in Example 2.
    Solution For this **A**, $\lambda_1 = 10$, $\lambda_2 = -4$, $\det(\mathbf{A}) = -55 + 15 = -40 = \lambda_1 \lambda_2$. ■

Problems 6.4

1. One eigenvalue of the matrix

$$\mathbf{A} = \begin{bmatrix} 8 & 2 \\ 3 & 3 \end{bmatrix}$$

is known to be 2. Determine the second eigenvalue by inspection.

2. One eigenvalue of the matrix

$$\mathbf{A} = \begin{bmatrix} 8 & 3 \\ 3 & 2 \end{bmatrix}$$

is known to be 0.7574, rounded to four decimal places. Determine the second eigenvalue by inspection.

3. Two eigenvalues of a *3 × 3* matrix are known to be 5 and 8. What can be said about the remaining eigenvalue if the trace of the matrix is −4?

4. Redo Problem 3 if the determinant of the matrix is −4 instead of its trace.

5. The determination of a *4 × 4* matrix **A** is 144 and two of its eigenvalues are known to be −3 and 2. What can be said about the remaining eigenvalues?

6. A *2 × 2* matrix **A** is known to have the eigenvalues −3 and 4. What are the eigenvalues of (a) 2**A**, (b) 5**A**, (c) **A** − 3**I**, and (d) **A** + 4**I**?

7. A *3 × 3* matrix **A** is known to have the eigenvalues −2, 2, and 4. What are the eigenvalues of (a) $\mathbf{A}^2$, (b) $\mathbf{A}^3$, (c) −3**A**, and (d) **A** + 3**I**?

8. A *2 × 2* matrix **A** is known to have the eigenvalues −1 and 1. Find a matrix in terms of **A** that has for its eigenvalues:
    (a) −2 and 2,
    (b) −5 and 5,
    (c) 1 and 1,
    (d) 2 and 4.

9. A *3 × 3* matrix **A** is known to have the eigenvalues 2, 3, and 4. Find a matrix in terms of **A** that has for its eigenvalues:
    (a) 4, 6, and 8,
    (b) 4, 9, and 16,
    (c) 8, 27, and 64,
    (d) 0, 1, and 2.

10. Verify Property 1 for

$$A = \begin{bmatrix} 12 & 16 \\ -3 & -7 \end{bmatrix}.$$

11. Verify Property 2 for

$$A = \begin{bmatrix} 1 & 3 & 6 \\ -1 & 2 & -1 \\ 2 & 1 & 7 \end{bmatrix}.$$

12. Show that if λ is an eigenvalue of A, then it is also an eigenvalue for $S^{-1} A S$ for any nonsingular matrix S.

13. Show by example that, in general, an eigenvalue of $A + B$ is not the sum of an eigenvalue of A with an eigenvalue of B.

14. Show by example that, in general, an eigenvalue of AB is not the product of an eigenvalue of A with an eigenvalue of B.

15. Prove Property 3.

16. Use mathematical induction to complete the proof of Property 6.

17. The determinant of $A - \lambda I$ is known as the characteristic polynomial of A. For an $n \times n$ matrix A, it has the form

$$\det(A - \lambda I) = (-1)^n \left(\lambda^n + a_{n-1}\lambda^{n-1} + a_{n-2}\lambda^{n-2} + \cdots + a_2\lambda^2 + a_1\lambda + a_0 \right),$$

where $a_{n-1}, a_{n-2}, \ldots, a_2, a_1, a_0$ are constants that depend on the elements of A. Show that $(-1)^n a_0 = \det(A)$.

18. (Problem 17 continued) Convince yourself by considering arbitrary 3×3 and 4×4 matrices that $a_{n-1} = \mathrm{tr}(A)$.

19. Assume that A is an $n \times n$ matrix with eigenvalues $\lambda_2, \lambda_2, \ldots, \lambda_n$, where some or all of the eigenvalues may be equal. Since each eigenvalue λ_i $(i = 1, 2, \ldots, n)$ is a root of the characteristic polynomial, $(\lambda - \lambda_i)$ must be a factor of that polynomial. Deduce that

$$\det(A - \lambda I) = (-1)^n (\lambda - \lambda_1)(\lambda - \lambda_2) \ldots (\lambda - \lambda_n).$$

20. Use the results of Problems 18 and 19 to prove Property 1.

21. Use the results of Problems 17 and 19 to prove Property 9.

22. Show, by example, that an eigenvector of A need not be an eigenvector of A^T.

23. Prove that an eigenvector of A is a left eigenvector of A^T.

6.5 Linearly independent eigenvectors

Since every eigenvalue has an infinite number of eigenvectors associated with it (recall that if $\mathbf{x}$ is an eigenvector, then any scalar multiple of $\mathbf{x}$ is also an eigenvector), it becomes academic to ask how many different eigenvectors can a matrix have? The answer is clearly an infinite number. A more revealing question is how many linearly independent eigenvectors can a matrix have? Theorem 4 of Section 2.6 provides us with a partial answer.

Theorem 1. *In an n-dimensional vector space, every set of $n + 1$ vectors is linearly dependent.*

    Therefore, since all of the eigenvectors of an $n \times n$ matrix must be n-dimensional (why?), it follows from Theorem 1 that an $n \times n$ matrix can have *at most n* linearly independent eigenvectors. The following three examples shed more light on the subject.

Example 1. Find the eigenvectors of

$$\mathbf{A} = \begin{bmatrix} 2 & 1 & 0 \\ 0 & 2 & 1 \\ 0 & 0 & 2 \end{bmatrix}.$$

    Solution The eigenvalues of $\mathbf{A}$ are $\lambda_1 = \lambda_2 = \lambda_3 = 2$; therefore, $\lambda = 2$ is an eigenvalue of multiplicity 3. If we designate the unknown eigenvector $\mathbf{x}$ by

$$\begin{bmatrix} x \\ y \\ z \end{bmatrix},$$

then Eq. (6) gives rise to the three equations

$$y = 0,$$
$$z = 0,$$
$$0 = 0.$$

Thus, $y = z = 0$ and x is arbitrary; hence,

$$\mathbf{x} = \begin{bmatrix} x \\ y \\ z \end{bmatrix} = \begin{bmatrix} x \\ 0 \\ 0 \end{bmatrix} = x \begin{bmatrix} 1 \\ 0 \\ 0 \end{bmatrix}.$$

Setting $x = 1$, we see that $\lambda = 2$ generates only one linearly independent eigenvector,

$$\mathbf{x} = \begin{bmatrix} 1 \\ 0 \\ 0 \end{bmatrix}. \blacksquare$$

Example 2. Find the eigenvectors of

$$A = \begin{bmatrix} 2 & 1 & 0 \\ 0 & 2 & 0 \\ 0 & 0 & 2 \end{bmatrix}.$$

Solution Again, the eigenvalues are $\lambda_1 = \lambda_2 = \lambda_3 = 2$; therefore, $\lambda = 2$ is an eigenvalue of multiplicity 3. Designate the unknown eigenvector $\mathbf{x}$ by

$$\begin{bmatrix} x \\ y \\ z \end{bmatrix}.$$

Eq. (6) now gives rise to the equations

$$y = 0,$$
$$0 = 0,$$
$$0 = 0.$$

Thus, $y = 0$ and both x and z are arbitrary; hence,

$$\mathbf{x} = \begin{bmatrix} x \\ y \\ z \end{bmatrix} = \begin{bmatrix} x \\ 0 \\ z \end{bmatrix} = \begin{bmatrix} x \\ 0 \\ 0 \end{bmatrix} + \begin{bmatrix} 0 \\ 0 \\ z \end{bmatrix} = x \begin{bmatrix} 1 \\ 0 \\ 0 \end{bmatrix} + z \begin{bmatrix} 0 \\ 0 \\ 1 \end{bmatrix}.$$

Since x and z can be chosen arbitrarily, we can first choose $x = 1$ and $z = 0$ to obtain

$$\mathbf{x}_1 = \begin{bmatrix} 1 \\ 0 \\ 0 \end{bmatrix}$$

and then choose $x = 0$ and $z = 1$ to obtain

$$\mathbf{x}_2 = \begin{bmatrix} 0 \\ 0 \\ 1 \end{bmatrix}.$$

$\mathbf{x}_1$ and $\mathbf{x}_2$ can easily be shown to be linearly independent vectors; hence, we see that $\lambda = 2$ generates the two linearly independent eigenvectors

$$\begin{bmatrix} 1 \\ 0 \\ 0 \end{bmatrix} \quad \text{and} \quad \mathbf{x}_1 = \begin{bmatrix} 0 \\ 0 \\ 1 \end{bmatrix}. \blacksquare$$

Example 3. Find the eigenvectors of

$$A = \begin{bmatrix} 2 & 0 & 0 \\ 0 & 2 & 0 \\ 0 & 0 & 2 \end{bmatrix}.$$

Solution Again the eigenvalues are $\lambda_1 = \lambda_2 = \lambda_3 = 2$; so again, $\lambda = 2$ is an eigenvalue of multiplicity 3. Designate the unknown eigenvector $\mathbf{x}$ by

$$\begin{bmatrix} x \\ y \\ z \end{bmatrix}.$$

Eq. (6) gives rise to the equations

$$0 = 0,$$
$$0 = 0,$$
$$0 = 0.$$

Thus, x, y, and z are all arbitrary; hence,

$$\mathbf{x} = \begin{bmatrix} x \\ y \\ z \end{bmatrix} \begin{bmatrix} x \\ 0 \\ 0 \end{bmatrix} + \begin{bmatrix} 0 \\ y \\ 0 \end{bmatrix} + \begin{bmatrix} 0 \\ 0 \\ z \end{bmatrix} = x \begin{bmatrix} 1 \\ 0 \\ 0 \end{bmatrix} + y \begin{bmatrix} 0 \\ 1 \\ 0 \end{bmatrix} + z \begin{bmatrix} 0 \\ 0 \\ 1 \end{bmatrix}.$$

Since x, y, and z can be chosen arbitrarily, we can first choose $x = 1$, $y = z = 0$, then choose $x = z = 0$, $y = 1$, and finally choose $y = x = 0$, $z = 1$ to generate the three linearly independent eigenvectors

$$\begin{bmatrix} 1 \\ 0 \\ 0 \end{bmatrix}, \quad \begin{bmatrix} 0 \\ 1 \\ 0 \end{bmatrix}, \quad \begin{bmatrix} 0 \\ 0 \\ 1 \end{bmatrix}.$$

In this case, we see that three linearly independent eigenvectors are generated by $\lambda = 2$. (Note that, from Theorem 1, this is the maximal number that could be generated.) ∎

The preceding examples are illustrations of

Theorem 2. *If λ is an eigenvalue of multiplicity k of an $n \times n$ matrix $\mathbf{A}$, then the number of linearly independent eigenvectors of $\mathbf{A}$ associated with λ is given by $\rho = n - r(\mathbf{A} - \lambda \mathbf{I})$. Furthermore, $1 \le \rho \le k$.*

Proof. Let $\mathbf{x}$ be an n-dimensional vector. If $\mathbf{x}$ is an eigenvector, then it must satisfy the vector equation $\mathbf{A}\mathbf{x} = \lambda\mathbf{x}$ or, equivalently, $(\mathbf{A} - \lambda\mathbf{I})\mathbf{x} = 0$. This system is homogeneous, hence consistent; so by Theorem 2 of Section 2.7, we have that the solution vector $\mathbf{x}$ will be in terms of $n - r(\mathbf{A} - \lambda\mathbf{I})$ arbitrary unknowns. Since these unknowns can be picked independently of each other, it follows that the number of linearly independent eigenvectors of $\mathbf{A}$ associated with λ is also $\rho = n - r(\mathbf{A} - \lambda\mathbf{I})$. We defer a proof that $1 \le \rho \le k$ until Chapter 9.

In Example 1, $\mathbf{A}$ is 3×3; hence, $n = 3$, and $r(\mathbf{A} - 2\mathbf{I}) = 2$. Thus, there should be $3 - 2 = 1$ linearly independent eigenvectors associated with $\lambda = 2$, which is indeed the case. In Example 2, once again $n = 3$ but $r(\mathbf{A} - 2\mathbf{I}) = 1$. Thus, there should be $3 - 1 = 2$ linearly independent eigenvectors associated with $\lambda = 2$, which also is the case.

The next theorem gives the relationship between eigenvectors that correspond to different eigenvalues.

Theorem 3. Eigenvectors corresponding to distinct (that is, different) eigenvalues are linearly independent.

Proof. For the sake of clarity, we consider the case of three distinct eigenvectors and leave the more general proof as an exercise (see Problem 17). Therefore, let λ_1, λ_2, λ_3, be distinct eigenvalues of the matrix $\mathbf{A}$ and let x_1, x_2, x_3 be the associated eigenvectors. That is

$$\begin{aligned}
\mathbf{A}\mathbf{x}_1 &= \lambda_1\mathbf{x}_1, \\
\mathbf{A}\mathbf{x}_2 &= \lambda_2\mathbf{x}_2, \\
\mathbf{A}\mathbf{x}_3 &= \lambda_3\mathbf{x}_3,
\end{aligned} \tag{8}$$

and $\lambda_1 \neq \lambda_2 \neq \lambda_3 \neq \lambda_1$.

Since we want to show that x_1, x_2, x_3 are linearly independent, we must show that the only solution to

$$c_1\mathbf{x}_1 = c_2\mathbf{x}_2 + c_3\mathbf{x}_3 = 0 \tag{9}$$

is $c_1 = c_2 = c_3 = 0$. By premultiplying (9) by $\mathbf{A}$, we obtain

$$c_1\mathbf{A}\mathbf{x}_1 + c_2\mathbf{A}\mathbf{x}_2 + c_3\mathbf{A}\mathbf{x}_3 = \mathbf{A}{\cdot}0 = 0.$$

It follows from (8), therefore, that

$$c_1\lambda_1\mathbf{x}_1 = c_2\lambda_2\mathbf{x}_2 + c_3\lambda_3\mathbf{x}_3 = 0. \tag{10}$$

By premultiplying (10) by $\mathbf{A}$ and again using (8), we obtain

$$c_1\lambda_1^1\mathbf{x}_1 + c_2\lambda_2\mathbf{x}_2 + c_3\lambda_3^2\mathbf{x}_3 = 0. \tag{11}$$

Eqs. 9−11 can be written in the matrix form

$$\begin{bmatrix} 1 & 1 & 1 \\ \lambda_1 & \lambda_2 & \lambda_3 \\ \lambda_1^2 & \lambda_2^2 & \lambda_3^2 \end{bmatrix} \begin{bmatrix} c_1\mathbf{x}_1 \\ c_2\mathbf{x}_2 \\ c_3\mathbf{x}_3 \end{bmatrix} = \begin{bmatrix} \mathbf{0} \\ \mathbf{0} \\ \mathbf{0} \end{bmatrix}.$$

Define

$$\mathbf{B} = \begin{bmatrix} 1 & 1 & 1 \\ \lambda_1 & \lambda_2 & \lambda_3 \\ \lambda_1^2 & \lambda_2^2 & \lambda_3^2 \end{bmatrix}.$$

It can be shown that $\det(\mathbf{B}) = (\lambda_2 - \lambda_1)(\lambda_3 - \lambda_2)(\lambda_3 - \lambda_1)$. Thus, since all the eigenvalues are distinct, $\det(\mathbf{B}) \neq 0$ and $\mathbf{B}$ is invertible. Therefore,

$$\begin{bmatrix} c_1\mathbf{x}_1 \\ c_2\mathbf{x}_2 \\ c_3\mathbf{x}_3 \end{bmatrix} = \mathbf{B}^{-1}\begin{bmatrix} \mathbf{0} \\ \mathbf{0} \\ \mathbf{0} \end{bmatrix} = \begin{bmatrix} \mathbf{0} \\ \mathbf{0} \\ \mathbf{0} \end{bmatrix}$$

or

$$\begin{bmatrix} c_1\mathbf{x}_1 = 0 \\ c_2\mathbf{x}_2 = 0 \\ c_3\mathbf{x}_3 = 0 \end{bmatrix} \tag{12}$$

But since x_1, x_2, x_3 are eigenvectors, they are nonzero; therefore, it follows from (12) that $c_1 = c_2 = c_3 = 0$. This result together with (9) implies Theorem 3.

Theorems 2 and 3 together completely determine the number of linearly independent eigenvectors of a matrix.

Example 4.   Find a set of linearly independent eigenvectors for

$$A = \begin{bmatrix} 1 & 0 & 0 \\ 4 & 3 & 2 \\ 4 & 2 & 3 \end{bmatrix}.$$

Solution The eigenvalues of A are $\lambda_1 = \lambda_2 = 1$ and $\lambda_3 = 5$. For this matrix, $n = 3$ and $r(A - 1I) = 1$, hence $n - r(A - 1I) = 2$. Thus, from Theorem 2, we know that A has two linearly independent eigenvectors corresponding to $\lambda = 1$ and one linearly independent eigenvector corresponding to $\lambda = 5$ (why?). Furthermore, Theorem 3 guarantees that the two eigenvectors corresponding to $\lambda = 1$ will be linearly independent of the eigenvector corresponding to $\lambda = 5$ and vice versa. It only remains to produce these vectors.

For $\lambda = 1$, the unknown vector

$$x_1 = \begin{bmatrix} x_1 \\ y_1 \\ z_1 \end{bmatrix}$$

must satisfy the vector equation $(A - 1I)x_1 = 0$, or equivalently, the set of equations

$$0 = 0,$$
$$4x_1 + 2y_1 + 2z_1 = 0,$$
$$4x_1 + 2y_1 + 2z_1 = 0.$$

A solution to this equation is $z_1 = -2x_1 - y_1$, x_1, and y_1 arbitrary. Thus,

$$x_1 = \begin{bmatrix} x_1 \\ y_1 \\ z_1 \end{bmatrix} = \begin{bmatrix} x_1 \\ y_1 \\ -2x_1 - y_1 \end{bmatrix} = x_1 \begin{bmatrix} 1 \\ 0 \\ -2 \end{bmatrix} + y_1 \begin{bmatrix} 0 \\ 1 \\ -1 \end{bmatrix}.$$

By first choosing $x_1 = 1$, $y_1 = 0$ and then $x_1 = 0$, $y_1 = 1$, we see that $\lambda = 1$ generates the two linearly independent eigenvectors

$$\begin{bmatrix} 1 \\ 0 \\ -2 \end{bmatrix}, \begin{bmatrix} 0 \\ 1 \\ -1 \end{bmatrix}.$$

An eigenvector corresponding to $\lambda_3 = 5$ is found to be

$$\begin{bmatrix} 0 \\ 1 \\ 1 \end{bmatrix}.$$

Therefore, A possesses the three linearly independent eigenvectors

$$\begin{bmatrix} 1 \\ 0 \\ -2 \end{bmatrix}, \begin{bmatrix} 0 \\ 1 \\ -1 \end{bmatrix}, \begin{bmatrix} 0 \\ 1 \\ 1 \end{bmatrix}. \blacksquare$$

Problems 6.5

In Problems 1–16, find a set of linearly independent eigenvectors for the given matrices.

1. $\begin{bmatrix} 2 & -1 \\ 1 & 4 \end{bmatrix}$,

2. $\begin{bmatrix} 3 & 1 \\ 0 & 3 \end{bmatrix}$,

3. $\begin{bmatrix} 3 & 0 \\ 0 & 3 \end{bmatrix}$,

4. $\begin{bmatrix} 2 & 1 & 1 \\ 0 & 1 & 0 \\ 1 & 1 & 2 \end{bmatrix}$,

5. $\begin{bmatrix} 2 & 1 & 1 \\ 0 & 1 & 0 \\ 1 & 2 & 2 \end{bmatrix}$,

6. $\begin{bmatrix} 2 & 0 & -1 \\ 2 & 1 & -2 \\ -1 & 0 & 2 \end{bmatrix}$,

7. $\begin{bmatrix} 1 & 1 & -1 \\ 0 & 0 & 0 \\ 1 & 2 & 3 \end{bmatrix}$,

8. $\begin{bmatrix} 1 & 2 & 3 \\ 2 & 4 & 6 \\ 3 & 6 & 9 \end{bmatrix}$,

9. $\begin{bmatrix} 3 & -1 & 1 \\ -1 & 3 & -1 \\ 1 & -1 & 3 \end{bmatrix}$,

10. $\begin{bmatrix} 0 & 1 & 0 \\ 0 & 0 & 1 \\ 27 & -27 & 9 \end{bmatrix}$,

11. $\begin{bmatrix} 0 & 1 & 0 \\ 0 & 0 & 1 \\ 1 & -3 & 3 \end{bmatrix}$,

12. $\begin{bmatrix} 4 & 2 & 1 \\ 2 & 7 & 2 \\ 1 & 2 & 4 \end{bmatrix}$,

13. $\begin{bmatrix} 0 & 1 & 0 & 0 \\ 0 & 0 & 1 & 0 \\ 0 & 0 & 0 & 1 \\ -1 & 4 & -6 & 4 \end{bmatrix}$,

14. $\begin{bmatrix} 1 & 0 & 0 & 0 \\ 0 & 0 & 1 & 0 \\ 0 & 0 & 0 & 1 \\ 0 & 1 & -3 & 3 \end{bmatrix}$,

15. $\begin{bmatrix} 1 & 0 & 0 & 0 \\ 1 & 1 & 1 & 1 \\ 1 & 1 & 2 & 1 \\ 1 & 1 & 1 & 2 \end{bmatrix}$,

16. $\begin{bmatrix} 3 & 1 & 1 & 2 \\ 0 & 3 & 1 & 1 \\ 0 & 0 & 2 & 0 \\ 0 & 0 & 0 & 2 \end{bmatrix}$.

17. The Vandermonde determinant

$$\begin{vmatrix} 1 & 1 & \cdots & 1 \\ x_1 & x_2 & \cdots & x_n \\ x_1^2 & x_2^2 & \cdots & x_n^2 \\ \vdots & \vdots & \vdots & \vdots \\ x_1^{n-1} & x_2^{n-1} & \cdots & x_n^{n-1} \end{vmatrix}$$

is known to equal the product

$$(x_2 - x_1)(x_3 - x_2)(x_3 - x_1)(x_4 - x_3)(x_4 - x_2)\cdots(x_n - x_1).$$

Using this result, prove Theorem 3 for n distinct eigenvalues.

6.6 **Power methods**

The analytic methods described in Sections 6.2 and 6.3 are impractical for calculating the eigenvalues and eigenvectors of matrices of large order. Determining
the characteristic equations for such matrices involves enormous effort, while

finding its roots algebraically is usually impossible. Instead, iterative methods that lend themselves to computer implementation are used. Ideally, each iteration yields a new approximation, which converges to an eigenvalue and the corresponding eigenvector.

The *dominant* eigenvalue of a matrix is the one having largest absolute values. Thus, if the eigenvalues of a matrix are 2, 5, and -13, then -13 is the dominant eigenvalue because it is the largest in absolute value. The *power method* is an algorithm for locating the dominant eigenvalue and a corresponding eigenvector for a matrix of real numbers when the following two conditions exist:

Condition 1. The dominant eigenvalue of a matrix is real (not complex) and is strictly greater in absolute values than all other eigenvalues.

Condition 2. If the matrix has order $n \times n$, then it possesses n linearly independent eigenvectors.

Denote the eigenvalues of a given square matrix $\mathbf{A}$ satisfying Conditions 1 and 2 by $\lambda_1, \lambda_2, \ldots, \lambda_2$ and a set of corresponding eigenvectors by $\mathbf{v}_1, \mathbf{v}_2, \ldots, \mathbf{v}_2$, respectively. Assume the indexing is such that

$$|\lambda_1| + > |\lambda_2| \geq |\lambda_3| \geq \cdots \geq |\lambda_n|.$$

Any vector $\mathbf{x}_0$ can be expressed as a linear combination of the eigenvectors of $\mathbf{A}$, so we may write

$$\mathbf{x}_0 = c_1\mathbf{v}_1 + c_2\mathbf{v}_2 + \cdots + c_n\mathbf{v}_n.$$

Multiplying this equation by $\mathbf{A}^k$, for some large, positive integer k, we get

$$\mathbf{A}^k\mathbf{x}_0 = \mathbf{A}^k(c_1\mathbf{v}_1 + c_2\mathbf{v}_2 + \cdots + c_n\mathbf{v}_n)$$
$$= c_1\mathbf{A}^k\mathbf{v}_1 + c_2\mathbf{A}^k\mathbf{v}_2 + \cdots + c_n\mathbf{A}^k\mathbf{v}_n$$

It follows from Property 6 and Observation 3 of Section 6.4 that

$$\mathbf{A}^k\mathbf{x}_0 = c_1\lambda_1^k\mathbf{v}_1 + c_2\lambda_2^k\mathbf{v}_2 + \cdots + c_n\lambda_n^k\mathbf{v}_n$$

$$= \lambda_1^k\left[c_1\mathbf{v}_1 + c_2\left(\frac{\lambda_2}{\lambda_1}\right)^k\mathbf{v}_2 + \cdots + c_n\left(\frac{\lambda_n}{\lambda_1}\right)^k\mathbf{v}_n\right]$$

$$\approx \lambda_1^k c_1\mathbf{v}_1 \quad \text{for large } k.$$

This last pseudoequality follows from noting that each quotient of eigenvalues is less than unity in absolute value, as a result of indexing the first eigenvalue as the dominant one, and therefore tends to zero as that quotient is raised to successively higher powers.

Thus, $\mathbf{A}^k\mathbf{x}_0$ approaches a scalar multiple of $\mathbf{v}_1$. But any nonzero scalar multiple of an eigenvector is itself an eigenvector, so $\mathbf{A}^k\mathbf{x}_0$ approaches an eigenvector of $\mathbf{A}$ corresponding to the dominant eigenvalue, providing c_1 is not zero. The scalar c_1 will be zero only if $\mathbf{x}_0$ is a linear combination of $\{\mathbf{v}_2, \mathbf{v}_3, \ldots, \mathbf{v}_n\}$.

The power method begins with an initial vector $\mathbf{x}_0$, usually the vector having all ones for its components, and then iteratively calculates the vectors

$$\mathbf{x}_1 = \mathbf{A}\mathbf{x}_0,$$

$$\mathbf{x}_2 = \mathbf{A}\mathbf{x}_1 = \mathbf{A}^2\mathbf{x}_0,$$

$$\mathbf{x}_3 = \mathbf{A}\mathbf{x}_2 = \mathbf{A}^3\mathbf{x}_0,$$

$$\vdots$$

$$\mathbf{x}_k = \mathbf{A}\mathbf{x}_{k-1} = \mathbf{A}^k\mathbf{x}_0.$$

As k gets larger, $\mathbf{x}_k$ approaches an eigenvector of $\mathbf{A}$ corresponding to its dominant eigenvalue.

We can even determine the dominant eigenvalue by scaling appropriately. If k is large enough so that $\mathbf{x}_k$ is a good approximation to the eigenvector, say to within acceptable roundoff error, then it follows from Eq. (1) that

$$\mathbf{A}\mathbf{x}_k = \lambda_1 \mathbf{x}_k.$$

If $\mathbf{x}_k$ is scaled so that its largest component is unity, then the component of $\mathbf{x}_{k+1} = \mathbf{A}\mathbf{x}_k = \lambda_1\mathbf{x}_k$ having the largest absolute value must be λ_1.

We can now formalize the power method. Begin with an initial guess $\mathbf{x}_0$ for the eigenvector, having the property that its largest component in absolute value is unity. Iteratively, calculate $\mathbf{x}_1$, $\mathbf{x}_2$, $\mathbf{x}_3$,... by multiplying each successive iterate by $\mathbf{A}$, the matrix of interest. Each time $\mathbf{x}_k$ ($k = 1, 2, 3, \ldots$) is computed, identify its dominant component and divide each component by it. Redefine this scaled vector as the new $\mathbf{x}_k$. Each $\mathbf{x}_k$ is an estimate of an eigenvector for $\mathbf{A}$ and each dominant component is an estimate for the associated eigenvalue.

Example 1.   Find the dominant eigenvalue and a corresponding eigenvector for

$$\mathbf{A} = \begin{bmatrix} 1 & 2 \\ 4 & 3 \end{bmatrix}.$$

Solution We initialize $\mathbf{x}_0 = [1\ 1]^T$. Then
First iteration

$$\mathbf{x}_1 = \mathbf{A}\mathbf{x}_0 = \begin{bmatrix} 1 & 2 \\ 4 & 3 \end{bmatrix}\begin{bmatrix} 1 \\ 1 \end{bmatrix} = \begin{bmatrix} 3 \\ 7 \end{bmatrix},$$

$$\lambda \approx 7,$$

$$\mathbf{x}_1 \leftarrow \frac{1}{7}[3\ \ 7]^T = [0.428571\ \ 1]^T.$$

Second iteration

$$\mathbf{x}_2 = \mathbf{A}\mathbf{x}_1 = \begin{bmatrix} 1 & 2 \\ 4 & 3 \end{bmatrix}\begin{bmatrix} 0.428571 \\ 1 \end{bmatrix} = \begin{bmatrix} 2.428571 \\ 4.714286 \end{bmatrix},$$

$$\lambda \approx 4.714286,$$

$$\mathbf{x}_2 \leftarrow \frac{1}{4.714286}[2.428571\ \ 4.714286]^T = [0.515152\ \ 1]^T.$$

Third iteration

$$\mathbf{x}_3 = \mathbf{A}\mathbf{x}_2 = \begin{bmatrix} 1 & 2 \\ 4 & 3 \end{bmatrix} \begin{bmatrix} 0.515152 \\ 1 \end{bmatrix} = \begin{bmatrix} 2.515152 \\ 5.060606 \end{bmatrix},$$

$$\lambda \approx 5.060606,$$

$$\mathbf{x}_3 \leftarrow \frac{1}{5.060606} \begin{bmatrix} 2.515152 & 5.060606 \end{bmatrix}^{\mathrm{T}} = \begin{bmatrix} 0.497006 & 1 \end{bmatrix}^{\mathrm{T}}.$$

Fourth iteration

$$\mathbf{x}_4 = \mathbf{A}\mathbf{x}_3 = \begin{bmatrix} 1 & 2 \\ 4 & 3 \end{bmatrix} \begin{bmatrix} 0.497006 \\ 1 \end{bmatrix} = \begin{bmatrix} 2.497006 \\ 4.988024 \end{bmatrix},$$

$$\lambda \approx 4.988024,$$

$$\mathbf{x}_4 \leftarrow \frac{1}{4.988024} \begin{bmatrix} 2.497006 & 4.988024 \end{bmatrix}^{\mathrm{T}} = \begin{bmatrix} 0.500600 & 1 \end{bmatrix}^{\mathrm{T}}.$$

The method is converging to the eigenvalue 5 and its corresponding eigenvector $[0.5 \ 1]^{\mathrm{T}}$. ■

Example 2.   Find the dominant eigenvalue and a corresponding eigenvector for

$$\mathbf{A} = \begin{bmatrix} 0 & 1 & 0 \\ 0 & 0 & 1 \\ 18 & -1 & -7 \end{bmatrix}.$$

Solution We initialize $\mathbf{x}_0 = [1 \ 1 \ 1]^{\mathrm{T}}$. Then

First iteration

$$\mathbf{x}_1 = \mathbf{A}\mathbf{x}_0 = [1 \ 1 \ 10]^{\mathrm{T}},$$

$$\lambda \approx 10,$$

$$\mathbf{x}_1 \leftarrow \frac{1}{10} [1 \ 1 \ 10]^{\mathrm{T}} = [0.1 \ 0.1 \ 1]^{\mathrm{T}}.$$

Second iteration

$$\mathbf{x}_1 = \mathbf{A}\mathbf{x}_0 = \begin{bmatrix} 0 & 1 & 0 \\ 0 & 0 & 1 \\ 18 & -1 & -7 \end{bmatrix} \begin{bmatrix} 0.1 \\ 0.1 \\ 1 \end{bmatrix} = \begin{bmatrix} 0.1 \\ 1 \\ -5.3 \end{bmatrix},$$

$$\lambda \approx -5.3,$$

$$\mathbf{x}_2 \leftarrow \frac{1}{-5.3} [0.1 \ 1 \ -5.3]^{\mathrm{T}},$$

$$= [-0.018868 \ -0.188679 \ 1]^{\mathrm{T}}.$$

Third iteration

$$\mathbf{x}_3 = \mathbf{A}\mathbf{x}_2 = \begin{bmatrix} 0 & 1 & 0 \\ 0 & 0 & 1 \\ 18 & -1 & -7 \end{bmatrix} \begin{bmatrix} -0.018868 \\ -0.188679 \\ 1 \end{bmatrix} = \begin{bmatrix} -0.188679 \\ 1 \\ -7.150943 \end{bmatrix},$$

$$\lambda \approx -7.150943,$$

$$\mathbf{x}_3 \leftarrow \frac{1}{-7.150943} \begin{bmatrix} -0.188679 & 1 & -7.150943 \end{bmatrix}^{\mathrm{T}}$$

$$= \begin{bmatrix} 0.026385 & -0.139842 & 1 \end{bmatrix}^{\mathrm{T}}.$$

Continuing in this manner, we generate Table 6.1, where all entries are rounded to four decimal places. The algorithm is converging to the eigenvalue -6.405125 and its corresponding eigenvector

$$\begin{bmatrix} 0.024376 & -0.1561240 & 1 \end{bmatrix}^{\mathrm{T}}. \blacksquare$$

Although effective when it converges, the power method has deficiencies. It does not converge to the dominant eigenvalue when that eigenvalue is complex, and it may not converge when there are more than one equally dominant eigenvalues (see Problem 12). Furthermore, the method, in general, cannot be used to locate all the eigenvalues.

A more powerful numerical method is the *inverse power method*, which is the power method applied to the inverse of a matrix. See Tables 6.1, 6.2 and 6.3 for illustrations of the *inverse power method*. This, of course, adds another assumption: the inverse must exist, or equivalently, the matrix must not have any zero eigenvalues. Since a nonsingular matrix and its inverse share identical eigenvectors and reciprocal eigenvalues (see Property 4 and Observation 1 of Section 6.4), once we know the eigenvalues and eigenvectors of the inverse of a matrix, we have the analogous information about the matrix itself.

The power method applied to the inverse of a matrix $\mathbf{A}$ will generally converge to the dominant eigenvalue of $\mathbf{A}^{-1}$. Its reciprocal will be the eigenvalue of $\mathbf{A}$ having the smallest absolute value. The advantages of the inverse power method are that it

Table 6.1

Iteration	Eigenvector components			Eigenvalue
0	1.0000	1.0000	1.0000	
1	0.1000	0.1000	1.0000	10.0000
2	−0.0189	−0.1887	1.0000	−5.3000
3	0.0264	−0.1398	1.0000	−7.1509
4	0.0219	−0.1566	1.0000	−6.3852
5	0.0243	−0.1551	1.0000	−6.4492
6	0.0242	−0.1561	1.0000	−6.4078
7	0.0244	−0.1560	1.0000	−6.4084
8	0.0244	−0.1561	1.0000	−6.4056

converges more rapidly than the power method, and it often can be used to find all real eigenvalues of $\mathbf{A}$; a disadvantage is that it deals with $\mathbf{A}^{-1}$, which is laborious to calculate for matrices of large order. Such a calculation, however, can be avoided using **LU** decomposition.

The power method generates the sequence of vectors

$$\mathbf{x}_k = \mathbf{A}\mathbf{x}_{k-1}.$$

The inverse power method will generate the sequence

$$\mathbf{x}_k = \mathbf{A}^{-1}\mathbf{x}_{k-1},$$

which may be written as

$$\mathbf{A}\mathbf{x}_k = \mathbf{x}_{k-1}.$$

We solve for the unknown vector $\mathbf{x}_k$ using **LU** decomposition (see Section 3.5).

Example 3. Use the inverse power method to find an eigenvalue for

$$\mathbf{A} = \begin{bmatrix} 2 & 1 \\ 2 & 3 \end{bmatrix}.$$

Solution We initialize $\mathbf{x}_0 = [1\ 1]^T$. The **LU** decomposition for $\mathbf{A}$ has $\mathbf{A} = \mathbf{LU}$ with

$$\mathbf{L} = \begin{bmatrix} 1 & 0 \\ 1 & 1 \end{bmatrix} \quad \text{and} \quad \mathbf{U} = \begin{bmatrix} 2 & 1 \\ 0 & 2 \end{bmatrix}.$$

First iteration. We solve the system $\mathbf{LU}\mathbf{x}_1 = \mathbf{x}_0$ by first solving the system $\mathbf{Ly} = \mathbf{x}_0$ for $\mathbf{y}$, and then solving the system $\mathbf{U}\mathbf{x}_1 = \mathbf{y}$ for $\mathbf{x}_1$. Set $\mathbf{y} = [y_1\ y_2]^T$ and $\mathbf{x}_1 = [a\ b]^T$. The first system is

$$y_1 + 0y_2 = 1,$$
$$y_1 + \ y_2 = 1,$$

which has as its solution $y_1 = 1$ and $y_2 = 0$. The system $\mathbf{U}\mathbf{x}_1 = \mathbf{y}$ becomes

$$2a + b = 1,$$
$$2b = 0,$$

which admits the solution $a = 0.5$ and $b = 0$. Thus,

$$\mathbf{x}_1 = \mathbf{A}^{-1}\mathbf{x}_0 = [0.5\ 0]^T,$$

$$\lambda \approx 0.5 \quad \text{(an approximation to an eigenvalue for } \mathbf{A}^{-1}\text{)},$$

$$\mathbf{x}_1 \leftarrow \frac{1}{0.5}\ [0.5\ 0]^T = [1\ 0]^T.$$

Second iteration. We solve the system $\mathbf{LU}\mathbf{x}_2 = \mathbf{x}_1$ by first solving the system $\mathbf{Ly} = \mathbf{x}_1$ for $\mathbf{y}$, and then solving the system $\mathbf{U}\mathbf{x}_2 = \mathbf{y}$ for $\mathbf{x}_2$. Set $\mathbf{y} = [y_1\ y_2]^T$ and $\mathbf{x}_2 = [a\ b]^T$. The first system is

$$y_1 + 0y_2 = 1,$$
$$y_1 + \ y_2 = 0,$$

which has as its solution $y_1 = 1$ and $y_2 = -1$. The system $\mathbf{U}\mathbf{x}_2 = \mathbf{y}$ becomes

$$2a + b = 1,$$
$$2b = -1,$$

which admits the solution $a = 0.75$ and $b = -0.5$. Thus,

$$\mathbf{x}_2 = \mathbf{A}^{-1}\mathbf{x}_1 = [0.75 \quad -0.5]^T,$$

$$\lambda \approx 0.75,$$

$$\mathbf{x}_2 \leftarrow \frac{1}{0.75} [0.75 \quad -0.5]^T = [1 \quad -0.666667]^T.$$

Third iteration. We first solve $\mathbf{Ly} = \mathbf{x}_2$ to obtain $\mathbf{y} = [1 \ -1.666667]^T$, and then $\mathbf{Ux}_3 = \mathbf{y}$ to obtain $\mathbf{x}_3 = [0.916667 \ -0.833333]^T$. Then,

$$\lambda \approx 0.916667$$

$$\mathbf{x}_3 \leftarrow \frac{1}{0.916667} [0.916667 \quad -0.833333]^T = [1 \quad -0.909091]^T.$$

Continuing, we converge to the eigenvalue 1 for $\mathbf{A}^{-1}$ and its reciprocal $1/1 = 1$ for $\mathbf{A}$. The vector approximations are converging to $[1 \ -1]^T$, which is an eigenvector for both $\mathbf{A}^{-1}$ and $\mathbf{A}$. ∎

Example 4.   Use the inverse power method to find an eigenvalue for

$$\mathbf{A} = \begin{bmatrix} 7 & 2 & 0 \\ 2 & 1 & 6 \\ 0 & 6 & 7 \end{bmatrix}.$$

Solution We initialize $\mathbf{x}_0 = [1 \ 1 \ 1]^T$. The **LU** decomposition for **A** has $\mathbf{A} = \mathbf{LU}$ with

$$\mathbf{L} = \begin{bmatrix} 1 & 0 & 0 \\ 0.285714 & 1 & 0 \\ 0 & 14 & 1 \end{bmatrix} \quad \text{and} \quad \mathbf{U} = \begin{bmatrix} 7 & 2 & 0 \\ 0 & 0.428571 & 6 \\ 0 & 0 & -77 \end{bmatrix}.$$

First iteration set $\mathbf{y} = [y_1 \ y_2 \ y_3]^T$ and $\mathbf{x}_1 = [a \ b \ c]^T$. The first system is

$$y_1 + 0y_2 + 0y_3 = 1,$$

$$0.285714y_1 + y_2 + 0y_3 = 1,$$

$$0y_1 + 14y_2 + y_3 = 1,$$

which has as its solution $y_1 = 1$, $y_2 = 0.714286$, and $y_3 = -9$. The system $\mathbf{Ux}_1 = \mathbf{y}$ becomes

$$7a + 2b = 1,$$

$$0.428571b + 6c = 0.714286,$$

$$-77c = -9,$$

which admits the solution $a = 0.134199$, $b = 0.030303$, and $c = 0.116883$. Thus,

$$\mathbf{x}_1 = \mathbf{A}^{-1}\mathbf{x}_0 = [0.134199 \quad 0.030303 \quad 0.116833]^T,$$

$$\lambda \approx 0.134199 \quad \text{(an approximation to an eigenvalue for } \mathbf{A}^{-1}\text{)},$$

$$\mathbf{x}_1 \leftarrow \frac{1}{0.134199} [0.134199 \quad 0.030303 \quad 0.116833]^T$$

$$= [1 \quad 0.225806 \quad 0.870968]^T.$$

Second iteration solving the system $\mathbf{Ly} = \mathbf{x}_1$ for **y**, we obtain

$$y = [1 \quad -0.059908 \quad 1.709677]^{\text{T}}.$$

Then, solving the system $\mathbf{U}\mathbf{x}_2 = \mathbf{y}$ for $\mathbf{x}_2$, we get

$$\mathbf{x}_2 = [0.093981 \quad 0.171065 \quad -0.022204]^{\text{T}}.$$

Therefore,

$$\lambda \approx 0.171065,$$

$$\mathbf{x}_2 \leftarrow \frac{1}{0.171065} [0.093981 \quad 0.171065 \quad -0.022204]^{\text{T}},$$

$$= [0.549388 \quad 1 \quad -0.129796]^{\text{T}}.$$

Third iteration solving the system $\mathbf{L}\mathbf{y} = \mathbf{x}_2$ for $\mathbf{y}$, we obtain

$$\mathbf{y} = [0.549388 \quad 0.843032 \quad -11.932245]^{\text{T}}.$$

Then, solving the system $\mathbf{U}\mathbf{x}_3 = \mathbf{y}$ for $\mathbf{x}_3$, we get

$$\mathbf{x}_3 = [0.136319 \quad -0.202424 \quad 0.154964]^{\text{T}}.$$

Therefore,

$$\lambda \approx -0.202424,$$

$$\mathbf{x}_3 \leftarrow \frac{1}{-0.202424} [0.136319 \quad -0.202424 \quad 0.154964]^{\text{T}}$$

$$= [-0.673434 \quad 1 \quad -0.765542]^{\text{T}}.$$

Continuing in this manner, we generate Table 6.2, where all entries are rounded to four decimal places. The algorithm is converging to the eigenvalue $-1/3$ for $\mathbf{A}^{-1}$ and its reciprocal -3 for $\mathbf{A}$. The vector approximations are converging to $[-0.2 \ 1 {-} 0.6]^{\text{T}}$, which is an eigenvector for both $\mathbf{A}^{-1}$ and $\mathbf{A}$. ■

Table 6.2

Iteration	Eigenvector components			Eigenvalue
0	1.0000	1.0000	1.0000	
1	1.0000	0.2258	0.8710	0.1342
2	0.5494	1.0000	−0.1298	0.1711
3	−0.6734	1.0000	−0.7655	−0.2024
4	−0.0404	1.0000	−0.5782	−0.3921
5	−0.2677	1.0000	−0.5988	−0.3197
6	−0.1723	1.0000	−0.6035	−0.3372
7	−0.2116	1.0000	−0.5977	−0.3323
8	−0.1951	1.0000	−0.6012	−0.3336
9	−0.2021	1.0000	−0.5994	−0.3333
10	−0.1991	1.0000	−0.6003	−0.3334
11	−0.2004	1.0000	−0.5999	−0.3333
12	−0.1998	1.0000	−0.6001	−0.3333

We can use Property 7 and Observation 4 of Section 6.4 in conjunction with the inverse power method to develop a procedure for finding all eigenvalues and a set of corresponding eigenvectors for a matrix, providing that the eigenvalues are real and distinct, and estimates of their locations are known. The algorithm is known as the *shifted inverse power method.*

If c is an estimate for an eigenvalue of $\mathbf{A}$, then $\mathbf{A} - c\mathbf{I}$ will have an eigenvalue near zero and its reciprocal will be the dominant eigenvalue of $(\mathbf{A} - c\mathbf{I})^{-1}$. We use the inverse power method with an **LU** decomposition of $\mathbf{A} - c\mathbf{I}$ to calculate the dominant eigenvalue λ and its corresponding eigenvector $\mathbf{x}$ for $(\mathbf{A} - c\mathbf{I})^{-1}$. Then $1/\lambda$ and $\mathbf{x}$ are an eigenvalue and eigenvector for $\mathbf{A} - c\mathbf{I}$, while $1/\lambda + c$ and $\mathbf{x}$ are an eigenvalue and eigenvector for $\mathbf{A}$.

Example 5.   Find a second eigenvalue for the matrix given in Example 4.

Solution Since we do not have an estimate for any of the eigenvalues, we arbitrarily choose $c = 15$. Then

$$\mathbf{A} = c\mathbf{I} = \begin{bmatrix} -8 & 2 & 0 \\ 2 & -14 & 6 \\ 0 & 6 & -8 \end{bmatrix},$$

which has an **LU** decomposition with

$$\mathbf{L} = \begin{bmatrix} 1 & 0 & 0 \\ 0.25 & 1 & 0 \\ 0 & -0.444444 & 1 \end{bmatrix} \quad \text{and} \quad \mathbf{U} = \begin{bmatrix} -8 & 2 & 0 \\ 0 & -13.5 & 6 \\ 0 & 0 & -5.333333 \end{bmatrix}.$$

Applying the inverse power method to $\mathbf{A} - 15\mathbf{I}$, we generate Table 6.3, which is converging to $\lambda = -0.25$ and $\mathbf{x} = \begin{bmatrix} \frac{1}{3} & \frac{2}{3} & 1 \end{bmatrix}^{\mathrm{T}}$. The corresponding eigenvalue of $\mathbf{A}$ is $1/- 0.25 + 15 = 11$, with the same eigenvector.

Using the results of Examples 4 and 5, we have two eigenvalues, $\lambda_1 = -3$ and $\lambda_2 = 11$, of the 3×3 matrix defined in Example 4. Since the trace of a matrix equals the sum of the eigenvalues (Property 1 of Section 6.4), we know $7 + 1 + 7 = -3 + 11 + \lambda_3$, so the last eigenvalue is $\lambda_3 = 7$. ∎

Table 6.3

Iteration	Eigenvector components			Eigenvalue
0	1.0000	1.0000	1.0000	
1	0.6190	0.7619	1.0000	−0.2917
2	0.4687	0.7018	1.0000	−0.2639
3	0.3995	0.6816	1.0000	−0.2557
4	0.3661	0.6736	1.0000	−0.2526
5	0.3496	0.6700	1.0000	−0.2513
6	0.3415	0.6683	1.0000	−0.2506
7	0.3374	0.6675	1.0000	−0.2503
8	0.3354	0.6671	1.0000	−0.2502
9	0.3343	0.6669	1.0000	−0.2501
10	0.3338	0.6668	1.0000	−0.2500
11	0.3336	0.6667	1.0000	−0.2500

Problems 6.6

In Problems 1 through 10, use the power method to locate the dominant eigenvalue and a corresponding eigenvector for the given matrices. Stop after five iterations.

1. $\begin{bmatrix} 2 & 1 \\ 2 & 3 \end{bmatrix}$,

2. $\begin{bmatrix} 2 & 3 \\ 4 & 6 \end{bmatrix}$,

3. $\begin{bmatrix} 3 & 6 \\ 9 & 6 \end{bmatrix}$,

4. $\begin{bmatrix} 0 & 1 \\ -4 & 6 \end{bmatrix}$,

5. $\begin{bmatrix} 8 & 2 \\ 3 & 3 \end{bmatrix}$,

6. $\begin{bmatrix} 8 & 3 \\ 3 & 2 \end{bmatrix}$,

7. $\begin{bmatrix} 3 & 0 & 0 \\ 2 & 6 & 4 \\ 2 & 3 & 5 \end{bmatrix}$,

8. $\begin{bmatrix} 7 & 2 & 0 \\ 2 & 1 & 6 \\ 0 & 6 & 7 \end{bmatrix}$,

9. $\begin{bmatrix} 3 & 2 & 3 \\ 2 & 6 & 6 \\ 3 & 6 & 11 \end{bmatrix}$,

10. $\begin{bmatrix} 2 & -17 & 7 \\ -17 & -4 & 1 \\ 7 & 1 & -14 \end{bmatrix}$.

11. Use the power method on

$$A = \begin{bmatrix} 2 & 0 & -1 \\ 2 & 2 & 2 \\ -1 & 0 & 2 \end{bmatrix},$$

and explain why it does not converge to the dominant eigenvalue $\lambda = 3$.

12. Use the power method on

$$\mathbf{A} = \begin{bmatrix} 3 & 5 \\ 5 & -3 \end{bmatrix},$$

and explain why it does not converge.

13. Shifting can also be used with the power method to locate the next most dominant eigenvalue, if it is real and distinct, once the dominant eigenvalue has been determined. Construct $\mathbf{A} - \lambda\mathbf{I}$, where λ is the dominant eigenvalue of $\mathbf{A}$, and apply the power method to the shifted matrix. If the algorithm converges to μ and $\mathbf{x}$, then $\mu + \lambda$ is an eigenvalue of $\mathbf{A}$ with the corresponding eigenvector $\mathbf{x}$. Apply this shifted power method algorithm to the matrix in Problem 1. Use the results of Problem 1 to determine the appropriate shift.

14. Use the shifted power method as described in Problem 13 to the matrix in Problem 9. Use the results of Problem 9 to determine the appropriate shift.

15. Use the inverse power method on the matrix defined in Example 1. Stop after five iterations.

16. Use the inverse power method on the matrix defined in Problem 3. Take $\mathbf{x}_0 = [1 \ -0.5]^T$ and stop after five iterations.

17. Use the inverse power method on the matrix defined in Problem 5. Stop after five iterations.

18. Use the inverse power method on the matrix defined in Problem 6. Stop after five iterations.

19. Use the inverse power method on the matrix defined in Problem 9. Stop after five iterations.

20. Use the inverse power method on the matrix defined in Problem 10. Stop after five iterations.

21. Use the inverse power method on the matrix defined in Problem 11. Stop after five iterations.

22. Use the inverse power method on the matrix defined in Problem 4. Explain the difficulty and suggest a way to avoid it.

23. Use the inverse power method on the matrix defined in Problem 2. Explain the difficulty and suggest a way to avoid it.

24. Can the power method converge to a dominant eigenvalue if that eigenvalue is not distinct?

25. Apply the shifted inverse power method to the matrix defined in Problem 9, with a shift constant of 10.

26. Apply the shifted inverse power method to the matrix defined in Problem 10, with a shift constant of -25.

Matrix calculus

Chapter outline

7.1 Well-defined functions

The student should be aware of the vast importance of polynomials and exponentials to calculus and differential equations. One should not be surprised to find, therefore, that polynomials and exponentials of matrices play an equally important role in matrix calculus and matrix differential equations. Since we will be interested in using matrices to solve linear differential equations, we shall devote this entire chapter to defining matrix functions, specifically polynomials and exponentials, developing techniques for calculating these functions, and discussing some of their important properties.

Let $p_k(x)$ denote an arbitrary polynomial in x of degree k,

$$p_k(x) = a_k x^k + a_{k-1} x^{k-1} + \cdots + a_1 x + a_0, \tag{1}$$

where the coefficients $a_k, a_{k-1}, \ldots, a_1, a_0$ are real numbers. We then define

$$p_k(\mathbf{A}) = a_k \mathbf{A}^k + a_{k-1} \mathbf{A}^{k-1} + \cdots + a_1 \mathbf{A} + a_0 \mathbf{I}. \tag{2}$$

Recall from Chapter 1 that $\mathbf{A}^2 = \mathbf{A} \cdot \mathbf{A}$, $\mathbf{A}^3 = \mathbf{A}^2 \cdot \mathbf{A} = \mathbf{A} \cdot \mathbf{A} \cdot \mathbf{A}$ and, in general, $\mathbf{A}^k = \mathbf{A}^{k-1} \cdot \mathbf{A}$. Also $\mathbf{A}^0 = \mathbf{I}$.

Two observations are now immediate. Whereas a_0 in (1) is actually multiplied by $x^0 = 1$, a_0 in (2) is multiplied by $\mathbf{A}^0 = \mathbf{I}$. Also, if $\mathbf{A}$ is an $n \times n$ matrix, then $p_k(\mathbf{A})$ is an $n \times n$ matrix since the right-hand side of (2) may be summed.

Example 1. Find $p_2(\mathbf{A})$ for

$$\mathbf{A} = \begin{bmatrix} 0 & 1 & 0 \\ 0 & 0 & 1 \\ 0 & 0 & 0 \end{bmatrix}$$

if $p_2(x) = 2x^2 + 3x + 4$.

Solution In this case, $p_2(\mathbf{A}) = 2\mathbf{A}^2 + 3\mathbf{A} + 4\mathbf{I}$. Thus,

$$p_2(\mathbf{A}) = 2\begin{bmatrix} 0 & 1 & 0 \\ 0 & 0 & 1 \\ 0 & 0 & 0 \end{bmatrix}^2 + 3\begin{bmatrix} 0 & 1 & 0 \\ 0 & 0 & 1 \\ 0 & 0 & 0 \end{bmatrix} + \begin{bmatrix} 1 & 0 & 0 \\ 0 & 1 & 0 \\ 0 & 0 & 1 \end{bmatrix}$$

$$= 2\begin{bmatrix} 0 & 0 & 1 \\ 0 & 0 & 0 \\ 0 & 0 & 0 \end{bmatrix} + 3\begin{bmatrix} 0 & 1 & 0 \\ 0 & 0 & 1 \\ 0 & 0 & 0 \end{bmatrix} + 4\begin{bmatrix} 1 & 0 & 0 \\ 0 & 1 & 0 \\ 0 & 0 & 1 \end{bmatrix} = \begin{bmatrix} 4 & 3 & 2 \\ 0 & 4 & 3 \\ 0 & 0 & 4 \end{bmatrix}.$$

Note that had we defined $p_2(\mathbf{A}) = 2\mathbf{A}^2 + 3\mathbf{A} + 4$ (that is, without the $\mathbf{I}$ term), we could not have performed the addition since addition of a matrix and a scalar is undefined. ∎

Since a matrix commutes with itself, many of the properties of polynomials (addition, subtraction, multiplication, and factoring but *not* division) are still valid for polynomials of a matrix. For instance, if $f(x)$, $d(x)$, $q(x)$, and $r(x)$ represent polynomials in x and if

$$f(x) = d(x)q(x) + r(x) \tag{3}$$

then it must be the case that

$$f(\mathbf{A}) = d(\mathbf{A})q(\mathbf{A}) + r(\mathbf{A}). \tag{4}$$

Eq. (4) follows from (3) only because $\mathbf{A}$ commutes with itself; thus, we multiply together two polynomials in $\mathbf{A}$ precisely in the same manner that we multiply together two polynomials in x.

If we recall from calculus that many functions can be written as a Maclaurin series, then we can define functions of matrices quite easily. For instance, the Maclaurin series for e^x is

$$e^x = \sum_{k=0}^{\infty} \frac{x^k}{k!} = 1 + \frac{x}{1!} + \frac{x^2}{2!} + \frac{x^3}{3!} + \cdots. \tag{5}$$

Thus, we define the exponential of a matrix $\mathbf{A}$ as

$$e^{\mathbf{A}} = \sum_{k=0}^{\infty} \frac{\mathbf{A}^k}{k!} = \mathbf{I} + \frac{\mathbf{A}}{1!} + \frac{\mathbf{A}^2}{2!} + \frac{\mathbf{A}^3}{3!} + \cdots. \tag{6}$$

The question of convergence now arises. For an infinite series of matrices, we define convergence as follows:

Definition 1. A sequence $\{\mathbf{B}_k\}$ of matrices $\mathbf{B}_k = \left[b_{ij}^k \right]$ is said to *converge* to a matrix $\mathbf{B} = [b_{ij}]$, if the elements b_{ij}^k converge to b_{ij} for every i and j.

Definition 2. The infinite series $\sum_{n=0}^{\infty} \mathbf{B}_n$ converges to $\mathbf{B}$, if the sequence $\{\mathbf{S}_k\}$ of partial sums, where $\mathbf{S}_k = \sum_{n=0}^{k} \mathbf{B}_n$, converges to $\mathbf{B}$.

It can be shown (see Theorem 1, this section) that the infinite series given in (6) converges for any matrix $\mathbf{A}$. Thus, $e^{\mathbf{A}}$ is defined for every matrix.

Example 2. Find $e^{\mathbf{A}}$ if

$$\mathbf{A} = \begin{bmatrix} 2 & 0 \\ 0 & 0 \end{bmatrix}.$$

Solution

$$e^{\mathbf{A}} = e^{\begin{bmatrix} 2 & 0 \\ 0 & 0 \end{bmatrix}} = \begin{bmatrix} 1 & 0 \\ 0 & 1 \end{bmatrix} + \frac{1}{1!}\begin{bmatrix} 2 & 0 \\ 0 & 2 \end{bmatrix} + \frac{1}{2!}\begin{bmatrix} 2 & 0 \\ 0 & 0 \end{bmatrix}^2 + \frac{1}{3!}\begin{bmatrix} 2 & 0 \\ 0 & 0 \end{bmatrix}^3 + \cdots$$

$$= \begin{bmatrix} 1 & 0 \\ 0 & 1 \end{bmatrix} + \begin{bmatrix} 2/1! & 0 \\ 0 & 0 \end{bmatrix} + \begin{bmatrix} 2^2/2! & 0 \\ 0 & 0 \end{bmatrix} + \begin{bmatrix} 2^3/3! & 0 \\ 0 & 0 \end{bmatrix} + \cdots$$

$$= \begin{bmatrix} \sum_{k=0}^{\infty} 2^k/k! & 0 \\ 0 & 1 \end{bmatrix} = \begin{bmatrix} e^2 0 & 0 \\ 0 & e^0 \end{bmatrix}. \quad \blacksquare$$

In general, if $\mathbf{A}$ is the diagonal matrix

$$\mathbf{A} = \begin{bmatrix} \lambda_1 & 0 & \cdots & 0 \\ 0 & \lambda_2 & \cdots & 0 \\ \vdots & \vdots & & \vdots \\ 0 & 0 & \cdots & \lambda_n \end{bmatrix},$$

then we can show (see Problem 12) that

$$e^{\mathbf{A}} = \begin{bmatrix} e^{\lambda_1} & 0 & \cdots & 0 \\ 0 & e^{\lambda_2} & \cdots & 0 \\ \vdots & \vdots & & \vdots \\ 0 & 0 & \cdots & e^{\lambda_n} \end{bmatrix}. \tag{7}$$

If **A** is not a diagonal matrix, then it is very difficult to find $e^{\mathbf{A}}$ directly from the definition given in (6). For an arbitrary **A**, $e^{\mathbf{A}}$ does not have the form exhibited in (7). For example, if

$$\mathbf{A} = \begin{bmatrix} 1 & 2 \\ 4 & 3 \end{bmatrix},$$

it can be shown (however, not yet by us) that

$$e^{\mathbf{A}} = \frac{1}{6}\begin{bmatrix} 2e^5 + 4e^{-1} & 2e^5 - 2e^{-1} \\ 4e^5 - 4e^{-1} & 4e^5 + 2e^{-1} \end{bmatrix}.$$

For the purposes of this book, the exponential is the only function that is needed. However, it may be of some value to know how other functions of matrices, sines, cosines, etc., are defined. The following theorem, the proof of which is beyond the scope of this book, provides this information.

Theorem 1. *Let z represent the complex variable $x + iy$. If $f(z)$ has the Taylor series $\sum_{k=0}^{\infty} a_k z^k$, which converges for $|z| < R$, and if the eigenvalues $\lambda_1, \lambda_2, \ldots, \lambda_n$ of an $n \times n$ matrix **A** have the property that $|\lambda_i| < R (i = 1, 2, \ldots, n)$, then $\sum_{k=0}^{\infty} a_k \mathbf{A}^k$ will converge to an $n \times n$ matrix, which is defined to be $f(\mathbf{A})$. In such a case, $f(\mathbf{A})$ is said to be well defined.*

Example 3. Define sin **A**.
Solution A Taylor series for $\sin z$ is

$$\sin z = \sum_{k=0}^{\infty} \frac{(-1)^k z^{2k+1}}{(2k+1)!}$$

$$= z - \frac{z^3}{3!} + \frac{z^5}{5!} - \frac{z^7}{7!} + \cdots$$

This series can be shown to converge for all z (that is, $R = \infty$). Hence, since any eigenvalue λ of **A** must have the property $|\lambda| < \infty$ (that is, λ is finite), sin **A** can be defined for every **A** as

$$\sin \mathbf{A} = \sum_{k=0}^{\infty} \frac{(-1)^k \mathbf{A}^{2k+1}}{(2k+1)!} = \mathbf{A} - \frac{\mathbf{A}^3}{3!} + \frac{\mathbf{A}^5}{5!} - \frac{\mathbf{A}^7}{7!} + \cdots . \blacksquare \tag{8}$$

Problems 7.1

1. Let $q(x) = x - 1$. Find $p_k(\mathbf{A})$ and $q(\mathbf{A})p_k(\mathbf{A})$ if

(a) $\mathbf{A} = \begin{bmatrix} 1 & 2 & 3 \\ 0 & -1 & 4 \\ 0 & 0 & 1 \end{bmatrix}$, $k = 2$, and $p_2(x) = x^2 - 2x + 1$,

(b) $\mathbf{A} = \begin{bmatrix} 1 & 2 \\ 3 & 4 \end{bmatrix}$, $k = 3$, and $p_3(x) = 2x^3 - 3x^2 + 4$.

2. If $p_k(x)$ is defined by (1), find $p_k(\mathbf{A})$ for the diagonal matrix

$$\mathbf{A} = \begin{bmatrix} \lambda_1 & 0 & 0 \\ 0 & \lambda_2 & 0 \\ 0 & 0 & \lambda_3 \end{bmatrix}. \text{ Can you generalize?}$$

3. By actually computing both sides of the following equation separately, verify that $(\mathbf{A} - 3\mathbf{I})(\mathbf{A} + 2\mathbf{I}) = \mathbf{A}^2 - \mathbf{A} - 6\mathbf{I}$ for

(a) $\mathbf{A} = \begin{bmatrix} 1 & 2 \\ 3 & 4 \end{bmatrix}$,

(b) $\mathbf{A} = \begin{bmatrix} 1 & 0 & -2 \\ 3 & 1 & 1 \\ -2 & -2 & 3 \end{bmatrix}$.

The above equation is an example of matrix factoring.

4. Although $x^2 - y^2 = (x - y)(x + y)$ whenever x and y denote real-valued variables, show by example that $\mathbf{A}^2 - \mathbf{B}^2$ need not equal the product $(\mathbf{A} - \mathbf{B})(\mathbf{A} + \mathbf{B})$ whenever $\mathbf{A}$ and $\mathbf{B}$ denote 2×2 real matrices. Why?

5. It is known that $x^2 - 5x + 6$ factors into the product $(x - 2)(x - 3)$ whenever x denotes a real-valued variable. Is it necessarily true that $\mathbf{A}^2 - 5\mathbf{A} + 6\mathbf{I} = (\mathbf{A} - 2\mathbf{I})(\mathbf{A} - 3\mathbf{I})$ whenever $\mathbf{A}$ represents a square real matrix? Why?

6. Determine $\lim_{\to \infty} \mathbf{B}_k$ when

$$\mathbf{B}_k = \begin{bmatrix} \dfrac{1}{k} & 2 - \dfrac{2}{k^2} \\ 3 & (0.5)^k \end{bmatrix}.$$

7. Determine $\lim_{k\to\infty} \mathbf{B}_k$ when

$$\mathbf{B}_k = \begin{bmatrix} \dfrac{2k}{k+1} \\[2ex] \dfrac{k+3}{k^2-2k+1} \\[2ex] \dfrac{3k^2+2k}{2k^2} \end{bmatrix}.$$

8. Determine $\lim_{k\to\infty} \mathbf{D}_k$ when

$$\mathbf{D}_k = \begin{bmatrix} (0.2)^k & 1 & (0.1)^k \\ 4 & 3^k & 0 \end{bmatrix}.$$

9. It is known that $\arctan(z) = \sum_{n=0}^{\infty}[(-1)^n/(2n+1)]z^{2n+1}$ converges for all $|z| < \pi/2$. Determine for which of the following matrices $\mathbf{A}$, $\arctan(\mathbf{A}) = \sum_{n=0}^{\infty}[(-1)^n/2n+1]\mathbf{A}^{2n+1}$ is well defined:

(a) $\begin{bmatrix} -3 & 6 \\ -2 & 4 \end{bmatrix}$.

(b) $\begin{bmatrix} 5 & -4 \\ 6 & -5 \end{bmatrix}$.

(c) $\begin{bmatrix} 6 & -5 \\ 2 & -1 \end{bmatrix}$.

(d) $\begin{bmatrix} 0 & 1 & 0 \\ 0 & 0 & -1 \\ 0 & 1 & 0 \end{bmatrix}$.

(e) $\begin{bmatrix} 1 & 2 & 1 \\ 0 & 3 & 5 \\ 0 & -1 & -3 \end{bmatrix}$.

(f) $\begin{bmatrix} 0 & 1 & 0 \\ 0 & 0 & 1 \\ 0 & -\dfrac{1}{8} & \dfrac{3}{4} \end{bmatrix}$.

10. It is known that $\ln(1 + z) = \sum_{n=0}^{\infty}\left[(-1)^{n+1}/n\right]z^n$ converges for all $|z| < 1$. Determine for which of the matrices given in Problem 9 $\ln(\mathbf{I} + \mathbf{A}) = \sum_{n=0}^{\infty}\left[(-1)^{n+1/n}\right]\mathbf{A}^n$ is well defined.

11. It is known that $f(z) = \sum_{n=0}^{\infty}z^n/3^n$ converges for all $|z| < 3$. Determine for which of the matrices given in Problem 9 $f(\mathbf{A}) = \sum_{n=0}^{\infty}\mathbf{A}^n/3^n$ is well defined.

12. Derive Eq. (7).

13. Find $e^{\mathbf{A}}$ when

$$\mathbf{A} = \begin{bmatrix} 1 & 0 \\ 0 & 2 \end{bmatrix}.$$

14. Find $e^{\mathbf{A}}$ when

$$\mathbf{A} = \begin{bmatrix} -1 & 0 \\ 0 & 28 \end{bmatrix}.$$

15. Find $e^{\mathbf{A}}$ when

$$\mathbf{A} = \begin{bmatrix} 2 & 0 & 0 \\ 0 & -2 & 0 \\ 0 & 0 & 0 \end{bmatrix}.$$

16. Derive an expression for $\sin(\mathbf{A})$ similar to Eq. (7) when $\mathbf{A}$ is a square diagonal matrix.

17. Find $\sin(\mathbf{A})$ for the matrix given in Problem 13.

18. Find $\sin(\mathbf{A})$ for the matrix given in Problem 14.

19. Using Theorem 1, give a definition for $\cos \mathbf{A}$ and use this definition to find

$$\cos\begin{bmatrix} 1 & 0 \\ 0 & 2 \end{bmatrix}.$$

20. Find $\cos(\mathbf{A})$ for the matrix given in Problem 15.

7.2 Cayley–Hamilton theorem

We now state one of the most powerful theorems of matrix theory, the proof of which is given in the Final Comments at the end of this chapter.

Cayley–Hamilton theorem. *A matrix satisfies its own characteristic equation. That is, if the characteristic equation of an $n \times n$ matrix $\mathbf{A}$ is $\lambda^n + a_{n-1}\lambda^{n-1} + \ldots + a_1\lambda + a_0 = 0$, then*

$$\mathbf{A}^n + a_{n-1}\mathbf{A}^{n-1} + \cdots + a_1\mathbf{A} + a_0\mathbf{I} = 0.$$

Note once again that when we change a scalar equation to a matrix equation, the unity element 1 is replaced by the identity matrix $\mathbf{I}$.

Example 1. Verify the Cayley–Hamilton theorem for

$$\mathbf{A} = \begin{bmatrix} 1 & 2 \\ 4 & 3 \end{bmatrix}.$$

Solution The characteristic equation for $\mathbf{A}$ is $\lambda^2 - 4\lambda - 5 = 0$.

$$\mathbf{A}^2 - 4\mathbf{A} - 5\mathbf{I} = \begin{bmatrix} 1 & 2 \\ 4 & 3 \end{bmatrix}\begin{bmatrix} 1 & 2 \\ 4 & 3 \end{bmatrix} - 4\begin{bmatrix} 1 & 2 \\ 4 & 3 \end{bmatrix} - 5\begin{bmatrix} 1 & 0 \\ 0 & 1 \end{bmatrix}$$

$$= \begin{bmatrix} 9 & 8 \\ 16 & 47 \end{bmatrix} - \begin{bmatrix} 4 & 8 \\ 16 & 12 \end{bmatrix} - \begin{bmatrix} 5 & 0 \\ 0 & 5 \end{bmatrix}$$

$$= \begin{bmatrix} 9 - 4 - 5 & 8 - 8 - 0 \\ 16 - 16 - 0 & 17 - 12 - 5 \end{bmatrix} = \begin{bmatrix} 0 & 0 \\ 0 & 0 \end{bmatrix} = 0. \quad \blacksquare$$

Example 2. Verify the Cayley–Hamilton theorem for

$$\mathbf{A} = \begin{bmatrix} 3 & 0 & -1 \\ 2 & 0 & 1 \\ 0 & 0 & 4 \end{bmatrix}.$$

Solution The characteristic equation of $\mathbf{A}$ is $(3 - \lambda)(-\lambda)(4 - \lambda) = 0$.

$$(3\mathbf{I} - \mathbf{A})(-\mathbf{A})(4\mathbf{I} - \mathbf{A}) = \left(\begin{bmatrix} 3 & 0 & 0 \\ 0 & 3 & 0 \\ 0 & 0 & 3 \end{bmatrix} - \begin{bmatrix} 3 & 0 & -1 \\ 2 & 0 & 1 \\ 0 & 0 & 4 \end{bmatrix}\right)\left(-\begin{bmatrix} 3 & 0 & -1 \\ 2 & 0 & 1 \\ 0 & 0 & 4 \end{bmatrix}\right)$$

$$\left(\begin{bmatrix} 4 & 0 & 0 \\ 0 & 4 & 0 \\ 0 & 0 & 4 \end{bmatrix} - \begin{bmatrix} 3 & 0 & -1 \\ 2 & 0 & 1 \\ 0 & 0 & 4 \end{bmatrix}\right)$$

$$= \begin{bmatrix} 0 & 0 & 1 \\ -2 & 3 & -1 \\ 0 & 0 & -1 \end{bmatrix}\begin{bmatrix} -3 & 0 & 1 \\ -2 & 0 & -1 \\ 0 & 0 & -4 \end{bmatrix}\begin{bmatrix} 1 & 0 & 1 \\ -2 & 4 & -1 \\ 0 & 0 & 0 \end{bmatrix}$$

$$= \begin{bmatrix} 0 & 0 & 1 \\ -2 & 3 & -1 \\ 0 & 0 & -1 \end{bmatrix}\begin{bmatrix} -3 & 0 & -3 \\ -2 & 0 & -2 \\ 0 & 0 & 0 \end{bmatrix} = \begin{bmatrix} 0 & 0 & 0 \\ 0 & 0 & 0 \\ 0 & 0 & 0 \end{bmatrix} = 0. \quad \blacksquare$$

One immediate consequence of the Cayley–Hamilton theorem is a new method for finding the inverse of a nonsingular matrix. If

$$\lambda^n + a_{n-1}\lambda^{n-1} + \cdots + a_1\lambda + a_0 = 0$$

is the characteristic equation of a matrix $\mathbf{A}$, it follows from Problem 17 of Section 6.4 that $\det(\mathbf{A}) = (-1)^n a_0$. Thus, $\mathbf{A}$ is invertible if and only if $a_0 \neq 0$.

Now assume that $a_0 \neq 0$. By the Cayley–Hamilton theorem, we have

$$\mathbf{A}^n + a_{n-1}\mathbf{A}^{n-1} + \cdots + a_1\mathbf{A} + a_0\mathbf{I} = \mathbf{0},$$

$$\mathbf{A}\left[\mathbf{A}^{n-1} + a_{n-1}\mathbf{A}^{n-2} + \cdots + a_1\mathbf{I}\right] = \mathbf{0} - a_0\mathbf{I},$$

or

$$\mathbf{A}\left[-\frac{1}{a_0}\left(\mathbf{A}^{n-1} + a_{n-1}\mathbf{A}^{n-2} + \cdots + a_1\mathbf{I}\right)\right] = \mathbf{I}.$$

Thus, $(-1/a_0)\,(\mathbf{A}^{n-1} + a_{n-1}\mathbf{A}^{n-2} + \ldots + a_1\,\mathbf{I})$ is an inverse of $\mathbf{A}$. But since the inverse is unique (see Theorem 2 of Section 3.4), we have that

$$\mathbf{A}^{-1} = \frac{-1}{a_0}\left(\mathbf{A}^{n-1} + a_{n-1}\mathbf{A}^{n-2} + \cdots + a_1\mathbf{I}\right). \tag{9}$$

Example 3.   Using the Cayley–Hamilton theorem, find $\mathbf{A}^{-1}$ for

$$\mathbf{A} = \begin{bmatrix} 1 & -2 & 4 \\ 0 & -1 & 2 \\ 2 & 0 & 3 \end{bmatrix}.$$

Solution The characteristic equation for $\mathbf{A}$ is $\lambda^3 - 3\lambda^2 - 9\lambda + 3 = 0$. Thus, by the Cayley–Hamilton theorem,

$$\mathbf{A}^3 - 3\mathbf{A}^2 - 9\mathbf{A} + 3\mathbf{I} = \mathbf{0}.$$

Hence,

$$\mathbf{A}^3 - 3\mathbf{A}^2 - 9\mathbf{A} = -3\mathbf{I},$$
$$\mathbf{A}(\mathbf{A}^2 - 3\mathbf{A} - 9\mathbf{I}) = -3\mathbf{I},$$

or

$$\mathbf{A}\left(\frac{1}{3}\right)(-\mathbf{A}^2 + 3\mathbf{A} + 9\mathbf{I}) = \mathbf{I}.$$

Thus,

$$\mathbf{A}^{-1} = \left(\frac{1}{3}\right)(-\mathbf{A}^2 + 3\mathbf{A} + 9\mathbf{I})$$

$$= \frac{1}{3}\left(\begin{bmatrix} -9 & 0 & -12 \\ -4 & -1 & -4 \\ -8 & 4 & -17 \end{bmatrix} + \begin{bmatrix} 3 & -6 & 12 \\ 0 & -3 & 6 \\ 6 & 0 & 9 \end{bmatrix} + \begin{bmatrix} 9 & 0 & 0 \\ 0 & 9 & 0 \\ 0 & 0 & 9 \end{bmatrix}\right)$$

$$= \frac{1}{3}\begin{bmatrix} 3 & -6 & 0 \\ -4 & 5 & 2 \\ -2 & 4 & 1 \end{bmatrix}. \quad \blacksquare$$

Problems 7.2

Verify the Cayley–Hamilton theorem and use it to find $\mathbf{A}^{-1}$, where possible, for

1. $\mathbf{A} = \begin{bmatrix} 1 & 2 \\ 3 & 4 \end{bmatrix}$,

2. $\mathbf{A} = \begin{bmatrix} 1 & 2 \\ 2 & 4 \end{bmatrix}$,

3. $\mathbf{A} = \begin{bmatrix} 2 & 0 & 1 \\ 4 & 0 & 2 \\ 0 & 0 & -1 \end{bmatrix}$,

4. $\mathbf{A} = \begin{bmatrix} 1 & -1 & 2 \\ 0 & 3 & 2 \\ 2 & 1 & 2 \end{bmatrix}$,

5. $\mathbf{A} = \begin{bmatrix} 1 & 0 & 0 & 0 \\ 0 & -1 & 0 & 0 \\ 0 & 0 & -1 & 0 \\ 0 & 0 & 0 & 1 \end{bmatrix}$.

7.3 Polynomials of matrices—distinct eigenvalues

In general, it is very difficult to compute functions of matrices from their definition as infinite series (one exception is the diagonal matrix). The Cayley–Hamilton theorem, however, provides a starting point for the development of an alternate, straightforward method for calculating these functions. In this section, we shall develop the method for polynomials of matrices having distinct eigenvalues. In the ensuing sections, we shall extend the method to functions of matrices having arbitrary eigenvalues.

Let $\mathbf{A}$ represent an $n \times n$ matrix. Define $d(\lambda) = \det(\mathbf{A} - \lambda \mathbf{I})$. Thus, $d(\lambda)$ is an nth degree polynomial in λ and the characteristic equation of $\mathbf{A}$ is $d(\lambda) = 0$. From Chapter 6, we know that if λ_i is an eigenvalue of $\mathbf{A}$, then λ_i is a root of the characteristic equation, hence

$$d(\lambda_i) = 0. \tag{10}$$

From the Cayley–Hamilton theorem, we know that a matrix must satisfy its own characteristic equation, hence

$$d(\mathbf{A}) = \mathbf{0}. \tag{11}$$

Let $f(\mathbf{A})$ be any matrix polynomial of arbitrary degree that we wish to compute. $f(\lambda)$ represents the corresponding polynomial of λ. A theorem of algebra states that there exist polynomials $q(\lambda)$ and $r(\lambda)$ such that

$$f(\lambda) = d(\lambda)q(\lambda) + r(\lambda), \tag{12}$$

where $r(\lambda)$ is called the remainder. The degree of $r(\lambda)$ is less than that of $d(\lambda)$, which is n, and must be less than or equal to the degree of $f(\lambda)$ (why?).

Example 1.   Find $q(\lambda)$ and $r(\lambda)$ if $f(\lambda) = \lambda^4 + 2\lambda^3 - 1$ and $d(\lambda) = \lambda^2 - 1$.
Solution For $\lambda \neq \pm 1$, $d(\lambda) \neq 0$. Dividing $f(\lambda)$ by $d(\lambda)$, we obtain

$$\frac{f(\lambda)}{d(\lambda)} = \frac{\lambda^4 + 2\lambda^3 - 1}{\lambda^2 - 1} = (\lambda^2 + 2\lambda + 1) + \frac{2\lambda}{\lambda^2 - 1},$$

$$\frac{f(\lambda)}{d(\lambda)} = (\lambda^2 + 2\lambda + 1) + \frac{2\lambda}{d(\lambda)},$$

or

$$f(\lambda) = d(\lambda)\left(\lambda^2 + 2\lambda + 1\right)(2\lambda). \tag{13}$$

If we define $q(\lambda) = \lambda^2 + 2\lambda + 1$ and $r(\lambda) = 2\lambda$, (13) has the exact form of (12) for all λ except possibly $\lambda = \pm 1$. However, by direct substitution, we find that (13) is also valid for $\lambda = \pm 1$; hence, (13) is an identity for all (λ). ∎

From (12), (3), and (4), we have

$$f(\mathbf{A}) = d(\mathbf{A})q(\mathbf{A}) + r(\mathbf{A}). \tag{14}$$

Using (11), we obtain

$$f(\mathbf{A}) = r(\mathbf{A}). \tag{15}$$

Therefore, it follows that any polynomial in $\mathbf{A}$ may be written as a polynomial of degree $n - 1$ or less. For example, if $\mathbf{A}$ is a 4×4 matrix and if we wish to compute $f(\mathbf{A}) = \mathbf{A}^{957} - 3\mathbf{A}^{59} + 2\mathbf{A}^3 - 4\mathbf{I}$, then (15) implies that $f(\mathbf{A})$ can be written as a polynomial of degree three or less in $\mathbf{A}$, that is,

$$\mathbf{A}^{957} - 3\mathbf{A}^{59} + 2\mathbf{A}^3 - 4\mathbf{I} = \alpha_3 \mathbf{A}^3 + \alpha_2 \mathbf{A}^2 + \alpha_1 \mathbf{A} + \alpha_0 \mathbf{I}, \tag{16}$$

where $\alpha_3, \alpha_2, \alpha_1, \alpha_0$ are scalars that still must be determined. Once $\alpha_3, \alpha_2, \alpha_1, \alpha_0$ are computed, the student should observe that it is much easier to calculate the right side rather than the left side of (16).

If $\mathbf{A}$ is an $n \times n$ matrix, then $r(\lambda)$ will be a polynomial having the form

$$r(\lambda) = \alpha_{n-1}\lambda^{n-1} + \alpha_{n-2}\lambda^{n-2} + \cdots + \alpha_1\lambda + \alpha_0. \tag{17}$$

If λ_i is an eigenvalue of $\mathbf{A}$, then we have, after substituting (10) into (12), that

$$f(\lambda_i) = r(\lambda_i). \tag{18}$$

Thus, using (17), Eq. (18) may be rewritten as

$$f(\lambda_i) = \alpha_{n-1}(\lambda_i)^{n-1} + \alpha_{n-2}(\lambda_i)^{n-2} + \cdots + \alpha_1(\lambda_i) + \alpha_0 \tag{19}$$

if λ_i is an eigenvalue.

If we now assume that $\mathbf{A}$ has distinct eigenvalues, $\lambda_1, \lambda_2, \ldots, \lambda_n$ (note that if the eigenvalues are distinct, there must be n of them), then (19) may be used to generate n simultaneous linear equations for the n unknowns $\alpha_{n-1}, \alpha_{n-2}, \ldots, \alpha_1, \alpha_0$:

$$
\begin{aligned}
f(\lambda_1) &= r(\lambda_1) = \alpha_{n-1}(\lambda_1)^{n-1} + \alpha_{n-2}(\lambda_1)^{n-2} + \cdots + \alpha_1(\lambda_1) + \alpha_0, \\
f(\lambda_2) &= r(\lambda_2) = \alpha_{n-1}(\lambda_2)^{n-1} + \alpha_{n-2}(\lambda_2)^{n-2} + \cdots + \alpha_1(\lambda_2) + \alpha_0, \\
&\;\vdots \\
f(\lambda_n) &= r(\lambda_n) = \alpha_{n-1}(\lambda_n)^{n-1} + \alpha_{n-2}(\lambda_n)^{n-2} + \cdots + \alpha_1(\lambda_n) + \alpha_0.
\end{aligned}
\tag{20}
$$

Note that $f(\lambda)$ and the eigenvalues $\lambda_1, \lambda_2, \ldots, \lambda_n$ are assumed known; hence, $f(\lambda_1)$, $f(\lambda_2), \ldots, f(\lambda_n)$ are known and the only unknowns in (20) are $\alpha_{n-1}, \alpha_{n-2}, \ldots, \alpha_1, \alpha_0$.

Example 2.   Find $\mathbf{A}^{593}$ if

$$\mathbf{A} = \begin{bmatrix} -3 & -4 \\ 2 & 3 \end{bmatrix}.$$

Solution The eigenvalues of $\mathbf{A}$ are $\lambda_1 = 1$, $\lambda_2 = -1$. For this example, $f(\mathbf{A}) = \mathbf{A}^{593}$; thus, $f(\lambda) = \lambda^{593}$. Since $\mathbf{A}$ is a 2×2 matrix, $r(\mathbf{A})$ will be a polynomial of degree $(2 - 1) = 1$ or less; hence $r(\mathbf{A}) = \alpha_1 \mathbf{A} + \alpha_0 \mathbf{I}$ and $r(\lambda) = \alpha_1 \lambda + \alpha_0$. From (15), we have that $f(\mathbf{A}) = r(\mathbf{A})$, thus, for this example,

$$\mathbf{A}^{593} = \alpha_1 \mathbf{A} + \alpha_0 \mathbf{I}. \tag{21}$$

From (18), we have that $f(\lambda_i) = r(\lambda_i)$ if λ_i is an eigenvalue of $\mathbf{A}$; thus, for this example, $(\lambda_i)^{593} = \alpha_1 \lambda_1 + \alpha_0$. Substituting the eigenvalues of $\mathbf{A}$ into this equation, we obtain the following system for α_1 and α_0:

$$
\begin{aligned}
(1)^{593} &= \alpha_1(1) + \alpha_0, \\
(-1)^{593} &= \alpha_1(-1) + \alpha_0,
\end{aligned}
$$

or

$$
\begin{aligned}
1 &= \alpha_1 + \alpha_0, \\
-1 &= -\alpha_1 + \alpha_0.
\end{aligned}
\tag{22}
$$

Solving (22), we obtain $\alpha_0 = 0$, $\alpha_1 = 1$. Substituting these values into (21), we obtain $\mathbf{A}^{593} = 1$. $\mathbf{A} + 0. \mathbf{I}$ or

$$\begin{bmatrix} -3 & -4 \\ 2 & 3 \end{bmatrix}^{593} = \begin{bmatrix} -3 & -4 \\ 2 & 3 \end{bmatrix}. \blacksquare$$

Example 3.   Find $\mathbf{A}^{39}$ if

$$\mathbf{A} = \begin{bmatrix} 4 & 1 \\ 2 & 3 \end{bmatrix}.$$

Solution The eigenvalues of $\mathbf{A}$ are $\lambda_1 = 5$, $\lambda_2 = 2$. For this example, $f(\mathbf{A}) = \mathbf{A}^{39}$, thus $f(\lambda) = \lambda^{39}$. Since $\mathbf{A}$ is a 2×2 matrix, $r(\mathbf{A})$ will be a polynomial of degree 1 or less, hence $r(\mathbf{A}) = \alpha_1 \mathbf{A} + \alpha_0 \mathbf{I}$ and $r(\lambda) = \alpha_1 \lambda + \alpha_0$. From (15), we have that $f(\mathbf{A}) = r(\mathbf{A})$; thus, for this example,

$$\mathbf{A}^{39} = \alpha_1 \mathbf{A} + \alpha_0 \mathbf{I}. \tag{23}$$

From (18), we have that $f(\lambda_i) = r(\lambda_i)$, if λ_i is an eigenvalue of $\mathbf{A}$, thus for this example, $(\lambda_i)^{39} = \alpha_1 \lambda_i + \alpha_0$. Substituting the eigenvalues of $\mathbf{A}$ into this equation, we obtain the following system for α_1 and α_0:

$$
\begin{aligned}
5^{39} &= 5\alpha_1 + \alpha_0, \\
2^{39} &= 2\alpha_1 + \alpha_0.
\end{aligned}
\tag{24}
$$

Solving (24), we obtain

$$\alpha_1 = \frac{5^{39} - 2^{39}}{3}, \qquad \alpha_0 = \frac{-2(5)^{39} + 5(2)^{39}}{3}.$$

Substituting these values into (23), we obtain

$$
\begin{aligned}
\mathbf{A}^{39} &= \frac{5^{39} - 2^{39}}{3} \begin{bmatrix} 4 & 1 \\ 2 & 3 \end{bmatrix} + \frac{-2(5)^{39} + 5(2)^{39}}{3} \begin{bmatrix} 1 & 0 \\ 0 & 1 \end{bmatrix}, \\[2mm]
&= \frac{1}{2} \begin{bmatrix} 2(5)^{39} + 2^{39} & 5^{39} - 2^{39} \\ 2(5)^{39} - 2(2)^{39} & 5^{39} + 2(2)^{39} \end{bmatrix}.
\end{aligned}
\tag{25}
$$

The numbers 5^{39} and 2^{39} can be determined on a calculator. For our purposes, however, the form of (25) is sufficient and no further simplification is required. ∎

Example 4. Find $\mathbf{A}^{602} - 3\mathbf{A}^3$ if

$$\mathbf{A} = \begin{bmatrix} 1 & 4 & -2 \\ 0 & 0 & 0 \\ 0 & -3 & 3 \end{bmatrix}.$$

Solution The eigenvalues of $\mathbf{A}$ are $\lambda_1 = 0$, $\lambda_2 = 1$, $\lambda_3 = 3$.

$$
\begin{aligned}
f(\mathbf{A}) &= \mathbf{A}^{602} - 3\mathbf{A}^3, & r(\mathbf{A}) &= \alpha_2 \mathbf{A}^2 + \alpha_1 \mathbf{A} + \alpha_0 \mathbf{I}, \\
f(\lambda) &= \lambda^{602} - 3\lambda^3, & r(\lambda) &= \alpha_2 \lambda^2 + \alpha_1 \lambda + \alpha_0.
\end{aligned}
$$

Note that since $\mathbf{A}$ is a 3×3 matrix, $r(\mathbf{A})$ must be no more than a second degree polynomial. Now

$$f(\mathbf{A}) = r(\mathbf{A});$$

thus,

$$\mathbf{A}^{602} - 3\mathbf{A}^3 = \alpha_2 \mathbf{A}^2 + \alpha_1 \mathbf{A} + \alpha_0 \mathbf{I}. \tag{26}$$

If λ_i is an eigenvalue of $\mathbf{A}$, then $f(\lambda_i) = r(\lambda_i)$. Thus,

$$(\lambda_i)^{602} - 3(\lambda_i)^3 = \alpha_2 (\lambda_i)^2 + \alpha_1 \lambda_i + \alpha_0;$$

hence,

$$
\begin{aligned}
(0)^{602} - 3(0)^3 &= \alpha_2 (0)^2 + \alpha_1 (0) + \alpha_0, \\
(1)^{602} - 3(1)^3 &= \alpha_2 (1)^2 + \alpha_1 (1) + \alpha_0, \\
(3)^{602} - 3(3)^3 &= \alpha_2 (3)^2 + \alpha_1 (3) + \alpha_0,
\end{aligned}
$$

or

$$0 = \alpha_0,$$
$$-2 = \alpha_2 + \alpha_1 + \alpha_0,$$
$$3^{602} - 81 = 9\alpha_2 + 3\alpha_1 - \alpha_0.$$

Thus,

$$\alpha_2 = \frac{3^{602} - 75}{6}, \quad \alpha_1 = \frac{-(3)^{302} + 63}{6}, \quad \alpha_0 = 0. \qquad (27)$$

Substituting (27) into (26), we obtain

$$\mathbf{A}^{602} - 3\mathbf{A}^3 = \frac{3^{602} - 75}{6} \begin{bmatrix} 1 & 10 & -8 \\ 0 & 0 & 0 \\ 0 & -9 & 9 \end{bmatrix} + \frac{-(3)^{602} + 63}{6} \begin{bmatrix} 1 & 4 & -2 \\ 0 & 0 & 0 \\ 0 & -3 & 3 \end{bmatrix}$$

$$= \frac{1}{6} \begin{bmatrix} -12 & 6(3)^{602} - 498 & -6(3)^{602} + 474 \\ 0 & 0 & 0 \\ 0 & -6(3)^{602} + 486 & 6(3)^{602} - 486 \end{bmatrix}. \blacksquare$$

Finally, the student should note that if the polynomial to be calculated is already of a degree less than or equal to $n - 1$, then this method affords no simplification and the polynomial must still be computed directly.

Problems 7.3

1. Specialize system (20) for $f(\mathbf{A}) = \mathbf{A}^7$ and

$$\mathbf{A} = \begin{bmatrix} -2 & 3 \\ -1 & 2 \end{bmatrix}.$$

Solve this system and use the results to determine $\mathbf{A}^7$. Check your answer by direct calculations.

2. Find $\mathbf{A}^{50}$ for the matrix $\mathbf{A}$ given in Problem 1.

3. Specialize system (20) for $f(\mathbf{A}) = \mathbf{A}^{735}$ and

$$\mathbf{A} = \begin{bmatrix} 0 & 1 \\ 0 & -1 \end{bmatrix}.$$

Solve this system and use the results to determine $\mathbf{A}^{735}$. (What do you notice about $\mathbf{A}^3$?)

4. Specialize system (20) for $f(\mathbf{A}) = \mathbf{A}^{20}$ and

$$\mathbf{A} = \begin{bmatrix} -3 & 6 \\ -1 & 2 \end{bmatrix}.$$

Solve this system and use the results to determine $\mathbf{A}^{20}$.

5. Find $\mathbf{A}^{97}$ for the matrix $\mathbf{A}$ given in Problem 4.
6. Find $\mathbf{A}^{50}$ for the matrix $\mathbf{A}$ given in Example 3.
7. Specialize system (20) for $f(\mathbf{A}) = \mathbf{A}^{78}$ and

$$\mathbf{A} = \begin{bmatrix} 2 & -1 \\ 2 & 5 \end{bmatrix}.$$

Solve this system and use the results to determine $\mathbf{A}^{78}$.

8. Find $\mathbf{A}^{41}$ for the matrix $\mathbf{A}$ given in Problem 7.
9. Specialize system (20) for $f(\mathbf{A}) = \mathbf{A}^{222}$ and

$$\mathbf{A} = \begin{bmatrix} 1 & -1 & 2 \\ 0 & -1 & 2 \\ 0 & 0 & 2 \end{bmatrix}.$$

Solve this system and use the results to determine $\mathbf{A}^{222}$.

10. Specialize system (20) for $f(\mathbf{A}) = \mathbf{A}^{17}$, when $\mathbf{A}$ is a 3×3 matrix having 3, 5, and 10 as its eigenvalues.
11. Specialize system (20) for $f(\mathbf{A}) = \mathbf{A}^{25}$, when $\mathbf{A}$ is a 4×4 matrix having 2, -2, 3, and 4 as its eigenvalues.
12. Specialize system (20) for $f(\mathbf{A}) = \mathbf{A}^{25}$, when $\mathbf{A}$ is a 4×4 matrix having 1, -2, 3, and -4 as its eigenvalues.
13. Specialize system (20) for $f(\mathbf{A}) = \mathbf{A}^{8}$, when $\mathbf{A}$ is a 5×5 matrix having 1, -1, 2, -2, and 3 as its eigenvalues.
14. Specialize system (20) for $f(\mathbf{A}) = \mathbf{A}^{8} - 3\mathbf{A}^{5} + 5\mathbf{I}$, when $\mathbf{A}$ is the matrix described in Problem 10.
15. Specialize system (20) for $f(\mathbf{A}) = \mathbf{A}^{8} - 3\mathbf{A}^{5} + 5\mathbf{I}$, when $\mathbf{A}$ is the matrix described in Problem 11.
16. Specialize system (20) for $f(\mathbf{A}) = \mathbf{A}^{8} - 3\mathbf{A}^{5} + 5\mathbf{I}$, when $\mathbf{A}$ is the matrix described in Problem 12.
17. Specialize system (20) for $f(\mathbf{A}) = \mathbf{A}^{10} + 6\mathbf{A}^{3} + 8\mathbf{A}$, when $\mathbf{A}$ is the matrix described in Problem 12.
18. Specialize system (20) for $f(\mathbf{A}) = \mathbf{A}^{10} + 6\mathbf{A}^{3} + 8\mathbf{A}$, when $\mathbf{A}$ is the matrix described in Problem 13.

19. Find $\mathbf{A}^{202} - 3\mathbf{A}^{147} + 2\mathbf{I}$ for the $\mathbf{A}$ of Problem 1.

20. Find $\mathbf{A}^{1025} - 4\mathbf{A}^5$ for the $\mathbf{A}$ of Problem 1.

21. Find $\mathbf{A}^8 - 3\mathbf{A}^5 - \mathbf{I}$ for the matrix given in Problem 7.

22. Find $\mathbf{A}^{13} - 12\mathbf{A}^9 + 5\mathbf{I}$ for

$$\mathbf{A} = \begin{bmatrix} 3 & -5 \\ 1 & -3 \end{bmatrix}.$$

23. Find $\mathbf{A}^{10} - 2\mathbf{A}^5 + 10\mathbf{I}$ for the matrix given in Problem 22.

24. Find $\mathbf{A}^{593} - 2\mathbf{A}^{15}$ for

$$\mathbf{A} = \begin{bmatrix} -2 & 4 & 3 \\ 0 & 0 & 0 \\ -1 & 5 & 2 \end{bmatrix}.$$

25. Specialize system (20) for $f(\mathbf{A}) = \mathbf{A}^{12} - 3\mathbf{A}^9 + 2\mathbf{A} + 5\mathbf{I}$ and

$$\mathbf{A} = \begin{bmatrix} 0 & 1 & 0 \\ 0 & 0 & 1 \\ -4 & 4 & 1 \end{bmatrix}.$$

Solve this system, and use the results to determine $f(\mathbf{A})$.

26. Specialize system (20) for $f(\mathbf{A}) = \mathbf{A}^9 - 3\mathbf{A}^4 + 1$ and

$$\mathbf{A} = \begin{bmatrix} 0 & 1 & 0 \\ 0 & 0 & 1 \\ -\dfrac{1}{16} & \dfrac{1}{4} & \dfrac{1}{4} \end{bmatrix}.$$

Solve this system and use the results to determine $f(\mathbf{A})$.

7.4 Polynomials of matrices—general case

The only restriction in the previous section was that the eigenvalues of $\mathbf{A}$ had to be distinct. The following theorem suggests how to obtain n equations for the unknown α's in (15), even if some of the eigenvalues are identical.

Theorem 1. *Let $f(\lambda)$ and $r(\lambda)$ be defined as in Eq. (12). If λ_i is an eigenvalue of multiplicity k, then*

$$f(\lambda_1) = r(\lambda_i),$$

$$\frac{df(\lambda_i)}{d\lambda} = \frac{dr(\lambda_i)}{d\lambda},$$

$$\frac{d^2 f(\lambda_i)}{d\lambda^2} = \frac{d^2 r(\lambda_i)}{d\lambda^2}, \qquad (28)$$

$$\vdots$$

$$\frac{d^{k-1} f(\lambda_i)}{d\lambda^{k-1}} = \frac{d^{k-1} r(\lambda_i)}{d\lambda^{k-1}},$$

where the notation $d^n f(\lambda_i)/d\lambda^n$ denotes the nth derivative of $f(\lambda)$ with respect to λ evaluated at $\lambda = \lambda_i$.*

Thus, for example, if λ_i is an eigenvalue of multiplicity 3, Theorem 1 implies that $f(\lambda)$ and its first two derivatives evaluated at $\lambda = \lambda_i$ are equal, respectively, to $r(\lambda)$ and its first two derivatives also evaluated at $\lambda = \lambda_i$. If λ_i is an eigenvalue of multiplicity 5, then $f(\lambda)$ and the first four derivatives of $f(\lambda)$ evaluated at $\lambda = \lambda_i$ are equal respectively to $r(\lambda)$ and the first four derivatives of $r(\lambda)$ evaluated at $\lambda = \lambda_i$. Note, furthermore, that if λ_i is an eigenvalue of multiplicity 1, then Theorem 1 implies that $f(\lambda_i) = r(\lambda_i)$, which is Eq. (18).

Example 1. Find $\mathbf{A}^{24} - 3\mathbf{A}^{15}$ if

$$\mathbf{A} = \begin{bmatrix} 3 & 2 & 4 \\ 0 & 1 & 0 \\ -1 & -3 & -1 \end{bmatrix}.$$

Solution The eigenvalues of $\mathbf{A}$ are $\lambda_1 = \lambda_2 = \lambda_3 = 1$; hence, $\lambda = 1$ is an eigenvalue of multiplicity 3:

$f(\mathbf{A}) = \mathbf{A}^{24} - 3\mathbf{A}^{15}$	$r(\mathbf{A}) = \alpha_2 \mathbf{A}^2 + \alpha_1 \mathbf{A} + \alpha_0 \mathbf{I}$
$f(\lambda) = \lambda^{24} - 3\lambda^{15}$	$r(\lambda) = \alpha_2 \lambda^2 + \alpha_1 \lambda + \alpha_0$
$f'(\lambda) = 24\lambda^{23} - 45\lambda^{14}$	$r'(\lambda) = 2\alpha_2 \lambda + \alpha_1$
$f''(\lambda) = 552\lambda^{22} - 630\lambda^{13}$	$r''(\lambda) = 2\alpha_2.$

---

* Theorem 1 is proved by differentiating Eq. (12) $k - 1$ times and noting that if λ_i is an eigenvalue of multiplicity k, then

$$d(\lambda_i) = \frac{d[d(\lambda_i)]}{d\lambda} = \cdots = \frac{d^{(k-1)} d(\lambda_i)}{d\lambda^{k-1}} = 0.$$

Now $f(\mathbf{A}) = r(\mathbf{A})$, hence

$$\mathbf{A}^{24} - 3\mathbf{A}^{15} = \alpha_2 \mathbf{A}^2 + \alpha_1 \mathbf{A} + \alpha_0 \mathbf{I}. \tag{29}$$

Also, since $\lambda = 1$ is an eigenvalue of multiplicity 3, it follows from Theorem 1 that

$$f(1) = r(1),$$
$$f'(1) = r'(1),$$
$$f''(1) = r''(1).$$

Hence,

$$(1)^{24} - 3(1)^{15} = \alpha_2(1)^2 + \alpha_1(1) + \alpha_0,$$
$$24(1)^{23} - 45(1)^{14} = 2\alpha_2(1) + \alpha_1,$$
$$552(1)^{22} - 630(1)^{13} = 2\alpha_2,$$

or

$$-2 = \alpha_2 + \alpha_1 + \alpha_0,$$
$$-21 = 2\alpha_2 + \alpha_1,$$
$$-78 = 2\alpha_2.$$

Thus, $\alpha_2 = -39$, $\alpha_1 = 57$, $\alpha_0 = -20$, and from Eq. (29)

$$\mathbf{A}^{24} = 3\mathbf{A}^{15} = -39\mathbf{A}^2 + 57\mathbf{A} - 20\mathbf{I} = \begin{bmatrix} -44 & 270 & -84 \\ 0 & -2 & 0 \\ 21 & -93 & 40 \end{bmatrix}. \blacksquare$$

Example 2.   Set up the necessary equation to find $\mathbf{A}^{15} - 6\mathbf{A}^2$ if

$$\mathbf{A} = \begin{bmatrix} 1 & 4 & 3 & 2 & 1 & -7 \\ 0 & 0 & 2 & 11 & 1 & 0 \\ 0 & 0 & 1 & -1 & 0 & 1 \\ 0 & 0 & 0 & -1 & 2 & 1 \\ 0 & 0 & 0 & 0 & -1 & 17 \\ 0 & 0 & 0 & 0 & 0 & 1 \end{bmatrix}.$$

Solution The eigenvalues of $\mathbf{A}$ are $\lambda_1 = \lambda_2 = \lambda_3 = 1$, $\lambda_4 = \lambda_5 = -1$, $\lambda_6 = 0$.

$$f(\mathbf{A}) = \mathbf{A}^{15} - 6\mathbf{A}^2 \qquad r(\mathbf{A}) = \alpha_5 \mathbf{A}^5 + \alpha_4 \mathbf{A}^4 + \alpha_3 \mathbf{A}^3 + \alpha_2 \mathbf{A}^2 + \alpha_1 \mathbf{A} + \alpha_0 \mathbf{I}$$
$$f(\lambda) = \lambda^{15} - 6\lambda^2 \qquad r(\lambda) = \alpha_5 \lambda^5 + 4\alpha_4 \lambda^4 + 3\alpha_3 \lambda^3 + 2\alpha_2 \lambda^2 + \alpha_1 \lambda^1 + \alpha_0$$
$$f'(\lambda) = 15\lambda^{14}12\lambda \qquad r(\lambda) = \alpha_5 \lambda^4 + 4\alpha_4 \lambda^3 + 3\alpha_3 \lambda^2 + 2\alpha_2 \lambda + \alpha_1$$
$$f''(\lambda) = 210\lambda^{13} - 12 \qquad r''(\lambda) = 20\alpha_5 \lambda^3 + 12\alpha_4 \lambda^2 + 6\alpha_3 \lambda + 2\alpha_2.$$

Since $f(\mathbf{A}) = r(\mathbf{A})$,

$$\mathbf{A}^{15} - 6\mathbf{A}^2 = a_5 \mathbf{A}^5 + \alpha_4 \mathbf{A}^4 + \alpha_3 \mathbf{A}^3 + \alpha_2 \mathbf{A}^2 + \alpha_1 \mathbf{A} + \alpha_0 \mathbf{I}. \tag{30}$$

Since $\lambda = 1$ is an eigenvalue of multiplicity 3, $\lambda = -1$ is an eigenvalue of multiplicity 2, and $\lambda = 0$ is an eigenvalue of multiplicity 1, it follows from Theorem 1 that

$$f(1) = r(1),$$
$$f'(1) = r'(1),$$
$$f''(1) = r''(1),$$
$$f(-1) = r(-1),$$
$$f'(-1) = r'(-1),$$
$$f(0) = r(0).$$

(31)

Hence,

$$(1)^{15} - 6(1)^2 = \alpha_5(1)^5 + \alpha_4(1)^4 + \alpha_3(1)^3 + \alpha_2(1)^2 + \alpha_1(1) + \alpha_0$$
$$15(1)^{14} - 12(1) = 5\alpha_5(1)^4 + 4\alpha_4(1)^3 + 3\alpha_3(1)^2 + 2\alpha_2(1)^2 + \alpha_1$$
$$210(1)^{13} - 12 = 20\alpha_5(1)^3 + 12\alpha_4(1)^2 + 6\alpha_3(1) + 2\alpha_2$$
$$(-1)^{15} - 6(-1)^2 = \alpha_5(-1)^5 + \alpha_4(-1)^4 + \alpha_3(-1)^3 + \alpha_2(-1)^2 + \alpha_1(-1) + \alpha_0$$
$$15(-1)^{14} - 12(-1) = 5\alpha_5(-1)^4 + 4\alpha_4(-1)^3 + 3\alpha_3(-1)^2 + 2\alpha_2(-1) + \alpha_1$$
$$(0)^{15} - 12(0)^2 = \alpha_5(0)^5 + \alpha_4(0)^4 + \alpha_3(0)^3 + \alpha_2(0)^2 + \alpha_1(0) + \alpha_0$$

or

$$-5 = \alpha_5 + \alpha_4 + \alpha_3 + \alpha_2 + \alpha_1 + \alpha_0$$
$$3 = 5\alpha_5 + 4\alpha_4 + 3\alpha_3 + 2\alpha_2 + \alpha_1$$
$$198 = 20\alpha_5 + 12\alpha_4 + 6\alpha_3 + 2\alpha_2$$
$$-7 = -\alpha_5 + \alpha_4 - \alpha_3 + \alpha_2 - \alpha_1 + \alpha_0$$
$$27 = 5\alpha_5 - 4\alpha_4 + 3\alpha_3 - 2\alpha_2 + \alpha_1$$
$$0 = \alpha_0.$$

(32)

System (32) can now be solved uniquely for $\alpha_5, \alpha_4, \ldots, \alpha_0$; the results are then substituted into (30) to obtain $f(\mathbf{A})$. ∎

Problems 7.4

1. Using Theorem 1, establish the equations that are needed to find $\mathbf{A}^7$ if $\mathbf{A}$ is a 2×2 matrix having 2 and 2 as multiple eigenvalues.
2. Using Theorem 1, establish the equations that are needed to find $\mathbf{A}^7$ if $\mathbf{A}$ is a 3×3 matrix having 2 as an eigenvalue of multiplicity 3.
3. Redo Problem 2 if instead the eigenvalues are 2, 2, and 1.
4. Using Theorem 1, establish the equations that are needed to find $\mathbf{A}^{10}$ if $\mathbf{A}$ is a 2×2 matrix having 3 as an eigenvalue of multiplicity 2.
5. Redo Problem 4 if instead the matrix has order 3×3 with 3 as an eigenvalue of multiplicity 3.
6. Redo Problem 4 if instead the matrix has order 4×4 with 3 as an eigenvalue of multiplicity 4.

7. Using Theorem 1, establish the equations that are needed to find $\mathbf{A}^9$ if $\mathbf{A}$ is a 4×4 matrix having 2 as an eigenvalue of multiplicity 4.

8. Redo Problem 7 if instead the eigenvalues are 2, 2, 2, and 1.

9. Redo Problem 7 if instead the eigenvalues are 2 and 1, both with multiplicity 2.

10. Set up (but do not solve) the necessary equations to find $\mathbf{A}^{10} - 3\mathbf{A}^5$ if

$$\mathbf{A} = \begin{bmatrix} 5 & -2 & 1 & 1 & 5 & -7 \\ 0 & 5 & 2 & 1 & -1 & 1 \\ 0 & 0 & 5 & 0 & 1 & -3 \\ 0 & 0 & 0 & 2 & 1 & 2 \\ 0 & 0 & 0 & 0 & 2 & 0 \\ 0 & 0 & 0 & 0 & 0 & 5 \end{bmatrix}.$$

11. Find $\mathbf{A}^6$ in two different ways if

$$\mathbf{A} = \begin{bmatrix} 5 & 8 \\ -2 & -5 \end{bmatrix}.$$

(First find $\mathbf{A}^6$ using Theorem 1 and then by direct multiplication.)

12. Find $\mathbf{A}^{521}$ if

$$\mathbf{A} = \begin{bmatrix} 4 & 1 & -3 \\ 0 & -1 & 0 \\ 5 & 1 & 4 \end{bmatrix}.$$

13. Find $\mathbf{A}^{14} - 3\mathbf{A}^{13}$ if

$$\mathbf{A} = \begin{bmatrix} 4 & 1 & 2 \\ 0 & 0 & 0 \\ -8 & 1 & -4 \end{bmatrix}.$$

7.5 Functions of a matrix

Once the student understands how to compute polynomials of a matrix, computing exponentials and other functions of a matrix is easy because the methods developed in the previous two sections remain valid for more general functions.

Let $f(\lambda)$ represent a *function* of λ and suppose we wish to compute $f(\mathbf{A})$. It can be shown, for a large class of problems, that there exists a function $q(\lambda)$ and an $n - 1$ degree polynomial $r(\lambda)$ (we assume $\mathbf{A}$ is of order $n \times n$) such that

$$f(\lambda) = q(\lambda)d(\lambda) + r(\lambda), \tag{33}$$

where $d(\lambda) = \det(\mathbf{A} - \lambda \mathbf{I})$. Hence, it follows that

$$f(\mathbf{A}) = q(\mathbf{A})d(\mathbf{A}) + r(\mathbf{A}). \tag{34}$$

Since (33) and (34) are exactly Eqs. (12) and (14), where $f(\lambda)$ is now understood to be a general function and not restricted to polynomials, the analysis of Sections 7.3 and 7.4 can again be applied. It then follows that

(a) $f(\mathbf{A}) = r(\mathbf{A})$, and
(b) Theorem 1 of Section 7.4 remains valid.

Thus, the methods used to compute a polynomial of a matrix can be generalized and used to compute arbitrary functions of a matrix.

Example 1.   Find $e^{\mathbf{A}}$ if

$$\mathbf{A} = \begin{bmatrix} 1 & 2 \\ 4 & 3 \end{bmatrix}.$$

Solution The eigenvalues of $\mathbf{A}$ are $\lambda_1 = 5$, $\lambda_2 = -1$; thus,

$$f(\mathbf{A}) = e^{\mathbf{A}} \quad r(\mathbf{A}) = \alpha_1 \mathbf{A} + \alpha_0 \mathbf{I}$$
$$f(\lambda) = e^{\lambda} \quad r(\lambda) = \alpha_1 \lambda + \alpha_0.$$

Now $f(\mathbf{A}) = r(\mathbf{A})$; hence,

$$e^{\mathbf{A}} = \alpha_1 \mathbf{A} + \alpha_0 \mathbf{I}. \tag{35}$$

Also, since Theorem 1 of Section 7.4 is still valid,

$$f(5) = r(5),$$

and

$$f(-1) = r(-1);$$

hence,

$$e^5 = 5\alpha_1 + \alpha_0,$$
$$e^{-1} = -\alpha_1 + \alpha_0.$$

Thus,

$$\alpha_1 = \frac{e^5 - e^{-1}}{6} \quad \text{and} \quad \alpha_0 = \frac{e^5 + 5e^{-1}}{6}.$$

Substituting these values into (35), we obtain

$$e^{\mathbf{A}} = \frac{1}{6} \begin{bmatrix} 2e^5 + 4e^{-1} & 2e^5 - 2e^{-1} \\ 4e^5 - 4e^{-1} & 4e^5 + 2e^{-1} \end{bmatrix}. \blacksquare$$

Example 2.   Find $e^{\mathbf{A}}$ if

$$\mathbf{A} = \begin{bmatrix} 2 & 1 & 0 \\ 0 & 2 & 1 \\ 0 & 0 & 2 \end{bmatrix}.$$

Solution The eigenvalues of $\mathbf{A}$ are $\lambda_1 = \lambda_2 = \lambda_3 = 2$; thus,

$$f(\mathbf{A}) = e^{\mathbf{A}} \qquad\qquad r(\mathbf{A}) = \alpha_2 \mathbf{A}^2 + \alpha_1 \mathbf{A} + \alpha_0 \mathbf{I}$$
$$f(\lambda) = e^{\lambda} \qquad\qquad r(\lambda) = \alpha_2 \lambda^2 + \alpha_1 \lambda + \alpha_0$$
$$f'(\lambda) = e^{\lambda} \qquad\qquad r'(\lambda) = 2\alpha_2 \lambda + \alpha_1$$
$$f''(\lambda) = e^{\lambda} \qquad\qquad r''(\lambda) = 2\alpha_2.$$

Since $f(\mathbf{A}) = r(\mathbf{A})$,

$$e^{\mathbf{A}} = \alpha_2 \mathbf{A}^2 + \alpha_1 \mathbf{A} + \alpha_0 \mathbf{I}. \tag{36}$$

Since $\lambda = 2$ is an eigenvalue of multiplicity 3,

$$f(2) = r(2),$$
$$f'(2) = r'(2),$$
$$f''(2) = r''(2);$$

hence,

$$e^2 = 4\alpha_2 + 2\alpha_1 + \alpha_0,$$
$$e^2 = 4\alpha_2 + \alpha_1,$$
$$e^2 = 2\alpha_2,$$

or

$$\alpha_2 = \frac{e^2}{2}, \quad \alpha_1 = -e^2, \quad \alpha_0 = e^2.$$

Substituting these values into (36), we obtain

$$e^{\mathbf{A}} = \frac{e^2}{2} \begin{bmatrix} 4 & 4 & 1 \\ 0 & 4 & 4 \\ 0 & 0 & 4 \end{bmatrix} - e^2 \begin{bmatrix} 2 & 1 & 0 \\ 0 & 2 & 1 \\ 0 & 0 & 2 \end{bmatrix} + e^2 \begin{bmatrix} 1 & 0 & 0 \\ 0 & 1 & 0 \\ 0 & 0 & 1 \end{bmatrix} = \begin{bmatrix} e^2 & e^2 & e^2/2 \\ 0 & e^2 & e^2 \\ 0 & 0 & e^2 \end{bmatrix}. \blacksquare$$

Example 3. Find $\sin \mathbf{A}$ if

$$\mathbf{A} = \begin{bmatrix} \pi & 1 & 0 \\ 0 & \pi & 0 \\ 4 & 1 & \pi/2 \end{bmatrix}.$$

Solution The eigenvalues of $\mathbf{A}$ are $\lambda_1 = \pi/2$, $\lambda_2 = \lambda_3 = \pi$; thus,

$$f(\mathbf{A}) = \sin \mathbf{A} \qquad r(\mathbf{A}) = \alpha_2 \mathbf{A}^2 + \alpha_1 \mathbf{A} + \alpha_0 \mathbf{I}$$
$$f(\lambda) = \sin \lambda \qquad r(\lambda) = \alpha_2 \lambda^2 + \alpha_1 \lambda + \alpha_0$$
$$f'(\lambda) = \cos \lambda \qquad r'(\lambda) = 2\alpha_2 \lambda + \alpha_1.$$

But $f(\mathbf{A}) = r(\mathbf{A})$, hence

$$\sin \mathbf{A} = \alpha_2 \mathbf{A}^2 + \alpha_1 \mathbf{A} + \alpha_0 \mathbf{I}. \tag{37}$$

Since $\lambda = \pi/2$ is an eigenvalue of multiplicity 1 and $\lambda = \pi$ is an eigenvalue of multiplicity 2, it follows that

$$f(\pi/2) = r(\pi/2),$$
$$f(\pi) = r(\pi),$$
$$f'(\pi) = r'(\pi);$$

hence,

$$\sin \pi/2 = \alpha_2(\pi/2)^2 + \alpha_1(\pi/2) + \alpha_0,$$
$$\sin \pi = \alpha_2(\pi)^2 + \alpha_1(\pi) + \alpha_0,$$
$$\cos \pi = 2\alpha_2\pi + \alpha_1,$$

or simplifying

$$4 = \alpha_2\pi^2 + 2\alpha_1\pi + 4\alpha_0,$$
$$0 = \alpha_2\pi^2 + \alpha_1\pi + \alpha_0,$$
$$-1 = 2\alpha_2\pi + \alpha_1.$$

Thus, $\alpha_2 = (1/\pi^2)\,(4 - 2\pi)$, $\alpha = (1/\pi^2)\,(-8\pi + 3\pi^2)$, $\alpha_0 = (1/\pi^2)\,(4\pi^2 - \pi^3)$. Substituting these values into (37), we obtain

$$\sin \mathbf{A} = 1/\pi^2 \begin{bmatrix} 0 & -\pi^2 & 0 \\ 0 & 0 & 0 \\ -8\pi & 16 - 10\pi & \pi^2 \end{bmatrix}. \blacksquare$$

In closing, we point out that although exponentials of any square matrix can always be computed by the above methods, not all functions of all matrices can; $f(\mathbf{A})$ must first be "well defined," whereby "well defined" (see Theorem 1 of Section 7.1) we mean that $f(z)$ has a Taylor series, which converges for $|z| < R$ and all eigenvalues of $\mathbf{A}$ have the property that their absolute values are also less than R.

Problems 7.5

1. Establish the equations necessary to find $e^{\mathbf{A}}$ if $\mathbf{A}$ is a 2×2 matrix having 1 and 2 as its eigenvalues.

2. Establish the equations necessary to find $e^{\mathbf{A}}$ if $\mathbf{A}$ is a 2×2 matrix having 2 and 2 as multiple eigenvalues.

3. Establish the equations necessary to find $e^{\mathbf{A}}$ if $\mathbf{A}$ is a 3×3 matrix having 2 as an eigenvalue of multiplicity 3.

4. Establish the equations necessary to find $e^{\mathbf{A}}$ if $\mathbf{A}$ is a 3×3 matrix having $1, -2$, and 3 as its eigenvalues.

5. Redo Problem 4 if instead the eigenvalues are $-2, -2$, and 1.

6. Establish the equations necessary to find $\sin(\mathbf{A})$ if $\mathbf{A}$ is a 3×3 matrix having 1, 2, and 3 as its eigenvalues.

7. Redo Problem 6 if instead the eigenvalues are $-2, -2$, and 1.

8. Establish the equations necessary to find $e^{\mathbf{A}}$ if $\mathbf{A}$ is a 4×4 matrix having 2 as an eigenvalue of multiplicity 4.

9. Establish the equations necessary to find $e^{\mathbf{A}}$ if $\mathbf{A}$ is a 4×4 matrix having both 2 and -2 as eigenvalues of multiplicity 2.

10. Redo Problem 9 if instead the function of interest is $\sin(\mathbf{A})$.

11. Establish the equations necessary to find $e^{\mathbf{A}}$ if $\mathbf{A}$ is a 4×4 matrix having 3, 3, 3, and -1 as its eigenvalues.

12. Redo Problem 11 if instead the function of interest is $\cos(\mathbf{A})$.

13. Find $e^{\mathbf{A}}$ for

$$\mathbf{A} = \begin{bmatrix} 1 & 3 \\ 4 & 2 \end{bmatrix}.$$

14. Find $e^{\mathbf{A}}$ for

$$\mathbf{A} = \begin{bmatrix} 4 & -1 \\ 1 & 2 \end{bmatrix}.$$

15. Find $e^{\mathbf{A}}$ for

$$\mathbf{A} = \begin{bmatrix} 1 & 1 & 2 \\ -1 & 3 & 4 \\ 0 & 0 & 2 \end{bmatrix}.$$

16. Find $e^{\mathbf{A}}$ for

$$\mathbf{A} = \begin{bmatrix} 1 & 1 & 2 \\ 3 & -1 & 4 \\ 0 & 0 & 2 \end{bmatrix}.$$

17. Find $\cos \mathbf{A}$ if

$$\mathbf{A} = \begin{bmatrix} \pi & 3\pi \\ 2\pi & 2\pi \end{bmatrix}.$$

18. The function $f(z) = \log(1 + z)$ has the Taylor series

$$\sum_{k=1}^{\infty} \frac{(-1)^{k-1} z^k}{k},$$

which converges for $|z| < 1$. For the following matrices, $\mathbf{A}$, determine whether or not $\log(\mathbf{A} + \mathbf{I})$ is well defined and, if so, find it:

(a) $\begin{bmatrix} \dfrac{1}{2} & 1 \\ 0 & -\dfrac{1}{2} \end{bmatrix}$,

(b) $\begin{bmatrix} -6 & 9 \\ -2 & 3 \end{bmatrix}$,

(c) $\begin{bmatrix} 3 & 5 \\ -1 & -3 \end{bmatrix}$,

(d) $\begin{bmatrix} 0 & 0 \\ 0 & 0 \end{bmatrix}$.

7.6 The function e^{At}

A very important function in the matrix calculus is e^{At}, where $\mathbf{A}$ is a square constant matrix (that is, all of its entries are constants) and t is a variable. This function may be calculated by defining a new matrix $\mathbf{B} = \mathbf{A}t$ and then computing $e^{\mathbf{B}}$ by the methods of the previous section.

Example 1. Find e^{At} if

$$\mathbf{A} = \begin{bmatrix} 1 & 2 \\ 4 & 3 \end{bmatrix}.$$

Solution Define

$$\mathbf{B} = \mathbf{A}t = \begin{bmatrix} t & 2t \\ 4t & 3t \end{bmatrix}.$$

The problem then reduces to finding $e^{\mathbf{B}}$. The eigenvalues of $\mathbf{B}$ are $\lambda_1 = 5t$, $\lambda_2 = -t$. Note that the eigenvalues now depend on t.

$$f(\mathbf{B}) = e^{\mathbf{B}} \quad r(\mathbf{B}) = \alpha_1 \mathbf{B} + \alpha_0 \mathbf{I}$$
$$f(\lambda) = e^{\lambda} \quad r(\lambda) = \alpha_1 \lambda + \alpha_0.$$

Since $f(\mathbf{B}) = r(\mathbf{B})$,

$$e^{\mathbf{B}} = \alpha_1 \mathbf{B} + \alpha_0 \mathbf{I}. \tag{38}$$

Also, $f(\lambda_i) = r(\lambda_i)$; hence,

$$e^{5t} = \alpha_1(5t) + \alpha_0,$$
$$e^{-t} = \alpha_1(-t) + \alpha_0.$$

Thus, $\alpha_1 = (1/6t)(e^{5t} - e^{-t})$ and $\alpha_0 = (1/6)(e^{5t} + 5e^{-t})$. Substituting these values into (38), we obtain

$$e^{\mathbf{A}t} = e^{\mathbf{B}} = \left(\frac{1}{6t}\right)(e^{5t} - e^{-t})\begin{bmatrix} t & 2t \\ 4t & 3t \end{bmatrix} + \left(\frac{1}{6}\right)(e^{5t} + 5e^{-1})\begin{bmatrix} 1 & 0 \\ 0 & 1 \end{bmatrix}$$

$$= \frac{1}{6}\begin{bmatrix} 2e^{5t} + 4e^{-t} & 2e^{5t} - 2e^{-t} \\ 4e^{5t} - 4e^{-t} & 4e^{5t} + 2e^{-t} \end{bmatrix}. \blacksquare$$

Example 2. Find $e^{\mathbf{A}t}$ if

$$\mathbf{A} = \begin{bmatrix} 3 & 1 & 0 \\ 0 & 3 & 1 \\ 0 & 0 & 3 \end{bmatrix}.$$

Solution Define

$$\mathbf{B} = \mathbf{A}t = \begin{bmatrix} 3t & t & 0 \\ 0 & 3t & t \\ 0 & 0 & 3t \end{bmatrix}.$$

The problem reduces to finding $e^{\mathbf{B}}$. The eigenvalues of $\mathbf{B}$ are

$$\lambda_1 = \lambda_2 = \lambda_3 = 3t$$

thus,

$$\begin{aligned} f(\mathbf{B}) &= e^{\mathbf{B}} & r(\mathbf{B}) &= \alpha_2\mathbf{B}^2 + \alpha_1\mathbf{B} + \alpha_0\mathbf{I} \\ f(\lambda) &= e^{\lambda} & r(\lambda) &= \alpha_2\lambda^2 + \alpha_1\lambda + \alpha_0 \end{aligned} \tag{39}$$

$$f'(\lambda) = e^{\lambda} \quad r'(\lambda) = 2\alpha_2\lambda + \alpha_1 \tag{40}$$

$$f''(\lambda) = e^{\lambda} \quad r''(\lambda) = 2\alpha_2. \tag{41}$$

Since $f(\mathbf{B}) = r(\mathbf{B})$,

$$e^{\mathbf{B}} = \alpha_2\mathbf{B}^2 + \alpha_1\mathbf{B} + \alpha_0\mathbf{I}. \tag{42}$$

Since $\lambda = 3t$ is an eigenvalue of multiplicity 3,

$$f(3t) = r(3t), \tag{43}$$

$$f'(3t) = r'(3t), \tag{44}$$

$$f''(3t) = r''(3t). \tag{45}$$

Thus, using (39)–(41), we obtain

$$e^{3t} = (3t)^2\alpha_2 + (3t)\alpha_1 + \alpha_0,$$

$$e^{3t} = 2(3t)\alpha_2 + \alpha_1,$$

$$e^{3t} = 2\alpha_2$$

or

$$e^{3t} = 9t^2\alpha_2 + 3t\alpha_1 + \alpha_0, \tag{46}$$

$$e^{3t} = 6t\alpha_2 + \alpha_1, \tag{47}$$

$$e^{3t} = 2\alpha_2. \tag{48}$$

Solving (46)–(48) simultaneously, we obtain

$$\alpha_2 = \frac{1}{2}e^{3t}, \quad \alpha_1 = (1 - 3t)e^{3t}, \quad \alpha_0 = \left(1 - 3t + \frac{9}{2}t^2\right)e^{3t}.$$

From (42), it follows that

$$e^{\mathbf{A}t} = e^{\mathbf{B}} = \frac{1}{2}e^{3t}\begin{bmatrix} 9t^2 & 6t^2 & t^2 \\ 0 & 9t^2 & 6t^2 \\ 0 & 0 & 9t^2 \end{bmatrix} + (1 - 3t)e^{3t}\begin{bmatrix} 3t & t & 0 \\ 0 & 3t & t \\ 0 & 0 & 3t \end{bmatrix}$$

$$+ \left(1 - 3t + \frac{9}{2}t^2\right)e^{3t}\begin{bmatrix} 1 & 0 & 0 \\ 0 & 1 & 0 \\ 0 & 0 & 1 \end{bmatrix}$$

$$= e^{3t}\begin{bmatrix} 1 & t & t^2/2 \\ 0 & 1 & t \\ 0 & 0 & 1 \end{bmatrix}. \blacksquare$$

Problems 7.6

Find $e^{\mathbf{A}t}$ if $\mathbf{A}$ is given by:

1. $\begin{bmatrix} 4 & 4 \\ 3 & 5 \end{bmatrix}$.

2. $\begin{bmatrix} 2 & 1 \\ -1 & -2 \end{bmatrix}$.

3. $\begin{bmatrix} 4 & 1 \\ -1 & 2 \end{bmatrix}$.

4. $\begin{bmatrix} 0 & 1 \\ -14 & -9 \end{bmatrix}$.

5. $\begin{bmatrix} -3 & 2 \\ 2 & -6 \end{bmatrix}$.

6. $\begin{bmatrix} -10 & 5 \\ 36 & -10 \end{bmatrix}$.

7. $\begin{bmatrix} 0 & 1 & 0 \\ 0 & 0 & 1 \\ 0 & 0 & 0 \end{bmatrix}$.

8. $\begin{bmatrix} 1 & 0 & 0 \\ 4 & 1 & 2 \\ -1 & 4 & -1 \end{bmatrix}$.

7.7 Complex eigenvalues

When computing $e^{\mathbf{A}t}$, it is often the case that the eigenvalues of $\mathbf{B} = \mathbf{A}t$ are complex. If this occurs, the complex eigenvalues will appear in conjugate pairs, assuming the elements of $\mathbf{A}$ to be real, and these can be combined to produce real functions.

Let z represent a complex variable. Define e^z by

$$e^z = \sum_{k=0}^{\infty} \frac{z^k}{k!} = 1 + z + \frac{z^2}{2!} + \frac{z^3}{3!} + \frac{z^4}{4!} + \frac{z^5}{5!} + \cdots \qquad (49)$$

(see Eq. 5). Setting $z = i\theta$, θ real, we obtain

$$e^{i\theta} = 1 + i\theta + \frac{(i\theta)}{2!} + \frac{(i\theta)^3}{3!} + \frac{(i\theta)^4}{4!} + \frac{(i\theta)^5}{5!} + \cdots$$

$$= 1 + i\theta - \frac{\theta^2}{2!} - \frac{i\theta^3}{3!} + \frac{\theta^4}{4!} + \frac{i\theta^5}{5!} - \cdots.$$

Combining real and imaginary terms, we obtain

$$e^{i\theta} = \left(1 - \frac{\theta^2}{2!} + \frac{\theta^4}{4!} - \cdots\right) + i\left(\theta - \frac{\theta^3}{3!} + \frac{\theta^5}{5!} - \cdots\right). \qquad (50)$$

But the Maclaurin series expansions for $\sin\theta$ and $\cos\theta$ are

$$\sin\theta = \frac{\theta}{1!} - \frac{\theta^3}{3!} + \frac{\theta^5}{5!} - \cdots$$

$$\cos\theta = 1 - \frac{\theta^2}{2!} + \frac{\theta^4}{4!} + \frac{\theta^6}{6!} + \cdots;$$

hence, Eq. (50) may be rewritten as

$$e^{i\theta} = \cos\theta + i\sin\theta. \qquad (51)$$

Eq. (51) is referred to as De Moivre's formula. If the same analysis is applied to $z = -i\theta$, it follows that

$$e^{-i\theta} = \cos\theta - i\sin\theta. \qquad (52)$$

Adding (51) and (52), we obtain

$$\cos \theta = \frac{e^{i\theta} + e^{-i\theta}}{2},$$ (53)

while subtracting (52) from (51), we obtain

$$\sin \theta = \frac{e^{i\theta} - e^{-i\theta}}{2i}.$$ (54)

Eqs. (53) and (54) are Euler's relations and can be used to reduce complex exponentials to expressions involving real numbers.

Example 1.   Find $e^{\mathbf{A}t}$ if

$$\mathbf{A} = \begin{bmatrix} -1 & 5 \\ -2 & 1 \end{bmatrix}.$$

Solution

$$\mathbf{B} = \mathbf{A}t = \begin{bmatrix} -t & 5t \\ -2t & t \end{bmatrix}.$$

Hence, the eigenvalues of $\mathbf{B}$ are $\lambda_1 = 3ti$ and $\lambda_2 = -3ti$; thus

$$f(\mathbf{B}) = e^{\mathbf{B}} \quad r(\mathbf{B}) = \alpha_1 \mathbf{B} + \alpha_0 \mathbf{I}$$
$$f(\lambda) = e^{\lambda} \quad r(\lambda) = \alpha_1 \lambda + \alpha_0.$$

Since $f(\mathbf{B}) = r(\mathbf{B})$,

$$e^{\mathbf{B}} = \alpha_1 \mathbf{B} + \alpha_0 \mathbf{I},$$ (55)

and since $f(\lambda_i) = r(\lambda_i)$,

$$e^{3ti} = \alpha_1(3ti) + \alpha_0,$$
$$e^{-3ti} = \alpha_1(-3ti) + \alpha_0.$$

Thus,

$$\alpha_0 = \frac{e^{3ti} + e^{-3ti}}{2} \quad \text{and} \quad \alpha_1 = \frac{1}{3t}\left(\frac{e^{3ti} - e^{-3ti}}{2i}\right).$$

If we now use (53) and (54), where in this case $\theta = 3t$, it follows that

$$\alpha_0 = \cos \quad \text{and} \quad \alpha_1 = (1/3t)\sin 3t.$$

Substituting these values into (55), we obtain

$$e^{\mathbf{A}t} = e^{\mathbf{B}} = \begin{bmatrix} -\dfrac{1}{3}\sin 3t + \cos 3t & \dfrac{5}{3}\sin 3t \\ -\dfrac{2}{3}\sin 3t & \dfrac{1}{3}\sin 3t + \cos 3t \end{bmatrix}.$$

In Example 1, the eigenvalues of $\mathbf{B}$ are pure imaginary permitting the application of (53) and (54) in a straightforward manner. In the general case, where the eigenvalues are complex numbers, we can still use Euler's relations providing we note the following:

$$\frac{e^{\beta+i\theta} + e^{\beta-i\theta}}{2} = \frac{e^{\beta}e^{i\theta} + e^{\beta}e^{-i\theta}}{2} = \frac{e^{\beta}\left(e^{i\theta} + e^{-i\theta}\right)}{2} = e^{\beta}\cos \theta,$$ (56)

and

$$\frac{e^{\beta+i\theta} - e^{\beta-i\theta}}{2i} = \frac{e^{\beta}e^{i\theta} - e^{\beta}e^{-i\theta}}{2i} = \frac{e^{\beta}\left(e^{i\theta} - e^{-i\theta}\right)}{2i} = e^{\beta}\sin\theta. \blacksquare \tag{57}$$

Example 2.   Find $e^{\mathbf{A}t}$ if

$$\mathbf{A} = \begin{bmatrix} 2 & -1 \\ 4 & 1 \end{bmatrix}.$$

Solution

$$\mathbf{B} = \mathbf{A}t = \begin{bmatrix} 2t & -t \\ 4t & t \end{bmatrix};$$

hence, the eigenvalues of $\mathbf{B}$ are

$$\lambda_1 = \left(\frac{3}{2} + i\frac{\sqrt{15}}{2}\right)t, \quad \lambda_2 = \left(\frac{3}{2} + i\frac{\sqrt{15}}{2}\right)t.$$

Thus,

$$f(\mathbf{B}) = e^{\mathbf{B}} \quad r(\mathbf{B}) = \alpha_1\mathbf{B} + \alpha_0\mathbf{I}$$
$$f(\lambda) = e^{\lambda} \quad r(\lambda) = \alpha_1\lambda + \alpha_0.$$

Since $f(\mathbf{B}) = r(\mathbf{B})$,

$$e^{\mathbf{B}} = \alpha_1\mathbf{B} + \alpha_0\mathbf{I}, \tag{58}$$

and since $f(\lambda_i) = r(\lambda_i)$,

$$e^{[3/2+i(\sqrt{15}/2)]t} = \alpha_1\left[\frac{3}{2} + i\left(\sqrt{15}/2\right)\right]t + \alpha_0,$$

$$e^{[3/2-i(\sqrt{15}/2)]t} = \alpha_1\left[\frac{3}{2} - i\left(\sqrt{15}/2\right)\right]t + \alpha_0.$$

Putting this system into matrix form and solving for α_1 and α_0 by inversion, we obtain

$$\alpha_1 = \frac{2}{\sqrt{15}t}\left[\frac{e^{[(3/2)t+(\sqrt{15}/2)ti]} - e^{[(3/2)t-(\sqrt{15}/2)ti]}}{2i}\right]$$

$$\alpha_0 = \frac{-3}{\sqrt{15}}\left(\frac{e^{[(3/2)t+(\sqrt{15}/2)ti]} - e^{[(3/2)t-(\sqrt{15}/2)ti]}}{2i}\right)$$

$$+ \left(\frac{e^{[(3/2)t+(\sqrt{15}/2)ti]} + e^{[(3/2)t-(\sqrt{15}/2)ti]}}{2i}\right).$$

Using (56) and (57), where $\beta = \frac{3}{2}t$ and $\theta = (15/2)t$, we obtain

$$\alpha_1 = \frac{2}{\sqrt{15}t}e^{3t/2}\sin\frac{\sqrt{15}t}{2}$$

$$\alpha_0 = -\frac{3}{\sqrt{15}}e^{3t/2}\sin\frac{\sqrt{15}}{2}t + e^{3t/2}\cos\frac{\sqrt{15}}{2}t.$$

Substituting these values into (58), we obtain

$$e^{At} = e^{3t/2} \begin{bmatrix} \dfrac{2}{\sqrt{15}}\sin\dfrac{\sqrt{15}}{2}t + \cos\dfrac{\sqrt{15}}{2}t & \dfrac{-2}{\sqrt{15}}\sin\dfrac{\sqrt{15}}{2}t \\[3mm] \dfrac{8}{\sqrt{15}}\sin\dfrac{\sqrt{15}}{2}t & \dfrac{-1}{\sqrt{15}}\sin\dfrac{\sqrt{15}}{2}t + \cos\dfrac{\sqrt{15}}{2}t \end{bmatrix}. \quad \blacksquare$$

Problems 7.7

Find e^{At} if A is given by

1. $\begin{bmatrix} 1 & -1 \\ 5 & -1 \end{bmatrix}$.

2. $\begin{bmatrix} 2 & -2 \\ 3 & -2 \end{bmatrix}$.

3. $\begin{bmatrix} 0 & 1 \\ -64 & 0 \end{bmatrix}$.

4. $\begin{bmatrix} 4 & -8 \\ 10 & -4 \end{bmatrix}$.

5. $\begin{bmatrix} 2 & 5 \\ -1 & -2 \end{bmatrix}$.

6. $\begin{bmatrix} 0 & 1 \\ -25 & -8 \end{bmatrix}$.

7. $\begin{bmatrix} 3 & 1 \\ -2 & 5 \end{bmatrix}$.

8. $\begin{bmatrix} 0 & 1 & 0 \\ 0 & -2 & -5 \\ 0 & 1 & 2 \end{bmatrix}$.

7.8 Properties of e^A

Since the scalar function e^x and the matrix function e^A are defined similarly (see Eqs. 5 and 6), it should not be surprising to find that they possess some similar properties. What might be surprising, however, is that *not* all properties of e^x are common to e^A.

For example, while it is always true that $e^x e^y = e^{x+y} = e^y e^x$, the same cannot be said for matrices $e^{\mathbf{A}}$ and $e^{\mathbf{B}}$ *unless* $\mathbf{A}$ *and* $\mathbf{B}$ *commute.*

Example 1.   Find $e^{\mathbf{A}} e^{\mathbf{B}}$, $e^{\mathbf{A+B}}$, and $e^{\mathbf{B}} e^{\mathbf{A}}$ if

$$\mathbf{A} = \begin{bmatrix} 1 & 1 \\ 0 & 0 \end{bmatrix} \quad \text{and} \quad \mathbf{B} = \begin{bmatrix} 0 & 0 \\ 0 & 1 \end{bmatrix}.$$

Solution Using the methods developed in Section 7.5, we find

$$e^{\mathbf{A}} = \begin{bmatrix} e & e-1 \\ 0 & 1 \end{bmatrix}, \quad e^{\mathbf{B}} = \begin{bmatrix} 1 & 0 \\ 0 & e \end{bmatrix}, \quad e^{\mathbf{A+B}} = \begin{bmatrix} e & e \\ 0 & e \end{bmatrix}.$$

Therefore,

$$e^{\mathbf{A}} e^{\mathbf{B}} = \begin{bmatrix} e & e-1 \\ 0 & 1 \end{bmatrix} \begin{bmatrix} 1 & 0 \\ 0 & e \end{bmatrix} = \begin{bmatrix} e & e^2 - e \\ 0 & e \end{bmatrix}$$

and

$$e^{\mathbf{B}} e^{\mathbf{A}} = \begin{bmatrix} 1 & 0 \\ 0 & e \end{bmatrix} \begin{bmatrix} e & e-1 \\ 0 & 1 \end{bmatrix} = \begin{bmatrix} e & e-1 \\ 0 & e \end{bmatrix};$$

hence,

$$e^{\mathbf{A+B}} \neq e^{\mathbf{A}} e^{\mathbf{B}}, \quad e^{\mathbf{A+B}} \neq e^{\mathbf{B}} e^{\mathbf{A}} \text{ and } e^{\mathbf{B}} e^{\mathbf{A}} \neq e^{\mathbf{A}} e^{\mathbf{B}}. \blacksquare$$

Two properties that both e^x and $e^{\mathbf{A}}$ do have in common are given by the following:

Property 1. $e^{\mathbf{0}} = \mathbf{I}$, *where* $\mathbf{0}$ *represents the zero matrix.*
   Proof. From (6) we have that

$$e^{\mathbf{A}} = \sum_{k=0}^{\infty} \left(\frac{\mathbf{A}^k}{k!} \right) = \mathbf{I} + \sum_{k=0}^{\infty} \left(\frac{\mathbf{A}^k}{k!} \right).$$

Hence,

$$e^{\mathbf{0}} = \mathbf{I} + \sum_{k=1}^{\infty} \frac{\mathbf{0}^k}{k!} = \mathbf{I}. \blacksquare$$

Property 2. $(e^A)^{-1} = e^{-A}$.

   Proof.

$$(e^A)(e^{-A}) = \left[\sum_{k=0}^{\infty}\left(\frac{A^k}{k!}\right)\right]\left[\sum_{k=0}^{\infty}\frac{(-A)^k}{k!}\right]$$

$$= \left[I + A + \frac{A^2}{2!} + \frac{A^3}{3!} + \cdots\right]\left[I + A + \frac{A^2}{2!} + \frac{A^3}{3!} + \cdots\right]$$

$$= II + A[1 - 1] + A^2\left[\frac{1}{2!} - 1 + \frac{1}{2!}\right] + A^3\left[-\frac{1}{3!} + \frac{1}{2!} - \frac{1}{2!} + \frac{1}{3!}\right] + \cdots$$

$$= I.$$

Thus, e^{-A} is an inverse of e^A. However, by definition, an inverse of e^A is $(e^A)^{-1}$; hence, from the uniqueness of the inverse (Theorem 2 of Section 3.4), we have that $e^{-A} = (e^A)^{-1}$. ■

Example 2. Verify Property 2 for

$$A = \begin{bmatrix} 0 & 1 \\ 0 & 0 \end{bmatrix}.$$

   Solution

$$-A = \begin{bmatrix} 0 & -1 \\ 0 & 0 \end{bmatrix},$$

$$e^A = \begin{bmatrix} 1 & 1 \\ 0 & 1 \end{bmatrix}, \text{ and } e^{-A} = \begin{bmatrix} 1 & -1 \\ 0 & 1 \end{bmatrix}.$$

   Thus,

$$(e^A)^{-1} = \begin{bmatrix} 1 & 1 \\ 0 & 1 \end{bmatrix}^{-1} = \begin{bmatrix} 1 & -1 \\ 0 & 1 \end{bmatrix} = e^{-A}. ■$$

Note that Property 2 implies that e^A is *always invertible* even if A itself is not.

Property 3. $(e^A)^T = e^{A^T}$.

   Proof. The proof of this property is left as an exercise for the reader (see Problem 7). ■

Example 3. Verify Property 3 for

$$A = \begin{bmatrix} 1 & 2 \\ 4 & 3 \end{bmatrix}.$$

Solution

$$\mathbf{A}^{\mathrm{T}} = \begin{bmatrix} 1 & 4 \\ 2 & 3 \end{bmatrix},$$

$$e^{\mathbf{A}^{\mathrm{T}}} = \frac{1}{6} \begin{bmatrix} 2e^5 + 4e^{-1} & 4e^5 - 4e^{-1} \\ 2e^5 - 2e^{-1} & 4e^5 + 2e^{-1} \end{bmatrix},$$

and

$$e^{\mathbf{A}} = \frac{1}{6} \begin{bmatrix} 2e^5 + 4e^{-1} & 2e^5 - 2e^{-1} \\ 2e^5 - 4e^{-1} & 4e^5 + 2e^{-1} \end{bmatrix}.$$

Hence, $\left(e^{\mathbf{A}}\right)^{\mathrm{T}} = e^{\mathbf{A}^{\mathrm{T}}}$. ∎

Problems 7.8

1. Verify Property 2 for

$$\mathbf{A} = \begin{bmatrix} 1 & 3 \\ 0 & 1 \end{bmatrix}.$$

2. Verify Property 2 for

$$\mathbf{A} = \begin{bmatrix} 0 & 1 \\ -64 & 0 \end{bmatrix}.$$

3. Verify Property 2 for

$$\mathbf{A} = \begin{bmatrix} 0 & 1 & 0 \\ 0 & 0 & 1 \\ 0 & 0 & 0 \end{bmatrix}.$$

What is the inverse of **A**?

4. Verify Property 3 for

$$\mathbf{A} = \begin{bmatrix} 2 & 1 & 0 \\ 0 & 2 & 0 \\ 1 & -1 & 1 \end{bmatrix}.$$

5. Verify Property 3 for the matrix given in Problem 2.
6. Verify Property 3 for the matrix given in Problem 3.

7. Prove Property 3. (*Hint:* Using the fact that the eigenvalues of $\mathbf{A}$ are identical to eigenvalues of $\mathbf{A}^{\mathrm{T}}$, show that if $e^{\mathbf{A}} = \alpha_{n-1}\mathbf{A}^{n-1} + \ldots + \alpha_1\mathbf{A} + \alpha_0\mathbf{I}$, and if

$$e^{\mathbf{A}^{\mathrm{T}}} = \beta_{n-1}\left(\mathbf{A}^{\mathrm{T}}\right)^{n-1} + \cdots + \beta_1\mathbf{A}^{\mathrm{T}} + \beta_0\mathbf{I},$$

8. Find $e^{\mathbf{A}}e^{\mathbf{B}}$, $e^{\mathbf{B}}e^{\mathbf{A}}$, and $e^{\mathbf{A}+\mathbf{B}}$ if

$$\mathbf{A} = \begin{bmatrix} 1 & 1 \\ 0 & 0 \end{bmatrix} \quad \text{and} \quad \mathbf{B} = \begin{bmatrix} 0 & 1 \\ 0 & 1 \end{bmatrix}.$$

9. Find two matrices $\mathbf{A}$ and $\mathbf{B}$ such that

$$e^{\mathbf{A}+\mathbf{B}} \neq e^{\mathbf{A}}e^{\mathbf{B}}, \quad e^{\mathbf{A}+\mathbf{B}} \neq e^{\mathbf{B}}e^{\mathbf{A}}, \quad \text{and} \quad e^{\mathbf{B}}e^{\mathbf{A}} \neq e^{\mathbf{A}}e^{\mathbf{B}}.$$

10. By using the definition of $e^{\mathbf{A}}$, prove that if A and B commute, then $e^{\mathbf{A}}e^{\mathbf{B}} = e^{\mathbf{A}+\mathbf{B}}$.

11. Show that if $\mathbf{A} = \mathbf{P}^{-1}\mathbf{BP}$ for some invertible matrix $\mathbf{P}$, then $e^{\mathbf{A}} = \mathbf{P}^{-1}e^{\mathbf{B}}\mathbf{P}$.

7.9 Derivatives of a matrix

Definition 1. An $n \times n$ matrix $\mathbf{A}(t) = [a_{ij}(t)]$ is *continuous* at $t = t_0$ if each of its elements $a_{ij}(t)$ $(i, j = 1, 2, \ldots, n)$ is continuous at $t = t_0$.

For example, the matrix given in (59) is continuous everywhere because each of its elements is continuous everywhere, while the matrix given in (60) is not continuous at $t = 0$ because the $(1, 2)$ element, $\sin(1/t)$, is not continuous at $t = 0$.

$$\begin{bmatrix} e^t & t^2 - 1 \\ 2 & \sin^2 t \end{bmatrix} \tag{59}$$

$$\begin{bmatrix} t^3 - 3t & \sin(1/t) \\ 2t & 45 \end{bmatrix}. \tag{60}$$

We shall use the notation $\mathbf{A}(t)$ to emphasize that the matrix $\mathbf{A}$ may depend on the variable t.

Definition 2. An $n \times n$ matrix $\mathbf{A}(t) = [a_{ij}(t)]$ is *differentiable* at $t = t_0$ if each of the elements $a_{ij}(t)$ $(i, j = 1, 2, \ldots, n)$ is differentiable at $t = t_0$ and

$$\frac{d\mathbf{A}(t)}{dt} = \left[\frac{da_{ij}(t)}{dt}\right]. \tag{61}$$

Generally we will use the notation $\dot{\mathbf{A}}(t)$ to represent $d\mathbf{A}(t)/dt$.

Example 1. Find $\dot{\mathbf{A}}(t)$ if

$$\mathbf{A}(t) = \begin{bmatrix} t^2 & \sin t \\ \ln t & e^{t^2} \end{bmatrix}.$$

Solution

$$\dot{\mathbf{A}}(t) = \frac{d\mathbf{A}(t)}{dt} = \begin{bmatrix} \dfrac{d(t^2)}{dt} & \dfrac{d(\sin t)}{dt} \\ \dfrac{d(\ln t)}{dt} & \dfrac{d\left(e^{t^2}\right)}{dt} \end{bmatrix} = \begin{bmatrix} 2t & \cos t \\ \dfrac{1}{t} & 2te^{t^2} \end{bmatrix}. \blacksquare$$

Example 2. Find $\dot{\mathbf{A}}(t)$ if

$$\mathbf{A}(t) = \begin{bmatrix} 3t \\ 45 \\ t^2 \end{bmatrix}.$$

Solution

$$\dot{\mathbf{A}}(t) = \begin{bmatrix} \dfrac{d(3t)}{dt} \\ \dfrac{d(45)}{dt} \\ \dfrac{d(t^2)}{dt} \end{bmatrix} = \begin{bmatrix} 3 \\ 0 \\ 2t \end{bmatrix}. \blacksquare$$

Example 3. Find $\dot{\mathbf{x}}(t)$ if

$$\mathbf{x}(t) = \begin{bmatrix} x_1(t) \\ x_2(t) \\ \vdots \\ x_n(t) \end{bmatrix}.$$

Solution

$$\dot{\mathbf{x}}(t) = \begin{bmatrix} \dot{x}_1(t) \\ \dot{x}_2(t) \\ \vdots \\ \dot{x}_n(t) \end{bmatrix}.$$

The following properties of the derivative can be verified:

(P1) $\dfrac{d(\mathbf{A}(t) + \mathbf{B}(t))}{dt} = \dfrac{d\mathbf{A}(t)}{dt} + \dfrac{d\mathbf{B}(t)}{dt}.$

(P2) $\dfrac{d[\alpha\mathbf{A}(t)]}{dt} = \alpha\dfrac{d\mathbf{A}(t)}{dt}$, where α is a constant.

(P3) $\dfrac{d[\beta(t)\mathbf{A}(t)]}{dt} = \left(\dfrac{d\beta(t)}{dt}\right)\mathbf{A}(t) + \beta(t)\left(\dfrac{d\mathbf{A}(t)}{dt}\right)$, when $\beta(t)$ is a scalar function of t.

(P4) $\dfrac{d[\mathbf{A}(t)\mathbf{B}(t)]}{dt} = \left(\dfrac{d\mathbf{A}(t)}{dt}\right)\mathbf{B}(t) + \mathbf{A}(t)\left(\dfrac{d\mathbf{B}(t)}{dt}\right)$.

We warn the student to be very careful about the order of the matrices in (P4). Any commutation of the matrices on the right side will generally yield a wrong answer. For instance, it generally is not true that

$$\frac{d}{dt}\left[\mathbf{A}(t)\mathbf{B}(t)\right] = \left(\frac{d\mathbf{A}(t)}{dt}\right)\mathbf{B}(t) + \left(\frac{d\mathbf{B}(t)}{dt}\right)\mathbf{A}(t).$$

Example 4.   Verify Property (P4) for

$$\mathbf{A}(t) = \begin{bmatrix} 2t & 3t^2 \\ 1 & t \end{bmatrix} \quad \text{and} \quad \mathbf{B}(t) = \begin{bmatrix} 1 & 2t \\ 3t & 2 \end{bmatrix}.$$

Solution

$$\frac{d}{dt}[\mathbf{A}(t)\mathbf{B}(t)] = \frac{d}{dt}\left(\begin{bmatrix} 2t & 3t^2 \\ 1 & t \end{bmatrix}\begin{bmatrix} 1 & 2t \\ 3t & 2 \end{bmatrix}\right)$$

$$= \frac{d}{dt}\begin{bmatrix} 2t + 9t^3 & 10t^2 \\ 1 + 3t^2 & 4t \end{bmatrix} = \begin{bmatrix} 2 + 27t^2 & 10t \\ 6t & 4 \end{bmatrix},$$

and

$$\left[\frac{d\mathbf{A}(t)}{dt}\right]\mathbf{B}(t) + \mathbf{A}(t)\left[\frac{d\mathbf{B}(t)}{dt}\right] = \begin{bmatrix} 2 & 6t \\ 0 & 1 \end{bmatrix}\begin{bmatrix} 1 & 2t \\ 3t & 2 \end{bmatrix} + \begin{bmatrix} 2t & 3t^2 \\ 1 & t \end{bmatrix}\begin{bmatrix} 0 & 2 \\ 3 & 0 \end{bmatrix}$$

$$= \begin{bmatrix} 2 + 27t^2 & 20t \\ 6t & 4 \end{bmatrix}$$

$$= \frac{d[\mathbf{A}(t)\mathbf{B}(t)]}{dt}. \blacksquare$$

We are now in a position to establish one of the more important properties of $e^{\mathbf{A}t}$. It is this property that makes the exponential so useful in differential equations (as we shall see in Chapter 8) and hence so fundamental in analysis.

Theorem 1.   *If* $\mathbf{A}$ *is a constant matrix then*

$$\frac{de^{\mathbf{A}t}}{dt} = \mathbf{A}e^{\mathbf{A}t} = e^{\mathbf{A}t}\mathbf{A}.$$

Proof. From (6) we have that

$$e^{\mathbf{A}t} = \sum_{k=0}^{\infty} \frac{(\mathbf{A}t)^k}{k!}$$

or

$$e^{\mathbf{A}t} = \mathbf{I} + t\mathbf{A} + \frac{t^2\mathbf{A}^2}{2!} + \frac{t^3\mathbf{A}^3}{3!} + \cdots + \frac{t^{n-1}\mathbf{A}^{n-1}}{(n-1)!} + \frac{t^n\mathbf{A}^n}{n!} + \frac{t^{n+1}\mathbf{A}^{n+1}}{(n+1)!} + \cdots.$$

Therefore,

$$\frac{de^{\mathbf{A}t}}{dt} = 0 + \frac{\mathbf{A}}{1!} + \frac{2t\mathbf{A}^2}{2!} + \frac{3t^2\mathbf{A}^3}{3!} + \cdots + \frac{nt^{n-1}\mathbf{A}^n}{n!} + \frac{(n+1)t^n\mathbf{A}^{n+1}}{(n+1)!} + \cdots$$

$$= \mathbf{A} + \frac{t\mathbf{A}^2}{1!} + \frac{t^2\mathbf{A}^3}{2!} + \cdots + \frac{t^{n-1}\mathbf{A}^n}{(n-1)!} + \frac{t^n\mathbf{A}^{n+1}}{n!} + \cdots$$

$$= \left[\mathbf{I} + \frac{t\mathbf{A}}{1!} + \frac{t^2\mathbf{A}^2}{2!} + \cdots + \frac{t^{n-1}\mathbf{A}^{n-1}}{(n-1)!} + \frac{t^n\mathbf{A}^n}{n!} + \cdots\right]\mathbf{A}$$

$$= e^{\mathbf{A}t}\mathbf{A}.$$

If we had factored $\mathbf{A}$ on the left, instead of on the right, we would have obtained the other identity,

$$\frac{de^{\mathbf{A}t}}{dt} = \mathbf{A}e^{\mathbf{A}t}. \blacksquare$$

Corollary 1. If $\mathbf{A}$ is a constant matrix, then

$$\frac{de^{-\mathbf{A}t}}{dt} = -\mathbf{A}e^{-\mathbf{A}t} = -e^{-\mathbf{A}t}\mathbf{A}.$$

Proof. Define $\mathbf{C} = -\mathbf{A}$. Hence, $e^{-\mathbf{A}t} = e^{\mathbf{C}t}$. Since $\mathbf{C}$ is a constant matrix, using Theorem 1, we have

$$\frac{de^{\mathbf{C}t}}{dt} = \mathbf{C}e^{\mathbf{C}t} = e^{\mathbf{C}t}\mathbf{C}.$$

If we now substitute for $\mathbf{C}$ its value, $-\mathbf{A}$, Corollary 1 is immediate. $\blacksquare$

Definition 3. An $n \times n$ matrix $\mathbf{A}(t) = [a_{ij}(t)]$ is *integrable* if each of its elements $a_{ij}(t)$ $(i, 1, 2, \ldots, n)$ is integrable, and if this is the case,

$$\int \mathbf{A}(t)\, dt = \left[\int a_{ij}(t)\, dt\right].$$

Example 5. Find $\int A(t) \, dt$ if

$$A(t) = \begin{bmatrix} 3t & 2 \\ t^2 & e^t \end{bmatrix}.$$

Solution

$$\int A(t) \, dt = \begin{bmatrix} \int 3t \, dt & \int 2 \, dt \\ \int t^2 \, dt & \int e^t \, dt \end{bmatrix} = \begin{bmatrix} \left(\dfrac{3}{2}\right)t^2 + c_1 & 2t + c_2 \\ \left(\dfrac{1}{3}\right)t^3 + c_3 & e^t + c_4 \end{bmatrix}. \blacksquare$$

Example 6. Find $\int_0^1 A(t) \, dt$ if

$$A(t) = \begin{bmatrix} 2t & 1 & 2 \\ e^t & 6t^2 & -1 \\ \sin \pi t & 0 & 1 \end{bmatrix}.$$

Solution

$$\int_0^1 A(t) \, dt = \begin{bmatrix} \int_0^1 2t \, dt & \int_0^1 1 \, dt & \int_0^1 2 \, dt \\ \int_0^1 e^t \, dt & \int_0^1 6t^2 \, dt & \int_0^1 -1 \, dt \\ \int_0^1 \sin \pi t \, dt & \int_0^1 0 \, dt & \int_0^1 1 \, dt \end{bmatrix} \blacksquare$$

$$= \begin{bmatrix} 1 & 1 & 2 \\ e-1 & 2 & -1 \\ 2/\pi & 0 & 1 \end{bmatrix}.$$

The following property of the integral can be verified:

$$(P5) \quad \int [\alpha A(t) + \beta B(t)] \, dt = \alpha \int A(t) \, dt + \beta \int B(t) \, dt,$$

where α and β are constants.

Problems 7.9

1. Find $\dot{A}(t)$ if

(a) $A(t) = \begin{bmatrix} \cos t & t^2 - 1 \\ 2t & e^{(t-1)} \end{bmatrix}.$

(b) $A(t) = \begin{bmatrix} 2e^{t3} & t(t-1) & 17 \\ t^2 + 3t - 1 & \sin 2t & t \\ \cos^3(3t^2) & 4 & \ln t \end{bmatrix}$.

2. Verify Properties (P1)–(P4) for

$$\alpha = 7, \quad \beta(t) = t^2, \quad A(t) = \begin{bmatrix} t^3 & 3t^2 \\ 1 & 2t \end{bmatrix}, \quad \text{and} \quad B(t) = \begin{bmatrix} t & -2t \\ t^3 & t^5 \end{bmatrix}.$$

3. Prove that if $dA(t)/dt = 0$, then $A(t)$ is a constant matrix. (That is, a matrix independent of t.)

4. Find $\int A(t)\, dt$ for the $A(t)$ given in Problem 1(a).

5. Verify Property (P5) for

$$\alpha = 2, \quad \beta = 10, \quad A(t) = \begin{bmatrix} 6t & t^2 \\ 2t & 1 \end{bmatrix}, \quad \text{and} \quad B(t) = \begin{bmatrix} t & 4t^2 \\ 1 & 2t \end{bmatrix}.$$

6. Using Property (P4), derive a formula for differentiating $A^2(t)$. Use this formula to find $dA^2(t)/dt$, where

$$A(t) = \begin{bmatrix} t & 2t^2 \\ 4t^3 & e^t \end{bmatrix},$$

and show that $dA^2(t)/dt \neq 2A(t)\, dA(t)/dt$. Therefore, the power rule of differentiation *does not hold* for matrices unless a matrix commutes with its derivative.

7.10 Final comments on Chapter 7

We begin a proof of the Cayley–Hamilton theorem by noting that if B is an $n \times n$ matrix having elements, which are polynomials in λ with constant coefficients, then B can be expressed as a matrix polynomial in λ whose coefficients are $n \times n$ constant matrices. As an example, consider the following decomposition:

$$\begin{bmatrix} \lambda^3 + 2\lambda^2 + 3\lambda + 4 & 2\lambda^3 + 3\lambda^2 + 4\lambda + 5 \\ 3\lambda^3 + 4\lambda^2 + 5\lambda & 2\lambda + 3 \end{bmatrix}$$

$$= \begin{bmatrix} 1 & 2 \\ 3 & 0 \end{bmatrix}\lambda^3 + \begin{bmatrix} 2 & 3 \\ 4 & 0 \end{bmatrix}\lambda^2 + \begin{bmatrix} 3 & 4 \\ 5 & 2 \end{bmatrix}\lambda + \begin{bmatrix} 4 & 5 \\ 0 & 3 \end{bmatrix}.$$

In general, if the elements of $\mathbf{B}$ are polynomials of degree k or less, then

$$\mathbf{B} = \mathbf{B}_k \lambda^k + \mathbf{B}_{k-1} \lambda^{k-1} + \cdots + \mathbf{B}_1 \lambda + \mathbf{B}_0,$$

where $\mathbf{B}_j (j = 0, 1, \ldots, k)$ is an $n \times n$ constant matrix.

Now, let $\mathbf{A}$ be any arbitrary $n \times n$ matrix. Define

$$\mathbf{C} = (\mathbf{A} - \lambda \mathbf{I}) \tag{62}$$

and let

$$d(\lambda) = \lambda_n + a_{n-1} \lambda^{n-1} + \cdots + a_1 \lambda + a_0 \tag{63}$$

represent the characteristic polynomial of $\mathbf{A}$. Thus,

$$d(\lambda) = \det(\mathbf{A} - \lambda \mathbf{I}) = \det \mathbf{C}. \tag{64}$$

Since $\mathbf{C}$ is an $n \times n$ matrix, it follows that the elements of $\mathbf{C}^a$ (see Definition 2 of Section 5.5) will be polynomials in λ of either degree $n - 1$ or $n - 2$. (Elements on the diagonal of $\mathbf{C}^a$ will be polynomials of degree $n - 1$, while all other elements will be polynomials of degree $n - 2$.) Thus, $\mathbf{C}^a$ can be written as

$$\mathbf{C}^a = \mathbf{C}_{n-1} \lambda^{n-1} + \mathbf{C}_{n-2} \lambda^{n-2} + \cdots + \mathbf{C}_1 \lambda + \mathbf{C}_0, \tag{65}$$

where $\mathbf{C}_j$ $(j = 0, 1, \ldots, n - 1)$ is an $n \times n$ constant matrix.

From Theorem 2 of Section 5.5 and (64), we have that

$$\mathbf{C}^a \mathbf{C} = [\det \mathbf{C}] \mathbf{I} = d(\lambda) \mathbf{I}. \tag{66}$$

From (62), we have that

$$\mathbf{C}^a \mathbf{C} = \mathbf{C}^a (\mathbf{A} - \lambda \mathbf{I}) = \mathbf{C}^a \mathbf{A} - \lambda \mathbf{C}^a. \tag{67}$$

Equating (66) and (67), we obtain

$$d(\lambda) \mathbf{I} = \mathbf{C}^a \mathbf{A} - \lambda \mathbf{C}^a. \tag{68}$$

Substituting (63) and (65) into (68), we find that

$$\mathbf{I} \lambda_n + a_{n-1} \mathbf{I} \lambda^{n-1} + \cdots + a_1 \mathbf{I} \lambda + a_0 \mathbf{I}$$
$$= \mathbf{C}_{n-1} \mathbf{A} \lambda^{n-1} + \mathbf{C}_{n-2} \mathbf{A} \lambda^{n-2} + \cdots + \mathbf{C}_1 \mathbf{A} \lambda + \mathbf{C}_0 \mathbf{A}$$
$$- \mathbf{C}_{n-1} \lambda^n - \mathbf{C}_{n-2} \lambda^{n-1} - \cdots - \mathbf{C}_1 \lambda^2 - \mathbf{C}_0 \lambda.$$

Both sides of this equation are matrix polynomials in λ of degree n. Since two polynomials are equal if and only if their corresponding coefficients are equal, we have

$$\begin{aligned} \mathbf{I} &= -\mathbf{C}_{n-1} \\ a_{n-1} \mathbf{I} &= -\mathbf{C}_{n-2} + \mathbf{C}_{n-1} \mathbf{A} \\ a_1 \mathbf{I} &= -\mathbf{C}_0 + \mathbf{C}_1 \mathbf{A} \\ a_0 \mathbf{I} &= \mathbf{C}_0 \mathbf{A}. \end{aligned} \tag{69}$$

Multiplying the first equation in (69) by $\mathbf{A}^n$, the second equation by $\mathbf{A}^{n-1}$, ..., and the last equation by $\mathbf{A}^0 = \mathbf{I}$ and adding, we obtain (note that the terms on the right-hand side cancel out)

$$\mathbf{A}^n + a_{n-1}\mathbf{A}^{n-1} + \cdots + a_1\mathbf{A} + \mathbf{A}_0\mathbf{I} = 0. \tag{70}$$

Eq. (70) is the Cayley–Hamilton theorem.

Linear differential equations

Chapter outline

8.1 Fundamental form

We are now ready to solve linear differential equations. The method that we shall use involves introducing new variables $x_1(t)$, $x_2(t)$, ..., $x_n(t)$ and then reducing a given system of differential equations to the system

$$\frac{dx_1(t)}{dt} = a_{11}(t)x_1(t) + a_{12}(t)x_2(t) + \cdots + a_{1n}(t)x_n(t) + f_1(t)$$

$$\frac{dx_2(t)}{dt} = a_{21}(t)x_1(t) + a_{22}(t)x_2(t) + \cdots + a_{2n}(t)x_n(t) + f_2(t)$$

$$\vdots \tag{1}$$

$$\frac{dx_n(t)}{dt} = a_{n1}(t)x_1(t) + a_{n2}(t)x_2(t) + \cdots + a_{nn}(t)x_n(t) + f_n(t).$$

Matrix Methods. https://doi.org/10.1016/B978-0-12-818419-6.00008-3

267

If we define

$$\mathbf{x}(t) = \begin{bmatrix} x_1(t) \\ x_2(t) \\ \vdots \\ x_n(t) \end{bmatrix},$$

$$\mathbf{A}(t) = \begin{bmatrix} a_{11}(t) & a_{12}(t) & \cdots & a_{1n}(t) \\ a_{21}(t) & a_{22}(t) & \cdots & a_{2n}(t) \\ \vdots & \vdots & & \vdots \\ a_{n1}(t) & a_{n2}(t) & \cdots & a_{nn}(t) \end{bmatrix}, \quad \text{and} \quad \mathbf{f}(t) = \begin{bmatrix} f_1(t) \\ f_2(t) \\ \vdots \\ f_n(t) \end{bmatrix}, \tag{2}$$

then, (1) can be rewritten in the matrix form

$$\frac{d\mathbf{x}(t)}{dt} = \mathbf{A}(t)\mathbf{x}(t) + \mathbf{f}(t). \tag{3}$$

Example 1.   Put the following system into matrix form:

$$\dot{y}(t) = t^2 y(t) + 3z(t) + \sin t,$$
$$\dot{z}(t) = -e^t y(t) + tz(t) - t^2 + 1.$$

Note that we are using the standard notation $\dot{y}(t)$ and $\dot{z}(t)$ to represent

$$\frac{dy(t)}{dt} \quad \text{and} \quad \frac{dz(t)}{dt}.$$

Solution Define $x_1(t) = y(t)$ and $x_2(t) = z(t)$. This system is then equivalent to the matrix equation

$$\begin{bmatrix} \dot{x}_1(t) \\ \dot{x}_2(t) \end{bmatrix} = \begin{bmatrix} t^2 & 3 \\ -e^t & t \end{bmatrix} \begin{bmatrix} x_1(t) \\ x_2(t) \end{bmatrix} + \begin{bmatrix} \sin t \\ -t^2 + 1 \end{bmatrix}. \tag{4}$$

If we define

$$\mathbf{x}(t) = \begin{bmatrix} x_1(t) \\ x_2(t) \end{bmatrix}, \quad \mathbf{A}(t) = \begin{bmatrix} t^2 & 3 \\ -e^t & t \end{bmatrix}, \quad \text{and} \quad \mathbf{f}(t) = \begin{bmatrix} \sin t \\ -t^2 + 1 \end{bmatrix},$$

then, (4) is in the required form, $\dot{\mathbf{x}}(t) = \mathbf{A}(t)\mathbf{x}(t) + \mathbf{f}(t)$. ∎

In practice, we are usually interested in solving an initial value problem; that is, we seek functions $x_1(t), x_2(t), \ldots, x_n(t)$ that satisfy not only the differential equations given by (1) but also a set of initial conditions of the form

$$x_1(t_0) = c_1, \quad x_2(t_0) = c_2, \ldots, x_n(t_0) = c_n, \tag{5}$$

where $c_1, c_2, \ldots, c_n$, and t_0 are known constants. Upon defining

$$\mathbf{c} = \begin{bmatrix} c_1 \\ c_2 \\ \vdots \\ c_n \end{bmatrix},$$

it follows from the definition of $\mathbf{x}(t)$ (see Eqs. 2 and 5) that

$$\mathbf{x}(t_0) = \begin{bmatrix} x_1(t_0) \\ x_2(t_0) \\ \vdots \\ x_n(t_0) \end{bmatrix} = \begin{bmatrix} c_1 \\ c_1 \\ \vdots \\ c_n \end{bmatrix} = \mathbf{c}.$$

Thus, the initial conditions can be put into the matrix form

$$\mathbf{x}(t_0) = \mathbf{c}. \tag{6}$$

Definition 1. A system of differential equations is in *fundamental form* if it is given by the matrix equations

$$\dot{\mathbf{x}}(t) = \mathbf{A}(t)\mathbf{x}(t) + \mathbf{f}(t)$$
$$\mathbf{x}(t_0) = \mathbf{c}. \tag{7}$$

Example 2. Put the following system into fundamental form:

$$\dot{x}(t) = 2x(t) - ty(t)$$
$$\dot{y}(t) = t^2 x(t) + e^t$$
$$x(2) = 3, \qquad y(2) = 1.$$

Solution Define $x_1(t) = x(t)$ and $x_2(t) = y(t)$. This system is then equivalent to the matrix equations

$$\begin{bmatrix} \dot{x}_1(t) \\ \dot{x}_2(t) \end{bmatrix} = \begin{bmatrix} 2 & -t \\ t^2 & 0 \end{bmatrix} \begin{bmatrix} x_1(t) \\ x_2(t) \end{bmatrix} + \begin{bmatrix} 0 \\ e^t \end{bmatrix}$$
$$\begin{bmatrix} x_1(2) \\ x_2(2) \end{bmatrix} = \begin{bmatrix} 3 \\ 1 \end{bmatrix}. \tag{8}$$

Consequently, if we define

$$\mathbf{x}(t) = \begin{bmatrix} x_1(t) \\ x_2(t) \end{bmatrix}, \qquad \mathbf{A}(t) = \begin{bmatrix} 2 & -t \\ t^2 & 0 \end{bmatrix},$$

$$\mathbf{f}(t) = \begin{bmatrix} 0 \\ e^t \end{bmatrix}, \qquad \mathbf{c} = \begin{bmatrix} 3 \\ 1 \end{bmatrix}, \quad \text{and} \quad t_0 = 2,$$

then, (8) is in fundamental form. ∎

Example 3. Put the following system into fundamental form:

$$\dot{l}(t) = 2l(t) + 3m(t) - n(t)$$
$$\dot{m}(t) = l(t) - m(t)$$
$$\dot{n}(t) = m(t) - n(t)$$
$$l(15) = 0, \quad m(15) = -170, \quad n(15) = 1007.$$

Solution Define $x_1(t) = l(t)$, $x_2(t) = m(t)$, $x_3(t) = n(t)$. This system is then equivalent to the matrix equations

$$\begin{bmatrix} \dot{x}_1(t) \\ \dot{x}_2(t) \\ \dot{x}_3(t) \end{bmatrix} = \begin{bmatrix} 2 & 3 & -1 \\ 1 & -1 & 0 \\ 0 & 1 & -1 \end{bmatrix} \begin{bmatrix} x_1(t) \\ x_2(t) \\ x_3(t) \end{bmatrix},$$

$$\begin{bmatrix} x_1(15) \\ x_2(15) \\ x_3(15) \end{bmatrix} = \begin{bmatrix} 0 \\ -170 \\ 1007 \end{bmatrix}.$$

(9)

Thus, if we define

$$\mathbf{x}(t) = \begin{bmatrix} x_1(t) \\ x_2(t) \\ x_3(t) \end{bmatrix}, \quad \mathbf{A}(t) = \begin{bmatrix} 2 & 3 & -1 \\ 1 & -1 & 0 \\ 0 & 1 & -1 \end{bmatrix}, \quad \mathbf{f}(t) = \begin{bmatrix} 0 \\ 0 \\ 0 \end{bmatrix},$$

$$\mathbf{c} = \begin{bmatrix} 0 \\ -170 \\ 1007 \end{bmatrix} \quad \text{and} \quad t_0 = 15,$$

then, (9) is in fundamental form. ■

Definition 2. A system in fundamental form is *homogeneous* if $\mathbf{f}(t) = 0$ (that is, if $f_1(t) = f_2(t) = \ldots = f_n(t) = 0$) and *nonhomogeneous* if $f(t) \neq 0$ (that is, if at least one component of $\mathbf{f}(t)$ differs from zero).

The system given in Examples 2 and 3 are nonhomogeneous and homogeneous, respectively.

Since we will be attempting to solve differential equations, it is important to know exactly what is meant by a solution.

Definition 3. $\mathbf{x}(t)$ is a *solution* of (7) if

(a) both $\mathbf{x}(t)$ and $\dot{\mathbf{x}}(t)$ are continuous in some neighborhood J of the initial time $t = t_0$,

(b) the substitution of $\mathbf{x}(t)$ into the differential equation

$$\dot{\mathbf{x}}(t) = \mathbf{A}(t)\mathbf{x}(t) + \mathbf{f}(t)$$

makes the equation an identity in t on the interval J; that is, the equation is valid for each t in J, and

(c) $\mathbf{x}(t_0) = \mathbf{c}$.

It would also seem advantageous, before trying to find the solutions, to know whether or not a given system has any solutions at all, and if it does, how many. The following theorem from differential equations answers both of these questions.

Theorem 1. *Consider a system given by (7). If $\mathbf{A}(t)$ and $\mathbf{f}(t)$ are continuous in some interval containing $t = t_0$, then this system possesses a unique continuous solution on that interval.*

Hence, to insure the applicability of this theorem, we assume for the remainder of the chapter that $\mathbf{A}(t)$ and $\mathbf{f}(t)$ are both continuous on some common interval containing $t = t_0$.

Problems 8.1

In Problems 8.1 through 8.8, put the given systems into fundamental form.

1. $\dfrac{dx(t)}{dt} = 2x(t) + 3y(t),$

$\dfrac{dy(t)}{dt} = 4x(t) + 5y(t),$

$x(0) = 6, \quad y(0) = 7.$

2. $\dot{y}(t) = 3y(t) + 2z(t),$

$\dot{z}(t) = 4y(t) + z(t),$

$y(0) = 1, \quad z(0) = 1.$

3. $\dfrac{dx(t)}{dt} = -3x(t) + 3y(t) + 1,$

$\dfrac{dy(t)}{dt} = 4x(t) - 4y(t) - 1,$

$x(0) = 0, \quad y(0) = 0.$

4. $\dfrac{dx(t)}{dt} = 3x(t) + t,$

$\dfrac{dy(t)}{dt} = 2x(t) + t + 1,$

$x(0) = 1, \quad y(0) = -1.$

5. $\dfrac{dx(t)}{dt} = 3t^2x(t) + 7y(t) + 2,$

$\dfrac{dx(t)}{dt} = x(t) + ty(t) + 2t,$

$x(1) = 2, \quad y(1) = -3.$

6. $\dfrac{du(t)}{dt} = e^t u(t) + tv(t) + w(t),$

$\dfrac{dv(t)}{dt} = t^2 u(t) - 3v(t) + (t+1)w(t),$

$\dfrac{dw(t)}{dt} = v(t) + e^{t^2} w(t),$

$u(4) = 0, \quad u(4) = 1, \quad z(4) = -1.$

7. $\dfrac{dx(t)}{dt} = 6y(t) + zt,$

$\dfrac{dy(t)}{dt} = x(t) - 3z(t),$

$\dfrac{dz(t)}{dt} = -2y(t),$

$x(0) = 10, \quad y(0) = 10, \quad z(0) = 20.$

8. $\dot{r}(t) = t^2 r(t) - 3s(t) - (\sin t)u(t) + \sin t,$

$\dot{s}(t) = r(t) - s(t) + t^2 - 1,$

$\dot{u}(t) = 2r(t) - e^t s(t) + (t^2 - 1)u(t) + \cos t,$

$r(1) = 4, \quad s(1) = -2, \quad u(1) = 5.$

9. Determine which of the following are solutions to the system

$$\begin{bmatrix} \dot{x}_1 \\ \dot{x}_2 \end{bmatrix} = \begin{bmatrix} 0 & 1 \\ -1 & 0 \end{bmatrix} \begin{bmatrix} x_1 \\ x_2 \end{bmatrix}, \qquad \begin{bmatrix} x_1(0) \\ x_2(0) \end{bmatrix} = \begin{bmatrix} 1 \\ 0 \end{bmatrix}:$$

(a) $\begin{bmatrix} \sin t \\ \cos t \end{bmatrix},$ (b) $\begin{bmatrix} e^t \\ 0 \end{bmatrix},$ (c) $\begin{bmatrix} \cos t \\ -\sin t \end{bmatrix}.$

10. Determine which of the following are solutions to the system:

$$\begin{bmatrix} \dot{x}_1 \\ \dot{x}_2 \end{bmatrix} = \begin{bmatrix} 1 & 2 \\ 4 & 3 \end{bmatrix} \begin{bmatrix} x_1 \\ x_2 \end{bmatrix}, \quad \begin{bmatrix} x_1(0) \\ x_2(0) \end{bmatrix} = \begin{bmatrix} 1 \\ 2 \end{bmatrix}:$$

(a) $\begin{bmatrix} e^{-t} \\ -e^{-t} \end{bmatrix}$,   (b) $\begin{bmatrix} e^{-t} \\ 2e^{-t} \end{bmatrix}$,   (c) $\begin{bmatrix} e^{5t} \\ 2e^{5t} \end{bmatrix}$.

11. Determine which of the following are solutions to the system:

$$\begin{bmatrix} \dot{x}_1 \\ \dot{x}_2 \end{bmatrix} = \begin{bmatrix} 0 & 1 \\ -2 & 3 \end{bmatrix} \begin{bmatrix} x_1 \\ x_2 \end{bmatrix}, \quad \begin{bmatrix} x_1(1) \\ x_2(1) \end{bmatrix} = \begin{bmatrix} 1 \\ 0 \end{bmatrix}:$$

(a) $\begin{bmatrix} -e^{2t} + 2e^t \\ -2e^{2t} + 2e^t \end{bmatrix}$,   (b) $\begin{bmatrix} -e^{2(t-1)} + 2e^{(t-1)} \\ -2e^{2(t-1)} + 2e^{(t-1)} \end{bmatrix}$,   (c) $\begin{bmatrix} e^{2(t-1)} \\ 0 \end{bmatrix}$.

8.2 **Reduction of an *n*th order equation**

Before seeking solutions to linear differential equations, we will first develop techniques for reducing these equations to fundamental form. In this section, we consider the initial value problems given by

$$a_n(t)\frac{d^n x(t)}{dt^n} + a_{n-1}(t)\frac{d^{n-1}x(t)}{dt^{n-1}} + \cdots + a_1(t)\frac{dx(t)}{dt} + a_0(t)x(t) = f(t) \tag{10}$$

$$x(t_0) = c_1, \frac{dx(t_0)}{dt} = c_2, \cdots, \frac{d^{n-1}x(t_0)}{dt^{n-1}} = c_n.$$

Eq. (10) is an *n*th order differential equation for $x(t)$, where $a_0(t)$, $a_1(t)$, ..., $a_n(t)$ and $f(t)$ are assumed known and continuous on some interval containing t_0. Furthermore, we assume that $a_n(t) \neq 0$ on this interval.

A method of reduction, particularly useful for differential equations defined by system (10), is the following:

Step 1. Rewrite (10) so that the *n*th derivative of $x(t)$ appears by itself:

$$\frac{d^n x(t)}{dt^n} = -\frac{a_{n-1}(t)}{a_n(t)}\frac{d^{n-1}x(t)}{dt^{n-1}} - \cdots - \frac{a_1(t)}{a_n(t)}\frac{dx(t)}{dt} - \frac{a_0(t)}{a_n(t)}x(t) + \frac{f(t)}{a_n(t)}. \tag{11}$$

Step 2. Define n new variables (the same number as the order of the differential equation), $x_1(t)$, $x_2(t)$, ..., $x_n(t)$ by the equations

$$x_1 = x(t),$$

$$x_2 = \frac{dx_1}{dt},$$

$$x_3 = \frac{dx_2}{dt},$$

$$\vdots$$

$$x_{n-1} = \frac{dx_{n-2}}{dt},$$

$$x_n = \frac{dx_{n-1}}{dt}.$$

(12)

Generally, we will write $x_j(t)$ ($j = 1, 2, \ldots, n$) simply as x_j when the dependence on the variable t is obvious from the context. It is immediate from system (12) that we also have the following relationships between $x_1, x_2, \ldots, x_n$ and the unknown $x(t)$:

$$x_1 = x,$$

$$x_2 = \frac{dx}{dt},$$

$$x_3 = \frac{d^2x}{dt^2},$$

$$\vdots$$

$$x_{n-1} = \frac{d^{n-2}x}{dt^{n-2}},$$

$$x_n = \frac{d^{n-1}x}{dt^{n-1}}.$$

(13)

Hence, by differentiating the last equation of (13), we have

$$\frac{dx_n}{dt} = \frac{d^nx}{dt^n}.$$

(14)

Step 3. Rewrite dx_n/dt in terms of the new variables $x_1, x_2, \ldots, x_n$.
Substituting (11) into (14), we have

$$\frac{dx_n}{dt} = -\frac{a_{n-1}(t)}{a_n(t)}\frac{d^{n-1}x}{dt^{n-1}} - \cdots - \frac{a_1(t)}{a_n(t)}\frac{dx}{dt} - \frac{a_0(t)}{a_n(t)}x + \frac{f(t)}{a_n(t)}.$$

Substituting (13) into this equation, we obtain

$$\frac{dx_n}{dt} = -\frac{a_{n-1}(t)}{a_n(t)}x_n - \cdots - \frac{a_1(t)}{a_n(t)}x_2 - \frac{a_0(t)}{a_n(t)}x_1 + \frac{f(t)}{a_n(t)}.$$

(15)

Step 4. Form a system of n first order differential equations for x_1, x_2, ..., x_n. Using (12) and (15), we obtain the system

$$\frac{dx_1}{dt} = x_2,$$

$$\frac{dx_2}{dt} = x_3,$$

$$\vdots$$

$$\frac{dx_{n-2}}{dt} = x_{n-1}, \tag{16}$$

$$\frac{dx_{n-1}}{dt} = x_n,$$

$$\frac{dx_n}{dt} = -\frac{a_0(t)}{a_n(t)}x_1 - \frac{a_1(t)}{a_n(t)}x_2 - \cdots - \frac{a_{n-1}(t)}{a_n(t)}x_n + \frac{f(t)}{a_n(t)}.$$

Note that in the last equation of (16), we have rearranged the order of (15) so that the x_1 term appears first, the x_2 term appears second, etc. This was done in order to simplify the next step.

Step 5. Put (16) into matrix form. Define

$$x(t) = \begin{bmatrix} x_1(t) \\ x_2(t) \\ \vdots \\ x_{n-1}(t) \\ x_n(t) \end{bmatrix},$$

$$\mathbf{A}(t) = \begin{bmatrix} 0 & 1 & 0 & 0 & \cdots & 0 \\ 0 & 0 & 1 & 0 & \cdots & 0 \\ 0 & 0 & 0 & 1 & \cdots & 0 \\ \vdots & \vdots & \vdots & \vdots & & \vdots \\ 0 & 0 & 0 & 0 & \cdots & 1 \\ -\dfrac{a_0(t)}{a_n(t)} & -\dfrac{a_1(t)}{a_n(t)} & -\dfrac{a_2(t)}{a_n(t)} & -\dfrac{a_3(t)}{a_n(t)} & \cdots & -\dfrac{a_{n-1}(t)}{a_n(t)} \end{bmatrix}, \tag{17}$$

and

$$f(t) = \begin{bmatrix} 0 \\ 0 \\ \vdots \\ 0 \\ \dfrac{f(t)}{a_n(t)} \end{bmatrix}.$$

Then (16) can be written as

$$\dot{\mathbf{x}}(t) = \mathbf{A}(t)\mathbf{x}(t) + \mathbf{f}(t). \tag{18}$$

Step 6. Rewrite the initial conditions in matrix form.
From (17), (13), and (10), we have that

$$\mathbf{x}(t_0) = \begin{bmatrix} x_1(t_0) \\ x_2(t_0) \\ \vdots \\ x_n(t_0) \end{bmatrix} = \begin{bmatrix} x(t_0) \\ \dfrac{dx(t_0)}{dt} \\ \vdots \\ \dfrac{d^{n-1}x(t_0)}{dt^{n-1}} \end{bmatrix} = \begin{bmatrix} c_1 \\ c_2 \\ \vdots \\ c_n \end{bmatrix}.$$

Thus, if we define

$$\mathbf{c} = \begin{bmatrix} c_1 \\ c_2 \\ \vdots \\ c_n \end{bmatrix},$$

the initial conditions can be put into matrix form

$$\mathbf{x}(t_0) = \mathbf{c}. \tag{19}$$

Eqs. (18) and (19) together represent the fundamental form for (10).

Since $\mathbf{A}(t)$ and $\mathbf{f}(t)$ are continuous (why?), Theorem 1 of the previous section guarantees that a unique solution exists to (18) and (19). Once this solution is obtained, $\mathbf{x}(t)$ will be known; hence, the components of $x(t)$, $x_1(t)$, ..., $x_n(t)$ will be known and, consequently, so will $x(t)$, the variable originally sought (from (12), $x_1(t) = x(t)$).

Example 1. Put the following initial value problem into fundamental form:

$$2\dddot{x} - 4\ddot{x} + 16t\ddot{x} - \dot{x} + 2t^2x = \sin t,$$
$$x(0) = 1, \quad \dot{x}(0) = 2, \quad \ddot{x}(0) = -1, \quad \dddot{x}(0) = 0.$$

Solution The differential equation may be rewritten as

$$\ddddot{x} = 2\dddot{x} - 8t\ddot{x} + \frac{1}{2}\dot{x} - t^2x + \left(\frac{1}{2}\right)\sin t.$$

Define

$$x_1 = x,$$
$$x_2 = \dot{x}_1 = \dot{x},$$
$$x_3 = \dot{x}_2 = \ddot{x},$$
$$x_4 = \dot{x}_3 = \dddot{x}$$

hence, $\dot{x}_4 = \ddddot{x}$. Thus,

$$\dot{x}_1 = x_2$$
$$\dot{x}_2 = x_3$$
$$\dot{x}_3 = x_4$$

$$\dot{x}_4 = \ddddot{x} = 2\dddot{x} - 8t\ddot{x} + \frac{1}{2}\dot{x} - t^2x + \frac{1}{2}\sin t$$

$$= 2x_4 - 8tx_3 + \frac{1}{2}x_2 - t^2x_1 + \frac{1}{2}\sin t,$$

or

$$\dot{x}_1 = \qquad\qquad x_2$$
$$\dot{x}_2 = \qquad\qquad\qquad x_3$$
$$\dot{x}_3 = \qquad\qquad\qquad\qquad x_4$$
$$\dot{x}_4 = -t^2x_1 + \frac{1}{2}x_2 - 8tx_3 + 2x_4 + \frac{1}{2}\sin t.$$

Define

$$\mathbf{x}(t) = \begin{bmatrix} x_1 \\ x_2 \\ x_3 \\ x_4 \end{bmatrix}, \qquad \mathbf{A}(t) = \begin{bmatrix} 0 & 1 & 0 & 0 \\ 0 & 0 & 1 & 0 \\ 0 & 0 & 0 & 1 \\ -t^2 & \frac{1}{2} & -8t & 2 \end{bmatrix},$$

$$\mathbf{f}(t) = \begin{bmatrix} 0 \\ 0 \\ 0 \\ \frac{1}{2}\sin t \end{bmatrix}, \qquad \mathbf{c} = \begin{bmatrix} 1 \\ 2 \\ -1 \\ 0 \end{bmatrix}, \qquad \text{and} \qquad t_0 = 0.$$

Thus, the initial value problem may be rewritten in the fundamental form

$$\dot{\mathbf{x}}(t) = \mathbf{A}(t)\mathbf{x}(t) + \mathbf{f}(t),$$
$$\mathbf{x}(t_0) = \mathbf{c}. \quad \blacksquare$$

Example 2. Put the following initial value problem into fundamental form:

$$e^t \frac{d^5 x}{dt^5} - 2e^{2t} \frac{d^4 x}{dt^4} + tx = 4e^t,$$

$$x(2) = 1, \quad \frac{dx(2)}{dt} = -1, \quad \frac{d^2 x(2)}{dt^2} = -1, \quad \frac{d^3 x(2)}{dt^3} = 2, \quad \frac{d^4 x(2)}{dt^4} = 3.$$

Solution The differential equation may be rewritten

$$\frac{d^5 x}{dt^5} = 2e^t \frac{d^4 x}{dt^4} - te^{-t} x + 4.$$

Define

$$x_1 = x$$
$$x_2 = \dot{x}_1 = \dot{x}$$
$$x_3 = \dot{x}_2 = \ddot{x}$$
$$x_4 = \dot{x}_3 = \dddot{x}$$
$$x_5 = \dot{x}_4 = \frac{d^4 x}{dt^4};$$

hence,

$$\dot{x}_5 = \frac{d^5 x}{dt^5}.$$

Thus,

$$\dot{x}_1 = x_2$$
$$\dot{x}_2 = x_3$$
$$\dot{x}_3 = x_4$$
$$\dot{x}_4 = x_5$$
$$\dot{x}_5 = \frac{d^5 x}{dt^5} = 2e^t \frac{d^4 x}{dt^4} - te^{-t} x + 4$$

$$= 2e^t x_5 - te^{-t} x_1 + 4,$$

or

$$\dot{x}_1 = \qquad x_2$$
$$\dot{x}_2 = \qquad x_3$$
$$\dot{x}_3 = \qquad x_4$$
$$\dot{x}_4 = \qquad x_5$$
$$\dot{x}_5 = -te^{-t} x_1 \qquad + 2e^t x_5 + 4.$$

Define

$$\mathbf{x}(t) = \begin{bmatrix} x_1 \\ x_2 \\ x_3 \\ x_4 \\ x_5 \end{bmatrix}, \quad \mathbf{A}(t) = \begin{bmatrix} 0 & 1 & 0 & 0 & 0 \\ 0 & 0 & 1 & 0 & 0 \\ 0 & 0 & 0 & 1 & 0 \\ 0 & 0 & 0 & 0 & 1 \\ -te^{-t} & 0 & 0 & 0 & 2e^t \end{bmatrix},$$

$$\mathbf{f}(t) = \begin{bmatrix} 0 \\ 0 \\ 0 \\ 0 \\ 4 \end{bmatrix}, \quad \mathbf{c} = \begin{bmatrix} 1 \\ -1 \\ -1 \\ 2 \\ 3 \end{bmatrix}, \quad \text{and} \quad t_0 = 2.$$

Thus, the initial value problem may be rewritten in the fundamental form

$$\dot{\mathbf{x}}(t) = \mathbf{A}(t)\mathbf{x}(t) + \mathbf{f}(t),$$
$$\mathbf{x}(t_0) = \mathbf{c}.$$

Problems 8.2

Put the following initial value problems into fundamental form:

1. $\dfrac{d^2x}{dt^2} - 2\dfrac{dx}{dt} - 3x = 0;$

   $x(0) = 4, \quad \dfrac{dx(0)}{dt} = 5.$

2. $\dfrac{d^2x}{dt^2} + e^t\dfrac{dx}{dt} - tx = 0;$

   $x(1) = 2, \quad \dfrac{dx(1)}{dt} = 0.$

3. $\dfrac{d^2x}{dt^2} - x = t^2;$

   $x(0) = -3, \quad \dfrac{dx(0)}{dt} = 3.$

4. $e^t\dfrac{d^2x}{dt^2} - 2e^{2t}\dfrac{dx}{dt} - 3e^t x = 2;$

   $x(0) = 0, \quad \dfrac{dx(0)}{dt} = 0.$

5. $\ddot{x} - 3\dot{x} + 2x = e^{-t}$,
   $x(1) = \dot{x}(1) = 2.$

6. $4\ddot{x} + t\ddot{x} - x = 0$,
   $x(-1) = 2, \quad \dot{x}(-1) = 1, \quad \ddot{x}(-1) = -205.$

7. $e^t\dfrac{d^4x}{dt^4} + t\dfrac{d^2x}{dt^2} = 1 + \dfrac{dx}{dt}$,

   $x(0) = 1, \quad \dfrac{dx(0)}{dt} = 2, \quad \dfrac{d^2x(0)}{dt^2} = \pi, \quad \dfrac{d^3x(0)}{dt^3} = e^3.$

8. $\dfrac{d^6x}{dt^6} + 4\dfrac{d^4x}{dt^4} = t^2 - t$,

   $x(\pi) = 2, \quad \dot{x}(\pi) = 1, \quad \ddot{x}(\pi) = 0, \quad \dddot{x}(\pi) = 2,$

   $\dfrac{d^4x(\pi)}{dt^4} = 1, \quad \dfrac{d^5x(\pi)}{dt^5} = 0.$

8.3 Reduction of a system

Based on our work in the preceding section, we are now able to reduce systems of higher order linear differential equations to fundamental form. The method, which is a straightforward extension of that used to reduce the nth order differential equation to fundamental form, is best demonstrated by examples.

Example 1. Put the following system into fundamental form:

$$\ddot{x} = 5\ddot{x} + \dot{y} - 7y + e^t,$$
$$\ddot{y} = \dot{x} - 2\dot{y} + 3y + \sin t, \tag{20}$$
$$x(1) = 2, \quad \dot{x}(1) = 3, \quad \ddot{x}(1) = -1, \quad y(1) = 0, \quad \dot{y}(1) = -2.$$

Step 1. Rewrite the differential equations so that the highest derivative of *each* unknown function appears by itself. For the above system, this has already been done.

Step 2. Define new variables $x_1(t)$, $x_2(t)$, $x_3(t)$, $y_1(t)$, and $y_2(t)$. (Since the highest derivative of $x(t)$ is of order 3, and the highest derivative of $y(t)$ is of order 2, we need 3 new variables for $x(t)$ and 2 new variables for $y(t)$. In general, for each unknown function, we define a set a k new variables, where k is the order of the highest derivative of the original function appearing in the system under consideration.) The new variables are defined in a manner analogous to that used in the previous section:

$$\begin{aligned} x_1 &= x, \\ x_2 &= \dot{x}_1, \\ x_3 &= \dot{x}_2, \\ y_1 &= y, \\ y_2 &= \dot{y}_1. \end{aligned} \tag{21}$$

From (21), the new variables are related to the functions $x(t)$ and $y(t)$ by the following:

$$
\begin{aligned}
x_1 &= x, \\
x_2 &= \dot{x}, \\
x_3 &= \ddot{x}, \\
y_1 &= y, \\
y_2 &= \dot{y}.
\end{aligned}
\tag{22}
$$

It follows from (22), by differentiating x_3 and y_2, that

$$
\begin{aligned}
\dot{x}_3 &= \dddot{x}, \\
\dot{y}_2 &= \ddot{y}.
\end{aligned}
\tag{23}
$$

Step 3. Rewrite $\dot{x}_3$ and $\dot{y}_2$ in terms of the new variables defined in (21).
Substituting (20) into (23), we have

$$
\begin{aligned}
\dot{x}_3 &= 5\ddot{x} + \dot{y} - 7y + e^t, \\
\dot{y}_2 &= \dot{x} - 2\dot{y} + 3y + \sin t.
\end{aligned}
$$

Substituting (22) into these equations, we obtain

$$
\begin{aligned}
\dot{x}_3 &= 5x_3 + y_2 - 7y_1 + e^t, \\
\dot{y}_2 &= x_2 - 2y_2 + 3y_1 + \sin t.
\end{aligned}
\tag{24}
$$

Step 4. Set up a system of first order differential equations for $x_1, x_2, x_3, y_1,$ and y_2.
Using (21) and (24), we obtain the system

$$
\begin{aligned}
\dot{x}_1 &= x_2, \\
\dot{x}_2 &= x_3, \\
\dot{x}_3 &= 5x_3 - 7y_1 + y_2 + e^t, \\
\dot{y}_1 &= y_2, \\
\dot{y}_2 &= x_2 + 3y_1 - 2y_2 + \sin t.
\end{aligned}
\tag{25}
$$

Note that for convenience we have rearranged terms in some of the equations to present them in their natural order.

Step 5. Write (25) in matrix form.
Define

$$
\mathbf{x}(t) = \begin{bmatrix} x_1(t) \\ x_2(t) \\ x_3(t) \\ y_1(t) \\ y_2(t) \end{bmatrix}, \quad
\mathbf{A}(t) = \begin{bmatrix} 0 & 1 & 0 & 0 & 0 \\ 0 & 0 & 1 & 0 & 0 \\ 0 & 0 & 5 & -7 & 1 \\ 0 & 0 & 0 & 0 & 1 \\ 0 & 1 & 0 & 3 & -2 \end{bmatrix}, \quad \text{and} \quad
\mathbf{f}(t) = \begin{bmatrix} 0 \\ 0 \\ e^t \\ 0 \\ \sin t \end{bmatrix}.
\tag{26}
$$

Thus, Eq. (25) can be rewritten in the matrix form

$$
\dot{\mathbf{x}}(t) = \mathbf{A}(t)\mathbf{x}(t) + \mathbf{f}(t).
\tag{27}
$$

Step 6. Rewrite the initial conditions in matrix form. From Eqs. (20), (22), and (26) we have

$$
\mathbf{x}(1) = \begin{bmatrix} x_1(1) \\ x_2(1) \\ x_3(1) \\ y_1(1) \\ y_2(1) \end{bmatrix} = \begin{bmatrix} x(1) \\ \dot{x}(1) \\ \ddot{x}(1) \\ y(1) \\ \dot{y}(1) \end{bmatrix} = \begin{bmatrix} 2 \\ 3 \\ -1 \\ 0 \\ -2 \end{bmatrix}.
$$

Thus, if we define

$$\mathbf{c} = \begin{bmatrix} 2 \\ 3 \\ -1 \\ 0 \\ -2 \end{bmatrix}$$

and $t_0 = 1$, then the initial conditions can be rewritten as

$$\mathbf{x}(1) = \mathbf{c}. \blacksquare \tag{28}$$

Since $\mathbf{A}(t)$ and $\mathbf{f}(t)$ are continuous, (27) and (28) possess a unique solution. Once $\mathbf{x}(t)$ is known, we immediately have the components of $\mathbf{x}(t)$, namely $x_1(t)$, $x_2(t)$, $x_3(t)$, $y_1(t)$, and $y_2(t)$. Thus, we have the functions $x(t)$ and $y(t)$ (from (21), $x_1(t) = x(t)$ and $y_1(t) = y(t)$).

All similar systems containing higher order derivatives may be put into fundamental form in exactly the same manner as that used here.

Example 2. Put the following system into fundamental form:

$$\ddot{x} = 2\dot{x} + t\dot{y} - 3z + t^2\dot{z} + t,$$
$$\ddot{y} = \dot{z} + (\sin t)y + x - t,$$
$$\dot{z} = \ddot{x} - \ddot{y} + t^2 + 1;$$
$$x(\pi) = 15, \quad \dot{x}(\pi) = 59, \quad \ddot{x}(\pi) = -117, \quad y(\pi = 2) \quad \dot{y}(\pi) = -19,$$
$$\ddot{y}(\pi) = 3, \quad z(\pi) = 36, \quad \dot{z}(\pi) = -212.$$

Solution Define

$$x_1 = x$$
$$x_2 = \dot{x}_1 = \dot{x}$$
$$x_3 = \dot{x}_2 = \ddot{x}; \quad \text{hence,} \quad \dot{x}_3 = \dddot{x}.$$
$$y_1 = y$$
$$y_2 = \dot{y}_1 = \dot{y}$$
$$y_3 = \dot{y}_2 = \ddot{y}; \quad \text{hence,} \quad \dot{y}_3 = \dddot{y}.$$
$$z_1 = x$$
$$z_2 = \dot{z}_1 = \dot{z}; \quad \text{hence,} \quad \dot{z}_2 = \ddot{z}.$$

Thus,

$$\dot{x}_1 = x_2$$
$$\dot{x}_2 = x_3$$
$$\dot{x}_3 = \dddot{x} = 2\dot{x} + t\dot{y} - 3z + t^2\dot{z} + t$$
$$\quad = 2x_2 + ty_2 - 3z_1 + t^2 z_2 + t;$$
$$\dot{y}_1 = y_2$$
$$\dot{y}_2 = y_3$$
$$\dot{y}_3 = \dddot{y} = \dot{z} + (\sin t)y + x - t$$
$$\quad = z_2 + (\sin t)y_1 + x_1 - t;$$
$$\dot{z}_1 = z_2$$
$$\dot{z}_2 = \dot{z} = \ddot{x} - \ddot{y} + t^2 + 1$$
$$\quad = x_3 - y_3 + t^2 + 1;$$

or

$$\dot{x}_1 = \quad x_2$$
$$\dot{x}_2 = \quad\quad x_3$$
$$\dot{x}_3 = \quad 2x_2 \quad\quad +ty_2 \quad\quad -3z_1 + t^2 z_2 + t$$
$$\dot{y}_1 = \quad\quad\quad y_2$$
$$\dot{y}_2 = \quad\quad\quad\quad y_3$$
$$\dot{y}_3 = x_1 \quad +(\sin t)y_1 \quad\quad z_2 - t$$
$$\dot{z}_1 = \quad\quad\quad\quad z_2$$
$$\dot{z}_2 = \quad\quad x_3 \quad\quad -y_3 \quad\quad +t^2 + 1.$$

Define

$$\mathbf{x} = \begin{bmatrix} x_1 \\ x_2 \\ x_3 \\ y_1 \\ y_2 \\ y_3 \\ z_1 \\ z_2 \end{bmatrix}, \qquad \mathbf{A}(t) = \begin{bmatrix} 0 & 1 & 0 & 0 & 0 & 0 & 0 & 0 \\ 0 & 0 & 1 & 0 & 0 & 0 & 0 & 0 \\ 0 & 2 & 0 & 0 & t & 0 & -3 & t^2 \\ 0 & 0 & 0 & 0 & 1 & 0 & 0 & 0 \\ 0 & 0 & 0 & 0 & 0 & 1 & 0 & 0 \\ 1 & 0 & 0 & \sin t & 0 & 0 & 0 & 1 \\ 0 & 0 & 0 & 0 & 0 & 0 & 0 & 1 \\ 0 & 0 & 1 & 0 & 0 & -1 & 0 & 0 \end{bmatrix},$$

$$\mathbf{f}(t) = \begin{bmatrix} 0 \\ 0 \\ t \\ 0 \\ 0 \\ -t \\ 0 \\ t^2 + 1 \end{bmatrix}, \qquad \mathbf{c} = \begin{bmatrix} 15 \\ 59 \\ -117 \\ 2 \\ -19 \\ 3 \\ 36 \\ -212 \end{bmatrix}, \quad \text{and} \quad t_0 = \pi.$$

Thus, the system can now be rewritten in the fundamental form

$$\dot{\mathbf{x}}(t) = \mathbf{A}(t)\mathbf{x}(t) + \mathbf{f}(t),$$
$$\mathbf{x}(t_0) = \mathbf{c}. \quad \blacksquare$$

Problems 8.3

Put the following initial value problems into fundamental form:

1. $\dfrac{d^2x}{dt^2} = 2\dfrac{dx}{dt} + 3x + 4y,$

$\dfrac{dy}{dt} = 5x - 6y,$

$x(0) = 7,$

$\dfrac{dx(0)}{dt} = 8,$

$y(0) = 9.$

2. $\dfrac{d^2x}{dt^2} = \dfrac{dx}{dt} + \dfrac{dy}{dt},$

$\dfrac{d^2y}{dt^2} = \dfrac{dy}{dt} - \dfrac{dx}{dt},$

$x(0) = 2, \quad \dfrac{dx(0)}{dt} = 3, \quad y(0) = 4, \quad \dfrac{dy(0)}{dt} = 4.$

3. $\dfrac{dx}{dt} = t^2\dfrac{dy}{dt} - 4x,$

$\dfrac{d^2y}{dt^2} = ty + t^2x,$

$x(2) = -1, \quad y(2) = 0, \quad \dfrac{dy(2)}{dt} = 0.$

4. $\dfrac{dx}{dt} = 2\dfrac{dy}{dt} - 4x + t,$

$\dfrac{d^2y}{dt^2} = ty + 3x - 1,$

$x(3) = 0, \quad y(3) = 0, \quad \dfrac{dy(3)}{dt} = 0.$

5. $\ddot{x} = 2\dot{x} + \ddot{y} - t,$
$\dddot{y} = tx - ty + \ddot{y} - e^t;$
$x(-1) = 2, \quad \dot{x}(-1) = 0, \quad y(-1) = 0, \quad \dot{y}(-1) = 3, \quad \ddot{y}(-1) = 9,$
$\dddot{y}(-1) = 4.$

6. $\ddot{x} = x - y + \dot{y},$
$\ddot{y} = \dot{x} - x + 2\dot{y};$
$x(0) = 21, \quad \dot{x}(0) = 4, \quad \ddot{x}(0) = -5, \quad y(0) = 5, \quad \dot{y}(0) = 7.$

7. $\dot{x} = y - 2,$
    $\dot{y} = z - 2,$
    $\dot{z} = x + y;$
    $x(\pi) = 1, \quad y(\pi) = 2, \quad \dot{y}(\pi) = 17, \quad z(\pi) = 0.$

8. $\ddot{x} = y + z + 2,$
    $\ddot{y} = x + y - 1,$
    $\ddot{z} = x - z + 1;$
    $x(20) = 4, \quad \dot{x}(20) = -4 \quad y(20) = 5, \quad \dot{y}(20) = -5, \quad z(20) = 9$
    $\dot{z}(20) = -9.$

8.4 Solutions of systems with constant coefficients

In general, when one reduces a system of differential equations to fundamental form, the matrix $\mathbf{A}(t)$ will depend explicitly on the variable t. For some systems, however, $\mathbf{A}(t)$ does not vary with t (that is, every element of $\mathbf{A}(t)$ is a constant). If this is the case, the system is said to have *constant coefficients*. For instance, in Section 8.1, Example 3 illustrates a system having constant coefficients, while Example 2 illustrates a system that does not have constant coefficients.

In this section, we only consider systems having constant coefficients; hence, we shall designate the matrix $\mathbf{A}(t)$ as $\mathbf{A}$ in order to emphasize its independence of t. We seek the solution to the initial value problem in the fundamental form

$$\dot{\mathbf{x}}(t) = \mathbf{A}\mathbf{x}(t) + \mathbf{f}(t), \tag{29}$$

$$\mathbf{x}(t_0) = \mathbf{c}.$$

The differential equation in (29) can be written as

$$\dot{\mathbf{x}}(t) - \mathbf{A}\mathbf{x}(t) = \mathbf{f}(t). \tag{30}$$

If we premultiply each side of (30) by $e^{-\mathbf{A}t}$, we obtain

$$e^{-\mathbf{A}t}[\dot{\mathbf{x}}(t) - \mathbf{A}\mathbf{x}(t)] = e^{-\mathbf{A}t}\mathbf{f}(t). \tag{31}$$

Using matrix differentiation and Corollary 1 of Section 7.9, we find that

$$\frac{d}{dt}\left[e^{-\mathbf{A}t}\mathbf{x}(t)\right] = e^{-\mathbf{A}t}(-\mathbf{A})\mathbf{x}(t) + e^{-\mathbf{A}r}\dot{\mathbf{x}}(t) \tag{32}$$

$$= e^{-\mathbf{A}t}[\dot{\mathbf{x}}(t) - \mathbf{A}\mathbf{x}(t)].$$

Substituting (32) into (31), we obtain

$$\frac{d}{dt}\left[e^{-\mathbf{A}t}\mathbf{x}(t)\right] = e^{-\mathbf{A}t}\mathbf{f}(t). \tag{33}$$

Integrating (33) between the limits $t = t_0$ and $t = t$, we have

$$\int_{t_0}^{t} \frac{d}{dt}\left[e^{-\mathbf{A}t}\mathbf{x}(t)\right]dt = \int_{t_0}^{t} e^{-\mathbf{A}t}\mathbf{f}(t)dt$$

$$= \int_{t_0}^{t} e^{-\mathbf{A}s}\mathbf{f}(s)ds. \tag{34}$$

Note that we have replaced the dummy variable t by the dummy variable s in the right-hand integral of (34), which *in no way* alters the definite integral (see Problem 1).

Upon evaluating the left-hand integral, it follows from (34) that

$$e^{-\mathbf{A}t}\mathbf{x}(t)\Big|_{t_0}^{t} = \int_{t_0}^{t} e^{-\mathbf{A}s}\mathbf{f}(s)ds$$

or that

$$e^{-\mathbf{A}t}\mathbf{x}(t) = e^{-\mathbf{A}t_0}\mathbf{x}(t_0) + \int_{t_0}^{t} e^{-\mathbf{A}s}\mathbf{f}(s)ds. \tag{35}$$

But $\mathbf{x}(t_0) = \mathbf{c}$; hence,

$$e^{-\mathbf{A}t}\mathbf{x}(t) = e^{-\mathbf{A}t_0}\mathbf{c} + \int_{t_0}^{t} e^{-\mathbf{A}s}\mathbf{f}(s)ds. \tag{36}$$

Premultiplying both sides of (36) by $(e^{-\mathbf{A}t})^{-1}$, we obtain

$$\mathbf{x}(t) = (e^{-\mathbf{A}t})^{-1}e^{-\mathbf{A}t_0}\mathbf{c} + (e^{-\mathbf{A}t})^{-1}\int_{t_0}^{t} e^{-\mathbf{A}s}\mathbf{f}(s)ds. \tag{37}$$

Using Property 2 of Section 7.8, we have

$$(e^{-\mathbf{A}t})^{-1} = e^{\mathbf{A}t}.$$

Whereupon we can rewrite (37) as

$$\mathbf{x}(t) = e^{-\mathbf{A}t}e^{-\mathbf{A}t_0}\mathbf{c} + e^{\mathbf{A}t}\int_{t_0}^{t} e^{-\mathbf{A}s}\mathbf{f}(s)ds. \tag{38}$$

Since $\mathbf{A}t$ and $-\mathbf{A}t_0$ commute (why?), we have from Problem 10 of Section 7.8.

$$e^{\mathbf{A}t}e^{-\mathbf{A}t_0} = e^{\mathbf{A}(t-t_0)}. \tag{39}$$

Finally, using (39), we can rewrite (38) as

$$\mathbf{x}(t) = e^{\mathbf{A}(t-t_0)}\mathbf{c} + e^{\mathbf{A}t}\int_{t_0}^{t} e^{-\mathbf{A}s}\mathbf{f}(s)ds. \tag{40}$$

Eq. (8.40) is the unique solution to the initial value problem given by (8.29).

A simple method for calculating the quantities $e^{\mathbf{A}\,(t-t_0)}$ and $e^{-\mathbf{A}s}$ is to first compute $e^{\mathbf{A}t}$ (see Section 7.6) and then replace the variable t wherever it appears by the variables $t - t_0$ and $(-s)$, respectively.

Example 1. Find $e^{\mathbf{A}\,(t-t_0)}$ and $e^{-\mathbf{A}s}$ for

$$\mathbf{A} = \begin{bmatrix} -1 & 1 \\ 0 & -1 \end{bmatrix}.$$

Solution Using the method of Section 7.6, we calculate $e^{\mathbf{A}t}$ as

$$e^{\mathbf{A}t} = \begin{bmatrix} e^{-t} & te^{-t} \\ 0 & e^{-t} \end{bmatrix}.$$

Hence,

$$e^{\mathbf{A}(t-t_0)} = \begin{bmatrix} e^{-(t-t_0)} & (t-t_0)e^{-(t-t_0)} \\ 0 & e^{-(t-t_0)} \end{bmatrix}$$

and

$$e^{-\mathbf{A}s} = \begin{bmatrix} e^{s} & -se^{s} \\ 0 & e^{s} \end{bmatrix}. \blacksquare$$

Note that when t is replaced by $(t - t_0)$ in e^{-t}, the result is $e^{-(t-t_0)} = e^{-t+t_0}$ and not e^{-t-t_0}. That is, we replaced the *quantity* t by the *quantity* $(t - t_0)$; we did not simply add $-t_0$ to the variable t wherever it appeared. Also note that the same result could have been obtained for $e^{-\mathbf{A}s}$ by first computing $e^{\mathbf{A}s}$ and then inverting by the method of cofactors (recall that $e^{-\mathbf{A}s}$ is the inverse of $e^{\mathbf{A}s}$) or by computing $e^{-\mathbf{A}s}$ directly (define $\mathbf{B} = -\mathbf{A}s$ and calculate $e^{\mathbf{B}}$). However, if $e^{\mathbf{A}t}$ is already known, the above method is by far the most expedient one for obtaining $e^{-\mathbf{A}s}$.

We can derive an alternate representation for the solution vector $\mathbf{x}(t)$ if we note that $e^{\mathbf{A}t}$ depends only on t and the integration is with respect to s. Hence, $e^{\mathbf{A}t}$ can be brought inside the integral, and (40) can be rewritten as

$$\mathbf{x}(t) = e^{\mathbf{A}(t-t_0)}\mathbf{c} + \int_{t_0}^{t} e^{\mathbf{A}t} e^{-\mathbf{A}s} \mathbf{f}(s)ds.$$

Since $\mathbf{A}t$ and $-\mathbf{A}s$ commute, we have that

$$e^{\mathbf{A}t} e^{-\mathbf{A}s} = e^{\mathbf{A}(t-s)}.$$

Thus, the solution to (29) can be written as

$$\mathbf{x}(t) = e^{\mathbf{A}(t-t_0)}\mathbf{c} + \int_{t_0}^{t} e^{\mathbf{A}(t-s)} \mathbf{f}(s)ds. \tag{41}$$

Again the quantity $e^{\mathbf{A}(t-s)}$ can be obtained by replacing the variable t in $e^{\mathbf{A}t}$ by the variable $(t - s)$.

In general, the solution $\mathbf{x}(t)$ may be obtained quicker by using (41) than by using (40), since there is one less multiplication involved. (Note that in (40) one must pre-multiply the integral by $e^{\mathbf{A}t}$, while in (41) this step is eliminated.) However, since the integration in (41) is more difficult than that in (40), the reader who is not confident of his or her integrating abilities will probably be more comfortable using (40).

If one has a homogeneous initial value problem with constant coefficients, that is, a system defined by

$$\dot{\mathbf{x}}(t) = \mathbf{A}\mathbf{x}(t),$$
$$\mathbf{x}(t_0) = \mathbf{c}, \tag{42}$$

a great simplification of (40) is effected. In this case, $\mathbf{f}(t) \equiv 0$. The integral in (40), therefore, becomes the zero vector and the solution to the system given by (42) is

$$\mathbf{x}(t) = e^{\mathbf{A}(t-t_0)}\mathbf{c}. \tag{43}$$

Occasionally, we are interested in just solving a differential equation and not an entire initial value problem. In this case, the general solution can be shown to be (see Problem 2)

$$\mathbf{x}(t) = e^{\mathbf{A}t}\mathbf{k} + e^{\mathbf{A}t}\int e^{-\mathbf{A}t}\mathbf{f}(t)dt, \tag{44}$$

where $\mathbf{k}$ is an arbitrary n-dimensional constant vector. The general solution to the homogeneous differential equation by itself is given by

$$\mathbf{x}(t) = e^{\mathbf{A}t}\mathbf{k}. \tag{45}$$

Example 2. Use matrix methods to solve

$$\dot{u}(t) = u(t) + 2v(t) + 1$$
$$\dot{v}(t) = 4u(t) + 3v(t) - 1$$
$$u(0) = 1, \qquad v(0) = 2.$$

Solution This system can be put into fundamental form if we define $t_0 = 0$,

$$\mathbf{x}(t) = \begin{bmatrix} u(t) \\ v(t) \end{bmatrix}, \quad \mathbf{A} = \begin{bmatrix} 1 & 2 \\ 4 & 3 \end{bmatrix}, \quad \mathbf{f}(t) = \begin{bmatrix} 1 \\ -1 \end{bmatrix}, \quad \text{and} \quad \mathbf{c} = \begin{bmatrix} 1 \\ 2 \end{bmatrix}. \tag{46}$$

Since $\mathbf{A}$ is independent of t, this is a system with constant coefficients and the solution is given by (40). For the $\mathbf{A}$ in (46), $e^{\mathbf{A}t}$ is found to be

$$e^{\mathbf{A}t} = \frac{1}{6}\begin{bmatrix} 2e^{5t} + 4e^{-t} & 2e^{5t} - 2e^{-t} \\ 4e^{5t} - 4e^{-t} & 4e^{5t} + 2e^{-t} \end{bmatrix}.$$

Hence,

$$e^{-\mathbf{A}s} = \frac{1}{6} \begin{bmatrix} 2e^{-5s} + 4e^s & 2e^{-5s} - 2e^s \\ 4e^{-5s} - 4e^s & 4e^{-5s} + 2e^s \end{bmatrix}$$

and

$$e^{\mathbf{A}(t-t_0)} = e^{\mathbf{A}t}, \quad \text{since} \quad t_0 = 0.$$

Thus,

$$
\begin{aligned}
e^{\mathbf{A}(t-t_0)}\mathbf{c} &= \frac{1}{6} \begin{bmatrix} 2e^{5t} + 4e^{-t} & 2e^{5t} - 2e^{-t} \\ 4e^{5t} - 4e^{-t} & 4e^{5t} + 2e^{-t} \end{bmatrix} \begin{bmatrix} 1 \\ 2 \end{bmatrix} \\
&= \frac{1}{6} \begin{bmatrix} 1\left[2e^{5t} + 4e^{-t}\right] + 2\left[2e^{5t} - 2e^{-t}\right] \\ 1\left[4e^{5t} - 4e^{-t}\right] + 2\left[4e^{5t} + 2e^{-t}\right] \end{bmatrix} \\
&= \begin{bmatrix} e^{5t} \\ 2e^{5t} \end{bmatrix},
\end{aligned}
\tag{47}
$$

and

$$
\begin{aligned}
e^{-\mathbf{A}s}\mathbf{f}(s) &= \frac{1}{6} \begin{bmatrix} 2e^{5t} + 4e^{-t} & 2e^{5t} - 2e^{-t} \\ 4e^{5t} - 4e^{-t} & 4e^{5t} + 2e^{-t} \end{bmatrix} \begin{bmatrix} 1 \\ -1 \end{bmatrix} \\
&= \frac{1}{6} \begin{bmatrix} 1\left[2e^{5t} + 4e^{-t}\right] - 1\left[2e^{5t} - 2e^{-t}\right] \\ 1\left[4e^{5t} - 4e^{-t}\right] - 1\left[4e^{5t} + 2e^{-t}\right] \end{bmatrix} = \begin{bmatrix} e^s \\ -e^s \end{bmatrix}.
\end{aligned}
$$

Hence,

$$e^{\mathbf{A}t} \int_{t_0}^{t} e^{-\mathbf{A}s}\mathbf{f}(s)\,ds = \begin{bmatrix} \int_0^t e^s\,ds \\ \int_0^t -e^s\,ds \end{bmatrix} = [e^s |]$$

and

$$
\begin{aligned}
e^{\mathbf{A}t} \int_{t_0}^{t} e^{-\mathbf{A}s}\mathbf{f}(s)\,ds &= \frac{1}{6} \begin{bmatrix} 2e^{5t} + 4e^{-t} & 2e^{5t} - 2e^{-t} \\ 4e^{5t} - 4e^{-t} & 4e^{5t} + 2e^{-t} \end{bmatrix} \begin{bmatrix} (e^t - 1) \\ (1 + e^t) \end{bmatrix} \\
&= \frac{1}{6} \begin{bmatrix} \left[2e^{5t} + 4e^{-t}\right]\left[e^t - 1\right] + \left[2e^{5t} - 2e^{-t}\right]\left[1 - e^t\right] \\ \left[4e^{5t} - 4e^{-t}\right]\left[e^t - 1\right] + \left[4e^{5t} + 2e^{-t}\right]\left[1 - e^t\right] \end{bmatrix} \\
&= \begin{bmatrix} (1 - e^{-t}) \\ (-1 + e^{-t}) \end{bmatrix}.
\end{aligned}
\tag{48}
$$

Substituting (47) and (48) into (40), we have

$$\begin{bmatrix} u(t) \\ v(t) \end{bmatrix} = \mathbf{x}(t) = \begin{bmatrix} e^{5t} \\ 2e^{5t} \end{bmatrix} + \begin{bmatrix} 1 - e^{-t} \\ -1 + e^{-t} \end{bmatrix} = \begin{bmatrix} e^{5t} + 1 - e^{-t} \\ 2e^{5t} - 1 + e^{-t} \end{bmatrix},$$

or

$$u(t) = e^{5t} - e^{-t} + 1,$$
$$v(t) = 2e^{5t} + e^{-t} - 1. \quad \blacksquare$$

Example 3. Use matrix methods to solve

$$\ddot{y} - 3\dot{y} + 2y = e^{-3t},$$
$$y(1) = 1, \quad \dot{y}(1) = 0.$$

Solution This system can be put into fundamental form, if we define $t_0 = 1$:

$$\mathbf{x}(t) = \begin{bmatrix} x_1(t) \\ x_2(t) \end{bmatrix}, \quad \mathbf{A} = \begin{bmatrix} 0 & 1 \\ -2 & 3 \end{bmatrix}, \quad \mathbf{f}(t) = \begin{bmatrix} 0 \\ e^{-3t} \end{bmatrix}, \quad \text{and} \quad \mathbf{c} = \begin{bmatrix} 1 \\ 0 \end{bmatrix}.$$

The solution to this system is given by (40). For this $\mathbf{A}$,

$$e^{\mathbf{A}t} = \begin{bmatrix} -e^{2t} + 2e^t & e^{2t} - e^t \\ -2e^{2t} + 2e^t & 2e^{2t} - e^t \end{bmatrix}.$$

Thus,

$$e^{\mathbf{A}(t-t_0)}\mathbf{c} = \begin{bmatrix} -e^{2(t-1)} + 2e^{(t-1)} & e^{2(t-1)} - e^{(t-1)} \\ -2e^{2(t-1)} + 2e^{(t-1)} & 2e^{2(t-1)} - e^{(t-1)} \end{bmatrix} \begin{bmatrix} 1 \\ 0 \end{bmatrix}$$

$$= \begin{bmatrix} -e^{2(t-1)} + 2e^{(t-1)} \\ -2e^{2(t-1)} + 2e^{(t-1)} \end{bmatrix}. \tag{49}$$

Now

$$\mathbf{f}(t) = \begin{bmatrix} 0 \\ e^{-3t} \end{bmatrix}, \quad \mathbf{f}(s) = \begin{bmatrix} 0 \\ e^{-3s} \end{bmatrix},$$

and

$$e^{-\mathbf{A}s}\mathbf{f}(s) = \begin{bmatrix} -e^{-2s} + 2e^{-s} & e^{-2s} - e^{-s} \\ -2e^{-2s} + 2e^{-s} & 2e^{-2s} - e^{-s} \end{bmatrix} \begin{bmatrix} 0 \\ e^{-3s} \end{bmatrix}$$

$$= \begin{bmatrix} e^{-5s} - e^{-4s} \\ 2e^{-5s} - e^{-4s} \end{bmatrix}.$$

Hence,

$$\int_{t_0}^{t} e^{-\mathbf{A}s}\mathbf{f}(s)\,ds = \begin{bmatrix} \int_1^t (e^{-5s} - e^{-4s})\,ds \\ \int_1^t (2e^{-5s} - e^{-4s})\,ds \end{bmatrix}$$

$$= \begin{bmatrix} \left(-\frac{1}{5}\right)e^{-5t} + \left(\frac{1}{4}\right)e^{-4t} + \left(\frac{1}{5}\right)e^{-5} - \left(\frac{1}{4}\right)e^{-4} \\ \left(-\frac{1}{5}\right)e^{-5t} + \left(\frac{1}{4}\right)e^{-4t} + \left(\frac{1}{5}\right)e^{-5} - \left(\frac{1}{4}\right)e^{-4} \end{bmatrix},$$

and

$$e^{\mathbf{A}t}\int_{t_0}^{t} e^{-\mathbf{A}s}\mathbf{f}(s)\,ds$$

$$= \begin{bmatrix} (-e^{2t}+2e^{t}) & (e^{2t}-e^{t}) \\ (-2e^{2t}+2e^{t}) & (2e^{2t}-e^{t}) \end{bmatrix} \begin{bmatrix} \left(-\dfrac{1}{5}e^{-5t}+\dfrac{1}{4}e^{-4t}+\dfrac{1}{5}e^{-5}-\dfrac{1}{4}e^{-4}\right) \\ \left(-\dfrac{2}{5}e^{-5t}+\dfrac{1}{4}e^{-4t}+\dfrac{2}{5}e^{-5}-\dfrac{1}{4}e^{-4}\right) \end{bmatrix} \qquad (50)$$

$$= \begin{bmatrix} \dfrac{1}{20}e^{-3t}+\dfrac{1}{5}e^{(2t-5)}-\dfrac{1}{4}e^{t-4} \\[2mm] -\dfrac{3}{20}e^{-3t}+\dfrac{2}{5}e^{(2t-5)}-\dfrac{1}{4}e^{t-4} \end{bmatrix}.$$

Substituting (49) and (50) into (40), we have that

$$\mathbf{x}(t) = \begin{bmatrix} x_1(t) \\ x_2(t) \end{bmatrix} = \begin{bmatrix} -e^{2(t-1)}+2e^{t-1} \\ -2e^{2(t-1)}+2e^{t-1} \end{bmatrix} + \begin{bmatrix} \dfrac{1}{20}e^{-3t}+\dfrac{1}{5}e^{(2t-5)}-\dfrac{1}{4}e^{t-4} \\[2mm] -\dfrac{3}{20}e^{-3t}+\dfrac{2}{5}e^{(2t-5)}-\dfrac{1}{4}e^{t-4} \end{bmatrix}$$

$$= \begin{bmatrix} -e^{2(t-1)}+2e^{t-1}+\dfrac{1}{20}e^{-3t}+\dfrac{1}{5}e^{(2t-5)}-\dfrac{1}{4}e^{t-4} \\[2mm] -2e^{2(t-1)}+2e^{t-1}-\dfrac{3}{20}e^{-3t}+\dfrac{2}{5}e^{(2t-5)}-\dfrac{1}{4}e^{t-4} \end{bmatrix}.$$

Thus, it follows that the solution to the initial value problem is given by

$$y(t) = x_1(t)$$

$$= -e^{2(t-1)}+2e^{t-1}+\frac{1}{20}e^{-3t}+\frac{1}{5}e^{(2t-5)}-\frac{1}{4}e^{t-4}. \blacksquare$$

The most tedious step in Example 3 was multiplying the matrix $e^{\mathbf{A}t}$ by the vector $\int_{t_0}^{t} e^{-\mathbf{A}s}\mathbf{f}(s)\,ds$. We could have eliminated this multiplication had we used (41) for the solution rather than (40). Of course, in using (41), we would have had to handle an integral rather more complicated than the one we encountered.

If $\mathbf{A}$ and $\mathbf{f}(t)$ are relatively simple (for instance, if $\mathbf{f}(t)$ is a constant vector), then the integral obtained in (41) may not be too tedious to evaluate, and its use can be a real savings in time and effort over the use of (40). We illustrate this point in the next example.

Example 4.   Use matrix methods to solve

$$\ddot{x}(t) + x(t) = 2,$$
$$x(\pi) = 0, \quad \dot{x}(\pi) = -1.$$

Solution This initial valued problem can be put into fundamental form if we define $t_0 = \pi$,

$$\mathbf{x}(t) = \begin{bmatrix} x_1(t) \\ x_2(t) \end{bmatrix}, \quad \mathbf{A} = \begin{bmatrix} 0 & 1 \\ -1 & 0 \end{bmatrix}, \quad \mathbf{f}(t) = \begin{bmatrix} 0 \\ 2 \end{bmatrix}, \quad \text{and} \quad \mathbf{c} = \begin{bmatrix} 0 \\ -1 \end{bmatrix}. \tag{51}$$

Here, $\mathbf{A}$ is again independent of the variable t; hence, the solution is given by either (40) or (41). This time we elect to use (41). For the $\mathbf{A}$ given in (51), $e^{\mathbf{A}t}$ is found to be

$$e^{\mathbf{A}t} = \begin{bmatrix} \cos t & \sin t \\ -\sin t & \cos t \end{bmatrix}.$$

Thus,

$$\begin{aligned} e^{\mathbf{A}(t-t_0)}\mathbf{c} &= \begin{bmatrix} \cos(t-\pi) & \sin(t-\pi) \\ -\sin(t-\pi) & \cos(t-\pi) \end{bmatrix} \begin{bmatrix} 0 \\ -1 \end{bmatrix} \\ &= \begin{bmatrix} -\sin(t-\pi) \\ -\cos(t-\pi) \end{bmatrix}, \end{aligned} \tag{52}$$

and

$$\begin{aligned} e^{\mathbf{A}(t-s)}\mathbf{f}(s) &= \begin{bmatrix} \cos(t-s) & \sin(t-s) \\ -\sin(t-s) & \cos(t-s) \end{bmatrix} \begin{bmatrix} 0 \\ 2 \end{bmatrix} \\ &= \begin{bmatrix} 2\sin(t-s) \\ 2\cos(t-s) \end{bmatrix}. \end{aligned}$$

Hence,

$$\begin{aligned} \int_{t_0}^{t} e^{\mathbf{A}(t-s)}\mathbf{f}(s)ds &= \begin{bmatrix} \int_{\pi}^{t} 2\sin(t-s)ds \\ \int_{\pi}^{t} 2\cos(t-s)ds \end{bmatrix} \\ &= \begin{bmatrix} 2 - 2\cos(t-\pi) \\ 2\sin(t-\pi) \end{bmatrix}. \end{aligned} \tag{53}$$

Substituting (52) and (53) into (41) and using the trigonometric identities $\sin(t-\pi) = -\sin t$ and $\cos(t-\pi) = -\cos t$, we have

$$\begin{aligned} \begin{bmatrix} x_1(t) \\ x_2(t) \end{bmatrix} = \mathbf{x}(t) &= \begin{bmatrix} -\sin(t-\pi) \\ -\cos(t-\pi) \end{bmatrix} + \begin{bmatrix} 2 - 2\cos(t-\pi) \\ 2\sin(t-\pi) \end{bmatrix} \\ &= \begin{bmatrix} \sin t + 2\cos t + 2 \\ \cos t - 2\sin t \end{bmatrix}. \end{aligned}$$

Thus, since $x(t) = x_1(t)$, it follows that the solution to the initial value problem is given by

$$x(t) = \sin t + 2\cos t + 2. \blacksquare$$

Example 5. Solve by matrix methods

$$\dot{u}(t) = u(t) + 2v(t),$$
$$\dot{v}(t) = 4u(t) + 3v(t).$$

Solution This system can be put into fundamental form, if we define

$$\mathbf{x}(t) = \begin{bmatrix} u(t) \\ v(t) \end{bmatrix}, \quad \mathbf{A} = \begin{bmatrix} 1 & 2 \\ 4 & 3 \end{bmatrix}, \quad \text{and} \quad \mathbf{f}(t) = \begin{bmatrix} 0 \\ 0 \end{bmatrix}.$$

This is a homogeneous system with constant coefficients and no initial conditions specified; hence, the general solution is given by (45).

As in Example 2, for this **A**, we have

$$e^{\mathbf{A}t} = \frac{1}{6} \begin{bmatrix} 2e^{5t} + 4e^{-t} & 2e^{5t} - 2e^{-t} \\ 4e^{5t} - 4e^{-t} & 4e^{5t} + 2e^{-t} \end{bmatrix}.$$

Thus,

$$
\begin{aligned}
e^{\mathbf{A}t}\mathbf{k} &= \frac{1}{6} \begin{bmatrix} 2e^{5t} + 4e^{-t} & 2e^{5t} - 2e^{-t} \\ 4e^{5t} - 4e^{-t} & 4e^{5t} + 2e^{-t} \end{bmatrix} \begin{bmatrix} k_1 \\ k_2 \end{bmatrix} \\
&= \frac{1}{6} \begin{bmatrix} k_1 [2e^{5t} + 4e^{-t}] + k_2 [2e^{5t} - 2e^{-t}] \\ k_1 [4e^{5t} - 4e^{-t}] + k_2 [4e^{5t} + 2e^{-t}] \end{bmatrix} \\
&= \frac{1}{6} \begin{bmatrix} e^{5t}(2k_1 + 2k_2) + e^{-t}(4k_1 - 2k_2) \\ e^{5t}(4k_1 + 4k_2) + e^{-t}(-4k_1 + 2k_2) \end{bmatrix}.
\end{aligned}
\tag{54}
$$

Substituting (54) into (45), we have that

$$\begin{bmatrix} u(t) \\ v(t) \end{bmatrix} = \mathbf{x}(t) = \frac{1}{6} \begin{bmatrix} e^{5t}(2k_1 + 2k_2) + e^{-t}(4k_1 - 2k_2) \\ e^{5t}(4k_1 + 4k_2) + e^{-t}(-4k_1 + 2k_2) \end{bmatrix}$$

or

$$
\begin{aligned}
u(t) &= \left(\frac{2k_1 + 2k_2}{6} \right) e^{5t} + \left(\frac{4k_1 - 2k_2}{6} \right) e^{-t} \\
v(t) &= 2 \left(\frac{2k_1 + 2k_2}{6} \right) e^{5t} + \left(\frac{-4k_1 + 2k_2}{6} \right) e^{-t}.
\end{aligned}
\tag{55}
$$

We can simplify the expressions for $u(t)$ and $v(t)$ if we introduce two new arbitrary constants k_3 and k_4 defined by

$$k_3 = \frac{2k_1 + 2k_2}{6}, \quad k_4 = \frac{4k_1 - 2k_2}{6}. \tag{56}$$

Substituting these values into (55), we obtain

$$
\begin{aligned}
u(t) &= k_3 e^{5t} + k_4 e^{-1} \\
v(t) &= 2k_3 e^{5t} - k_4 e^{-1}. \ \blacksquare
\end{aligned}
\tag{57}
$$

Problems 8.4

1. Show by direct integration that

$$\int_{t_0}^{t} t^2 dt = \int_{t_0}^{t} s^2 ds = \int_{t_0}^{t} p^2 dp.$$

In general, show that if $f(t)$ is integrable on $[a, b]$, then

$$\int_{a}^{b} f(t) \, dt = \int_{a}^{b} f(s) \, ds.$$

(*Hint*: Assume $\int f(t)dt = F(t) + c$. Hence, $\int f(s)ds = F(s) + c$. Use the fundamental theorem of integral calculus to obtain the result.)

2. Derive Eq. (44). (*Hint*: Follow steps (30)–(33). For step (34) use indefinite integration and note that

$$\int \frac{d}{dt} \left[e^{-\mathbf{A}t} \mathbf{x}(t) \right] dt = e^{-\mathbf{A}t} \mathbf{x}(t) + \mathbf{k},$$

where $\mathbf{k}$ is an arbitrary constant vector of integration.)

3. Find (a) $e^{-\mathbf{A}t}$, (b) $e^{\mathbf{A}(t-2)}$, (c) $e^{\mathbf{A}(t-s)}$, (d) $e^{-\mathbf{A}(t-2)}$ if

$$e^{\mathbf{A}t} = e^{3t} \begin{bmatrix} 1 & t & t^2/2 \\ 0 & 1 & t \\ 0 & 0 & 1 \end{bmatrix}.$$

4. For $e^{\mathbf{A}t}$ as given in Problem 3, invert by the method of cofactors to obtain $e^{-\mathbf{A}t}$ and hence verify part (a) of that problem.

5. Find (a) $e^{-\mathbf{A}t}$, (b) $e^{-\mathbf{A}s}$, (c) $e^{\mathbf{A}(t-3)}$ if

$$e^{\mathbf{A}t} = \frac{1}{6} \begin{bmatrix} 2e^{5t} + 4e^{-t} & 2e^{5t} - 2e^{-t} \\ 4e^{5t} - 4e^{-t} & 4e^{5t} + 2e^{-t} \end{bmatrix}.$$

6. Find (a) $e^{-\mathbf{A}t}$, (b) $e^{-\mathbf{A}s}$, (c) $e^{-\mathbf{A}(t-s)}$ if

$$e^{\mathbf{A}t} = \frac{1}{3} \begin{bmatrix} -\sin 3t + 3 \cos 3t & 5 \sin 3t \\ -2 \sin 3t & \sin 3t + 3 \cos 3t \end{bmatrix}.$$

Solve the systems given in Problems 7 through 14 by matrix methods. Note that Problems 7 through 10 have the same coefficient matrix.

7. $\dot{x}(t) = -2x(t) + 3y(t),$

$\dot{y}(t) = -x(t) + 2y(t);$

$x(2) = 2, \quad y(2) = 4.$

8. $\dot{x}(t) = -2x(t) + 3y(t) + 1,$

$\dot{y}(t) = -x(t) + 2y(t) + 1;$

$x(1) = 1, \quad y(1) = 1.$

9. $\dot{x}(t) = -2x(t) + 3y(t),$

$\dot{y}(t) = -x(t) + 2y(t).$

10. $\dot{x}(t) = -2x(t) + 3y(t) + 1,$
$\dot{y}(t) = -x(t) + 2y(t) + 1.$

11. $\ddot{x}(t) = -4x(t) + \sin t;$
$x(0) = 1, \quad \dot{x}(0) = 0.$

12. $\dddot{x}(t) = t;$
$x(1) = 1, \quad \dot{x}(1) = 2, \quad \ddot{x}(1) = 3$

13. $\ddot{x} - \dot{x} - 2x = e^{-t};$
$x(0) = 1, \quad \dot{x}(0) = 0.$

14. $\ddot{x} = 2\dot{x} + 5y + 3,$
$\dot{y} = -\dot{x} - 2y;$
$x(0) = 0, \quad \dot{x}(0) = 0, \quad y(0) = 1.$

8.5 Solutions of systems—general case

Having completely solved systems of linear differential equations with constant coefficients, we now turn our attention to the solutions of systems of the form

$$\dot{\mathbf{x}}(t) = \mathbf{A}(t)\mathbf{x}(t) + \mathbf{f}(t)$$
$$\mathbf{x}(t_0) = \mathbf{c}. \tag{58}$$

Note that $\mathbf{A}(t)$ may now depend on t; hence, the analysis of Section 8.4 does not apply. However, since we still require both $\mathbf{A}(t)$ and $\mathbf{f}(t)$ to be continuous in some interval about $t = t_0$, Theorem 1 of Section 8.1 still guarantees that (58) has a unique solution. Our aim in this section is to obtain a representation for this solution.

Definition 1. A *transition* (or fundamental) *matrix* of the homogeneous equation $\dot{\mathbf{x}}(t) = \mathbf{A}(t)\mathbf{x}(t)$ is an $n \times n$ matrix $\mathbf{\Phi}(t, t_0)$ having the properties that

$$\text{(a)} \quad \frac{d}{dt}\mathbf{\Phi}(t, t_0) = \mathbf{A}(t)\mathbf{\Phi}(t, t_0) \tag{59}$$

$$\text{(b)} \quad \mathbf{\Phi}(t_0, t_0) = \mathbf{I}. \tag{60}$$

Here t_0 is the initial time given in (58). In the Final Comments to this chapter, we show that $\mathbf{\Phi}(t, t_0)$ exists and is unique.

Example 1. Find $\mathbf{\Phi}$ (t, t_0) if $\mathbf{A}(t)$ is a constant matrix.

    Solution Consider the matrix $e^{\mathbf{A}(t-t_0)}$. From Property 1 of Section 7.8, we have that $e^{\mathbf{A}(t_0-t_0)} = e^0 = \mathbf{I}$, while from Theorem 1 of Section 7.9, we have that

$$\frac{d}{dt}e^{\mathbf{A}(t-t_0)} = \frac{d}{dt}\left(e^{\mathbf{A}t}e^{-\mathbf{A}t_0}\right)$$

$$= \mathbf{A}e^{\mathbf{A}t}e^{-\mathbf{A}t_0} = \mathbf{A}e^{\mathbf{A}(t-t_0)}.$$

    Thus, $e^{\mathbf{A}(t-t_0)}$ satisfies (59) and (60). Since $\mathbf{\Phi}(t, t_0)$ is unique, it follows for the case where $\mathbf{A}$ is a *constant* matrix that

$$\mathbf{\Phi}(t, t_0) = e^{\mathbf{A}(t-t_0)}. \tag{61}$$

Caution. Although $\mathbf{\Phi}(t, t_0) = e^{\mathbf{A}(t-t_0)}$, if $\mathbf{A}$ is a constant matrix, this equality is not valid if $\mathbf{A}$ actually depends on t. In fact, it is usually impossible to explicitly find $\mathbf{\Phi}(t, t_0)$ in the general time varying case. Usually, the best we can say about the transition matrix is that it exists, it is unique, and, of course, it satisfies (59) and (60).

    One immediate use of $\mathbf{\Phi}(t, t_0)$ is that it enables us to theoretically solve the general homogeneous initial value problem

$$\dot{\mathbf{x}}(t) = \mathbf{A}(t)\mathbf{x}(t)$$
$$\mathbf{x}(t_0) = \mathbf{c}. \tag{62}$$

Theorem 1. *The unique solution to* (62) *is*

$$\mathbf{x}(t) = \mathbf{\Phi}(t, t_0)\mathbf{c}. \tag{63}$$

    Proof. If $\mathbf{A}(t)$ is a constant matrix, (63) reduces to (43) (see (61)); hence, Theorem 1 is valid. In general, however, we have that

$$\frac{dx(t)}{dt} = \frac{d}{dt}[\mathbf{\Phi}(t, t_0)\mathbf{c}] = \frac{d}{dt}[\mathbf{\Phi}(t, t_0)]\mathbf{c},$$

$$= \mathbf{A}(t)\mathbf{\Phi}(t, t_0)\mathbf{c} \quad \{\text{from } (59),$$

$$= \mathbf{A}(t)\mathbf{x}(t) \quad\quad \{\text{from } (63),$$

and

$$\mathbf{x}(t_0) = \mathbf{\Phi}(t_0, t_0)\mathbf{c},$$
$$= \mathbf{Ic} \quad \{\text{from } (60),$$
$$= \mathbf{c}.\blacksquare$$

Example 2.   Find $x(t)$ and $y(t)$ if

$$\dot{x} = ty$$
$$\dot{y} = -tx$$
$$x(1) = 0, \qquad y(1) = 1.$$

Solution Putting this system into fundamental form, we obtain

$$t_0 = 1, \quad \mathbf{x}(t) = \begin{bmatrix} x(t) \\ y(t) \end{bmatrix}, \quad \mathbf{A}(t) = \begin{bmatrix} 0 & t \\ -t & 0 \end{bmatrix}, \quad \mathbf{f}(t) = \mathbf{0}, \quad \mathbf{c} = \begin{bmatrix} 0 \\ 1 \end{bmatrix},$$

and

$$\dot{\mathbf{x}}(t) = \mathbf{A}(t)\mathbf{x}(t),$$
$$\mathbf{x}(t_0) = \mathbf{c}.$$

The transition matrix for this system can be shown to be (see Problem 1)

$$\Phi(t, t_0) = \begin{bmatrix} \cos\left(\dfrac{t^2 - t_0^2}{2}\right) & \sin\left(\dfrac{t^2 - t_0^2}{2}\right) \\ -\sin\left(\dfrac{t^2 - t_0^2}{2}\right) & \cos\left(\dfrac{t^2 - t_0^2}{2}\right) \end{bmatrix}.$$

Thus, from (63), we have

$$\mathbf{x}(t) = \begin{bmatrix} \cos\left(\dfrac{t^2 - 1}{2}\right) & \sin\left(\dfrac{t^2 - 1}{2}\right) \\ -\sin\left(\dfrac{t^2 - 1}{2}\right) & \cos\left(\dfrac{t^2 - 1}{2}\right) \end{bmatrix} \begin{bmatrix} 0 \\ 1 \end{bmatrix}$$

$$= \begin{bmatrix} \sin\left(\dfrac{t^2 - 1}{2}\right) \\ \cos\left(\dfrac{t^2 - 1}{2}\right) \end{bmatrix}.$$

Consequently, the solution is

$$x(t) = \sin\left(\dfrac{t^2 - 1}{2}\right), \quad y(t) = \cos\left(\dfrac{t^2 - 1}{2}\right). \blacksquare$$

The transition matrix also enables us to give a representation for the solution of the general time-varying initial value problem

$$\dot{\mathbf{x}}(t) = \mathbf{A}(t)\mathbf{x}(t) + \mathbf{f}(t),$$
$$\mathbf{x}(t_0) = \mathbf{c}. \tag{58}$$

Theorem 2. *The unique solution to* (58) *is*

$$\mathbf{x}(t) = \mathbf{\Phi}(t, t_0)\mathbf{c} + \int_{t_0}^{t} \mathbf{\Phi}(t, s)\mathbf{f}(s)ds. \tag{64}$$

Proof. If $\mathbf{A}$ is a constant matrix, $\mathbf{\Phi}(t, t_0) = e^{\mathbf{A}(t-t_0)}$; hence, $\mathbf{\Phi}(t, s) = e^{\mathbf{A}(t-s)}$ and (64) reduces to (41). We defer the proof of the general case, where $\mathbf{A}(t)$ depends on t, until later in this section.

Eq. (64) is the solution to the general initial value problem given by (58). It should be noted, however, that since $\mathbf{\Phi}(t, t_0)$ is not explicitly known, $\mathbf{x}(t)$ will not be explicitly known either; hence, (64) is not as useful a formula as it might first appear. Unfortunately, (64) is the best solution that we can obtain for the general time-varying problem. The student should not despair, though. It is often the case that by knowing enough properties of $\mathbf{\Phi}(t, t_0)$, we can extract a fair amount of information about the **solution** from (64). In fact, we can sometimes even obtain the exact solution! ■

Before considering some important properties of the transition matrix, we state one lemma that we ask the student to prove (see Problem 3).

Lemma 1. If $\mathbf{B}(t)$ *is an* $n \times n$ *matrix having the property that* $\mathbf{B}(t)\mathbf{c} = 0$ *for every* n*-dimensional constant vector* $\mathbf{c}$*, then* $\mathbf{B}(t)$ *is the zero matrix.*

For the remainder of this section, we assume that $\mathbf{\Phi}(t, t_0)$ is the transition matrix for $\dot{\mathbf{x}}(t) = \mathbf{A}(t)\mathbf{x}(t)$.

Property 1. *(The transition property)*

$$\mathbf{\Phi}(t, \tau)\mathbf{\Phi}(\tau, t_0) = \mathbf{\Phi}(t, t_0). \tag{65}$$

Proof. If $\mathbf{A}(t)$ is a constant matrix, $\mathbf{\Phi}(t, t_0) = e^{\mathbf{A}(t-t_0)}$ hence,

$$\begin{aligned}
\mathbf{\Phi}(t, \tau)\mathbf{\Phi}(\tau, t_0) &= e^{\mathbf{A}(t-\tau)}e^{\mathbf{A}(\tau-t_0)} \\
&= e^{\mathbf{A}(t-\tau+\tau-t_0)} \\
&= e^{\mathbf{A}(t-t_0)} = \mathbf{\Phi}(t, t_0).\blacksquare
\end{aligned}$$

Thus, Property 1 is immediate. For the more general case, that in which $\mathbf{A}(t)$ depends on t, the argument runs as follows: Consider the initial value problem

$$\begin{aligned}
\dot{\mathbf{x}}(t) &= \mathbf{A}(t)\mathbf{x}(t) \\
\mathbf{x}(t_0) &= \mathbf{c}.
\end{aligned} \tag{66}$$

The unique solution of (66) is

$$\mathbf{x}(t) = \mathbf{\Phi}(t, t_0)\mathbf{c}. \tag{67}$$

Hence,

$$\mathbf{x}(t_1) = \mathbf{\Phi}(t_1, t_0)\mathbf{c} \tag{68}$$

and

$$\mathbf{x}(\tau) = \boldsymbol{\Phi}(\tau, t_0)\mathbf{c}, \tag{69}$$

where t_1 is any arbitrary time greater than τ. If we designate the vector $\mathbf{x}(t_1)$ by $\mathbf{d}$ and the vector $\mathbf{x}(\tau)$ by $\mathbf{b}$, then we can give the solution graphically by Fig. 8.1.

Consider an associated system governed by

$$\dot{\mathbf{x}}(t) = \mathbf{A}(t)\mathbf{x}(t),$$
$$\mathbf{x}(\tau) = \mathbf{b}. \tag{70}$$

We seek a solution to the above differential equation that has an initial value $\mathbf{b}$ at the initial time $t = \tau$. If we designate the solution by $\mathbf{y}(t)$, it follows from Theorem 1 that

$$\mathbf{y}(t) = \boldsymbol{\Phi}(t, \tau)\mathbf{b}, \tag{71}$$

hence,

$$\mathbf{y}(t_1) = \boldsymbol{\Phi}(t_1, \tau)\mathbf{b}. \tag{72}$$

But now we note that both $\mathbf{x}(t)$ and $\mathbf{y}(t)$ are governed by the same equation of motion, namely $\dot{\mathbf{x}}(t) = \mathbf{A}(t)\mathbf{x}(t)$, and both $\mathbf{x}(t)$ and $\mathbf{y}(t)$ go through the same point (τ, b). Thus, $\mathbf{x}(t)$ and $\mathbf{y}(t)$ must be the same solution. That is, the solution curve for $\mathbf{y}(t)$ looks exactly like that of $\mathbf{x}(t)$, shown in Fig. 8.1 except that it starts at $t = \tau$, while that of $\mathbf{x}(t)$ starts at $t = t_0$. Hence,

$$\mathbf{x}(t) = \mathbf{y}(t), \quad \mathbf{t} \geq \tau,$$

and, in particular,

$$\mathbf{x}(t_1) = \mathbf{y}(t_1). \tag{73}$$

Thus, substituting (68) and (72) into (73), we obtain

$$\boldsymbol{\Phi}(t_1, t_0)\mathbf{c} = \boldsymbol{\Phi}(t_1, \tau)\mathbf{b}. \tag{74}$$

However, $\mathbf{x}(\tau) = \mathbf{b}$, thus (74) may be rewritten as

$$\boldsymbol{\Phi}(t_1, t_0)\mathbf{c} = \boldsymbol{\Phi}(t_1, \tau)\mathbf{x}(\tau). \tag{75}$$

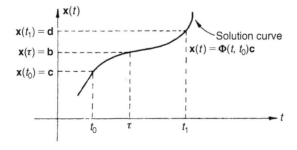

FIGURE 8.1

Substituting (69) into (75), we have

$$\Phi(t_1, t_0)\mathbf{c} = \Phi(t_1, \tau)\Phi(\tau, t_0)\mathbf{c}$$

or

$$[\Phi(t_1, t_0) - \Phi(t_1, \tau)\Phi(\tau, t_0)]\mathbf{c} = 0. \tag{76}$$

Since $\mathbf{c}$ may represent any n-dimensional initial state, it follows from Lemma 1 that

$$\Phi(t_1, t_0) - \Phi(t_1, \tau)\Phi(\tau, t_0) = \mathbf{0}$$

or

$$\Phi(t_1, t_0) = \Phi(t_1, \tau)\Phi(\tau, t_0). \tag{77}$$

Since t_1 is arbitrary, it can be replaced by t; Eq. (77), therefore, implies Eq. (65).

Property 2. $\Phi\ (t, t_0)$ *is invertible and*

$$[\Phi(t_1, t_0)]^{-1} = \Phi(t_0, t). \tag{78}$$

Proof. This result is obvious if $\mathbf{A}(t)$ is a constant matrix. We know from Section 7.8 that the inverse of $e^{\mathbf{A}t}$ is $e^{-\mathbf{A}t}$, hence,

$$[\Phi(t, t_0)]^{-1} = \left[e^{\mathbf{A}(t-t_0)}\right]^{-1} = e^{-\mathbf{A}(t-t_0)}$$
$$= e^{\mathbf{A}(t_0-t)} = \Phi(t_0, t).$$

In order to prove Property 2 for any $\mathbf{A}(t)$, we note that (65) is valid for any t; hence, it must be valid for $t = t_0$. Thus,

$$\Phi(t_0, \tau)\Phi(\tau, t_0) = \Phi(t_0, t_0).$$

It follows from (60) that

$$\Phi(t_0, \tau)\Phi(\tau, t_0) = \mathbf{I}.$$

Thus, from the definition of the inverse, we have

$$[\Phi(\tau, t_0)]^{-1} = \Phi(t_0, \tau),$$

which implies (78). ∎

Example 3. Find the inverse of

$$\begin{bmatrix} \cos\left(\dfrac{t^2 - t_0^2}{2}\right) & \sin\left(\dfrac{t^2 - t_0^2}{2}\right) \\ -\sin\left(\dfrac{t^2 - t_0^2}{2}\right) & \cos\left(\dfrac{t^2 - t_0^2}{2}\right) \end{bmatrix}.$$

Solution This matrix is a transition matrix. (See Problem 1.) Hence using (78), we find the inverse to be

$$
\begin{bmatrix} \cos\left(\frac{t_0^2-t^2}{2}\right) & \sin\left(\frac{t_0^2-t^2}{2}\right) \\ -\sin\left(\frac{t_0^2-t^2}{2}\right) & \cos\left(\frac{t_0^2-t^2}{2}\right) \end{bmatrix} = \begin{bmatrix} \cos\left(\frac{t^2-t_0^2}{2}\right) & -\sin\left(\frac{t^2-t_0^2}{2}\right) \\ -\sin\left(\frac{t^2-t_0^2}{2}\right) & \cos\left(\frac{t^2-t_0^2}{2}\right) \end{bmatrix}.
$$

Here we have used the identities $\sin(-\theta) = -\sin\theta$ and $\cos(-\theta) = \cos\theta$. ∎

Properties 1 and 2 enable us to prove Theorem 2, namely, that the solution of

$$
\dot{\mathbf{x}}(t) = \mathbf{A}(t)\mathbf{x}(t) + \mathbf{f}(t),
$$
$$
\mathbf{x}(t_0) = \mathbf{c}
$$

is

$$
\mathbf{x}(t) = \mathbf{\Phi}(t, t_0)\mathbf{c} + \int_{t_0}^{t} \mathbf{\Phi}(t, s)\mathbf{f}(s)ds. \tag{79}
$$

Using Property 1, we have that $\mathbf{\Phi}(t, s) = \mathbf{\Phi}(t, t_0)\, \mathbf{\Phi}(t_0, s)$; hence, (79) may be rewritten as

$$
\mathbf{x}(t) = \mathbf{\Phi}(t, t_0)\mathbf{c} + \mathbf{\Phi}(t, t_0) \int_{t_0}^{t} \mathbf{\Phi}(t_0, s)\mathbf{f}(s)ds. \tag{80}
$$

Now

$$
\mathbf{x}(t_0) = \mathbf{\Phi}(t_0, t_0)\mathbf{c} + \mathbf{\Phi}(t_0, t_0) \int_{t_0}^{t_0} \mathbf{\Phi}(t_0, s)\mathbf{f}(s)ds.
$$
$$
= \mathbf{I}\mathbf{c} + \mathbf{I}\mathbf{0} = \mathbf{c}.
$$

Thus, the initial condition is satisfied by (80). To show that the differential equation is also satisfied, we differentiate (80) and obtain

$$
\frac{d\mathbf{x}(t)}{dt} = \frac{d}{dt}\left[\mathbf{\Phi}(t, t_0)\mathbf{c} + \mathbf{\Phi}(t, t_0) \int_{t_0}^{t} \mathbf{\Phi}(t_0, s)\mathbf{f}(s)ds \right]
$$

$$
= \left[\frac{d}{dt}\mathbf{\Phi}(t, t_0) \right]\mathbf{c} + \left[\frac{d}{dt}\mathbf{\Phi}(t, t_0) \right] \int_{t_0}^{t} \mathbf{\Phi}(t_0, s)\mathbf{f}(s)ds
$$

$$
+ \mathbf{\Phi}(t, t_0)\left[\frac{d}{dt} \int_{t_0}^{t} \mathbf{\Phi}(t_0, s)\mathbf{f}(s)ds \right]
$$

$$
= \mathbf{A}(t)\mathbf{\Phi}(t, t_0)\mathbf{c} + \mathbf{A}(t)\mathbf{\Phi}(t, t_0) \int_{t_0}^{t} \mathbf{\Phi}(t_0, s)\mathbf{f}(s)ds
$$

$$
+ \mathbf{\Phi}(t, t_0)\mathbf{\Phi}(t_0, t)\mathbf{f}(t)
$$

$$
= \mathbf{A}(t)\left[\mathbf{\Phi}(t, t_0)\mathbf{c} + \mathbf{\Phi}(t, t_0) \int_{t_0}^{t} \mathbf{\Phi}(t_0, s)\mathbf{f}(s)ds \right]
$$

$$
+ \mathbf{\Phi}(t, t_0)\mathbf{\Phi}^{-1}(t, t_0)\mathbf{f}(t).
$$

The quantity inside the bracket is given by (80) to be $\mathbf{x}(t)$; hence,

$$\frac{d\mathbf{x}(t)}{dt} = \mathbf{A}(t)\mathbf{x}(t) + \mathbf{f}(t).$$

We conclude this section with one final property of the transition matrix, the proof of which is beyond the scope of this book.

Property 3.

$$\det \ \boldsymbol{\Phi}(t, t_0) = \exp\left\{ \int_{t_0}^{t} \text{tr}[\mathbf{A}(t)] dt \right\}. \tag{81}$$

Since the exponential is never zero, (81) establishes that $\det \boldsymbol{\Phi}(t, t_0) \neq 0$; *hence, we have an alternate proof that* $\boldsymbol{\Phi}(t, t_0)$ *is invertible.*

Problem 8.5

1. Use (59) and (60) to show that

$$\boldsymbol{\Phi}(t, t_0) = \begin{bmatrix} \cos\left(\dfrac{t^2 - t_0^2}{2}\right) & \sin\left(\dfrac{t^2 - t_0^2}{2}\right) \\ -\sin\left(\dfrac{t^2 - t_0^2}{2}\right) & \cos\left(\dfrac{t^2 - t_0^2}{2}\right) \end{bmatrix}$$

is a transition matrix for

$$\dot{x} = ty,$$
$$\dot{y} = -tx.$$

2. As a generalization of Problem 1, use (59) and (60) to show that

$$\boldsymbol{\Phi}(t, t_0) = \begin{bmatrix} \cos \displaystyle\int_{t_0}^{t} g(s)ds & \sin \displaystyle\int_{t_0}^{t} g(s)ds \\ -\sin \displaystyle\int_{t_0}^{t} g(s)ds & \cos \displaystyle\int_{t_0}^{t} g(s)ds \end{bmatrix}$$

is a transition matrix for

$$\dot{x} = g(t)y,$$
$$\dot{y} = -g(t)x.$$

3. Prove Lemma 1. (*Hint:* Consider the product $\mathbf{B}(t)\mathbf{c}$, where

$$\text{first, } \mathbf{c} = \begin{bmatrix} 1 \\ 0 \\ 0 \\ \vdots \\ 0 \end{bmatrix}, \quad \text{second, } \mathbf{c} = \begin{bmatrix} 0 \\ 1 \\ 0 \\ \vdots \\ 0 \end{bmatrix}, \quad \text{etc.)}$$

4. If $\boldsymbol{\Phi}(t, t_0)$ is a transition matrix, prove that

$$\boldsymbol{\Phi}^{\mathrm{T}}(t_1, t_0) \left[\int_{t_0}^{t_1} \boldsymbol{\Phi}(t_1, s) \boldsymbol{\Phi}^{\mathrm{T}}(t_1, s) ds \right]^{-1} \boldsymbol{\Phi}(t_1, t_0)$$

$$= \left[\int_{t_0}^{t_1} \boldsymbol{\Phi}(t_0, s) \boldsymbol{\Phi}^{\mathrm{T}}(t_0, s) ds \right]^{-1}.$$

8.6 Final comments on Chapter 8

We now prove that there exists a unique matrix $\boldsymbol{\Phi}(t, t_0)$ having properties (59) and (60).

Define n-dimensional unit vectors $\mathbf{e}_1, \mathbf{e}_2, \ldots, \mathbf{e}_n$ by

$$\mathbf{e}_1 = \begin{bmatrix} 1 \\ 0 \\ 0 \\ 0 \\ \vdots \\ 0 \end{bmatrix}, \quad \mathbf{e}_2 = \begin{bmatrix} 0 \\ 1 \\ 0 \\ 0 \\ \vdots \\ 0 \end{bmatrix}, \quad \mathbf{e}_3 = \begin{bmatrix} 0 \\ 0 \\ 1 \\ 0 \\ \vdots \\ 0 \end{bmatrix}, \quad \ldots, \quad \mathbf{e}_n = \begin{bmatrix} 0 \\ 0 \\ 0 \\ 0 \\ \vdots \\ 1 \end{bmatrix}. \quad (82)$$

Thus,

$$\begin{bmatrix} \mathbf{e}_1 & \mathbf{e}_2 & \mathbf{e}_3 & \cdots & \mathbf{e}_n \end{bmatrix} = \begin{bmatrix} 1 & 0 & 0 & \cdots & 0 \\ 0 & 1 & 0 & & 0 \\ 0 & 0 & 0 & & 0 \\ \vdots & \vdots & \vdots & & \vdots \\ 0 & 0 & 0 & \cdots & 1 \end{bmatrix} = \mathbf{I}. \quad (83)$$

Consider the homogeneous systems given by

$$\dot{\mathbf{x}}(t) = \mathbf{A}(t)\mathbf{x}(t)$$
$$\mathbf{x}(t) = \mathbf{e}_j \quad (j = 1, 2, ..., n), \tag{84}$$

where $\mathbf{A}(t)$ and t_0 are taken from (58). For each $j(j = 1, 2, ..., n)$, Theorem 1 of Section 8.1 guarantees the existence of a unique solution of (84); denote this solution by $\mathbf{x}_j(t)$. Thus, $\mathbf{x}_1(t)$ solves the system

$$\dot{\mathbf{x}}_1(t) = \mathbf{A}(t)\mathbf{x}_1(t)$$
$$\mathbf{x}_1(t_0) = \mathbf{e}_1, \tag{85}$$

$\mathbf{x}_2(t)$ satisfies the system

$$\dot{\mathbf{x}}_2(t) = \mathbf{A}(t)\mathbf{x}_2(t)$$
$$\mathbf{x}_2(t_0) = \mathbf{e}_2, \tag{86}$$

and $\mathbf{x}_n(t)$ satisfies the system

$$\dot{\mathbf{x}}_n(t) = \mathbf{A}(t)\mathbf{x}_n(t)$$
$$\mathbf{x}_n(t_0) = \mathbf{e}_n. \tag{87}$$

Define the matrix

$$\boldsymbol{\Phi}(t, t_0) = [\mathbf{x}_1(t)\mathbf{x}_2(t)...\mathbf{x}_n(t)].$$

Then,

$$\boldsymbol{\Phi}(t_0, t_0) = [\mathbf{x}_1(t_0)\mathbf{x}_2(t_0)...\mathbf{x}_n(t_0)].$$
$$= [\mathbf{e}_1\mathbf{e}_2...\mathbf{e}_n] \quad \{\text{from (85) to (87)}$$
$$= \mathbf{I} \qquad \{\text{from (83)}$$

and

$$\frac{d\boldsymbol{\Phi}(t, t_0)}{dt} = \frac{d}{dt}[\mathbf{x}_1(t) \; \mathbf{x}_2(t) \cdots \mathbf{x}_n(t)]$$
$$= [\dot{\mathbf{x}}_1(t) \; \dot{\mathbf{x}}_2(t) \cdots \dot{\mathbf{x}}_n(t)]$$
$$= [\mathbf{A}(t)\mathbf{x}_1(t) \; \mathbf{A}(t)\mathbf{x}_2(t) \cdots \mathbf{A}(t)\mathbf{x}_n(t)] \quad \{\text{from (85) to (87)}$$
$$= \mathbf{A}(t)[\mathbf{x}_1(t) \; \mathbf{x}_2(t) \cdots \mathbf{x}_n(t)]$$
$$= \mathbf{A}(t)\boldsymbol{\Phi}(t, t_0).$$

Thus, $\boldsymbol{\Phi}(t, t_0)$, as defined above, is a matrix that satisfies (59) and (60). Since this $\boldsymbol{\Phi}(t, t_0)$ always exists, it follows that there will always exist a matrix that satisfies these equations.

It only remains to be shown that $\boldsymbol{\Phi}(t, t_0)$ is unique. Let $\boldsymbol{\Psi}(t, t_0)$ be any matrix satisfying (59) and (60). Then the jth column $\boldsymbol{\Psi}(t, t_0)$ must satisfy the initial valued problem given by (84). However, the solution to (84) is unique by Theorem 1 of Section 8.1, hence, the jth column of $\boldsymbol{\Psi}(t, t_0)$ must be $\mathbf{x}_j(t)$. Thus,

$$\boldsymbol{\Psi}(t, t_0) = [\mathbf{x}_1(t)\mathbf{x}_2(t)...\mathbf{x}_n(t)] = \boldsymbol{\Phi}(t, t_0)$$

From this equation, it follows that the transition matrix is unique.

Probability and Markov chains

Chapter outline

9.1 Probability: an informal approach

Our approach to probability will be very basic in this section; we will be more formal in the next section.

We begin by considering a *set*; recall that a *set* can be thought of as a *collection of objects*. For example, consider a deck of regular playing cards, consisting of 52 cards. This set will be called the *sample space* or the *universe*. Now suppose we shuffle the deck a number of times and, at random, pick out a card. Assume that the card is the King of Diamonds. The action of selecting this card is called an *event*. And we might ask the following question: "How likely are we to pick out the King of Diamonds?"

Before attempting to answer *this* question, let us consider the following:

- How many times have we shuffled the deck?
- What do we mean by random?
- What do we mean by likely?

The answers to these three "simple" questions touch on a number of advanced mathematical concepts and can even go into philosophical areas. For our purposes, we will, by and large, appeal to our intuition when quantifying certain concepts beyond the scope of this section.

However, we can give a *reasonable* answer to our original question. We note that the *size* of our sample space (the deck of cards) is 52. We also observe that there is

Matrix Methods. https://doi.org/10.1016/B978-0-12-818419-6.00009-5

only one way to draw a King of Diamonds, since there is *only one* King of Diamonds in the deck; hence, the *size* of the desired event is 1. So we make the following statement, which should seem plausible to the reader: we will say that *the probability of the desired event is, simply, 1/52.*

Using mathematical notation, we can let S be the set that represents the sample space and let E be the set that represents the desired event. Since the number of objects (size) in any set is called the *cardinal number*, we can write $N(S) = 52$ and $N(E) = 1$, to represent the cardinal number of each set. So we now write

$$P(E) = \frac{N(E)}{N(S)} = \frac{1}{52} \tag{1}$$

to denote the *probability of event E*.

What does this mean? It does *not* mean that, should we make exactly 52 drawings of a card—returning it to the deck after each drawing and reshuffling the deck each time—we would draw the King of Diamonds exactly once. (Try it!)

A better interpretation is that *over a very large number of trials*, the *proportion* of times for which a King of Diamonds would be drawn would get closer and closer to 1 out of 52.

Continuing with this example, the probability of drawing a Spade (event F) is one-fourth, since there are 13 Spades in the deck,

$$P(F) = \frac{N(F)}{N(S)} = \frac{13}{52} = \frac{1}{4}. \tag{2}$$

Another example would be to consider a fair die; let's call it D. Since there are six faces, there are six equally likely outcomes (1, 2, 3, 4, 5, or 6) for every roll of the die, $N(D) = 6$. If the event G is to roll a "3," then

$$P(G) = \frac{N(G)}{N(D)} = \frac{1}{6}. \tag{3}$$

Experiment by rolling a die "many" times. You will find that the proportion of times a "3" occurs is close to one-sixth. In fact, if this is *not* the case, the die is most probably "not fair".

Remark 1. From this example it is clear that the probability of any of the six outcomes is one-sixth. Note, too, that the *sum* of the six probabilities is 1. Also, the probability of rolling a "7" is *zero*, simply because there are no 7s on the any of the six faces of the die.

Because of the above examples, it is most natural to think of probability as a *number*. This number will always be between 0 and 1. We say that an event is *certain* if the probability is 1, and that it is *impossible* if the probability is 0. Most probabilities will be strictly between 0 and 1.

To *compute* the number we call the *probability of an event*, we will adopt the following convention. *We will divide the number of ways the desired event can occur by the total number of possible outcomes. We always assume that each member of the sample space is "just as likely" to occur as any other member.* We call this a *relative frequency* approach.

Example 1. Consider a fair die. Find the probability of rolling a number that is a perfect square.
As before, the size of the sample space, D, is 6. The number of ways a perfect square can occur is two: only "1" or "4" are perfect squares out of the first six positive integers. Therefore, the desired probability is

$$P(K) = \frac{N(K)}{N(D)} = \frac{2}{6} = \frac{1}{3}. \blacksquare \qquad (4)$$

Example 2. Consider a pair of fair dice. What is the probability of rolling a "7"?
To solve this problem, we first have to find the cardinal number of the sample space, R. To do this, it may be helpful to consider the dice as composed of one red die and one green die, and to think of a "roll" as tossing the red die first, followed by the green die. Then $N(R) = 36$ because there are 36 possible outcomes. To see this, consider Fig. 9.1. Here, the first column represents the outcome of the red die, the first row represents the outcome of the green die, and the body is the *sum* of the two dice—the actual number obtained by the dice roll.
Notice, too, that if we label a "7" roll event Z, then $N(Z) = 6$, because there are six distinct ways of rolling a "7"; again, see Fig. 9.1.
So our answer is

$$P(Z) = \frac{N(Z)}{N(R)} = \frac{6}{36} = \frac{1}{6}. \blacksquare \qquad (5)$$

Example 3. Suppose a random number generator generates numbers ranging from 1 through 1000. Find the probability that a given number is divisible by 5.
Elementary number theory teaches that for a number to be divisible by 5, the number must end in either a "5" or a "0". By sheer counting, we know that there are 200 numbers between 1 and 1000 that satisfy this condition. Therefore, the required probability is $\frac{200}{1000}$. $\blacksquare$

In the next section, we will give some rules that pertain to probabilities and investigate the meaning of probability more fully.

R\G	1	2	3	4	5	6
1	2	3	4	5	6	7
2	3	4	5	6	7	8
3	4	5	6	7	8	9
4	5	6	7	8	9	10
5	6	7	8	9	10	11
6	7	8	9	10	11	12

FIGURE 9.1

Problems 9.1

1. Find the sample space, its cardinal number, and the probability of the desired event for each of the following scenarios:

   (a) Pick a letter, at random, out of the English alphabet. Desired event: choosing a vowel.

   (b) Pick a date, at random, for the Calendar Year 2008. Desired event: choosing December 7th.

   (c) Pick a US president, at random, from a list of all the presidents. Desired event: choosing Abraham Lincoln.

   (d) Pick a US president, at random, from a list of all the presidents. Desired event: choosing Grover Cleveland.

   (e) Pick a card, at random, from a well-shuffled deck of regular playing cards. Desired event: choosing the Ace of Spades.

   (f) Pick a card, at random, from a well-shuffled deck of Pinochle playing cards. Desired event: choosing the Ace of Spades.

   (g) Roll a pair of fair dice. Desired event: getting a roll of "2" (snake eyes).

   (h) Roll a pair of fair dice. Desired event: getting a roll of "12" (boxcars).

   (i) Roll a pair of fair dice. Desired event: getting a roll of "8".

   (j) Roll a pair of fair dice. Desired event: getting a roll of "11".

   (k) Roll a pair of fair dice. Desired event: getting a roll of an even number.

   (l) Roll a pair of fair dice. Desired event: getting a roll of a number that is a perfect square.

   (m) Roll a pair of fair dice. Desired event: getting a roll of a number that is a perfect cube.

   (n) Roll a pair of fair dice. Desired event: getting a roll of a number that is a multiple of 3.

   (o) Roll a pair of fair dice. Desired event: getting a roll of a number that is divisible by 3.

   (p) Roll a pair of fair dice. Desired event: getting a roll of "13".

2. Suppose we were to roll three fair dice: a red one first, followed by a white die, followed by a blue die. Describe the sample space and find its cardinal number.

3. Suppose the probability for event A is known to be 0.4. Find the cardinal number of the sample space if $N(A) = 36$.

4. Suppose the probability for event B is known to be 0.65. Find the cardinal number of B, if the cardinal number of the sample space, S, is $N(S) = 3000$.

9.2 Some laws of probability

In this section, we will continue our discussion of probability from a more theoretical and formal perspective.

Recall that the probability of event A, given a sample space S, is given by

$$P(A) = \frac{N(A)}{N(S)}, \tag{6}$$

where the numerator and denominator are the respective cardinal numbers of A and S.

We will now give a number of definitions and rules, which we will follow regarding the computations of probabilities. This list "formalizes" our approach. We note that the reader can find the mathematical justification in any number of sources devoted to more advanced treatments of this topic. We assume that A, B, $C \ldots$ are any events and that S is the sample space. We also use Φ to denote an impossible event.

- $P(\Phi) = 0$; that is, the probability of an impossible event is zero.
- $P(S) = 1$; that is, the probability of a certain event is one.
- For any two events, A and B, then

$$P(A \cup B) = P(A) + P(B) - P(A \cap B). \tag{7}$$

Remark 1.   Here we use both the *union* ($\cup$) and *intersection* ($\cap$) notation from set theory. For this rule, we subtract off the probability of the "common" even in order not to "count it twice". For example, if A is the event of drawing a *King* from a deck of regular playing cards, and B is the event of drawing a Diamond, clearly the *King of Diamonds* is both a King *and* a Diamond. Since there are 4 Kings in the deck, the probability of drawing a King is $\frac{4}{52}$. And since there are 13 Diamonds in the deck, the probability of drawing a Diamond is $\frac{13}{52}$. Since there is only one King of Diamonds in the deck, the probability of drawing this card is clearly $\frac{1}{52}$. We note that

$$\frac{4}{52} + \frac{13}{52} - \frac{1}{52} = \frac{16}{52}, \tag{8}$$

which is the probability of drawing a King *or* a Diamond because the deck contains *only* 16 Kings or Diamonds.

- If A and B are disjoint events, then

$$P(A \cup B) = P(A) + P(B). \tag{9}$$

Remark 2.   Two events are *disjoint* if they are mutually exclusive; that is, they cannot happen simultaneously. For example, the events of drawing a *King* from a deck of regular playing cards and, at the same time, drawing a *Queen* are disjoint events. In this case, we merely add the individual probabilities. Note, also, that since A and B are disjoint, we can write $A \cap B = \Phi$; hence, $P(A \cap B) = P(\Phi) = 0$. The reader will also see that Eq. (9) is merely a special case of Eq. (7).

- Consider event A; if A^C represents the *complement of A*, then

$$P(A^C) = 1 - P(A). \tag{10}$$

Remark 3. This follows from the fact that either an event occurs or it doesn't. Therefore, $P(A \cup A^C) = 1$; but since these two events are disjoint, $P(A \cup A^C) = P(A) + P(A^C) = 1$. Eq. (10) follows directly.

For example, if the probability of rolling a "3" on a fair die is $\frac{1}{6}$, then the probability of *not* rolling a "3" is $1 - \frac{1}{6} = \frac{5}{6}$.

In the next section, we will introduce the idea of *independent events*, along with associated concepts. For the rest of this section, we give a number of examples regarding the above rules of probability.

Example 1. Given a pair of fair dice, find the probability of rolling a "3" or a "4".

Since these events are disjoint, we use Eq. (9) and refer to Fig. 9.1 and obtain the desired probability: $\frac{2}{36} + \frac{3}{36} = \frac{5}{36}$. ∎

Example 2. Given a pair of fair dice, find the probability of not rolling a "3" or a "4".

From the previous example, we know that the probability of rolling a "3" or a "4" is $\frac{5}{36}$, therefore, using Eq. (10), we find that the probability of the complementary event is: $1 - \frac{5}{36} = \frac{31}{36}$. ∎

Remark 4. Note that we could have computed this probability directly by counting the number of ways—31—in which the rolls 2, 5, 6, 7, 8, 9, 10, 11, or 12 can occur. However, using Eq. (10) is the preferred method because it is quicker.

Example 3. Pick a card at random out of a well-shuffled deck of regular playing cards. What is the probability of drawing a picture card (that is, a King, Queen, or Jack)?

Since there are four suits (Spades, Hearts, Clubs, and Diamonds) and there are three picture cards for each suit, the desired event can occur 12 ways; these can be thought of as 12 disjoint events. Hence, the required probability is $\frac{12}{52}$. ∎

Example 4. Pick a card at random out of a well-shuffled deck of regular playing cards. Find the probability of drawing a red card or a picture card.

We know there are 12 picture cards as well as 26 red cards (Hearts or Diamonds). But our events are not disjoint, since six of the picture cards are red. Therefore, we apply Eq. (7) and compute the desired probability as $\frac{12}{52} + \frac{26}{52} - \frac{6}{52} = \frac{32}{52}$. ∎

Example 5.   Suppose events A and B are not disjoint. Find $P(A \cap B)$, if it is given that $P(A) = 0.4$, $P(B) = 0.3$, and $P(A \cup B) = 0.6$.

Recall Eq. (7): $P(A \cup B) = P(A) + P(B) - P(A \cap B)$. Therefore, $0.6 = 0.4 + 0.3 - P(A \cap B)$. Therefore, $P(A \cap B) = 0.1$. ∎

Example 6.   Extend formula (7) for three nondisjoint events. That is, consider $P(A \cup B \cup C)$. By using parentheses to group two events, we have the following equation:

$$P(A \cup B \cup C) = P((A \cup B) \cup C) = P(A \cup B) + P(C) - P((A \cup B)) \cap C \qquad (11)$$

From set theory, we know that the last term of (11) can be written as $P((A \cup B) \cap C) = P((A \cap C) \cup (B \cap C))$. Hence, applying (7) to (11) yields

$$P(A \cap C) + P(B \cap C) - P((A \cap C) \cap (B \cap C)). \qquad (12)$$

But the last term of (12) is equivalent to $P(A \cap B \cap C)$. After applying (7) to the $P(A \cup B)$ term in (11), we have

$$\begin{aligned} P(A \cup B \cup C) = P(A) + P(B) + P(C) - P(A \cap B) \\ - [P(A \cap C) + P(B \cap C) - P((A \cap C) \cap (B \cap C))]. \end{aligned} \qquad (13)$$

This simplifies to

$$\begin{aligned} P(A \cup B \cup C) = P(A) + P(B) + P(C) - P(A \cap B) - P(A \cap C) \\ - P(B \cap C) + P(A \cap B \cap C). \blacksquare \end{aligned} \qquad (14)$$

Remark 5.   Eq. (14) can be extended for any finite number of events and it holds even if some events are pairwise disjoint. For example, if events A and B are disjoint, we merely substitute $P(A \cap B) = 0$ into (14).

Problems 9.2

1. Pick a card at random from a well-shuffled deck of regular playing cards. Find the probabilities of
   (a) Picking an Ace or a King.
   (b) Picking an Ace or a picture card.
   (c) Picking an Ace or a black card.
   (d) Picking the Four of Diamonds or the Six of Clubs.
   (e) Picking a red card or a Deuce.
   (f) Picking a Heart or a Spade.
   (g) Not choosing a Diamond.
   (h) Not choosing a Queen.
   (i) Not choosing an Ace or a Spade.

2. Roll a pair of fair dice. Find the probabilities of
   (a) Getting an odd number.
   (b) Rolling a prime number.
   (c) Rolling a number divisible by four.
   (d) Not rolling a "7".
   (e) Not rolling a "6" or an "8".
   (f) Not rolling a "1".
3. See Problem 2 of Section 9.1. Roll the three dice. Find the probabilities of
   (a) Getting an odd number.
   (b) Rolling a "3".
   (c) Rolling an "18".
   (d) Rolling a "4".
   (e) Rolling a "17".
   (f) Rolling a "25".
   (g) Not rolling a "4".
   (h) Not rolling a "5".
   (i) Not rolling a "6".
4. Consider events A and B. Given $P(A) = 0.7$ and $P(B) = 0.2$, find the probability of "A or B" if $P(A \cap B) = 0.15$.
5. Suppose events A and B are equally likely. Find their probabilities if $P(A \cup B) = 0.46$ and $P(A \cap B) = 0.34$.
6. Extend Eq. (14) for any four events A, B, C, and D.

9.3 Bernoulli trials and combinatorics

In the previous section, we considered single events. For example, rolling dice *once* or drawing *one* card out of a deck. In this section, we consider multiple events that neither affect nor are affected by preceding or succeeding events.

For this section, we will consider events with only *two* outcomes. For example, flipping a coin, which can result in only "heads" or "tails". The coin cannot land on its edge. We do not insist that the probability of a "head" equals the probability of a "tail", but we will assume that the probabilities remain constant. Each one of the "flips" will be called a *Bernoulli trial*, in honor of Jakob Bernoulli (1654–1705).

Remark 1. A closely related underlying mathematical structure for these trials is known as a *binomial distribution*.

Remark 2. The Bernoulli family had a number of great mathematicians and scientists spanning several generations. This family is to mathematics what the Bach family is to music.

As we have indicated, we will assume that the events are *independent*. Hence, the probabilities are unaffected at all times. So, if we tossed a coin 10 times in a row, each of the tosses would be called a *Bernoulli trial* and the probability of getting a head on each toss would remain constant.

We will assume the following rule. If two events, *A* and *B*, are independent, then the probability of "*A and B*" or the probability of "*A followed by B*" is given by

$$P(AB) = P(A \cap B) = P(A)P(B). \tag{15}$$

Notice that we use the intersection ($\cap$) notation. This simple rule is called the *multiplication rule*.

Remark 3.   The reader must be careful not to confuse *disjoint* events with *independent* events. The former means that "nothing is in common" or that the events "cannot happen simultaneously". The latter means that the probabilities do not influence one another. Often, but not always, independent events are *sequential*, like flipping a coin 10 times in a row.

It is clear that probabilities depend on *counting*, as in determining the size of the sample space. We assume the following result from an area of mathematics known as *Combinatorics*:

- The number of ways we can choose *k* objects from a given collection of *n* objects is given by

$$\binom{n}{k} = \frac{n!}{k!(n-k)!}. \tag{16}$$

Remark 4.   This is equivalent to determining the number of *subsets of size k* given a *set of size n*, where $k \leq n$.

Remark 5.   We saw "factorials" in Chapter 7. Recall that *3!*, for example, is read "three factorial" and it is evaluated 3. 2. 1. = 6. Hence, "n − factorial" is given by $n! = n(n-1)(n-2) \dots 3. 2. 1.$ By convention, we define $0! = 1$. Finally, we only consider cases where *n* is a nonnegative integer.

Remark 6.   For these "number of ways" we are not concerned about the *order* of selection. We are merely interested in the number of *combinations* (as opposed to the number of *permutations*).

We will provide the reader with a number of examples that illustrate the salient points of this section.

Example 1.   Evaluate $\binom{5}{2}$. Using (16) we see that $\binom{5}{2} = \frac{5!}{2!(5-2)!} = \frac{5!}{2!3!}$.

Since $5! = 120$, $2! = 2$, and $3! = 6$, $\binom{5}{2} = \frac{120}{12} = 10$. ∎

Example 2.   Given a committee of five people, in how many ways can a subcommittee of size two be formed? The number is precisely what we computed in the previous example: 10. The reader can verify this as follows. Suppose the people are designated: A, B, C, D, and E. Then, the 10 subcommittees of size two are given by: AB, AC, AD, AE, BC, BD, BE, CD, CE, and DE. ■

Example 3.   Given a committee of five people, in how many ways can a subcommittee of size three be formed? We can use formula (16) to compute the answer; however, the answer must be 10. This is because a subcommittee of three is the *complement* of a subcommittee or two. That is, consider the three people *not* on a particular subcommittee of two as constituting a subcommittee of three. For example, if A and B are on one subcommittee of size two, put C, D, and E on a subcommittee of size three. Clearly, there are 10 such pairings. ■

Example 4.   Suppose we flip a fair coin twice. What is the probability of getting *exactly* one "head"?

Let H represent getting a "head" and T represent getting a "tail". Since the coin is fair, $P(H) = P(T) = \frac{1}{2}$. The only way we can obtain exactly *one* head in *two* tosses is if the order of outcomes is either HT or TH. Note that the events are *disjoint* or *mutually exclusive*; that is, we cannot get these two outcomes at the same time. Hence, Eq. (9) will come into play. And because of the *independence* of the coin flips (each toss is a *Bernoulli trial*), we will use Eq. (15) to determine the probability of obtaining HT and TH.

Therefore, the probability of getting exactly one H is equal to

$$P(HT \cup TH) = P(HT) + P(TH) = P(H)P(T) + P(T)P(H)$$

$$= \frac{1}{2} \cdot \frac{1}{2} + \frac{1}{2} \cdot \frac{1}{2} = \frac{1}{2}. \blacksquare$$

(17)

Remark 7. Note that (17) could have been obtained by finding the probability of HT—in that order—and then multiplying it by the *number of times (combinations)* we could get exactly one H in two tosses.

Example 5.   Now, suppose we flip a fair coin 10 times. What is the probability of getting exactly one "head"?

Suppose we get the H on the first toss. Then, the probability of getting $HTTTTTTTTT$—in that order—is equal to $\left(\frac{1}{2}\right) \cdot \left(\frac{1}{2}\right)^9 = \frac{1}{1024}$ because the tosses are all independent. Note that if the H occurs "in the second slot", the probability is also $\frac{1}{1024}$. In fact, we get the same number for all 10 possible "slots". Hence, the final answer to our question is $\frac{10}{1024}$. ■

Remark 8. Note that the previous example could have been answered by the following computation

$$\binom{10}{1} \left(\frac{1}{2}\right) \left(\frac{1}{2}\right)^9 = \frac{10}{1024}. \tag{18}$$

Here, the first factor gives the number of ways we can get exactly one H in 10 tosses; this is where mutual exclusivity comes in. The second factor is the probability of getting one H, and the third factor is the probability of getting nine Ts; the independence factor is employed here.

Example 6. Suppose we flip a fair coin 10 times. Find the probability of getting *exactly* five Hs.

Since there are $\binom{10}{5} = 252$ ways of getting five Hs in 10 tosses, the desired probability is given by

$$\binom{10}{5} \left(\frac{1}{2}\right)^5 \left(\frac{1}{2}\right)^5 = \frac{252}{1024}. \blacksquare \tag{19}$$

Example 7. Suppose we flip an *unfair* coin 10 times. If $P(H) = 0.3$ and the $P(T) = 0.7$, what is the probability of getting *exactly* five Hs?

As we learned from the previous problem, there are 252 ways of getting exactly five Hs in 10 tosses. Hence, our desired probability is given by

$$\binom{10}{5} (.3)^5 (.7)^5 \approx 0.103. \blacksquare$$

Remark 9. A calculator is useful for numerical computations. We will address the issues of calculations and technology in both Section 9.5 and in the Appendix. Note, too, that individual probabilities $P(H)$ and $P(T)$ must add to 1, and that the exponents in the formula must add to the total number of tosses; in this case, 10.

Example 8. Consider the previous example. Find the probability of getting *at least five Hs*.

Note that *at least five Hs* means *exactly five Hs* plus *exactly six Hs* plus ... etc. Note also that *exactly five Hs* and *exactly six Hs* are disjoint, so we will use Eq. (9). Therefore, the desired probability is given by

$$\binom{10}{5}(.3)^5(.7)^5 + \binom{10}{6}(.3)^7(.7)^3 + \binom{10}{8}(.3)^8(.7)^2 + \binom{10}{9}(.3)^9(.7)^1$$

$$+ \binom{10}{10}(.3)^{10}(.7)^{10} \approx 0.150. \blacksquare \tag{20}$$

Example 9. Consider the previous examples. Find the probability of getting *at least one H*.

While we could follow the approach in Example 9, there is a simpler way to answer this question. If we realize that the *complement* of the desired event is *getting no Hs* in 10 tosses, we can apply Eq. (10). That is, the probability of getting at *least one H* is equal to

$$1 - \binom{10}{0}(.3)^0(.7)^{10} \approx 0.972. \blacksquare \tag{21}$$

We summarize the results for the probability in this section as follows:

- Given n successive Bernoulli trials, the probability of getting exactly k successes, where $k \leq n$, is equal to

$$\frac{n!}{k!(n-k!)}p^k(1-p)^{n-k}, \tag{22}$$

where the probability of a success is p and the probability of a failure is $(1-p)$.

Problems 9.3

1. Evaluate the following:

(a) $\binom{6}{2}$;

(b) $\binom{7}{1}$;

(c) $\binom{5}{5}$;

(d) $\binom{20}{18}$;

(e) $\binom{20}{2}$;

(f) $\binom{1000}{1000}$;

(g) $\binom{1000}{0}$;

(h) $\begin{pmatrix} 100 \\ 99 \end{pmatrix};$

(i) $\begin{pmatrix} 1000 \\ 999 \end{pmatrix};$

(j) $\begin{pmatrix} 0 \\ 0 \end{pmatrix}.$

2. How many different nine-player lineups can the New York Yankees produce, if there are 25 players on the roster and every player can play every position?

3. Suppose 15 women comprised a club and a committee of 6 members was needed. How many different committees would be possible?

4. Toss a fair die eight times. Find the probability of
   (a) Rolling exactly one "5".
   (b) Rolling exactly three "5s".
   (c) Rolling at least three "5s".
   (d) Rolling at most three "5s".
   (e) Rolling at least one "5".

5. Suppose event A has a probability of occurring equal to 65. Without evaluating the expressions, find the following probabilities, given 500 independent Bernoulli trials:
   (a) Event A occurs 123 times.
   (b) Event A occurs 485 times.
   (c) Event A occurs at least 497 times.
   (d) Event A occurs at most 497 times.
   (e) Event A occurs any nonzero multiple of 100 times.

6. An urn contains 10 golf balls, three of which are white, with the remaining seven painted orange. A blindfolded golfer picks a golf ball from the urn and then replaces it. The process is repeated nine times, making a total of 10 trials. What is the probability of the golfer picking a white golf ball exactly three times?

7. An urn contains 10 golf balls colored as follows: three are white; two are green; one is red; four are orange. A blindfolded golfer picks a golf ball from the urn and then replaces it. The process is repeated nine times, making a total of 10 trials. What is the probability of the golfer picking a white golf ball exactly three times?

9.4 Modeling with Markov chains: an introduction

In Chapters 1 and 6, we mentioned the concept of Markov chains. We return to this idea, further formalizing it from the perspective of probability. Consider the following example, which we will call the moving situation.

Example 1. Suppose we have two families. Family (1) lives in state A and Family (2) lives in state B. Let us further assume that the matrix

$$P = \begin{bmatrix} .7 & .3 \\ .9 & .1 \end{bmatrix} \tag{23}$$

represents the following probabilities. The element in the first row and first column represents the probability of Family (1) originally residing in state A remaining in state A, while the element in the first row and second column represents the probability of starting in state A and then moving to state B. Note that these two probabilities add to one.

Similarly, let the element in the second row and first column, represent the probability of Family (2) starting in state B and moving to state A, while the element in the second row and second column represents the probability of starting in state B and remaining in state B. Here, too, these two probabilities add to one.

Note that we can consider the process as "time conditioned" in the sense that there is a *present* and a *future* (for example, 1 year from the present).

Such a matrix is called a *transition matrix* and the elements are called *transitional probabilities*. Let us consider the matrix in (23) and let us compute P^2. We find that

$$P^2 = \begin{bmatrix} .76 & .24 \\ .72 & .28 \end{bmatrix}. \tag{24}$$

What does P^2 represent? To answer this question, let us ask another question: From the perspective of Family (1), what is the probability of being in state A after 2 years?

There are two ways Family (1) can be in state A after 2 years:

• Scenario 1: Either the family stayed 2 years in a row.
• Scenario 2: The family moved to state B after 1 year and then moved back to state A after the second year.

The probability of the first scenario is $.7(0.7) = 0.49$ because these events can be considered *independent*. The probability of the second is $.3(0.7) = 0.27$.

Because these events are *disjoint*, we add the probabilities to get .76.

Note that this is the element in the first row and first column of P^2.

By similar analyses, we find that P^2 is indeed the transitional matrix of our moving situation after two time periods.

Matrix P is the transition matrix for a *Markov chain*. The sum of the probabilities of each row must add to one, and by the very nature of the process, the matrix must be square. We assume that at any time, each object is in one and only one state (although different objects can be in the same state). We also assume that the probabilities remain constant over the given time period. ∎

Remark 1. The notation $p^{(n)}{}_{ij}$ is used to signify the transitional probability of moving from state i to state j over n time periods.

Example 2. Suppose Moe, Curly, and Larry live in the same neighborhood. Let the transition matrix

$$P = \begin{bmatrix} .7 & .1 & .2 \\ .5 & .3 & .2 \\ .8 & .1 & .1 \end{bmatrix} \tag{25}$$

represent the respective probabilities of Moe, Curly, and Larry staying at home on Monday and either visiting one of their two neighbors or staying home on Tuesday. We ask the following questions regarding Thursday:

(a) What is the probability of Moe going to visit Larry at his home, $p^{(3)}_{13}$?

(b) What is the probability of Curly being at his own home, $p^{(3)}_{22}$?

To answer both of these questions, we must compute P^3 because three time periods would have elapsed. We find that

$$P^3 = \begin{bmatrix} .694 & .124 & .182 \\ .124 & .132 & .182 \\ .695 & .124 & .181 \end{bmatrix}. \tag{26}$$

So, our answers are as follows: (a) the probability is .182, the entry in the first row and third column; (b) the probability is .132, the entry in the second row and second column. ■

Example 3. Consider the transitional matrix K that represents respective probabilities of Republicans, Democrats, and Independents either remaining within their political parties or changing their political parties over a 2-year period:

$$K = \begin{bmatrix} .7 & .1 & .2 \\ .15 & .75 & .1 \\ .3 & .2 & .5 \end{bmatrix}. \tag{27}$$

What is the probability of a Republican becoming an Independent after 4 years? And what is the probability of a Democrat becoming a Republican after 4 years?

Both of these questions require *two* time periods; hence, we need $p^{(2)}_{13}$ and $p^{(2)}_{21}$, which can be obtained from K^2 below:

$$K^2 = \begin{bmatrix} .565 & .185 & .25 \\ .2475 & .5975 & .155 \\ .39 & .28 & .33 \end{bmatrix}. \tag{28}$$

Hence, $p^{(2)}_{13} = 0.25$ and $p^{(2)}_{21} = 0.2475$.

We close this discussion with the observation that sometimes transitional matrices have special properties. For example, consider the matrix

$$A = \begin{bmatrix} .5 & .5 \\ .3 & .7 \end{bmatrix}. \tag{29}$$

We find that if we raise this matrix to the 9th and 10th powers, our result is the same. That is,

$$A^9 = A^{10} = \begin{bmatrix} .375 & .625 \\ .375 & .625 \end{bmatrix}. \tag{30}$$

and the *same result occurs for any higher power of A*. This *absorbing* quality implies that "sooner or later" the transitional probabilities will stabilize.

Markov processes are used in many areas including decision theory, economics, and political science. ■

Problems 9.4

1. Why are the following matrices not transitional matrices?

(a) $\begin{bmatrix} 0 & 1 \\ -1 & 2 \end{bmatrix}$;

(b) $\begin{bmatrix} .6 & .5 \\ .4 & .5 \end{bmatrix}$;

(c) $\begin{bmatrix} .1 & .2 & .7 \\ 1 & 0 & 0 \\ 0 & 0 & 0 \end{bmatrix}$;

(d) $\begin{bmatrix} .1 & .5 & .4 \\ .2 & .6 & .2 \end{bmatrix}$

2. Consider the following transitional matrices. Construct scenarios for which these matrices might represent the transitional probabilities:

(a) $\begin{bmatrix} .5 & .5 \\ .7 & .3 \end{bmatrix}$;

(b) $\begin{bmatrix} .95 & .05 \\ .02 & .98 \end{bmatrix}$;

(c) $\begin{bmatrix} .5 & .5 \\ .5 & .5 \end{bmatrix}$;

(d) $\begin{bmatrix} 1 & 0 \\ 0 & .1 \end{bmatrix}$;

(e) $\begin{bmatrix} 0 & 1 \\ 1 & 0 \end{bmatrix}$;

(f) $\begin{bmatrix} .1 & .2 & .7 \\ .5 & .25 & .25 \\ .3 & .3 & .4 \end{bmatrix}$

3. Consider the (c) and (d) matrices in the previous problem; show that these matrices are "absorbing" matrices.
4. Consider the (e) matrix in Problem (2). Raise the matrix to the powers of 2, 3, 4, 5, 6, 7, and 8. What do you notice? Can you construct a scenario for which this transitional matrix could be a model?
5. Consider the following transitional matrix: $\begin{bmatrix} .6 & .4 \\ .1 & .9 \end{bmatrix}$. Find $p^{(2)}{}_{11}$, $p^{(2)}{}_{21}$, $p^{(3)}{}_{12}$, and $p(3)_{22}$.

6. Consider a game called Red—Blue. The rules state that after one "turn" Red can become Blue or remain Red. The same is true with Blue. Suppose you make a bet that after five turns, Red will be in the Red category. You are told that the following probabilities are valid:

   - Given Red, the probability of remaining Red after one turn is .7
   - Given Red, the probability of going to Blue is .3
   - Given Blue, the probability of remaining Blue is .6
   - Given Blue, the probability of going to Red is .4
     - (a) Give the transition matrix.
     - (b) What is the probability of you winning your bet?
     - (c) Does the probability increase, decrease, or stay the same if you bet six turns instead of five?

9.5 **Final comments on Chapter 9**

Probability is a fascinating area. For numbers that necessarily range between 0 and 1, inclusively, a lot can happen.

When using probability, we must understand exactly what is being asked and give precise answers without misrepresenting our conclusions. Concepts such as randomness and independence must be present before certain laws can be applied. While the mathematical underpinnings are rock solid, probabilities generally deal with "trends" and "likelihoods".

Regarding Bernoulli trials, if the number of experiments is *large*, the calculations can be overwhelming. In these cases, the use of computers and other technological aids is essential. From a theoretical perspective, there is a very good approximation that can be employed, known as the *Normal Approximation to the Binomial Distribution*. This technique is explored in basic courses on probability and statistics.

One final point: With the exception of the section on Markov chains, all the probabilities in this chapter were *theoretically* assigned. That is, we made assumptions, applied definitions, and then made our computations. For example, *if* a die was *fair, then* we assigned a probability of $\frac{1}{6}$ to the event of rolling a "3", based on our definition, which dealt with relative frequency.

However, there are many times when probabilities are obtained by *observation* and *empirical evidence*. For example, the greatest baseball player of all time, Babe Ruth, had a lifetime batting average of .342. Since batting average is defined as successful hits divided by total at-bats, we can interpret this as Ruth getting 342 hits for every 1000 at-bats *over a long period of time* (See Chapters 11 and 12).

There are many other occurrences of *empirical probabilities* in research areas such as medicine, psychology, economics, and sociology, to name but a few.

Real inner products and least squares

10

Chapter outline

10.1 Introduction

To any two vectors $\mathbf{x}$ and $\mathbf{y}$ of the same dimension having real components (as distinct from complex components), we associate a scalar called the *inner product*, denoted as $\langle \mathbf{x}, \mathbf{y} \rangle$, by multiplying together the corresponding elements of $\mathbf{x}$ and $\mathbf{y}$, and then summing the results. Students already familiar with the dot product of two- and three-dimensional vectors will undoubtedly recognize the inner product as an extension of the dot product to real vectors of all dimensions.

Example 1. Find $\langle \mathbf{x}, \mathbf{y} \rangle$ if

$$\mathbf{x} = \begin{bmatrix} 1 \\ 2 \\ 3 \end{bmatrix} \quad \text{and} \quad \mathbf{y} = \begin{bmatrix} 4 \\ -5 \\ 6 \end{bmatrix}.$$

Solution $\langle \mathbf{x}, \mathbf{y} \rangle = 1(4) + 2(-5) + 3(6) = 12.$ ∎

Example 2. Find $\langle \mathbf{u}, \mathbf{v} \rangle$ if $\mathbf{u} = [20\ {-4}\ 30\ 10]$ and $\mathbf{v} = [10\ {-5}\ {-8}\ {-6}]$.
Solution $\langle \mathbf{u}, \mathbf{v} \rangle = 20(10) + (-4)\,(-5) + 30(-8) + 10(-6) = -80.$ ∎

It follows immediately from the definition that the inner product of real vectors satisfies the following properties:

(I1) $\langle \mathbf{x}, \mathbf{x} \rangle$ is positive if $\mathbf{x} \neq \mathbf{0}$; $\langle \mathbf{x}, \mathbf{x} \rangle = 0$ if and only if $\mathbf{x} = \mathbf{0}$.
(I2) $\langle \mathbf{x}, \mathbf{y} \rangle = \langle \mathbf{y}, \mathbf{x} \rangle$.
(I3) $\langle \lambda \mathbf{x}, \mathbf{y} \rangle = \lambda \langle \mathbf{x}, \mathbf{y} \rangle$, for any real scalar λ.
(I4) $\langle \mathbf{x} + \mathbf{z}, \mathbf{y} \rangle = \langle \mathbf{x}, \mathbf{y} \rangle + \langle \mathbf{z}, \mathbf{y} \rangle$.
(I5) $\langle \mathbf{0}, \mathbf{y} \rangle = 0$.

We will only prove **(I1)** here and leave the proofs of the other properties as exercises for the students (see Problems 29 through 32). Let $\mathbf{x} = [x_1 \; x_2 \; x_3 \; \dots \; x_n]$ be an n-dimensional row vector whose components $x_1, x_2, x_3, \dots, x_n$ are all real. Then,

$$\langle \mathbf{x}, \mathbf{x} \rangle = (x_1)^2 + (x_2)^2 + (x_3)^2 + \cdots + (x_n)^2.$$

This sum of squares is zero if and only if $x_1 = x_2 = x_3 = \dots = x_n = 0$, which in turn implies $\mathbf{x} = \mathbf{0}$. If any one component is not zero, that is, if $\mathbf{x}$ is not the zero vector, then the sum of squares must be positive.

The inner product of real vectors is related to the magnitude of a vector as defined in Section 1.6. In particular,

$$\|\mathbf{x}\| = \sqrt{\langle \mathbf{x}, \mathbf{x} \rangle}.$$

Example 3. Find the magnitude of $\mathbf{x} = [2 \; -3 \; -4]$.
Solution $\langle \mathbf{x}, \mathbf{x} \rangle = 2(2) + (-3)(-3) + (-4)(-4) = 29$, so the magnitude of $\mathbf{x}$ is

$$\|\mathbf{x}\| = \sqrt{29}. \blacksquare$$

The concepts of a normalized vector and a unit vector are identical to the definitions given in Section 1.6. A nonzero vector is *normalized* if it is divided by its magnitude. A *unit vector* is a vector whose magnitude is unity. Thus, if $\mathbf{x}$ is any nonzero vector, then $(1/\|\mathbf{x}\|)\mathbf{x}$ is normalized. Furthermore,

$$\left\langle \frac{1}{\|\mathbf{x}\|}\mathbf{x}, \frac{1}{\|\mathbf{x}\|}\mathbf{x} \right\rangle = \frac{1}{\|\mathbf{x}\|} \left\langle \mathbf{x}, \frac{1}{\|\mathbf{x}\|}\mathbf{x} \right\rangle \qquad \text{(Property I3)}$$

$$= \frac{1}{\|\mathbf{x}\|} \left\langle \mathbf{x}, \frac{1}{\|\mathbf{x}\|}\mathbf{x}, \mathbf{x} \right\rangle \qquad \text{(Property I2)}$$

$$= \left(\frac{1}{\|\mathbf{x}\|}\right)^2 \langle \mathbf{x}, \mathbf{x} \rangle \qquad \text{(Property I3)}$$

$$= \left(\frac{1}{\|\mathbf{x}\|}\right)^2 \|\mathbf{x}\|^2 = 1,$$

so a normalized vector is always a unit vector.

Problems 10.1

In Problems 1 through 17, find (a) $\langle \mathbf{x}, \mathbf{y} \rangle$ and (b) $\langle \mathbf{x}, \mathbf{x} \rangle$ for the given vectors.

1. $\mathbf{x} = \begin{bmatrix} 1 \\ 2 \end{bmatrix}$ and $\mathbf{y} = \begin{bmatrix} 3 \\ 4 \end{bmatrix}$.

2. $\mathbf{x} = \begin{bmatrix} 2 \\ 0 \end{bmatrix}$ and $\mathbf{y} = \begin{bmatrix} 4 \\ -5 \end{bmatrix}$.

3. $\mathbf{x} = \begin{bmatrix} -5 \\ 7 \end{bmatrix}$ and $\mathbf{y} = \begin{bmatrix} 3 \\ -5 \end{bmatrix}$.

4. $\mathbf{x} = \begin{bmatrix} 3 & 14 \end{bmatrix}$ and $\mathbf{y} = \begin{bmatrix} 7 & 3 \end{bmatrix}$.

5. $\mathbf{x} = \begin{bmatrix} -2 & -8 \end{bmatrix}$ and $\mathbf{y} = \begin{bmatrix} -4 & -7 \end{bmatrix}$.

6. $\mathbf{x} = \begin{bmatrix} 2 \\ 0 \\ 1 \end{bmatrix}$ and $\mathbf{y} = \begin{bmatrix} 1 \\ 2 \\ 4 \end{bmatrix}$.

7. $\mathbf{x} = \begin{bmatrix} -2 \\ 2 \\ -4 \end{bmatrix}$ and $\mathbf{y} = \begin{bmatrix} -4 \\ 3 \\ -3 \end{bmatrix}$.

8. $\mathbf{x} = \begin{bmatrix} -3 \\ -2 \\ -5 \end{bmatrix}$ and $\mathbf{y} = \begin{bmatrix} 6 \\ -4 \\ -4 \end{bmatrix}$.

9. $\mathbf{x} = \begin{bmatrix} \dfrac{1}{2} & \dfrac{1}{3} & \dfrac{1}{6} \end{bmatrix}$ and $\mathbf{y} = \begin{bmatrix} \dfrac{1}{3} & \dfrac{3}{2} & 1 \end{bmatrix}$.

10. $\mathbf{x} = \begin{bmatrix} 1/\sqrt{2} & 1/\sqrt{3} & 1/\sqrt{6} \end{bmatrix}$ and $\mathbf{y} = \begin{bmatrix} 1/\sqrt{3} & 3/\sqrt{2} & 1 \end{bmatrix}$.

11. $\mathbf{x} = \begin{bmatrix} \dfrac{1}{3} & \dfrac{1}{3} & \dfrac{1}{3} \end{bmatrix}$ and $\mathbf{y} = \begin{bmatrix} \dfrac{1}{4} & \dfrac{1}{2} & \dfrac{1}{8} \end{bmatrix}$.

12. $\mathbf{x} = \begin{bmatrix} 10 & 20 & 30 \end{bmatrix}$ and $\mathbf{y} = \begin{bmatrix} 5 & -7 & 3 \end{bmatrix}$.

13. $\mathbf{x} = \begin{bmatrix} 1 \\ 0 \\ 1 \\ 1 \end{bmatrix}$ and $\mathbf{y} = \begin{bmatrix} 1 \\ 1 \\ 0 \\ 1 \end{bmatrix}$.

14. $\mathbf{x} = \begin{bmatrix} \frac{1}{2} \\ \frac{1}{2} \\ \frac{1}{2} \\ \frac{1}{2} \end{bmatrix}$ and $\mathbf{y} = \begin{bmatrix} 1 \\ 2 \\ 3 \\ -4 \end{bmatrix}$.

15. $\mathbf{x} = \begin{bmatrix} 3 \\ 5 \\ -7 \\ -8 \end{bmatrix}$ and $\mathbf{y} = \begin{bmatrix} 1 \\ -6 \\ -9 \\ 8 \end{bmatrix}$.

16. $\mathbf{x} = \begin{bmatrix} \frac{1}{5} & \frac{1}{5} & \frac{1}{5} & \frac{1}{5} & \frac{1}{5} \end{bmatrix}$ and $\mathbf{y} = \begin{bmatrix} 1 & 2 & -3 & 4 & -5 \end{bmatrix}$.

17. $\mathbf{x} = \begin{bmatrix} 1 & 1 & 1 & 1 & 1 \end{bmatrix}$ and $\mathbf{y} = \begin{bmatrix} -3 & 8 & 11 & -4 & 7 \end{bmatrix}$.
18. Normalize $\mathbf{y}$ as given in Problem 1.
19. Normalize $\mathbf{y}$ as given in Problem 2.
20. Normalize $\mathbf{y}$ as given in Problem 4.
21. Normalize $\mathbf{y}$ as given in Problem 7.
22. Normalize $\mathbf{y}$ as given in Problem 8.
23. Normalize $\mathbf{y}$ as given in Problem 11.
24. Normalize $\mathbf{y}$ as given in Problem 15.
25. Normalize $\mathbf{y}$ as given in Problem 16.
26. Normalize $\mathbf{y}$ as given in Problem 17.
27. Find $\mathbf{x}$ if $\langle \mathbf{x}, \mathbf{a} \rangle \mathbf{b} = \mathbf{c}$, where

$$\mathbf{a} = \begin{bmatrix} 1 \\ 3 \\ -1 \end{bmatrix}, \quad \mathbf{b} = \begin{bmatrix} 2 \\ 1 \\ 1 \end{bmatrix}, \quad \text{and} \quad \mathbf{c} = \begin{bmatrix} 3 \\ 0 \\ -1 \end{bmatrix}.$$

28. Determine whether it is possible for two nonzero vectors to have an inner product that is zero.

29. Prove Property I2.

30. Prove Property I3.

31. Prove Property I4.

32. Prove Property I5.

33. Prove that $\|\mathbf{x} + \mathbf{y}\|^2 = \|\mathbf{x}\|^2 + 2\langle\mathbf{x}, \mathbf{y}\rangle + \|\mathbf{y}\|^2$.

34. Prove the *parallelogram law*:

$$\|\mathbf{x} + \mathbf{y}\|^2 + \|\mathbf{x} - \mathbf{y}\|^2 = 2\|\mathbf{x}\|^2 + 2\|\mathbf{y}\|^2.$$

35. Prove that, for any scalar λ,

$$0 \le \|\lambda\mathbf{x} + \mathbf{y}\|^2 = \lambda^2\|\mathbf{x}\|^2 - 2\lambda\langle\mathbf{x}, \mathbf{y}\rangle + \|\mathbf{y}\|^2.$$

36. (Problem 35 continued) Take $\lambda = \langle\mathbf{x}, \mathbf{y}\rangle/\|\mathbf{x}\|^2$ and show that

$$0 \le \frac{-\langle\mathbf{x}, \mathbf{y}\rangle^2}{\|\mathbf{x}\|^2} + \|\mathbf{y}\|^2.$$

From this, deduce that

$$\langle\mathbf{x}, \mathbf{y}\rangle^2 \le \|\mathbf{x}\|^2\|\mathbf{y}\|^2$$

and that

$$|\langle\mathbf{x}, \mathbf{y}\rangle| \le \|\mathbf{x}\|\ \|\mathbf{y}\|.$$

This last inequality is known as the *Cauchy–Schwarz inequality.*

37. Using the results of Problem 33 and the Cauchy–Schwarz inequality, show that

$$\|\mathbf{x} + \mathbf{y}\|^2 \le \|\mathbf{x}\|^2 + 2\|\mathbf{x}\|\ \|\mathbf{y}\| + \|\mathbf{y}\|^2$$
$$= (\|\mathbf{x}\| + \|\mathbf{y}\|)^2.$$

From this, deduce that

$$\|\mathbf{x} + \mathbf{y}\| \le \|\mathbf{x}\| + \|\mathbf{y}\|.$$

38. Determine whether there exists a relationship between $\langle\mathbf{x}, \mathbf{y}\rangle$ and $\mathbf{x}^T\mathbf{y}$, when both $\mathbf{x}$ and $\mathbf{y}$ are column vectors of identical dimension with real components.

39. Use the results of Problem 38 to prove that $\langle\mathbf{A}\mathbf{x}, \mathbf{y}\rangle = \langle\mathbf{x}, \mathbf{A}^T\mathbf{y}\rangle$, when $\mathbf{A}$, $\mathbf{x}$, and $\mathbf{y}$ are real matrices of dimensions $n \times n$, $n \times 1$, and $n \times 1$, respectively.

40. A generalization of the inner product for n-dimensional column vectors with real components is $\langle\mathbf{x}, \mathbf{y}\rangle_{\mathbf{A}} = \langle\mathbf{A}\mathbf{x}, \mathbf{A}\mathbf{y}\rangle$ for any real $n \times n$ nonsingular matrix $\mathbf{A}$. This definition reduces to the usual one when $\mathbf{A} = \mathbf{I}$.

Compute $\langle \mathbf{x}, \mathbf{y} \rangle_{\mathbf{A}}$ for the vectors given in Problem 1 when

$$\mathbf{A} = \begin{bmatrix} 2 & 3 \\ 1 & -1 \end{bmatrix}.$$

41. Compute $\langle \mathbf{x}, \mathbf{y} \rangle_{\mathbf{A}}$ for the vectors given in Problem 6 when

$$\mathbf{A} = \begin{bmatrix} 1 & 1 & 0 \\ 1 & 0 & 1 \\ 0 & 1 & 1 \end{bmatrix}.$$

42. Redo Problem 41 with

$$\mathbf{A} = \begin{bmatrix} 1 & -1 & 1 \\ 0 & 1 & -1 \\ 1 & 1 & 1 \end{bmatrix}.$$

10.2 Orthonormal vectors

Definition 1. Two vectors $\mathbf{x}$ and $\mathbf{y}$ are *orthogonal* (or perpendicular) if $\langle \mathbf{x}, \mathbf{y} \rangle = 0$. Thus, given the vectors

$$\mathbf{x} = \begin{bmatrix} 1 \\ 1 \\ 1 \end{bmatrix}, \quad \mathbf{y} = \begin{bmatrix} -1 \\ 1 \\ 0 \end{bmatrix}, \quad \mathbf{z} = \begin{bmatrix} 1 \\ 1 \\ 0 \end{bmatrix},$$

we see that $\mathbf{x}$ is orthogonal to $\mathbf{y}$ and $\mathbf{y}$ is orthogonal to $\mathbf{z}$ since $\langle \mathbf{x}, \mathbf{y} \rangle = \langle \mathbf{y}, \mathbf{z} \rangle = 0$; but the vectors $\mathbf{x}$ and $\mathbf{z}$ are not orthogonal since $\langle \mathbf{x}, \mathbf{z} \rangle = 1 + 1 \neq 0$. In particular, as a direct consequence of Property (**I5**) of Section 10.1, we have that the zero vector is orthogonal to every vector.

A set of vectors is called an *orthogonal set* if each vector in the set is orthogonal to every other vector in the set. The set given above is not an orthogonal set since $\mathbf{z}$ is not orthogonal to $\mathbf{x}$, whereas the set given by $\{\mathbf{x}, \mathbf{y}, \mathbf{z}\}$,

$$\mathbf{x} = \begin{bmatrix} 1 \\ 1 \\ 1 \end{bmatrix}, \quad \mathbf{y} = \begin{bmatrix} 1 \\ 1 \\ -2 \end{bmatrix}, \quad \mathbf{z} = \begin{bmatrix} 1 \\ -1 \\ 0 \end{bmatrix},$$

is an orthogonal set because each vector is orthogonal to every other vector.

Definition 2. A set of vectors is *orthonormal* if it is an orthogonal set having the property that every vector is a unit vector (a vector of magnitude 1).

The set of vectors

$$\left\{ \begin{bmatrix} 1/\sqrt{2} \\ 1/\sqrt{2} \\ 0 \end{bmatrix}, \begin{bmatrix} 1/\sqrt{2} \\ -1/\sqrt{2} \\ 0 \end{bmatrix}, \begin{bmatrix} 1 \\ 1 \\ 0 \end{bmatrix} \right\}$$

is an example of an orthonormal set.

Definition 2 can be simplified if we make use of the Kronecker delta, δ_{ij}, defined by

$$\delta_{ij} = \begin{cases} 1 & \text{if} \quad i = j, \\ 0 & \text{if} \quad i \neq j. \end{cases} \tag{1}$$

A set of vectors $\{\mathbf{x}_1, \mathbf{x}_2, \ldots, \mathbf{x}_n\}$ is an orthonormal set if and only if

$$\langle \mathbf{x}_i, \mathbf{x}_j \rangle = \delta_{ij} \quad \text{for all } i \text{ and } j, \quad i, j = 1, 2, \ldots, n. \tag{2}$$

The importance of orthonormal sets is that they are almost equivalent to linearly independent sets. However, since orthonormal sets have associated with them the additional structure of an inner product, they are often more convenient. We devote the remaining portion of this section to showing the equivalence of these two concepts. The utility of orthonormality will become self-evident in later sections.

Theorem 1. *An orthonormal set of vectors is linearly independent.*

    Proof. Let $\{\mathbf{x}_1, \mathbf{x}_2, \ldots, \mathbf{x}_n\}$ be an orthonormal set and consider the vector equation

$$c_1 \mathbf{x}_1 + c_2 \mathbf{x}_2 + \cdots + c_n \mathbf{x}_n = \mathbf{0}, \tag{3}$$

where the c_j's ($j = 1, 2, \ldots, n$) are constants. The set of vectors will be linearly independent if the only constants that satisfy (3) are $c_1 = c_2 = \ldots = c_n = 0$. Take the inner product of both sides of (3) with $\mathbf{x}_1$. Thus,

$$\langle c_1 \mathbf{x}_1 + c_2 \mathbf{x}_2 + \cdots + c_n \mathbf{x}_n, \mathbf{x}_1 \rangle = \langle \mathbf{0}, \mathbf{x}_1 \rangle.$$

Using properties (**I3**), (**I4**), and (**I5**) of Section 10.1, we have

$$c_1 \langle \mathbf{x}_1, \mathbf{x}_1 \rangle + c_2 \langle \mathbf{x}_2, \mathbf{x}_1 \rangle + \cdots + c_n \langle \mathbf{x}_n, \mathbf{x}_1 \rangle = 0.$$

    Finally, noting that $\langle \mathbf{x}_i, \mathbf{x}_1 \rangle = \delta_{i1}$, we obtain $c_1 = 0$. Now taking the inner product of both sides of (3) with $\mathbf{x}_2, \mathbf{x}_3, \ldots, \mathbf{x}n$, successively, we obtain $c_2 = 0$, $c_3 = 0$, ..., $c_n = 0$. Combining these results, we find that $c_1 = c_2 = \ldots c_n = 0$, which implies the theorem.

Theorem 2. *For every linearly independent set of vectors* $\{\mathbf{x}_1, \mathbf{x}_2, \ldots, \mathbf{x}_n\}$, *there exists an orthonormal set of vectors* $\{\mathbf{q}_1, \mathbf{q}_2, \ldots, \mathbf{q}_n\}$ *such that each* $\mathbf{q}_j$ ($j = 1, 2, \ldots, n$) *is a linear combination of* $\mathbf{x}_1, \mathbf{x}_2, \ldots, \mathbf{x}_j$.

Proof. First define new vectors $\mathbf{y}_1, \mathbf{y}_2, \ldots, \mathbf{y}_n$ by

$$\mathbf{y}_1 = \mathbf{x}_1$$

$$\mathbf{y}_1 = \mathbf{x}_1 - \frac{\langle \mathbf{x}_2, \mathbf{y}_1 \rangle}{\langle \mathbf{y}_1, \mathbf{y}_1 \rangle} \mathbf{y}_1$$

$$\mathbf{y}_3 = \mathbf{x}_3 - \frac{\langle \mathbf{x}_3, \mathbf{y}_1 \rangle}{\langle \mathbf{y}_1, \mathbf{y}_1 \rangle} \mathbf{y}_1 - \frac{\langle \mathbf{x}_3, \mathbf{y}_2 \rangle}{\langle \mathbf{y}_2, \mathbf{y}_2 \rangle} \mathbf{y}_2$$

and, in general,

$$\mathbf{y}_j = \mathbf{x}_j - \sum_{k=1}^{j-1} \frac{\langle \mathbf{x}_j, \mathbf{y}_k \rangle}{\langle \mathbf{y}_k, \mathbf{y}_k \rangle} \mathbf{y}_k \quad (j = 2, 3, \ldots, n). \tag{4}$$

Each $\mathbf{y}_j$ is a linear combination of $\mathbf{x}_1, \mathbf{x}_2, \ldots, \mathbf{x}_j$ ($j = 1, 2, \ldots, n$). Since the $\mathbf{x}$'s are linearly independent, and the coefficient of the $\mathbf{x}_j$ term in (4) is unity, it follows that $\mathbf{y}_j$ is not the zero vector (see Problem 19). Furthermore, it can be shown that the $\mathbf{y}_j$ terms form an orthogonal set (see Problem 20), hence the only property that the $\mathbf{y}_j$ terms lack in order to be the required set is that their magnitudes may not be one. We remedy this situation by defining

$$\mathbf{q}_j = \frac{\mathbf{y}_j}{\|\mathbf{y}_j\|}. \tag{5}$$

The desired set is $\{\mathbf{q}_1, \mathbf{q}_2, \ldots \mathbf{q}_n\}$.

The process used to construct the $\mathbf{q}_j$ terms is called the *Gram–Schmidt orthonormalization process.*

Example 1. Use the Gram–Schmidt orthonormalization process to construct an orthonormal set of vectors from the linearly independent set $\{\mathbf{x}_1, \mathbf{x}_2, \mathbf{x}_3\}$, where

$$\mathbf{x}_1 = \begin{bmatrix} 1 \\ 1 \\ 0 \end{bmatrix}, \quad \mathbf{x}_2 = \begin{bmatrix} 0 \\ 1 \\ 1 \end{bmatrix}, \quad \mathbf{x}_3 = \begin{bmatrix} 1 \\ 0 \\ 1 \end{bmatrix},$$

Solution

$$\mathbf{y}_1 = \mathbf{x}_1 = \begin{bmatrix} 1 \\ 1 \\ 0 \end{bmatrix}.$$

Now $\langle \mathbf{x}_2, \mathbf{y}_1 \rangle = 0(1) + 1(1) + 1(1) = 1$, and $\langle \mathbf{y}_1, \mathbf{y}_1 \rangle = 1(1) + 1(1) + 0(0) = 2$; hence,

$$\mathbf{y}_2 = \mathbf{x}_2 - \frac{\langle \mathbf{x}_2, \mathbf{y}_1 \rangle}{\langle \mathbf{y}_1, \mathbf{y}_1 \rangle} \mathbf{y}_1 = \mathbf{x}_2 - \frac{1}{2} \mathbf{y}_1 = \begin{bmatrix} 0 \\ 1 \\ 1 \end{bmatrix} - \frac{1}{2} \begin{bmatrix} 1 \\ 1 \\ 0 \end{bmatrix} = \begin{bmatrix} -1/2 \\ 1/2 \\ 1 \end{bmatrix}.$$

Then,

$$\langle \mathbf{x}_3, \mathbf{y}_1 \rangle = 1(1) + 0(1) + 1(0) = 1,$$
$$\langle \mathbf{x}_3, \mathbf{y}_2 \rangle = 1(-1/2) + 0(1/2) + 1(1) = 1/2,$$
$$\langle \mathbf{y}_2, \mathbf{y}_2 \rangle = (-1/2)^2 + (1/2)^2 + (1)^2 = 3/2,$$

so

$$\mathbf{y}_3 = \mathbf{x}_3 - \frac{\langle \mathbf{x}_3, \mathbf{y}_1 \rangle}{\langle \mathbf{y}_1, \mathbf{y}_1 \rangle} \mathbf{y}_1 - \frac{\langle \mathbf{x}_3, \mathbf{y}_2 \rangle}{\langle \mathbf{y}_2, \mathbf{y}_2 \rangle} \mathbf{y}_2 = \mathbf{x}_3 - \frac{1}{2} \mathbf{y}_1 - \frac{1/2}{3/2} \mathbf{y}_2$$

$$= \begin{bmatrix} 1 \\ 0 \\ 1 \end{bmatrix} - \frac{1}{2} \begin{bmatrix} 1 \\ 1 \\ 0 \end{bmatrix} - \frac{1}{3} \begin{bmatrix} -1/2 \\ 1/2 \\ 1 \end{bmatrix} = \begin{bmatrix} 2/3 \\ -2/3 \\ 2/3 \end{bmatrix}.$$

The vectors $\mathbf{y}_1$, $\mathbf{y}_2$, and $\mathbf{y}_3$ form an orthogonal set. To make this set orthonormal, we note that $\langle \mathbf{y}_1, \mathbf{y}_1 \rangle = 2$, $\langle \mathbf{y}_2, \mathbf{y}_2 \rangle = 3/2$, and $\langle \mathbf{y}_3, \mathbf{y}_3 \rangle = (2/3)(2/3) + (-2/3)(-2/3) + (2/3)(2/3) = 4/3$. Therefore,

$$\|\mathbf{y}_1\| = \sqrt{\langle \mathbf{y}_1, \mathbf{y}_1 \rangle} = \sqrt{2} \qquad \|\mathbf{y}_2\| = \sqrt{\langle \mathbf{y}_2, \mathbf{y}_2 \rangle} = \sqrt{3/2},$$
$$\|\mathbf{y}_3\| = \sqrt{\langle \mathbf{y}_3, \mathbf{y}_3 \rangle} = 2 \big/ \sqrt{3},$$

and

$$\mathbf{q}_1 = \frac{\mathbf{y}_1}{\|\mathbf{y}_1\|} = \frac{1}{\sqrt{2}} \begin{bmatrix} 1 \\ 1 \\ 0 \end{bmatrix} = \begin{bmatrix} 1/\sqrt{2} \\ 1/\sqrt{2} \\ 0 \end{bmatrix},$$

$$\mathbf{q}_2 = \frac{\mathbf{y}_2}{\|\mathbf{y}_2\|} = \frac{1}{\sqrt{3/2}} \begin{bmatrix} -12 \\ 1/2 \\ 1 \end{bmatrix} = \begin{bmatrix} -1/\sqrt{6} \\ 1/\sqrt{6} \\ 2/\sqrt{6} \end{bmatrix},$$

$$\mathbf{q}_3 = \frac{\mathbf{y}_3}{\|\mathbf{y}_3\|} = \frac{1}{2/\sqrt{3}} \begin{bmatrix} 2/3 \\ -2/3 \\ 2/3 \end{bmatrix} = \begin{bmatrix} 1/\sqrt{3} \\ -1/\sqrt{3} \\ 1/\sqrt{3} \end{bmatrix}. \blacksquare$$

Example 2. Use the Gram–Schmidt orthonormalization process to construct an orthonormal set of vectors from the linearly independent set $\{\mathbf{x}_1, \mathbf{x}_2, \mathbf{x}_3, \mathbf{x}_4\}$, where

$$\mathbf{x}_1 = \begin{bmatrix} 1 \\ 1 \\ 0 \\ 1 \end{bmatrix}, \quad \mathbf{x}_2 = \begin{bmatrix} 1 \\ 2 \\ 1 \\ 0 \end{bmatrix}, \quad \mathbf{x}_3 = \begin{bmatrix} 0 \\ 1 \\ 2 \\ 1 \end{bmatrix}, \quad \mathbf{x}_4 = \begin{bmatrix} 1 \\ 0 \\ 1 \\ 1 \end{bmatrix}.$$

Solution

$$\mathbf{y}_1 = \mathbf{x}_1 = \begin{bmatrix} 1 \\ 1 \\ 0 \\ 1 \end{bmatrix},$$

$$\langle \mathbf{y}_1, \mathbf{y}_1 \rangle = 1(1) + 1(1) + 0(0) + 1(1) = 3,$$

$$\langle \mathbf{x}_1, \mathbf{y}_1 \rangle = 1(1) + 2(1) + 1(0) + 0(1) = 3,$$

$$\mathbf{y}_2 = \mathbf{x}_2 - \frac{\langle \mathbf{x}_2, \mathbf{y}_1 \rangle}{\langle \mathbf{y}_1, \mathbf{y}_1 \rangle} \mathbf{y}_1 = \mathbf{x}_2 - \frac{3}{3} \mathbf{y}_1 = \begin{bmatrix} 0 \\ 1 \\ 1 \\ -1 \end{bmatrix};$$

$$\langle \mathbf{y}_2, \mathbf{y}_2 \rangle = 0(0) + 1(1) + 1(1) + (-1)(-1) = 3,$$

$$\langle \mathbf{x}_3, \mathbf{y}_1 \rangle = 0(1) + 1(1) + 1(0) + 1(1) = 2,$$

$$\langle \mathbf{x}_3, \mathbf{y}_2 \rangle = 0(0) + 1(1) + 2(1) + 1(-1) = 2,$$

$$\mathbf{y}_3 = \mathbf{x}_3 - \frac{\langle \mathbf{x}_3, \mathbf{y}_1 \rangle}{\langle \mathbf{y}_1, \mathbf{y}_1 \rangle} \mathbf{y}_1 - \frac{\langle \mathbf{x}_3, \mathbf{y}_2 \rangle}{\langle \mathbf{y}_2, \mathbf{y}_2 \rangle} \mathbf{y}_2$$

$$= \mathbf{x}_3 - \frac{2}{3} \mathbf{y}_1 - \frac{2}{3} \mathbf{y}_2 = \begin{bmatrix} -2/3 \\ -1/3 \\ 4/3 \\ 1 \end{bmatrix};$$

$$\langle \mathbf{y}_3, \mathbf{y}_3 \rangle = \left(\frac{-2}{3} \right)^2 + \left(\frac{-1}{3} \right)^2 + \left(\frac{4}{3} \right)^2 + (1)^2 = \frac{10}{3},$$

$$\langle \mathbf{x}_4, \mathbf{y}_1 \rangle = 1(1) + 0(1) + 1(0) + 1(1) = 2,$$
$$\langle \mathbf{x}_4, \mathbf{y}_2 \rangle = 1(0) + 0(1) + 1(1) + 1(.1) = 0,$$

$$\langle \mathbf{x}_4, \mathbf{y}_3 \rangle = 1 \left(\frac{-2}{3} \right) + 0 \left(\frac{-1}{3} \right) + 1 \left(\frac{4}{3} \right) + 1(1) = \frac{5}{3},$$

$$\mathbf{y}_4 = \mathbf{x}_4 - \frac{\langle \mathbf{x}_4, \mathbf{y}_1 \rangle}{\langle \mathbf{y}_1, \mathbf{y}_1 \rangle} \mathbf{y}_1 - \frac{\langle \mathbf{x}_4, \mathbf{y}_2 \rangle}{\langle \mathbf{y}_2, \mathbf{y}_2 \rangle} \mathbf{y}_2 - \frac{\langle \mathbf{x}_4, \mathbf{y}_3 \rangle}{\langle \mathbf{y}_3, \mathbf{y}_3 \rangle} \mathbf{y}_3$$

$$= \mathbf{x}_4 - \frac{2}{3} \mathbf{y}_1 - \frac{0}{3} \mathbf{y}_2 - \frac{5/3}{10/3} \mathbf{y}_3 = \begin{bmatrix} 2/3 \\ -1/2 \\ 1/3 \\ -1/6 \end{bmatrix}.$$

Then

$$\langle \mathbf{y}_4, \mathbf{y}_4 \rangle = (2/3)(2/3) + (-1/2)(-1/2) + (1/3)(1/3) + (-1/6)(-1/6)$$
$$= 5/6,$$

and

$$\mathbf{q}_1 = \frac{1}{\sqrt{3}} \begin{bmatrix} 1 \\ 1 \\ 0 \\ 1 \end{bmatrix} = \begin{bmatrix} 1/\sqrt{3} \\ 1/\sqrt{3} \\ 0 \\ 1/\sqrt{3} \end{bmatrix},$$

$$\mathbf{q}_2 = \frac{1}{\sqrt{3}} \begin{bmatrix} 0 \\ 1 \\ 1 \\ -1 \end{bmatrix} = \begin{bmatrix} 0 \\ 1/\sqrt{3} \\ 1/\sqrt{3} \\ -1/\sqrt{3} \end{bmatrix},$$

$$\mathbf{q}_3 = \frac{1}{\sqrt{10/3}} \begin{bmatrix} -2/3 \\ -1/3 \\ 4/3 \\ 1 \end{bmatrix} = \begin{bmatrix} -2/\sqrt{30} \\ -1/\sqrt{30} \\ 4/\sqrt{30} \\ 3/\sqrt{30} \end{bmatrix},$$

$$\mathbf{q}_4 = \frac{1}{\sqrt{5/6}} \begin{bmatrix} 2/3 \\ -1/2 \\ 1/3 \\ -1/6 \end{bmatrix} = \begin{bmatrix} 4/\sqrt{30} \\ -3/\sqrt{30} \\ 2/\sqrt{30} \\ -1/\sqrt{30} \end{bmatrix}. \blacksquare$$

Problems 10.2

1. Determine which of the following vectors are orthogonal:

$$\mathbf{x} = \begin{bmatrix} 1 \\ 2 \end{bmatrix}, \ \mathbf{y} = \begin{bmatrix} 2 \\ -1 \end{bmatrix}, \ \mathbf{z} = \begin{bmatrix} -2 \\ -1 \end{bmatrix}, \ \mathbf{u} = \begin{bmatrix} -4 \\ 2 \end{bmatrix}, \ \mathbf{v} = \begin{bmatrix} 3 \\ 6 \end{bmatrix}.$$

2. Determine which of the following vectors are orthogonal:

$$\mathbf{x} = \begin{bmatrix} 1 \\ 1 \\ 2 \end{bmatrix}, \ \mathbf{y} = \begin{bmatrix} 1 \\ 1 \\ 1 \end{bmatrix}, \ \mathbf{z} = \begin{bmatrix} 1 \\ 1 \\ -1 \end{bmatrix}, \ \mathbf{u} = \begin{bmatrix} 1 \\ -1 \\ 0 \end{bmatrix}, \ \mathbf{v} = \begin{bmatrix} -2 \\ 1 \\ 1 \end{bmatrix}.$$

3. Find x so that

$$\begin{bmatrix} 3 \\ 5 \end{bmatrix} \text{ is orthogonal to } \begin{bmatrix} x \\ 4 \end{bmatrix}.$$

4. Find x so that

$$\begin{bmatrix} -1 \\ x \\ 3 \end{bmatrix} \text{ is orthogonal to } \begin{bmatrix} 1 \\ 2 \\ 3 \end{bmatrix}.$$

5. Find x so that $[x \; x \; 2]$ is orthogonal to $[1 \; 3 \; -1]$.

6. Find x and y so that $[x \; y]$ is orthogonal to $[1 \; 3]$.

7. Find x and y so that

$$\begin{bmatrix} x \\ y \\ 1 \end{bmatrix} \text{ is orthogonal to both } \begin{bmatrix} 1 \\ 2 \\ 3 \end{bmatrix} \text{ and } \begin{bmatrix} 1 \\ 1 \\ 1 \end{bmatrix}.$$

8. Find x, y, and z so that

$$\begin{bmatrix} x \\ y \\ z \end{bmatrix} \text{ is orthogonal to both } \begin{bmatrix} 1 \\ 0 \\ 1 \end{bmatrix} \text{ and } \begin{bmatrix} 1 \\ 1 \\ 2 \end{bmatrix}.$$

9. Redo Problem 8 with the additional stipulation that $[x \; y \; z]^T$ be a unit vector.

In Problems 10 through 18, use the Gram–Schmidt orthonormalization process to construct an orthonormal set from the given set of linearly independent vectors.

10. $\mathbf{x}_1 = \begin{bmatrix} 1 \\ 2 \end{bmatrix}$, $\mathbf{x}_2 = \begin{bmatrix} 2 \\ 1 \end{bmatrix}$.

11. $\mathbf{x}_1 = \begin{bmatrix} 1 \\ 1 \end{bmatrix}$, $\mathbf{x}_2 = \begin{bmatrix} 3 \\ 5 \end{bmatrix}$.

12. $\mathbf{x}_1 = \begin{bmatrix} 3 \\ -2 \end{bmatrix}$, $\mathbf{x}_2 = \begin{bmatrix} 3 \\ 3 \end{bmatrix}$.

13. $\mathbf{x}_1 = \begin{bmatrix} 1 \\ 2 \\ 1 \end{bmatrix}$, $\mathbf{x}_2 = \begin{bmatrix} 1 \\ 0 \\ 1 \end{bmatrix}$, $\mathbf{x}_3 = \begin{bmatrix} 1 \\ 0 \\ 2 \end{bmatrix}$.

14. $\mathbf{x}_1 = \begin{bmatrix} 2 \\ 1 \\ 0 \end{bmatrix}, \quad \mathbf{x}_2 = \begin{bmatrix} 0 \\ 1 \\ 1 \end{bmatrix}, \quad \mathbf{x}_3 = \begin{bmatrix} 2 \\ 0 \\ 2 \end{bmatrix}.$

15. $\mathbf{x}_1 = \begin{bmatrix} 1 \\ 1 \\ 0 \end{bmatrix}, \quad \mathbf{x}_2 = \begin{bmatrix} 2 \\ 0 \\ 1 \end{bmatrix}, \quad \mathbf{x}_3 = \begin{bmatrix} 2 \\ 2 \\ 1 \end{bmatrix}.$

16. $\mathbf{x}_1 = \begin{bmatrix} 0 \\ 3 \\ 4 \end{bmatrix}, \quad \mathbf{x}_2 = \begin{bmatrix} 3 \\ 5 \\ 0 \end{bmatrix}, \quad \mathbf{x}_3 = \begin{bmatrix} 2 \\ 0 \\ 5 \end{bmatrix}.$

17. $\mathbf{x}_1 = \begin{bmatrix} 0 \\ 1 \\ 1 \\ 1 \end{bmatrix}, \quad \mathbf{x}_2 = \begin{bmatrix} 1 \\ 0 \\ 1 \\ 1 \end{bmatrix}, \quad \mathbf{x}_3 = \begin{bmatrix} 1 \\ 1 \\ 0 \\ 1 \end{bmatrix}, \quad \mathbf{x}_2 = \begin{bmatrix} 1 \\ 1 \\ 1 \\ 0 \end{bmatrix}.$

18. $\mathbf{x}_1 = \begin{bmatrix} 1 \\ 1 \\ 0 \\ 0 \end{bmatrix}, \quad \mathbf{x}_2 = \begin{bmatrix} 0 \\ 1 \\ -1 \\ 0 \end{bmatrix}, \quad \mathbf{x}_3 = \begin{bmatrix} 1 \\ 0 \\ -1 \\ 0 \end{bmatrix}, \quad \mathbf{x}_4 = \begin{bmatrix} 1 \\ 0 \\ 0 \\ -1 \end{bmatrix}.$

19. Prove that no **y**-vector in the Gram−Schmidt orthonormalization process is zero.

20. Prove that the **y**-vectors in the Gram−Schmidt orthonormalization process form an orthogonal set. (*Hint*: first show that $\langle \mathbf{y}_2, \mathbf{y}_1 \rangle = 0$, hence $\mathbf{y}_2$ must be orthogonal to $\mathbf{y}_1$. Then use induction.)

21. With $\mathbf{q}_j$ defined by Eq. (5), show that Eq. (4) can be simplified to $\mathbf{y}_j = \mathbf{x}_j - \sum_{k=1}^{j-1} \langle \mathbf{x}_j, \mathbf{q}_k \rangle \mathbf{q}_k.$

22. The vectors

$$\mathbf{x}_1 = \begin{bmatrix} 1 \\ 1 \\ 0 \end{bmatrix}, \quad \mathbf{x}_2 = \begin{bmatrix} 0 \\ 1 \\ 1 \end{bmatrix}, \quad \mathbf{x}_3 = \begin{bmatrix} 1 \\ 0 \\ -1 \end{bmatrix}.$$

are linearly dependent. Apply the Gram–Schmidt process to it and use the results to deduce what occurs whenever the process is applied to a linearly dependent set of vectors.

23. Prove that if $\mathbf{x}$ and $\mathbf{y}$ are orthogonal, then

$$\|\mathbf{x} - \mathbf{y}\|^2 = \|\mathbf{x}\|^2 + \|\mathbf{y}\|^2.$$

24. Prove that if $\mathbf{x}$ and $\mathbf{y}$ are orthonormal, then

$$\|s\mathbf{x} + t\mathbf{y}\|^2 = s^2 + t^2$$

for any two scalars s and t.

25. Let $\mathbf{Q}$ be any $n \times n$ matrix whose columns, when considered as n-dimensional vectors, form an orthonormal set. What can you say about the product $\mathbf{Q}^T\mathbf{Q}$?

26. Prove that if $\langle \mathbf{y}, \mathbf{x} \rangle = 0$ for every n-dimensional vector $\mathbf{y}$, then $\mathbf{x} = 0$.

27. Let $\mathbf{x}$ and $\mathbf{y}$ be any two vectors of the same dimension. Prove that $\mathbf{x} + \mathbf{y}$ is orthogonal to $\mathbf{x} - \mathbf{y}$ if and only if $\|\mathbf{x}\| = \|\mathbf{y}\|$.

28. Let $\mathbf{A}$ be an $n \times n$ real matrix and $\mathbf{p}$ be a real n-dimensional column vector. Show that if $\mathbf{p}$ is orthogonal to the columns of $\mathbf{A}$, then $\langle \mathbf{Ay}, \mathbf{p} \rangle = 0$ for any n-dimensional real column vector $\mathbf{y}$.

10.3 Projections and QR decompositions

As with other vector operations, the inner product has a geometrical interpretation in two or three dimensions. For simplicity, we consider two-dimensional vectors here; the extension to three dimensions is straightforward.

Let $\mathbf{u}$ and $\mathbf{v}$ be two nonzero vectors, considered as directed line segments (see Section 1.7), positioned so that their initial points coincide. The *angle between* $\mathbf{u}$ *and* $\mathbf{v}$ is the angle θ between the two line segments satisfying $0 \le \theta \le \pi$. See Fig. 10.1.

Definition 1. If $\mathbf{u}$ and $\mathbf{v}$ are two-dimensional vectors and θ is the angle between them, then the *dot product* of these two vectors is $\mathbf{u} \cdot \mathbf{v} = \|\mathbf{u}\| \, \|\mathbf{v}\| \cos \theta$.

To use Definition 1, we need the cosine of the angle between two vectors, which requires us to measure the angle. We shall take another approach.

The vectors $\mathbf{u}$ and $\mathbf{v}$ along with their difference $\mathbf{u} - \mathbf{v}$ form a triangle (see Fig. 10.2) having sides $\|\mathbf{u}\|$, $\|\mathbf{v}\|$ and $\|\mathbf{u} - \mathbf{v}\|$. It follows from the law of cosines that

$$\|\mathbf{u} - \mathbf{v}\|^2 = \|\mathbf{u}\|^2 + \|\mathbf{v}\|^2 - 2\|\mathbf{u}\| \, \|\mathbf{v}\| \cos \theta,$$

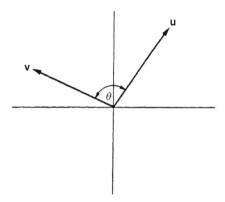

FIGURE 10.1

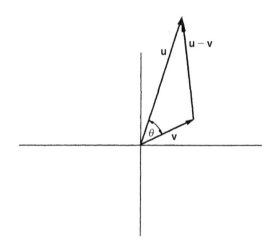

FIGURE 10.2

whereupon

$$\|\mathbf{u}\|\|\mathbf{v}\|\cos\theta = \frac{1}{2}\left[\|\mathbf{u}\|^2 + \|\mathbf{v}\|^2 - \|\mathbf{u} - \mathbf{v}\|^2\right]$$

$$= \frac{1}{2}[\langle\mathbf{u},\mathbf{u}\rangle + \langle\mathbf{v},\mathbf{v}\rangle - \langle\mathbf{u} - \mathbf{v}, \mathbf{u} - \mathbf{v}\rangle -]$$

$$= \langle\mathbf{u},\mathbf{v}\rangle.$$

Thus, the dot product of two-dimensional vectors is the inner product of those vectors. That is,

$$\mathbf{u} \cdot \mathbf{v} = \|\mathbf{u}\|\,\|\mathbf{v}\|\,\cos\theta = \langle\mathbf{u},\,\mathbf{v}\rangle. \tag{6}$$

The dot product of nonzero vectors is zero if and only if $\cos \theta = 0°$, or $\theta = 90°$. Consequently, the dot product of two nonzero vectors is zero if and only if the vectors are perpendicular. This, with Eq. (6), establishes the equivalence between orthogonality and perpendicularity for two-dimensional vectors. In addition, we may rewrite Eq. (6) as

$$\cos \theta = \frac{\langle \mathbf{u}, \mathbf{v} \rangle}{\|\mathbf{u}\| \, \|\mathbf{v}\|} \tag{7}$$

and use Eq. (7) to calculate the angle between two vectors.

Example 1. Find the angle between the vectors

$$\mathbf{u} = \begin{bmatrix} 2 \\ 5 \end{bmatrix} \quad \text{and} \quad \mathbf{v} = \begin{bmatrix} -3 \\ 4 \end{bmatrix}.$$

Solution $\langle \mathbf{u}, \mathbf{v} \rangle = 2(-3) + 5(4) = 14$, $\|\mathbf{u}\| = \sqrt{4 + 25} = \sqrt{29}$, $\|\mathbf{v}\| = \sqrt{9 + 16} = 5$, so $\cos \theta = 14/(5\sqrt{29}) = 0.1599$, and $\theta = 58.7°$. ∎

Eq. (7) is used to define the angle between any two vectors of the same but arbitrary dimension, even though the geometrical significance of an angle becomes meaningless for dimensions greater than three. (See Problems 9 and 10.)

A problem that occurs often in the applied sciences and that has important ramifications for us in matrices involves a given nonzero vector $\mathbf{x}$ and a nonzero reference vector $\mathbf{a}$. The problem is to decompose $\mathbf{x}$ into the sum of two vectors, $\mathbf{u} + \mathbf{v}$, where $\mathbf{u}$ is parallel to $\mathbf{a}$ and $\mathbf{v}$ is perpendicular to $\mathbf{a}$. This situation is illustrated in Fig. 10.3. In physics, $\mathbf{u}$ is called the parallel component of $\mathbf{x}$ and $\mathbf{v}$ is called the perpendicular component of $\mathbf{x}$, where parallel and perpendicular are understood to be with respect to the reference vector $\mathbf{a}$.

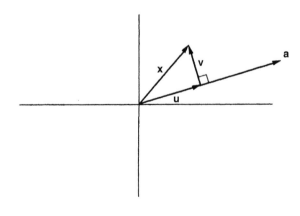

FIGURE 10.3

If $\mathbf{u}$ is to be parallel to $\mathbf{a}$, it must be a scalar multiple of $\mathbf{a}$, in particular $\mathbf{u} = \lambda\mathbf{a}$. Since we want $\mathbf{x} = \mathbf{u} + \mathbf{v}$, it follows that $\mathbf{v} = \mathbf{x} - \mathbf{u} = \mathbf{x} - \lambda\mathbf{a}$. Finally, if $\mathbf{u}$ and $\mathbf{v}$ are to be perpendicular, we require that

$$0 = \langle \mathbf{u}, \mathbf{v} \rangle = \langle \lambda\mathbf{a}, \mathbf{x} - \lambda\mathbf{a} \rangle$$
$$= \lambda\langle \mathbf{a}, \mathbf{x} \rangle = \lambda^2\langle \mathbf{a}, \mathbf{a} \rangle$$
$$= \lambda[\langle \mathbf{a}, \mathbf{x} \rangle = \lambda\langle \mathbf{a}, \mathbf{a} \rangle].$$

Thus, either $\lambda = 0$ or $\lambda = \langle \mathbf{a}, \mathbf{x} \rangle / \langle \mathbf{a}, \mathbf{a} \rangle$. If $\lambda = 0$, then $\mathbf{u} = \lambda\mathbf{a} = 0$ and $\mathbf{x} = \mathbf{u} + \mathbf{v} = \mathbf{v}$, which means that $\mathbf{x}$ and $\mathbf{a}$ are perpendicular. In such a case, $\langle \mathbf{a}, \mathbf{x} \rangle = 0$. Thus, we may always infer that $\lambda = \langle \mathbf{a}, \mathbf{x} \rangle / \langle \mathbf{a}, \mathbf{a} \rangle$ with

$$\mathbf{u} = \frac{\langle \mathbf{a}, \mathbf{x} \rangle}{\langle \mathbf{a}, \mathbf{a} \rangle}\mathbf{a} \quad \text{and} \quad \mathbf{v} = \mathbf{x} - \frac{\langle \mathbf{a}, \mathbf{x} \rangle}{\langle \mathbf{a}, \mathbf{x} \rangle}\mathbf{a}.$$

In this context, $\mathbf{u}$ is the *projection of $\mathbf{x}$ onto $\mathbf{a}$* and $\mathbf{v}$ is the *orthogonal complement*.

Example 2. Decompose the vector

$$\mathbf{x} = \begin{bmatrix} 2 \\ 7 \end{bmatrix}$$

into the sum of two vectors, one of which is parallel to

$$\mathbf{a} = \begin{bmatrix} -3 \\ 4 \end{bmatrix},$$

and one of which is perpendicular to $\mathbf{a}$.

Solution

$$\mathbf{u} = \frac{\langle \mathbf{a}, \mathbf{x} \rangle}{\langle \mathbf{a}, \mathbf{a} \rangle}\mathbf{a} = \frac{22}{25}\begin{bmatrix} -3 \\ 4 \end{bmatrix} = \begin{bmatrix} -2.64 \\ 3.52 \end{bmatrix},$$

$$\mathbf{v} = \mathbf{x} - \mathbf{u} = \begin{bmatrix} 2 \\ 7 \end{bmatrix} - \begin{bmatrix} -2.64 \\ 3.52 \end{bmatrix} = \begin{bmatrix} 4.64 \\ 3.48 \end{bmatrix}.$$

Then, $\mathbf{x} = \mathbf{u} + \mathbf{v}$, with $\mathbf{u}$ parallel to $\mathbf{a}$ and $\mathbf{v}$ perpendicular to $\mathbf{a}$. ∎

We now extend the relationships developed in two dimensions to vectors in higher dimensions. Given a nonzero vector $\mathbf{x}$ and another nonzero reference vector $\mathbf{a}$, we define the projection of $\mathbf{x}$ onto $\mathbf{a}$ as

$$\text{proj}_\mathbf{a}\mathbf{x} = \frac{\langle \mathbf{a}, \mathbf{x} \rangle}{\langle \mathbf{a}, \mathbf{a} \rangle}\mathbf{a}. \tag{8}$$

As a result, we obtain the very important relationship that

$$\mathbf{x} - \frac{\langle \mathbf{a}, \mathbf{x} \rangle}{\langle \mathbf{a}, \mathbf{x} \rangle} \mathbf{a} \text{ is orthogonal to } \mathbf{a}. \tag{9}$$

That is, if we subtract from a nonzero vector $\mathbf{x}$ its projection onto another nonzero vector $\mathbf{a}$, we are left with a vector that is orthogonal to $\mathbf{a}$. (See Problem 23.)

In this context, the Gram–Schmidt process, described in Section 10.2, is almost obvious. Consider Eq. (4) from that section:

$$\mathbf{y}_j = \mathbf{x}_j - \sum_{k=1}^{j-1} \frac{\langle \mathbf{x}_j, \mathbf{y}_k \rangle}{\langle \mathbf{y}_k, \mathbf{y}_k \rangle} \mathbf{y}_k \quad \text{(4 repeated)}.$$

The quantity inside the summation sign is the projection of $\mathbf{x}_j$ onto $\mathbf{y}_k$. Thus, for each k ($k = 1, 2, ..., j - 1$), we are sequentially subtracting from $\mathbf{x}_j$ its projection onto $\mathbf{y}_k$, leaving a vector that is orthogonal to $\mathbf{y}_k$.

We now propose to alter slightly the steps of the Gram–Schmidt orthonormalization process. First, we shall normalize the orthogonal vectors as soon as they are obtained, rather than waiting until the end. This will make for messier hand calculations, but for a more efficient computer algorithm. Observe that if the $\mathbf{y}_k$ vectors in Eq. (4) are unit vectors, then the denominator is unity and need not be calculated.

Once we have fully determined a $\mathbf{y}_k$ vector, we shall immediately subtract the various projections onto this vector from all succeeding $\mathbf{x}$ vectors. In particular, once $\mathbf{y}_1$ is determined, we shall subtract the projection of $\mathbf{x}_2$ onto $\mathbf{y}_1$ from $\mathbf{x}_2$, then we shall subtract the projection of $\mathbf{x}_3$ onto $\mathbf{y}_1$ from $\mathbf{x}_3$, and continue until we have subtracted the projection of $\mathbf{x}_n$ onto $\mathbf{y}_1$ from $\mathbf{x}_n$. Only then will we return to $\mathbf{x}_2$ and normalize it to obtain $\mathbf{y}_2$. Then, we shall subtract from $\mathbf{x}_3, \mathbf{x}_4, ..., \mathbf{x}_n$ the projections onto $\mathbf{y}_2$ from $\mathbf{x}_3, \mathbf{x}_4, ..., \mathbf{x}_n$, respectively, before returning to $\mathbf{x}_3$ and normalizing it, thus obtaining $\mathbf{y}_3$. As a result, once we have $\mathbf{y}_1$, we alter $\mathbf{x}_2, \mathbf{x}_3, ..., \mathbf{x}_n$ so each is orthogonal to $\mathbf{y}_1$; once we have $\mathbf{y}_2$, we alter again $\mathbf{x}_3, \mathbf{x}_4, ..., \mathbf{x}_n$ so each is also orthogonal to $\mathbf{y}_2$; and so on.

These changes are known as the *revised Gram–Schmidt algorithm*. Given a set of linearly independent vectors $\{\mathbf{x}_1, \mathbf{x}_2, ..., \mathbf{x}_n\}$, the algorithm may be formalized as follows: Begin with $k = 1$ and sequentially moving through $k = n$:

(i) calculate $r_{kk} = \sqrt{\langle \mathbf{x}_k, \mathbf{x}_k \rangle}$,
(ii) set $\mathbf{q}_k = (1/r_{kk}) \mathbf{x}_k$
(iii) for $j = k + 1, k + 2, ..., n$, calculate $r_{kj} = \langle \mathbf{x}_{j,k} \rangle$
(iv) for $j = k + 1, k + 2, ..., n$, replace $\mathbf{x}_j$ by $\mathbf{x}_j - r_{kj} \mathbf{q}_k$.

The first two steps normalize; the third and fourth steps subtract projections from vectors, thereby generating orthogonality.

Example 3.   Use the revised Gram–Schmidt algorithm to construct an orthonormal set of vectors from the linearly independent set $\{\mathbf{x}_1, \mathbf{x}_2, \mathbf{x}_3\}$, where

$$\mathbf{x}_1 = \begin{bmatrix} 1 \\ 1 \\ 0 \end{bmatrix}, \quad \mathbf{x}_2 = \begin{bmatrix} 0 \\ 1 \\ 1 \end{bmatrix}, \quad \mathbf{x}_3 = \begin{bmatrix} 1 \\ 0 \\ 1 \end{bmatrix}.$$

Solution
First iteration ($k = 1$)

$$r_{11} = \sqrt{\langle \mathbf{x}_1, \mathbf{x}_1 \rangle} = \sqrt{2},$$

$$\mathbf{q}_1 = \frac{1}{r_{11}}\mathbf{x}_1 = \frac{1}{\sqrt{2}}\begin{bmatrix} 1 \\ 1 \\ 0 \end{bmatrix} = \begin{bmatrix} 1/\sqrt{2} \\ 1/\sqrt{2} \\ 0 \end{bmatrix},$$

$$r_{12} = \langle \mathbf{x}_2, \mathbf{q}_1 \rangle = \frac{1}{\sqrt{2}},$$

$$r_{13} = \langle \mathbf{x}_3, \mathbf{q}_1 \rangle = \frac{1}{\sqrt{2}},$$

$$\mathbf{x}_2 \leftarrow \mathbf{x}_2 - r_{12}\mathbf{q}_1 = \begin{bmatrix} 0 \\ 1 \\ 1 \end{bmatrix} - \frac{1}{\sqrt{2}}\begin{bmatrix} 1/\sqrt{2} \\ 1/\sqrt{2} \\ 0 \end{bmatrix} = \begin{bmatrix} -1/2 \\ 1/2 \\ 1 \end{bmatrix},$$

$$\mathbf{x}_3 \leftarrow \mathbf{x}_3 - r_{12}\mathbf{q}_1 = \begin{bmatrix} 1 \\ 0 \\ 1 \end{bmatrix} - \frac{1}{\sqrt{2}}\begin{bmatrix} 1/\sqrt{2} \\ 1/\sqrt{2} \\ 0 \end{bmatrix} = \begin{bmatrix} 1/2 \\ -1/2 \\ 1 \end{bmatrix}.$$

Note that both $\mathbf{x}_2$ and $\mathbf{x}_3$ are now orthogonal to $\mathbf{q}_1$.
Second iteration ($k = 2$)
Using vectors from the first iteration, we compute

$$r_{22} = \sqrt{\langle \mathbf{x}_2, \mathbf{x}_2 \rangle} = \sqrt{3/2},$$

$$\mathbf{q}_2 = \frac{1}{r_{22}}\mathbf{x}_2 = \frac{1}{\sqrt{3/2}}\begin{bmatrix} -1/2 \\ 1/2 \\ 1 \end{bmatrix} = \begin{bmatrix} -1/\sqrt{6} \\ 1/\sqrt{6} \\ 2/\sqrt{6} \end{bmatrix},$$

$$r_{23} = \langle \mathbf{x}_3, \mathbf{q}_2 \rangle = \frac{1}{\sqrt{6}},$$

$$\mathbf{x}_3 \leftarrow \mathbf{x}_3 - r_{23}\mathbf{q}_2 = \begin{bmatrix} 1/2 \\ -1/2 \\ 1 \end{bmatrix} - \frac{1}{\sqrt{6}}\begin{bmatrix} -1/\sqrt{6} \\ 1/\sqrt{6} \\ 2/\sqrt{6} \end{bmatrix} = \begin{bmatrix} 2/3 \\ -2/3 \\ 2/3 \end{bmatrix}.$$

Third iteration ($k = 3$)
Using vectors from the second iteration, we compute

$$r_{33} = \sqrt{\langle x_3, x_3 \rangle} = \frac{2}{\sqrt{3}},$$

$$\mathbf{q}_3 = \frac{1}{r_{33}} \mathbf{x}_3 = \frac{1}{2/\sqrt{3}} \begin{bmatrix} 2/3 \\ -2/3 \\ 2/3 \end{bmatrix} = \begin{bmatrix} 1/\sqrt{3} \\ -1/\sqrt{3} \\ 1/\sqrt{3} \end{bmatrix}.$$

The orthonormal set is $\{\mathbf{q}_1, \mathbf{q}_2, \mathbf{q}_3\}$. Compare with Example 1 of Section 10.2. ∎

The revised Gram–Schmidt algorithm has two advantages over the Gram–Schmidt process developed in the previous section. First, it is less effected by roundoff errors, and second, the inverse process—recapturing the **x**-vectors from the **q**-vectors—becomes trivial. To understand this second advantage, let us redo Example 3 symbolically. In the first iteration, we calculated

$$\mathbf{q}_1 = \frac{1}{r_{11}} \mathbf{x}_1,$$

so, we immediately have

$$\mathbf{x}_1 = r_{11} \mathbf{q}_1. \tag{10}$$

We then replaced $\mathbf{x}_2$ and $\mathbf{x}_3$ with vectors that were orthogonal to $\mathbf{q}_1$. If we denote these replacement vectors as $\mathbf{x}'_2$ and $\mathbf{x}'_3$, respectively, we have

$$\mathbf{x}'_2 = \mathbf{x}_2 - r_{12} \mathbf{q}_1 \quad and \quad \mathbf{x}'_3 = \mathbf{x}_3 - r_{13} \mathbf{q}_1.$$

With the second iteration, we calculated

$$\mathbf{q}_2 = \frac{1}{r_{22}} \mathbf{x}'_2 = \frac{1}{r_{22}} (\mathbf{x}_2 - r_{12} \mathbf{q}_1).$$

Solving for $\mathbf{x}_2$, we get

$$\mathbf{x}_2 = r_{12} \mathbf{q}_1 + r_{22} \mathbf{q}_2. \tag{11}$$

We then replaced $\mathbf{x}_3$ with a vector that was orthogonal to $\mathbf{q}_2$. If we denote this replacement vector as $\mathbf{x}''_3$, we have

$$\mathbf{x}''_3 = \mathbf{x}'_3 - r_{23} \mathbf{q}_2 = (\mathbf{x}_3 - r_{13} \mathbf{q}_1) - r_{23} \mathbf{q}_2.$$

With the third iteration, we calculated

$$\mathbf{q}_3 = \frac{1}{r_{33}} \mathbf{x}''_3 = \frac{1}{r_{33}} (\mathbf{x}_3 - r_{13} \mathbf{q}_1 - r_{23} \mathbf{q}_2).$$

Solving for $\mathbf{x}_3$, we obtain

$$\mathbf{x}_3 = r_{13} \mathbf{q}_1 + r_{23} \mathbf{q}_2 + r_{33} \mathbf{q}_3. \tag{12}$$

Eqs. (10) through (12) form a pattern that is easily extended. Begin with linearly independent vectors $\mathbf{x}_1, \mathbf{x}_2, \ldots, \mathbf{x}_n$ and use the revised Gram$-$Schmidt algorithm to form $\mathbf{q}_1, \mathbf{q}_2, \ldots, \mathbf{q}_n$. Then, for any k ($k = 1, 2, \ldots, n$).

$$\mathbf{x}_k = r_{1k}\mathbf{q}_1 + r_{2k}\mathbf{q}_2 + r_{3k}\mathbf{q}_3 + \cdots + r_{kk}\mathbf{q}_k.$$

If we set $\mathbf{X} = [\mathbf{x}_1\ \mathbf{x}_2\ \ldots\ \mathbf{x}_n]$,

$$\mathbf{Q} = [\mathbf{q}_1\ \mathbf{q}_2 \cdots \mathbf{q}_n] \tag{13}$$

and

$$\mathbf{R} = \begin{bmatrix} r_{11} & r_{12} & r_{13} & \cdots & r_{1n} \\ 0 & r_{22} & r_{23} & \cdots & r_{2n} \\ 0 & 0 & r_{33} & \cdots & r_{3n} \\ \vdots & \vdots & \vdots & & \vdots \\ 0 & 0 & 0 & \cdots & r_{nn} \end{bmatrix} ; \tag{14}$$

we have the matrix representation

$$\mathbf{X} = \mathbf{QR},$$

which is known as the **QR** decomposition of the matrix $\mathbf{X}$. The columns of $\mathbf{Q}$ form an orthonormal set of column vectors, and $\mathbf{R}$ is upper (or right) triangular.

In general, we are given a matrix $\mathbf{X}$ and are asked to generate its **QR** decomposition. This is accomplished by applying the revised Gram$-$Schmidt algorithm to the columns of $\mathbf{X}$ providing those columns are linearly independent. Then, Eqs. (13) and (14) yield the desired factorization.

Example 4.   Construct a **QR** decomposition for

$$\mathbf{X} = \begin{bmatrix} 1 & 0 & 1 \\ 1 & 1 & 0 \\ 0 & 1 & 1 \end{bmatrix}.$$

Solution The columns of $\mathbf{X}$ are the vectors $\mathbf{x}_1, \mathbf{x}_2$, and $\mathbf{x}_3$ of Example 3. Using the results of that problem, we generate

$$\mathbf{Q} = \begin{bmatrix} 1/\sqrt{2} & -1/\sqrt{6} & 1/\sqrt{3} \\ 1/\sqrt{2} & 1/\sqrt{6} & -1/\sqrt{3} \\ 0 & 2/\sqrt{6} & 1/\sqrt{3} \end{bmatrix} \quad \text{and} \quad \begin{bmatrix} \sqrt{2} & 1/\sqrt{2} & 1/\sqrt{2} \\ 0 & \sqrt{3}/2 & 1/\sqrt{6} \\ 0 & 0 & 2/\sqrt{3} \end{bmatrix}. \blacksquare$$

Example 5.   Construct a **QR** decomposition for

$$\mathbf{X} = \begin{bmatrix} 1 & 1 & 0 & 1 \\ 1 & 2 & 1 & 0 \\ 0 & 1 & 2 & 1 \\ 1 & 0 & 1 & 1 \end{bmatrix}.$$

Solution The columns of $\mathbf{X}$ are the vectors

$$\mathbf{x}_1 \begin{bmatrix} 1 \\ 1 \\ 0 \\ 1 \end{bmatrix}, \quad \mathbf{x}_2 \begin{bmatrix} 1 \\ 2 \\ 1 \\ 0 \end{bmatrix}, \quad \mathbf{x}_3 \begin{bmatrix} 0 \\ 1 \\ 2 \\ 1 \end{bmatrix}, \quad \mathbf{x}_4 \begin{bmatrix} 1 \\ 0 \\ 1 \\ 1 \end{bmatrix}.$$

We apply the revised Gram–Schmidt algorithm to these vectors. Carrying eight significant figures through all computations but rounding to four decimals for presentation purposes, we get
First iteration $(k = 1)$

$$r_{11} = \sqrt{\langle \mathbf{x}_1, \mathbf{x}_1 \rangle} = \sqrt{3} = 1.7321,$$

$$\mathbf{q}_1 = \frac{1}{r_{11}} \mathbf{x}_1 = \frac{1}{\sqrt{3}} \begin{bmatrix} 1 \\ 1 \\ 0 \\ 1 \end{bmatrix} = \begin{bmatrix} 0.5774 \\ 0.5774 \\ 0.0000 \\ 0.5774 \end{bmatrix},$$

$$r_{12} = \langle \mathbf{x}_2, \mathbf{q}_1 \rangle = 1.7321,$$
$$r_{13} = \langle \mathbf{x}_3, \mathbf{q}_1 \rangle = 1.1547,$$
$$r_{14} = \langle \mathbf{x}_4, \mathbf{q}_1 \rangle = 1.1547,$$

$$\mathbf{x}_2 \leftarrow \mathbf{x}_2 - r_{12}\mathbf{q}_1 = \begin{bmatrix} 1 \\ 2 \\ 1 \\ 0 \end{bmatrix} - 1.7321 \begin{bmatrix} 0.5774 \\ 0.5774 \\ 0.0000 \\ 0.5774 \end{bmatrix} = \begin{bmatrix} 0.0000 \\ 1.0000 \\ 1.0000 \\ -1.0000 \end{bmatrix},$$

$$\mathbf{x}_3 \leftarrow \mathbf{x}_3 - r_{13}\mathbf{q}_1 = \begin{bmatrix} 0 \\ 1 \\ 2 \\ 1 \end{bmatrix} - 1.1547 \begin{bmatrix} 0.5774 \\ 0.5774 \\ 0.0000 \\ 0.5774 \end{bmatrix} = \begin{bmatrix} -0.6667 \\ 0.3333 \\ 2.0000 \\ 0.3333 \end{bmatrix},$$

$$\mathbf{x}_4 \leftarrow \mathbf{x}_4 - r_{14}\mathbf{q}_1 = \begin{bmatrix} 1 \\ 0 \\ 1 \\ 1 \end{bmatrix} - 1.1547 \begin{bmatrix} 0.5774 \\ 0.5774 \\ 0.0000 \\ 0.5774 \end{bmatrix} = \begin{bmatrix} 0.3333 \\ -0.6667 \\ 1.0000 \\ 0.3333 \end{bmatrix}.$$

Second iteration $(k = 2)$
Using vectors from the first iteration, we compute

$$r_{22} = \sqrt{\langle \mathbf{x}_2, \mathbf{x}_2 \rangle} = 1.7321,$$

$$\mathbf{q}_2 = \frac{1}{r_{22}} \mathbf{x}_2 = \frac{1}{1.7321} \begin{bmatrix} 0.0000 \\ 1.0000 \\ 1.0000 \\ -1.0000 \end{bmatrix} = \begin{bmatrix} 0.0000 \\ 0.5774 \\ 0.5774 \\ -0.5774 \end{bmatrix},$$

$$r_{23} = \langle \mathbf{x}_3, \mathbf{q}_2 \rangle = 1.1547,$$
$$r_{24} = \langle \mathbf{x}_4, \mathbf{q}_2 \rangle = 0.0000,$$

$$\mathbf{x}_3 \leftarrow \mathbf{x}_3 - r_{23}\mathbf{q}_2 = \begin{bmatrix} -0.6667 \\ 0.3333 \\ 2.0000 \\ 0.3333 \end{bmatrix} - 1.1547 \begin{bmatrix} 0.0000 \\ 0.5774 \\ 0.5774 \\ -0.5774 \end{bmatrix} = \begin{bmatrix} -0.6667 \\ -0.3333 \\ 1.3333 \\ 1.0000 \end{bmatrix},$$

$$\mathbf{x}_4 \leftarrow \mathbf{x}_4 - r_{24}\mathbf{q}_2 = \begin{bmatrix} 0.3333 \\ -0.6667 \\ 1.0000 \\ 0.3333 \end{bmatrix} - 0.0000 \begin{bmatrix} 0.0000 \\ 0.5774 \\ 0.5774 \\ -0.5774 \end{bmatrix} = \begin{bmatrix} 0.3333 \\ -0.6667 \\ 1.0000 \\ 0.3333 \end{bmatrix}.$$

Third iteration $(k = 3)$

Using vectors from the second iteration, we compute

$$r_{33} = \sqrt{\langle \mathbf{x}_3, \mathbf{x}_3 \rangle} = 1.8257,$$

$$\mathbf{q}_3 = \frac{1}{r_{33}}\mathbf{x}_3 = \frac{1}{1.8257} \begin{bmatrix} -0.6667 \\ -0.3333 \\ 1.3333 \\ 1.0000 \end{bmatrix} = \begin{bmatrix} -0.3651 \\ -0.1826 \\ 0.7303 \\ 0.5477 \end{bmatrix},$$

$$r_{34} = \langle \mathbf{x}_4, \mathbf{q}_3 \rangle = 0.9129,$$

$$\mathbf{x}_4 \leftarrow \mathbf{x}_4 - r_{34}\mathbf{q}_3 = \begin{bmatrix} 0.3333 \\ -0.6667 \\ 1.0000 \\ 0.3333 \end{bmatrix} - 0.9129 \begin{bmatrix} -0.3651 \\ -0.1826 \\ 0.7303 \\ 0.5477 \end{bmatrix} = \begin{bmatrix} 0.6667 \\ -0.5000 \\ 0.3333 \\ -0.1667 \end{bmatrix}.$$

Fourth iteration $(k = 4)$

Using vectors from the third iteration, we compute

$$r_{44} = \sqrt{\langle \mathbf{x}_4, \mathbf{x}_4 \rangle} = 0.9129,$$

$$\mathbf{q}_4 = \frac{1}{r_{44}}\mathbf{x}_4 = \frac{1}{0.9129} \begin{bmatrix} 0.6667 \\ -0.5000 \\ 0.3333 \\ -0.1667 \end{bmatrix} = \begin{bmatrix} 0.7303 \\ -0.5477 \\ 0.3651 \\ -0.1826 \end{bmatrix}.$$

With these entries calculated (compare with Example 2 of Section 10.2), we form

$$\mathbf{Q} = \begin{bmatrix} 0.5774 & 0.0000 & -0.3651 & 0.7303 \\ 0.5774 & 0.5774 & -0.1826 & -0.5477 \\ 0.0000 & 0.5774 & 0.7303 & 0.3651 \\ 0.5774 & -0.5774 & 0.5477 & -0.1826 \end{bmatrix}$$

and

$$\mathbf{R} = \begin{bmatrix} 1.7321 & 1.7321 & 1.1547 & 1.1547 \\ 0 & 1.7321 & 1.1547 & 0.0000 \\ 0 & 0 & 1.8257 & 0.9129 \\ 0 & 0 & 0 & 0.9129 \end{bmatrix}. \blacksquare$$

Finally, we note that in contrast to **LU** decompositions, **QR** decompositions are applicable to nonsquare matrices as well. In particular, if we consider a matrix containing just the first two columns of the matrix **X** in Example 5 and calculate $r11$, $r12$, $r22$, $\mathbf{q}1$, and $\mathbf{q}2$, as we did there, we have the decomposition

$$\begin{bmatrix} 1 & 1 \\ 1 & 2 \\ 0 & 1 \\ 1 & 0 \end{bmatrix} = \begin{bmatrix} 0.5774 & 0.0000 \\ 0.5774 & 0.5774 \\ 0.0000 & 0.5774 \\ 0.5774 & -0.5774 \end{bmatrix} \begin{bmatrix} 1.7321 & 1.7321 \\ 0 & 1.7321 \end{bmatrix}.$$

Problems 10.3

In Problems 1 through 10, determine (a) the angle between the given vectors, (b) the projection of $\mathbf{x}_1$ onto $\mathbf{x}_2$, and (c) its orthogonal component.

1. $\mathbf{x}_1 = \begin{bmatrix} 1 \\ 2 \end{bmatrix}$, $\mathbf{x}_2 = \begin{bmatrix} 2 \\ 1 \end{bmatrix}$.

2. $\mathbf{x}_1 = \begin{bmatrix} 1 \\ 1 \end{bmatrix}$, $\mathbf{x}_2 = \begin{bmatrix} 3 \\ 5 \end{bmatrix}$.

3. $\mathbf{x}_1 = \begin{bmatrix} 3 \\ -2 \end{bmatrix}$, $\mathbf{x}_2 = \begin{bmatrix} 3 \\ 3 \end{bmatrix}$.

4. $\mathbf{x}_1 = \begin{bmatrix} 4 \\ -1 \end{bmatrix}$, $\mathbf{x}_2 = \begin{bmatrix} 2 \\ 8 \end{bmatrix}$.

5. $\mathbf{x}_1 = \begin{bmatrix} -7 \\ -2 \end{bmatrix}$, $\mathbf{x}_2 = \begin{bmatrix} 2 \\ 9 \end{bmatrix}$.

6. $\mathbf{x}_1 = \begin{bmatrix} 2 \\ 1 \\ 0 \end{bmatrix}$, $\mathbf{x}_2 = \begin{bmatrix} 2 \\ 0 \\ 2 \end{bmatrix}$.

7. $\mathbf{x}_1 = \begin{bmatrix} 1 \\ 1 \\ 0 \end{bmatrix}$, $\mathbf{x}_2 = \begin{bmatrix} 2 \\ 2 \\ 1 \end{bmatrix}$.

8. $\mathbf{x}_1 = \begin{bmatrix} 0 \\ 3 \\ 4 \end{bmatrix}$, $\mathbf{x}_2 = \begin{bmatrix} 2 \\ 5 \\ 5 \end{bmatrix}$.

9. $\mathbf{x}_1 = \begin{bmatrix} 0 \\ 1 \\ 1 \\ 1 \end{bmatrix}$, $\mathbf{x}_2 = \begin{bmatrix} 1 \\ 1 \\ 1 \\ 0 \end{bmatrix}$.

10. $\mathbf{x}_1 = \begin{bmatrix} 1 \\ 2 \\ 3 \\ 4 \end{bmatrix}$, $\mathbf{x}_2 = \begin{bmatrix} 1 \\ -2 \\ 0 \\ -1 \end{bmatrix}$.

In Problems 11 through 21, determine **QR** decompositions for the given matrices.

11. $\begin{bmatrix} 1 & 2 \\ 2 & 1 \end{bmatrix}$.

12. $\begin{bmatrix} 1 & 3 \\ 1 & 5 \end{bmatrix}$.

13. $\begin{bmatrix} 3 & 3 \\ -2 & 3 \end{bmatrix}$.

14. $\begin{bmatrix} 1 & 2 \\ 2 & 2 \\ 2 & 1 \end{bmatrix}$.

15. $\begin{bmatrix} 1 & 1 \\ 1 & 0 \\ 3 & 5 \end{bmatrix}$.

16. $\begin{bmatrix} 3 & 1 \\ -2 & 1 \\ 1 & 1 \\ -1 & 1 \end{bmatrix}$.

17. $\begin{bmatrix} 2 & 0 & 2 \\ 1 & 1 & 0 \\ 0 & 1 & 2 \end{bmatrix}$.

18. $\begin{bmatrix} 1 & 2 & 2 \\ 1 & 0 & 2 \\ 0 & 1 & 1 \end{bmatrix}$.

19. $\begin{bmatrix} 0 & 3 & 2 \\ 3 & 5 & 5 \\ 4 & 0 & 5 \end{bmatrix}$.

20. $\begin{bmatrix} 0 & 1 & 1 \\ 1 & 0 & 1 \\ 1 & 1 & 0 \\ 1 & 1 & 1 \end{bmatrix}$.

21. $\begin{bmatrix} 1 & 0 & 1 \\ 1 & 1 & 0 \\ 0 & -1 & -1 \\ 0 & 0 & 0 \end{bmatrix}$.

22. Show that

$$\left\| \frac{\langle \mathbf{x}, \mathbf{a} \rangle}{\langle \mathbf{x}, \mathbf{a} \rangle} \mathbf{a} \right\| = \|\mathbf{x}\| |\cos \theta|,$$

where θ is the angle between $\mathbf{x}$ and $\mathbf{a}$.

23. Prove directly that

$$\mathbf{x} - \frac{\langle \mathbf{a}, \mathbf{x} \rangle}{\langle \mathbf{a}, \mathbf{x} \rangle} \mathbf{a}$$

is orthogonal to $\mathbf{a}$.

24. Discuss what is likely to occur in a **QR** decomposition if the columns are not linearly independent and all calculations are rounded.

10.4 The QR algorithm

The **QR** algorithm is one of the more powerful numerical methods developed for computing eigenvalues of real matrices. In contrast to the power methods described in Section 6.6, which converge only to a single dominant real eigenvalue of a matrix, the **QR** algorithm generally locates all eigenvalues, both real and complex, regardless of multiplicity.

Although a proof of the **QR** algorithm is beyond the scope of this book, the algorithm itself is deceptively simple. As its name suggests, the algorithm is based on **QR** decompositions. Not surprisingly then, the algorithm involves numerous arithmetic calculations, making it unattractive for hand computations but ideal for implementation on a computer.

Like many numerical methods, the **QR** algorithm is iterative. We begin with a square real matrix A_0. To determine its eigenvalues, we create a sequence of new matrices A_1, A_2, ..., A_{k-1}, A_k, ..., having the property that each new matrix has the same eigenvalues as A_0 and that these eigenvalues become increasingly obvious as the sequence progresses. To calculate A_k ($k = 1, 2, 3, ...$) once A_{k-1} is known, first construct a **QR** decomposition of A_{k-1}:

$$A_{k-1} = Q_{k-1} R_{k-1},$$

and then reverse the order of the product to define

$$A_k = R_{k-1} Q_{k-1}.$$

It can be shown that each matrix in the sequence $\{A_k\}$ ($k = 1, 2, 3, ...$) has identical eigenvalues. For now, we just note that the sequence generally converges to one of the following two partitioned forms:

$$\begin{array}{c|c} \mathbf{S} & T \\ \hline 0\ 0\ 0\ \cdots\ 0 & a \end{array} \tag{15}$$

or

$$\left[\begin{array}{c|c} \mathbf{U} & V \\ \hline 0\ 0\ 0\ \cdots\ 0 & bc \\ 0\ 0\ 0\ \cdots\ 0 & de \end{array}\right]. \tag{16}$$

If matrix (15) occurs, then the element a is an eigenvalue and the remaining eigenvalues are found by applying the **QR** algorithm anew to the submatrix **S**. If, on the other hand, matrix (16) occurs, then two eigenvalues are determined by solving for the roots of the characteristic equation of the 2×2 matrix in the lower right partition, namely

$$\lambda^2 - (b+e)\lambda + (be - cd) = 0.$$

The remaining eigenvalues are found by applying the **QR** algorithm anew to the submatrix **U**.

Convergence of the algorithm is accelerated by performing a shift at each iteration. If the orders of all matrices are $n \times n$, we denote the element in the (n, n)-position of the matrix A_{k-1} as w_{k-1}, and construct a **QR** decomposition for the shifted matrix $A_{k-1} - w_{k-1}I$. That is,

$$A_{k-1} - w_{k-1}I = Q_{k-1} R_{k-1}. \tag{17}$$

We define

$$A_k = R_{k-1} Q_{k-1} + w_{k-1}I. \tag{18}$$

Example 1. Find the eigenvalues of

$$\mathbf{A}_0 = \begin{bmatrix} 0 & 1 & 0 \\ 0 & 0 & 1 \\ 18 & -1 & -7 \end{bmatrix}.$$

Solution Using the **QR** algorithm with shifting, carrying all calculations to eight significant figures, but rounding to four decimals for presentation, we compute

$$\mathbf{A}_0 - (-7)\mathbf{I} = \begin{bmatrix} 7 & 1 & 0 \\ 0 & 7 & 1 \\ 18 & -1 & 0 \end{bmatrix}$$

$$= \begin{bmatrix} 0.3624 & 0.1695 & -0.9165 \\ 0.0000 & 0.9833 & 0.1818 \\ 0.9320 & -0.0659 & 0.3564 \end{bmatrix} \begin{bmatrix} 19.3132 & -0.5696 & -0.0000 \\ 0.0000 & 7.1187 & 0.9833 \\ 0.0000 & -0.0000 & 0.1818 \end{bmatrix}$$

$$= \mathbf{Q}_0 \mathbf{R}_0,$$

$$\mathbf{A}_1 = \mathbf{R}_0 \mathbf{Q}_0 + (-7)\mathbf{I}$$

$$= \begin{bmatrix} 19.3132 & -0.5696 & 0.0000 \\ 0.0000 & 7.1187 & 0.9833 \\ 0.0000 & 0.0000 & 0.1818 \end{bmatrix} \begin{bmatrix} 0.3624 & -0.1695 & 0.9165 \\ 0.0000 & 0.9833 & 0.1818 \\ 0.9320 & -0.0659 & 0.3564 \end{bmatrix}$$

$$+ \begin{bmatrix} -7 & 0 & 0 \\ 0 & -7 & 0 \\ 0 & 0 & -7 \end{bmatrix}$$

$$= \begin{bmatrix} 0.0000 & 2.7130 & -17.8035 \\ 0.9165 & -0.0648 & 1.6449 \\ 0.1695 & -0.0120 & -6.9352 \end{bmatrix},$$

$$\mathbf{A}_1 - (-6.9352)\mathbf{I} = \begin{bmatrix} 6.9352 & 2.7130 & -17.8035 \\ 0.9165 & 6.8704 & 1.6449 \\ 0.1695 & -0.0120 & 0.0000 \end{bmatrix}$$

$$= \begin{bmatrix} 0.9911 & -0.1306 & -0.0260 \\ 0.1310 & 0.9913 & 0.0120 \\ 0.0242 & -0.0153 & 0.9996 \end{bmatrix} \begin{bmatrix} 6.9975 & 3.5884 & -17.4294 \\ 0.0000 & 6.4565 & 3.9562 \\ 0.0000 & -0.0000 & 0.4829 \end{bmatrix}$$

$$= \mathbf{Q}_1 \mathbf{R}_1,$$

$$\mathbf{A}_2 = \mathbf{R}_1 \mathbf{Q}_1 + (-6.9352)\mathbf{I} = \begin{bmatrix} 0.0478 & 2.9101 & -17.5612 \\ 0.9414 & -0.5954 & 4.0322 \\ 0.0117 & -0.0074 & -6.4525 \end{bmatrix}.$$

Continuing in this manner, we generate sequentially

$$A_3 = \begin{bmatrix} 0.5511 & 2.7835 & -16.8072 \\ 0.7826 & -1.1455 & 6.5200 \\ 0.0001 & -0.0001 & -6.4056 \end{bmatrix}$$

and

$$A_4 = \begin{bmatrix} 0.9259 & 2.5510 & -15.9729 \\ 0.5497 & -1.5207 & 8.3583 \\ 0.0000 & -0.0000 & -6.4051 \end{bmatrix}.$$

A_4 has the form (15) with

$$S = \begin{bmatrix} 0.9259 & 2.5510 \\ 0.5497 & -1.5207 \end{bmatrix} \quad \text{and} \quad a = -6.4051.$$

One eigenvalue is -6.4051, which is identical to the value obtained in Example 2 of Section 6.6. In addition, the characteristic equation of R is $\lambda^2 + 0.5948\lambda - 2.8103 = 0$, which admits both -2 and 1.4052 as roots. These are the other two eigenvalues of A_0. ∎

Example 2.   Find the eigenvalues of

$$A_0 = \begin{bmatrix} 0 & 0 & 0 & -25 \\ 1 & 0 & 0 & 30 \\ 0 & 1 & 0 & -18 \\ 0 & 0 & 1 & 6 \end{bmatrix}.$$

Solution Using the **QR** algorithm with shifting, carrying all calculations to eight significant figures, but rounding to four decimals for presentation, we compute

$$A_0 - (6)I = \begin{bmatrix} -6 & 0 & 0 & -25 \\ 1 & -6 & 0 & 30 \\ 0 & 1 & -6 & -18 \\ 0 & 0 & 1 & 0 \end{bmatrix}$$

$$= \begin{bmatrix} 0.9864 & -0.1621 & -0.0270 & -0.0046 \\ 0.1644 & -0.9726 & -0.1620 & -0.0274 \\ 0.0000 & 0.1666 & -0.9722 & -0.1643 \\ 0.0000 & 0.0000 & 0.1667 & -0.9860 \end{bmatrix}$$

$$\times \begin{bmatrix} 6.0828 & -0.9864 & -0.0000 & -29.5918 \\ 0.0000 & 6.0023 & -0.9996 & -28.1246 \\ 0.0000 & 0.0000 & 6.0001 & 13.3142 \\ 0.0000 & 0.0000 & 0.0000 & 2.2505 \end{bmatrix}$$

$$= Q_0 R_0$$

$$\mathbf{A}_1 = \mathbf{R}_0\mathbf{Q}_0 + (6)\mathbf{I} = \begin{bmatrix} -0.1622 & -0.0266 & 4.9275 & -29.1787 \\ 0.9865 & -0.0044 & -4.6881 & 27.7311 \\ 0.0000 & 0.9996 & 2.3856 & -14.1140 \\ 0.0000 & 0.0000 & 0.3751 & 3.7810 \end{bmatrix},$$

$$\mathbf{A}_1 - (3.7810)\mathbf{I} = \begin{bmatrix} -3.9432 & -0.0266 & 4.9275 & -29.1787 \\ 0.9868 & -3.7854 & -4.6881 & 27.7311 \\ 0.0000 & 0.9996 & 1.3954 & -14.1140 \\ 0.0000 & 0.0000 & 0.3751 & 3.0000 \end{bmatrix}$$

$$= \begin{bmatrix} -0.9701 & -0.2343 & -0.0628 & -0.0106 \\ 0.2428 & -0.9361 & -0.2509 & -0.0423 \\ 0.0000 & 0.2622 & -0.9516 & -0.1604 \\ 0.0000 & 0.0000 & 0.1662 & -0.9861 \end{bmatrix}$$

$$\times \begin{bmatrix} 4.0647 & -0.8931 & -5.9182 & 35.0379 \\ 0.0000 & 3.8120 & 2.8684 & -22.8257 \\ 0.0000 & 0.0000 & 2.2569 & 8.3060 \\ 0.0000 & 0.0000 & 0.0000 & 1.3998 \end{bmatrix}$$

$$= \mathbf{Q}_1\mathbf{R}_1,$$

$$\mathbf{A}_2 = \mathbf{R}_1\mathbf{Q}_1 + (3.7810)\mathbf{I} = \begin{bmatrix} -0.3790 & -1.6681 & 11.4235 & -33.6068 \\ 0.9254 & 0.9646 & -7.4792 & 21.8871 \\ 0.0000 & 0.5918 & 3.0137 & -8.5524 \\ 0.0000 & 0.0000 & 0.2326 & 2.4006 \end{bmatrix}.$$

Continuing in this manner, we generate, after 25 iterations,

$$\mathbf{A}_{25} = \begin{bmatrix} 4.8641 & -4.4404 & 18.1956 & -28.7675 \\ 4.2635 & -2.8641 & 13.3357 & -21.3371 \\ 0.0000 & 0.0000 & 2.7641 & -4.1438 \\ 0.0000 & 0.0000 & 0.3822 & 1.2359 \end{bmatrix},$$

which has the form as in (16) with

$$\mathbf{U} = \begin{bmatrix} 4.8641 & -4.4404 \\ 4.2635 & -2.8641 \end{bmatrix} \quad \text{and} \quad \begin{bmatrix} b & c \\ d & e \end{bmatrix} = \begin{bmatrix} 2.7641 & -4.1438 \\ 0.3822 & 1.2359 \end{bmatrix}.$$

The characteristic equation of $\mathbf{U}$ is $\lambda^2 - 2\lambda + 5 = 0$, which has as its roots $1 \pm 2i$; the characteristic equation of the other 2×2 matrix is $\lambda^2 - 4\lambda + 4.9999 = 0$, which has as its roots $2 \pm i$. These roots are the four eigenvalues of $\mathbf{A}_0$. ∎

Problems 10.4

1. Use one iteration of the **QR** algorithm to calculate $\mathbf{A}_1$ when

$$\mathbf{A}_0 = \begin{bmatrix} 0 & 1 & 0 \\ 0 & 0 & 1 \\ 18 & -1 & 7 \end{bmatrix}.$$

Note that this matrix differs from the one in Example 1 by a single sign.

2. Use one iteration of the **QR** algorithm to calculate $\mathbf{A}_1$ when

$$\mathbf{A}_0 = \begin{bmatrix} 2 & -17 & 7 \\ -17 & -4 & 1 \\ 7 & 1 & -14 \end{bmatrix}.$$

3. Use one iteration of the **QR** algorithm to calculate $\mathbf{A}_1$ when

$$\mathbf{A}_0 = \begin{bmatrix} 0 & 0 & 0 & -13 \\ 1 & 0 & 0 & 4 \\ 0 & 1 & 0 & -14 \\ 0 & 0 & 1 & 4 \end{bmatrix}.$$

In Problems 4 through 14, use the **QR** algorithm to calculate the eigenvalues of the given matrices:

4. The matrix defined in Problem 1.

5. The matrix defined in Problem 2.

6. $\begin{bmatrix} 3 & 0 & 0 \\ 2 & 2 & 4 \\ 2 & 3 & 5 \end{bmatrix}.$

7. $\begin{bmatrix} 7 & 2 & 0 \\ 2 & 1 & 6 \\ 0 & 6 & 7 \end{bmatrix}.$

8. $\begin{bmatrix} 3 & 2 & 3 \\ 2 & 6 & 6 \\ 3 & 6 & 11 \end{bmatrix}.$

9. $\begin{bmatrix} 2 & 0 & -1 \\ 2 & 3 & 2 \\ -1 & 0 & 2 \end{bmatrix}.$

10. $\begin{bmatrix} 1 & 1 & 0 \\ 0 & 1 & 1 \\ 5 & -9 & 6 \end{bmatrix}$.

11. $\begin{bmatrix} 3 & 0 & 5 \\ 1 & 1 & 1 \\ -2 & 0 & -3 \end{bmatrix}$.

12. The matrix in Problem 3.

13. $\begin{bmatrix} 0 & 3 & 2 & -1 \\ 1 & 0 & 2 & -3 \\ 3 & 1 & 0 & -1 \\ 2 & -2 & 1 & 1 \end{bmatrix}$.

14. $\begin{bmatrix} 10 & 7 & 8 & 7 \\ 7 & 5 & 6 & 5 \\ 8 & 6 & 10 & 9 \\ 7 & 5 & 9 & 10 \end{bmatrix}$.

10.5 Least squares

Analyzing data for forecasting and predicting future events is common to business, engineering, and the sciences, both physical and social. If such data are plotted, as in Fig. 10.4, they constitute a *scatter diagram*, which may provide insight into the underlying relationship between system variables. For example, the data in Fig. 10.4 appear to follow a straight-line relationship reasonably well. The problem then is to determine the equation of the straight line that best fits the data.

A straight line in the variables x and y having the equation

$$y = mx + c, \tag{19}$$

where m and c are constants, will have one y-value on the line for each value of x. This y-value may or may not agree with the data at the same value of x. Thus, for values of x at which data are available, we generally have two values of y, one value from the data and a second value from the straight-line approximation to the data. This situation is illustrated in Fig. 10.5. The error at each x, designated as $e(x)$, is the difference between the y-value of the data and the y-value obtained from the straight-line approximation.

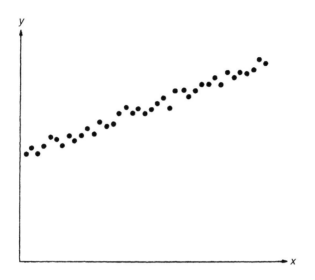

FIGURE 10.4

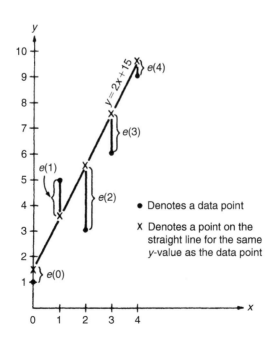

FIGURE 10.5

Example 1. Calculate the errors made in approximating the data given in Fig. 10.5 by the line $y = 2x + 1.5$.

Solution The line and the given data points are plotted in Fig. 10.5. There are errors at $x = 0$, $x = 1$, $x = 2$, $x = 3$, and $x = 4$. Evaluating the equation $y = 2x + 1.5$ at these values of x, we compute Table 10.1.

It now follows that

$$e(0) = 1 - 1.5 = -0.5,$$
$$e(1) = 5 - 3.5 = 1.5,$$
$$e(2) = 3 - 5.5 = -2.5,$$
$$e(3) = 6 - 7.5 = -1.5,$$

and

$$e(4) = 9 - 9.5 = -0.5.$$

Note that these errors could have been read directly from the graph. ■

We can extend this concept of error to the more general situation involving N data points. Let (x_1, y_1), (x_2, y_2), (x_3, y_3), ..., (x_N, y_N) be a set of N data points for a particular situation. Any straight-line approximation to these data generates errors $e(x_1)$, $e(x_2)$, $e(x_3)$, ..., $e(x_N)$, which individually can be positive, negative, or zero. The latter case occurs when the approximation agrees with the data at a particular point. We define the overall error as follows.

Definition 1. The *least squares error* E is the sum of the squares of the individual errors. That is,

$$E = [e(x_1)]^2 + [e(x_2)]^2 + [e(x_3)]^2 + \cdots + [e(x_N)]^2.$$

The only way the total error E can be zero is for each of the individual errors to be zero. Since each term of E is squared, an equal number of positive and negative individual errors cannot sum to zero.

Table 10.1

Given data		Evaluated from $y = 2x + 1.5$
X	Y	y
0	1	1.5
1	5	3.5
2	3	5.5
3	6	7.5
4	9	9.5

Example 2.   Compute the least squares error for the approximation used in Example 1.

Solution

$$E = [e(0)]^2 + [e(1)]^2 + [e(1)]^2 + [e(3)]^2 + [e(4)]^2$$
$$= (0.5)^2 + (1.5)^2 + (-2.5)^2 + (-1.5)^2 + (-0.5)^2$$
$$= 0.25 + 2.25 + 6.25 + 2.25 + 0.25$$
$$= 10.25. \blacksquare$$

Definition 2.   The *least squares straight line* is the line that minimizes the least squares error.

We seek values of m and c in (19) that minimize the least squares error. For such a line,

$$e(x_1) = y_i - (mx_i + c),$$

so we want the values for m and c that minimize

$$E = \sum_{i=1}^{N} (y_i - mx_i - c)^2.$$

This occurs when

$$\frac{\partial E}{\partial m} = \sum_{i=1}^{N} 2(y_i - mx_i - c)(-x_i) = 0$$

and

$$\frac{\partial E}{\partial c} = \sum_{i=1}^{N} 2(y_i - mx_i - c)(-1) = 0,$$

or, upon simplifying, when

$$\left(\sum_{i=1}^{N} x_i^2 \right) m + \left(\sum_{i=1}^{N} x_i \right) c = \sum_{i=1}^{N} x_i y_i, \tag{20}$$

$$\left(\sum_{i=1}^{N} x_i \right) m + Nc = \sum_{i=1}^{N} y_i.$$

System (20) makes up the *normal equations* for a least squares fit in two variables.

Example 3. Find the least squares straight line for the following $x-y$ data:

$$
\begin{array}{c|ccccc}
x & 0 & 1 & 2 & 3 & 4 \\
\hline
y & 1 & 5 & 3 & 6 & 9
\end{array}
$$

Solution Table 10.2 contains the required summations.
For these data, the normal equations become

$$30m + 10c = 65,$$
$$10m + 5c = 24,$$

which has as its solution $m = 1.7$ and $c = 1.4$. The least squares straight line is $y = 1.7x + 1.4$.

The normal equations have a simple matrix representation. Ideally, we would like to choose m and c for (19) so that

$$y_i = mx_i + c$$

for all data pairs (x_i, y_i), $i = 1, 2, \ldots, N$. That is, we want the constants m and c to solve the system

$$mx_1 + c = y_1,$$
$$mx_2 + c = y_2,$$
$$mx_3 + c = y_3,$$
$$\vdots$$
$$mx_N + c = y_N,$$

or, equivalently, the matrix equation

$$
\begin{bmatrix}
x_i & 1 \\
x_2 & 1 \\
x_3 & 1 \\
\vdots & \vdots \\
x_N & 1
\end{bmatrix}
\begin{bmatrix}
m \\
c
\end{bmatrix}
=
\begin{bmatrix}
y_1 \\
y_2 \\
y_3 \\
\vdots \\
y_N
\end{bmatrix}.
$$

Table 10.2

x_i	y_i	$(x_i)^2$	$x_i y_i$
0	1	0	0
1	5	1	5
2	3	4	6
3	6	9	18
4	9	16	36
Sum $\sum_{i=1}^{5} x_i = 10$	$\sum_{i=1}^{5} y_i = 24$	$\sum_{i=1}^{5} (x_i)^2 = 30$	$\sum_{i=1}^{5} x_i y_i = 65$

This system has the standard form $\mathbf{Ax} = \mathbf{b}$, where $\mathbf{A}$ is defined as a matrix having two columns, the first being the data vector $[x_1\, x_2\, x_3\, \ldots\, x_N]^T$, and the second containing all ones, $\mathbf{x} = [m\ c]^T$, and $\mathbf{b}$ is the data vector $[y_1\, y_2\, y_3\, \ldots\, y_N]^T$. In this context, $\mathbf{Ax} = \mathbf{b}$ has a solution for $\mathbf{x}$ if and only if the data fall on a straight line. If not, then the matrix system is inconsistent, and we seek the least squares solution. That is, we seek the vector $\mathbf{x}$ that minimizes the least squares error as stipulated in Definition 2, having the matrix form

$$E = \|\mathbf{Ax} - \mathbf{b}\|^2. \tag{21}$$

The solution is the vector $\mathbf{x}$ satisfying the normal equations, which take the matrix form

$$\mathbf{A}^T\mathbf{Ax} = \mathbf{A}^T\mathbf{b}. \tag{22}$$

System (22) is identical to system (20) when $\mathbf{A}$ and $\mathbf{b}$ are defined as above.

We now generalize to all linear systems of the form $\mathbf{Ax} = \mathbf{b}$. We are primarily interested in cases where the system is inconsistent (rendering the methods developed in Chapter 2 useless) and this generally occurs when $\mathbf{A}$ has more rows than columns. We shall place no restrictions on the number of columns in $\mathbf{A}$, but we will assume that *the columns are linearly independent*. We seek the vector $\mathbf{x}$ that minimizes the least squares error defined by Eq. (21).

Theorem 1. *If $\mathbf{x}$ has the property that $\mathbf{Ax} - \mathbf{b}$ is orthogonal to the columns of A, then $\mathbf{x}$ minimizes $\|\mathbf{Ax} - \mathbf{b}\|^2$.*

Proof. For any vector $\mathbf{x}_0$ of appropriate dimension,

$$\begin{aligned}
\|\mathbf{Ax}_0 - \mathbf{b}\|^2 &= \|(\mathbf{Ax}_0 - \mathbf{Ax}) + (\mathbf{Ax} - \mathbf{b})\|^2 \\
&= \langle (\mathbf{Ax}_0 - \mathbf{Ax}) + (\mathbf{Ax} - \mathbf{b}), (\mathbf{Ax}_0 - \mathbf{Ax}) + (\mathbf{Ax} - \mathbf{b}) \rangle \\
&= \langle (\mathbf{Ax}_0 - \mathbf{Ax}), (\mathbf{Ax}_0 - \mathbf{Ax}) \rangle + \langle (\mathbf{Ax} - \mathbf{b}), (\mathbf{Ax} - \mathbf{b}) \rangle \\
&= +2 \langle (\mathbf{Ax}_0 - \mathbf{Ax}), (\mathbf{Ax} - \mathbf{b}) \rangle \\
&= \|(\mathbf{Ax}_0 - \mathbf{Ax})\|^2 + \|(\mathbf{Ax} - \mathbf{b})\|^2 \\
&= +2 \langle \mathbf{Ax}_0 (\mathbf{Ax} - \mathbf{b}) \rangle - 2 \langle \mathbf{Ax}, (\mathbf{Ax} - \mathbf{b}) \rangle.
\end{aligned}$$

It follows directly from Problem 28 of Section 10.2 that the last two inner products are both zero (take $\mathbf{p} = \mathbf{Ax} - \mathbf{b}$). Therefore,

$$\|\mathbf{Ax}_0 - \mathbf{b}\|^2 = \|(\mathbf{Ax}_0 - \mathbf{Ax})\|^2 + \|(\mathbf{Ax} - \mathbf{b})\|^2$$

$$\geq \|(\mathbf{Ax} - \mathbf{b})\|^2$$

and $\mathbf{x}$ minimizes Eq. (21). ∎

As a consequence of Theorem 1, we seek a vector $\mathbf{x}$ having the property that $\mathbf{Ax} - \mathbf{b}$ is orthogonal to the columns of $\mathbf{A}$. Denoting the columns of $\mathbf{A}$ as $\mathbf{A}_1$, $\mathbf{A}_2$, ..., $\mathbf{A}n$, respectively, we require

$$\langle \mathbf{A}_i, \ \mathbf{A}\mathbf{x} - \mathbf{b} \rangle = 0 \quad (i = 1, 2, ..., n).$$

If $y = [y_1 \ y_2 \ ... \ y_n]^T$ denotes an arbitrary vector of appropriate dimension, then

$$\mathbf{A}\mathbf{y} = \mathbf{A}_1 y_1 + \mathbf{A}_2 y_2 + \cdots + \mathbf{A}_n y_n,$$

and

$$
\begin{aligned}
\langle \mathbf{A}\mathbf{y}, (\mathbf{A}\mathbf{x} - \mathbf{b}) \rangle &= \left\langle \sum_{i=1}^{n} \mathbf{A}_i y_i, \ (\mathbf{A}\mathbf{x} - \mathbf{b}) \right\rangle \\
&= \sum_{i=1}^{n} \langle \mathbf{A}_i y_i, (\mathbf{A}\mathbf{x} - \mathbf{b}) \rangle \qquad (23) \\
&= \sum_{i=1}^{n} y_i \langle \mathbf{A_i}, (\mathbf{A}\mathbf{x} - \mathbf{b}) \rangle \\
&= 0.
\end{aligned}
$$

It also follows from Problem 39 of Section 6.1 that

$$\langle \mathbf{A}\mathbf{y}, (\mathbf{A}\mathbf{x} - \mathbf{b}) \rangle = \langle \mathbf{y}, \ \mathbf{A}^T(\mathbf{A}\mathbf{x} - \mathbf{b}) \rangle = \langle \mathbf{y}, (\mathbf{A}^T\mathbf{A}\mathbf{x} - \mathbf{A}^T\mathbf{b}) \rangle. \qquad (24)$$

Eqs. (23) and (24) imply that $\langle \mathbf{y}, (\mathbf{A}^T \mathbf{A} - \mathbf{A}^T\mathbf{b}) \rangle = 0$ for any $\mathbf{y}$. We may deduce from Problem 26 of Section 10.2 that $\mathbf{A}^T\mathbf{A}\mathbf{x} - \mathbf{A}^T\mathbf{b} = 0$, or $\mathbf{A}^T\mathbf{A}\mathbf{x} = \mathbf{A}^T\mathbf{b}$, *which has the same form as* Eq. (22)! Therefore, a vector $\mathbf{x}$ is the least squares solution to $\mathbf{A}\mathbf{x} = \mathbf{b}$ if and only if it is the solution to $\mathbf{A}^T\mathbf{A}\mathbf{x} = \mathbf{A}^T\mathbf{b}$. This set of normal equations is guaranteed to have a unique solution whenever the columns of $\mathbf{A}$ are linearly independent, *and it may be solved using any of the methods described in the previous chapters!*

Example 4. Find the least squares solution to

$$x + 2y + z = 1,$$
$$3x - y = 2,$$
$$2x + y - z = 2,$$
$$x + 2y + 2z = 1.$$

Solution This system takes the matrix form $\mathbf{A}\mathbf{x} = \mathbf{b}$, with

$$
\mathbf{A} = \begin{bmatrix} 1 & 2 & 1 \\ 3 & -1 & 0 \\ 2 & 1 & -1 \\ 1 & 2 & 2 \end{bmatrix}, \quad \mathbf{x} = \begin{bmatrix} x \\ y \\ z \end{bmatrix}, \quad \text{and} \quad \mathbf{b} = \begin{bmatrix} 1 \\ 2 \\ 1 \\ 1 \end{bmatrix}.
$$

Then,

$$A^T A = \begin{bmatrix} 15 & 3 & 1 \\ 3 & 10 & 5 \\ 1 & 5 & 6 \end{bmatrix} \quad \text{and} \quad A^T b = \begin{bmatrix} 12 \\ 4 \\ 1 \end{bmatrix},$$

and the normal equations become

$$\begin{bmatrix} 15 & 3 & 1 \\ 3 & 10 & 5 \\ 1 & 5 & 6 \end{bmatrix} \begin{bmatrix} x \\ y \\ z \end{bmatrix} = \begin{bmatrix} 12 \\ 4 \\ 1 \end{bmatrix}.$$

Using Gaussian elimination, we obtain as the unique solution to this set of equations $x = 0.7597$, $y = 0.2607$, and $z = -0.1772$, which is also the least squares solution to the original system. ∎

Example 5.   Find the least squares solution to

$$0x + 3y = 80,$$
$$2x + 5y = 100,$$
$$5x - 2y = 60,$$
$$-x + 8y = 130,$$
$$10x - y = 150.$$

Solution This system takes the matrix form $Ax = b$, with

$$A = \begin{bmatrix} 1 & 3 \\ 2 & 5 \\ 5 & -2 \\ -1 & 8 \\ 10 & -1 \end{bmatrix}, \quad x = \begin{bmatrix} x \\ y \end{bmatrix}, \quad \text{and} \quad b = \begin{bmatrix} 80 \\ 100 \\ 60 \\ 130 \\ 150 \end{bmatrix}.$$

Then,

$$A^T A = \begin{bmatrix} 131 & -15 \\ -15 & 103 \end{bmatrix} \quad \text{and} \quad A^T b = \begin{bmatrix} 1950 \\ 1510 \end{bmatrix},$$

and the normal equations become

$$\begin{bmatrix} 131 & -15 \\ -15 & 103 \end{bmatrix} \begin{bmatrix} x \\ y \end{bmatrix} = \begin{bmatrix} 1950 \\ 1510 \end{bmatrix}.$$

The unique solution to this set of equations is $x = 16.8450$ and $y = 17.1134$, rounded to four decimals, which is also the least squares solution to the original system. ∎

Problems 10.5

In Problems 1 through 8, find the least squares solution to the given systems of equations:

1. $2x + 3y = 8,$
$3x - y = 5,$
$x + y = 6.$

2. $2x + y = 8,$
$x + y = 4,$
$-x + y = 0,$
$3x + y = 13.$

3. $x = 3y = 65,$
$2x - y = 0,$
$3x + y = 50,$
$2x + 2y = 55.$

4. $2x + y = \ 6,$
$x + y = \ 8,$
$-2x + y = \ 11,$
$-x + y = \ 8,$
$3x + y = \ 4.$

5. $2x + 3y - 4z = 1,$
$x - 2y + 3z = 3,$
$x + 4y + 2z = 6,$
$2x + y - 3z = 1.$

6. $2x + 3y + 2z = 25,$
$2x - y + 3z = 30,$
$3x + 4y - 2z = 20,$
$3x + 5y + 4z = 55.$

7. $x + y - z = 90,$
$2x + y + z = 200,$
$x + 2y + 2z = 320,$
$3x - 2y - 4z = 10,$
$3x + 2y - 3z = 220.$

8. $x + 2y + 2z = 1,$
$2x + 3y + 2z = 2,$
$2x + 4y + 4z = -2,$
$3x + 5y + 4z = 1,$
$x + 3y + 2z = -1.$

9. Which of the systems, if any, given in Problems 1 through 8 represent a least squares, straight line fit to data?

10. The monthly sales figures (in thousands of dollars) for a newly opened shoe store are

month	1	2	3	4	5
sales	9	16	14	15	21

(a) Plot a scatter diagram for these data.
(b) Find the least squares straight line that best fits these data.
(c) Use this line to predict sales revenue for month 6.

11. The number of new cars sold at a new **car** dealership over the first 8 weeks of the new season are

week	1	2	3	4	5	6	7	8
sales	51	50	45	46	43	39	35	34

(a) Plot a scatter diagram for these data.
(b) Find the least squares straight line that best fits these data.
(c) Use this line to predict sales for weeks 9 and 10.

12. Annual rainfall data (in inches) for a given town **over** the last 7 years are

year	1	2	3	4	5	6	7
rainfall	10.5	10.8	10.9	11.7	11.4	11.8	12.2

(a) Find the least squares straight line that best fits these data.
(b) Use this line to predict next year's rainfall.

13. Solve system (20) algebraically and explain why the solution would be susceptible to roundoff error.

14. **(Coding)** To minimize the roundoff error associated with solving the normal equations for a least squares straight line fit, the (x_i, y_i)-data are coded before using them in calculations. Each x_i-value is replaced by the difference between x_i and the average of all x_i-data. That is, if

$$\overline{X} = \frac{1}{N} \sum_{i=1}^{N} x_i, \quad \text{then set } x'_i = x_i - \overline{X},$$

and fit a straight line to the (x'_i, y_i)-data instead.

Explain why this coding scheme avoids the roundoff errors associated with uncoded data.

15. (a) Code the data given in Problem 10 using the procedure described in Problem 14.

    (b) Find the least squares straight line fit for these coded data.

16. (a) Code the data given in Problem 11 using the procedure described in Problem 14.

    (b) Find the least squares straight line fit for these coded data.

17. Census figures for the population (in millions of people) for a particular region of the country are as follows:

year	1950	1960	1970	1980	1990
population	25.3	23.5	20.6	18.7	17.8

(a) Code these data using the procedure described in Problem 14 and then find the least squares straight line that best fits it.

(b) Use this line to predict the population in 2000.

18. Show that if $\mathbf{A} = \mathbf{QR}$ is a $\mathbf{QR}$ decomposition of $\mathbf{A}$, then the normal equations given by Eq. (22) can be written as $\mathbf{R}^T\mathbf{Rx} = \mathbf{R}^T\mathbf{Q}^T\mathbf{b}$, which reduces to $\mathbf{Rx} = \mathbf{Q}^T\mathbf{b}$. This is a numerically stable set of equations to solve, not subject to the same roundoff errors associated with solving the normal equations directly.

19. Use the procedure described in Problem 18 to solve Problem 1.

20. Use the procedure described in Problem 18 to solve Problem 2.

21. Use the procedure described in Problem 18 to solve Problem 5.

22. Use the procedure described in Problem 18 to solve Problem 6.

23. Determine the error vector associated with the least squares solution of Problem 1 and then calculate the inner product of this vector with each of the columns of the coefficient matrix associated with the given set of equations.

24. Determine the error vector associated with the least squares solution of Problem 5 and then calculate the inner product of this vector with each of the columns of the coefficient matrix associated with the given set of equations.

Sabermetrics — An introduction

11

Chapter outline

We remind readers that it is assumed you have an abiding interest in baseball and are interested in deepening your understanding of "the national pastime". We hope you enjoy this special—and somewhat groundbreaking—approach to this fourth edition of *Matrix Methods*.

11.1 Introductory comments

In this chapter the authors will give a short overview of sabermetrics and why it has become so popular over the last several decades. In the following chapter, we will look at a good number of sabermetric measures, followed by a number of problems and questions. By way of sabermetric analysis, we will offer plausible solutions to our inquiries.

Since this chapter is an outline of the rudiments of sabermetrics and its fairly recent arrival into academia, and the next chapter is a module illustrating how one *does* sabermetrics, we will not derive many of the formulas or metrics. However, we will note certain aspects of specific instruments and comment on the reason for their value.

It is important to note that a "sabermetric proof" does not have the "absolute certainty" of a proof such as the Pythagorean theorem, for example. At the end of the day, "qualitative" factors are virtually always introduced. For example, Detroit Hall of Famer, Ty Cobb, never played a major league game under lights, nor did he ever face players of color during his career. Similarly, Angels' outfielder Mike Trout has benefited from technological and medical advances that never crossed the mind of the ballplayer of the 1920s. Is there a methodology that can at least assist us to compare players' numbers of different eras? That's where sabermetrics can offer *plausibility* to the discussion.

For those readers who are interested in the possibility of offering a course on the subject, we will also provide a "final exam" at the end of this chapter, to give potential educators an idea of a possible evaluation instrument.

The term *sabermetrics* was coined by Bill James, who popularized the concept in the late 1970s by way of his editions of *Baseball Abstract* (see the Informal Reference List at the end of this chapter). Basing it on the acronym for the Society for American Baseball Research, established in 1971, James used the abbreviation and defined sabermetrics as the "search for objective knowledge about baseball".

11.2 Some basic measures

The connection between numbers and Major League Baseball (MLB) is nothing new. Statistics on the national pastime have been kept from the inception of the game in the first half of the 19th century. Since the beginning of the National League (NL) in 1876 and the birth of the American League (AL) in 1901, statistics have been used to measure performance. In *The Hidden Game of Baseball*, Thorn and Palmer write: "Baseball may be loved without statistics, but it cannot be understood without them."

Some of the fundamental offensive categories that have been used in the past are as follows:

- Games played = G
- At bats = AB
- Plate Appearances = PA
- Hit by Pitch = HBP
- Hits = H
- Walks = Bases on Balls = BB
- Singles = 1B
- Doubles = 2B
- Triples = 3B
- Home Runs = HR
- Runs scored = R

- Runs batted in = RBI
- Batting Average = BA = H/AB
- Total Bases = TB = 1(1B) + 2(2B) + 3(3B) + 4(HR)
- Slugging Average = SLG = TB/AB
- Stolen Bases = SB
- Caught Stealing = CS

A few pitching abbreviations are:

- Wins = W
- Losses = L
- Innings Pitched = IP
- Strikeouts = K
- Bases on Ball = BB
- Earned Runs Allowed = ER
- Earned Run Average = ERA = 9(ER)/IP
- Shutouts = SHO

The following are basic fielding statistics:

- Putouts = PO
- Assists = A
- Errors = E
- Fielding Average = FA = (PO + A)/(PO + A + E)
- Range Factor/G = RF/G = (PO + A)/G

We will add more measures and instruments in Chapter 12.

11.3 Sabermetrics in the classroom

In 1988, sabermetrics made its way into the classroom at Seton Hall University, located in South Orange, New Jersey. A one-credit course (MATH 1011 − Sabermetrics) was approved by the administration. The junior author of this book was fortunate to have come up with the idea whose time had come.

The following matrix, dealing with Yankee Hall of Famer Babe Ruth, was to become the catalyst for the course:

Babe Ruth Home Run Totals Versus Team Home Run Totals 1918–1933

Year	HRS	Next	DIFF	>AL	>NL	HR titles
1918	11	11	0	4	1	AL ML
1919	29	11	18	4	6	AL ML
1920	54	19	35	7	7	AL ML
1921	59	24	35	5	3(a)	AL ML
1922	35	–	–	1	1	0
1923	41	41	0	2(b)	1	AL ML
1924	46	27	19	5	3	AL ML

Babe Ruth Home Run Totals Versus Team Home Run Totals 1918–1933
—*cont'd*

Year	HRS	Next	DIFF	>AL	>NL	HR titles
1925	25	—	—	—	—	0
1926	47	21	26	5	4	AL ML
1927	60	47	13	7	5	AL ML
1928	54	31	23	4	3	AL ML
1929	46	43	3	2(c)	2	AL ML
1930	49	—	—	1	0	AL
1931	46	46	0	3	3	AL ML
1932	41	—	—	1	0	0
1933	34	—	—	0	0(d)	0

- AL, ML ties in 1918, 1923, and 1931
- Tied Brooklyn Dodgers in 1921 (a)
- Tied Detroit Tigers in 1923 (b)
- Tied St. Louis Browns in 1929 (c)
- Tied Cincinnati Reds in 1933 (d)
- Summary:
  - Outhomered *teams* 90 times, excluding four ties
  - Outhomered *pairs of teams* 18 times:
    - 1918 (1)
    - 1920 (11)
    - 1921 (3)
    - 1927 (3)

Some clarification is needed. The fourth column in this matrix, DIFF, is column 2 (Ruth's seasonal HR total) minus column 3 (the ML runner-up). The fifth column gives the number of teams that Ruth outhomered (out of a possible seven other AL teams), and the sixth column is the counterpart for the eight NL teams).

Many questions come to mind:

- How significant was Ruth's outhomering teams 90 times?
- Was this *unique* to Ruth and his time?
- Did MLB have a *similar* occurrence in any other period?
- How about other sluggers like Lou Gehrig, Jimmie Foxx, Ted Williams, Mickey Mantle, Willie Mays, Hank Aaron, Mark McGwire, Sammy Sosa, Barry Bonds, and Alex Rodriguez?
- If Ruth was *this dominant*, how could/would it be measured … or did he *transcend* the game?

It was in the process of attempting to answer these questions that sabermetrics in the classroom came to be.

Over the years many other schools (the United States Military Academy, Quinnipiac University, University of Miami, etc.) offer related courses with a "sabermetric flavor".

11.4 **Run expectancy matrices**

A second example is given by the following two "Run Expectancy Matrices" for the first eight innings (see: http://tangotiger.net/re24.html).

This following matrix presents the *average number of runs that scored*, from that base/out state, to the end of that inning. Please use em (long dashes).

Base runners			2010—2015			1993—2009			1969—92			1950—68		
1B	2B	3B	0 outs	1 outs	2 outs	0 outs	1 outs	2 outs	0 outs	1 outs	2 outs	0 outs	1 outs	2 outs
—	—	—	0.481	0.254	0.098	0.547	0.293	0.113	0.477	0.252	0.094	0.476	0.256	0.098
1B	—	—	0.859	0.509	0.224	0.944	0.565	0.245	0.853	0.504	0.216	0.837	0.507	0.216
—	2B	—	1.100	0.664	0.319	1.175	0.723	0.349	1.102	0.678	0.325	1.094	0.680	0.330
1B	2B	—	1.437	0.884	0.429	1.562	0.966	0.471	1.476	0.902	0.435	1.472	0.927	0.441
—	—	3B	1.350	0.950	0.353	1.442	0.991	0.388	1.340	0.943	0.373	1.342	0.926	0.378
1B	—	3B	1.784	1.130	0.478	1.854	1.216	0.533	1.715	1.149	0.484	1.696	1.151	0.504
—	2B	3B	1.964	1.376	0.580	2.053	1.449	0.626	1.967	1.380	0.594	1.977	1.385	0.620
1B	2B	3B	2.292	1.541	0.752	2.390	1.635	0.815	2.343	1.545	0.752	2.315	1.540	0.747

The following matrix presents the *probability that a run will score* at some point in that inning, from each base/out state.

Base runners			2010—15			1993—2009			1969—92			1950—68		
1B	2B	3B	0 outs	1 outs	2 outs	0 outs	1 outs	2 outs	0 outs	1 outs	2 outs	0 outs	1 outs	2 outs
—	—	—	0.268	0.155	0.067	0.294	0.173	0.075	0.267	0.153	0.064	0.263	0.154	0.066
1B	—	—	0.416	0.265	0.127	0.442	0.285	0.135	0.426	0.269	0.125	0.410	0.264	0.122
—	2B	—	0.614	0.397	0.216	0.639	0.419	0.230	0.623	0.411	0.224	0.615	0.410	0.227
1B	2B	—	0.610	0.406	0.222	0.644	0.430	0.237	0.632	0.421	0.230	0.623	0.425	0.232
—	—	3B	0.843	0.660	0.257	0.854	0.675	0.271	0.840	0.664	0.274	0.818	0.650	0.278
1B	—	3B	0.860	0.634	0.270	0.868	0.652	0.289	0.855	0.647	0.280	0.849	0.648	0.287
—	2B	3B	0.852	0.676	0.260	0.867	0.698	0.280	0.855	0.678	0.275	0.839	0.664	0.285
1B	2B	3B	0.861	0.657	0.316	0.878	0.679	0.334	0.874	0.668	0.320	0.849	0.652	0.316

A close study of these matrices reveals that the advancement of a base is *virtually never worth* giving up an out. So, for example, it is generally more profitable for a team to have a man at first base with no outs than a man on second base with one out.

We also note that these matrices are based on the first eight innings. When the game is in the ninth inning (or later, for extra innings), a number of other factors must be taken into consideration, including issues such as:

- Is the game currently tied?
- Is the visiting team or the home team currently at bat?
- Is the batter a "contact" hitter? If so, should the batter bunt in this situation?
- Where are the fielders playing? Are they deep, shifted, or close?
- Are there any runners on base, and if so, do they possess base-running skills?

11.5 How to "do" sabermetrics

We now give a methodology—a blueprint of an algorithm, if you will—of a "sabermetric approach":

Sabermetrics — the search for objective knowledge about baseball (Bill James)

How to do sabermetrics

1. Carefully *identify* the question(s) or problem(s) to be discussed.
2. Get all *relevant* information.
3. Use as many *measures* and *instruments* as possible.
4. Use *relativity* and *normalization* arguments when comparing players of different eras. The effects of specific ballparks should be included in this part of the analysis as well.
5. Exercise *care* with regard to:
   - Labels (e.g., HR/AB vs. AB/HR)
   - Definitions (e.g., ratio vs. differential)
   - Interpolation versus extrapolation
   - Projections
6. *Interpret* the results. Do various approaches seem to *converge* to an objective and plausible conclusion? Can these results be *extended*?
7. Consider the historical *context* with regard to technological developments, medical advances, etc.
8. Consider any *nonsabermetric* factors germane to the analysis.
9. Be open to further discussions and developments.
10. Realize the *limitations* of a "sabermetric proof".

In the next chapter, we shall see various applications and illustrations of the above analytical technique. Using defined metrics, we give approximately two dozen sabermetric analyses of specific questions and issues. The reader will notice how essential both the *quantitative* and *qualitative* factors are in baseball "analytics".

11.6 Informal reference list

This resource guide is not meant to be exhaustive in any way. It is provided to give the reader a springboard for further studies in the area of sabermetrics. The reader is reminded to contact the junior author at Gabriel.costa@shu.edu, should they have any questions regarding sabermetrics.

(A) Websites

Baseball Reference: https://www.baseball-reference.com/
Society for American Baseball Research (SABR): https://sabr.org/

(B) Books

Curve Ball, Albert and Bennett (978-0-3879-8816-0)
Understanding Sabermetrics, Costa, Huber, and Saccoman (978-0-7864-3388-9)
Practicing Sabermetrics, Costa, Huber, and Saccoman (978-0-7864-4177-8)
Reasoning With Sabermetrics, Costa, Huber, and Saccoman (978-0-7864-6071-7)
Understanding Sabermetrics (2/e), Costa, Huber, and Saccoman (978-1-4766-6766-9)
The Bill James Baseball Abstract (1982), James (978-0-3453-0366-0)
The Bill James Historical Baseball Abstract, James (978-0-39,475-805-3)
Baseball's Ultimate Power, Jenkinson (978-1-59,921-544-0)
The Year Babe Ruth Hit 104 Home Runs, Jenkinson (978-0-78,671-906-8)
Moneyball, Lewis (978-0-39,305-765-2)
Leveling The Field, Scott (978-1-57,912-255-3)
The Numbers Game, Schwarz (978-0-31,232-222-9)
The Book, Tango, Lichtman, and Dolphin (978-1-59,797-129-4)
The Hidden Game of Baseball, Thorn and Palmer (978-0-38,518-284-3)
Ted Williams Hit List, Williams and Prime (978-1-57,028-078-9)

11.7 Testing

The "Sabermetrics Examination" below is an example of how one might evaluate a student taking a course on the subject. These "20 Questions" could serve, if amplified, as topics for term papers. Note that some of the metrics mentioned below are defined in the next chapter.

1. How does one reason *sabermetrically*? Include as many factors as possible. Also, comment on the validity of a "sabermetric proof".

2. Consider the following four sluggers from the late 1950s through the 1960s: Norm Cash, Rocky Colavito, Harmon Killebrew, and Frank Robinson. Using their career statistics and the eight measures below, find the best "power hitter" of the group.

	Cash	Colavito	Killebrew	Robinson
AB				
H				
BB				
TB				
HR				
RBI				
RC				
TPQ				

Conclusion:

3. Suppose we retitle Slugging Percentage (SLG) to be Modified Slugging Percentage (MSLG), and define it to be as follows:

$$MSLG = \frac{TB + (9)BB}{AB + (9)BB}$$

Complete the table below and determine who has the top career MSLG of the five Hall of Fame first basemen

	TB	BB	AB	MSLG
Jimmie Foxx				
Lou Gehrig				
Hank Greenberg				
Willie McCovey				
Willie Stargell				

Conclusion:

4. Procure the seminal article "Statistical Data Useful for the Operation of a Baseball Team" by George R. Lindsey, which appeared in the professional journal *Operations Research* in 1959. Write a two-page summary of the article, highlighting not only the conclusions that the author made but also the mathematical concepts that appeared in this paper.

5. *Who was the greatest pitcher ever?* Some suggestions might be Walter Johnson, Lefty Grove, Sandy Koufax, Bob Gibson, or Pedro Martinez. Give a thorough sabermetric analysis supported with as many measures as possible; consider "relativity" arguments as well.

6. Write a two-page reaction paper on the 1919 World Series.

7. Use the concept of the equivalence coefficient to predict the number of HR Don Mattingly would have accumulated if he had had 12,000 PA and was on a 5% decline for the final part of his career.

8. *Can we compare apples and oranges?* Ty Cobb won 12 batting titles (BA) in 13 years and Babe Ruth copped 13 slugging (SLG) crowns in a 14-year stretch. No players in history have come close to these two immortals with respect to these types of dominance. Yet, can we determine which one of the two Hall of Famers was *more dominant*? By looking at the league averages (LBA and LSLG), complete the table below and comment on the two lists of *ratios*. Give a short plausible conclusion based on your findings.

	BA	LBA	BA/LBA
Cobb,07			
Cobb,08			
Cobb,09			
Cobb,10			
Cobb,11			
Cobb,12			
Cobb,13			
Cobb,14			
Cobb,15			
Cobb,17			
Cobb,18			
Cobb,19			

	SLG	LSLG	SLG/LSLG
Ruth,18			
Ruth,19			
Ruth,20			
Ruth,21			
Ruth,22			
Ruth,23			
Ruth,24			
Ruth,26			
Ruth,27			
Ruth,28			
Ruth,29			
Ruth,30			
Ruth,31			

Conclusion:

9. Take a look at Hank Aaron's career and develop a regression line with HRs as the "independent variable" and RBIs as the "dependent variable". How "good of a fit" is this line vis-à-vis the data?

10. Using sabermetric analysis, argue either one of the following opposing positions:
    - Mike Trout will have greater lifetime statistics than Albert Pujols
    - Mike Trout will not have greater lifetime statistics than Albert Pujols

11. How does Met's ace Jacob deGrom compare with today's pitchers?

12. Considering both his playing career and his managerial record, should Gil Hodges be enshrined in the Baseball Hall of Fame?

13. In the 1940s, MLB was almost stunned with a proposed trade between the Red Sox and the Yankees: Ted Williams for Joe DiMaggio. Had the trade taken place, who would have received the better deal?

14. Determine the all-time leaders in (a) RC/AB; (b) MWPR; and (c) PF.

15. Determine the seasonal leaders in (a) WHIP; (b) TPQ; and (c) SLOB.

16. Honus Wagner versus Derek Jeter: who was the greater shortstop?

17. Tris Speaker or Joe DiMaggio: who was the superior centerfielder?

18. Who was the greatest hitter ever? Demonstrate your selection with a "sabermetric proof".

19. Who was the greatest player ever? Demonstrate your selection with a "sabermetric proof".

20. Critique "sabermetrics".

Peanuts! Popcorn! Crackerjacks! Cold beer!
We are now ready to PLAY BALL!

Sabermetrics — A module

12

Chapter outline

In this final chapter we will list a good number of "new" statistics—"new" in the sense that they have become more frequently invoked in this time of "analytics". We will give basic definitions, realizing that there are many versions of these models (such as runs created), which have been developed with further research. Following each formula will be a comment about specific instrumental nuances.

After these offensive, pitching, and defensive metrics are presented, we will delve into a good number of sabermetric questions.

12.1 Base stealing runs (BSRs)

$$BSRs = (.30)SB - (.60)CS$$

Comment: Note that if the runner does not steal at least two bases for each time he is caught stealing, then $BSR < 0$. Compare this to the run expectancy matrices in Chapter 11 and the importance of avoiding outs. In their "linear weights" system, approximately every 10 runs equals a win, which they use as the ultimate statistic to determine greatness.

12.2 Batting linear weights runs (BLWTS)

$$BR = (.47)1B + (0.78)2B + (1.09)3B + (1.40)HR$$
$$+ (.33)(BB + HBP) - (.25)(AB - H)$$

Comment: The "linear weights" system was devised by Thorn and Palmer and is, in some sense, a precursor to wins above replacement (WAR). WAR, incidentally, has been one of the buzzwords over the past decade or so. Currently, there are several definitions of WAR and the metrics have been modified somewhat. As time goes on, one can envision this instrument in a simplified form. Note that this number can be *negative*. A player with 0 BR is an *average* player and, in a sense, makes no positive or negative contribution to the team.

12.3 Equivalence coefficient (EC)

$$EC = 1 + \frac{\Delta AB}{AB}(k)$$

Comment: This instrument allows one to predict what might have happened over a normal career span if the player's time had not been truncated, such as with an illness (as for Lou Gehrig) or with military service (as for Ted Williams). The parameter "k" is a *kicker* and is used to factor in a "higher" career performance ($k > 1$) or a "lower" career performance ($k < 1$) for the additional playing time. If we are merely prorating, $k = 1$.

12.4 Isolated power (ISO)

$$ISO = SLG - BA = \frac{TB - H}{AB}$$

Comment: This instrument measures the "power beyond first base". Babe Ruth at 0.348 has the highest ISO in history.

12.5 On base average (OBA)

$$OBA = \frac{H + BB + HBP}{H + BB + HBP + SF}$$

Comment: "SF" denotes sacrifice flies. Ted Williams leads all others with a career OBA of 0.482.

12.6 On base plus slugging (OPS)

$$OPS = OBA + SLG$$

Comment: Ted Williams called this statistic the "bottom line in hitting". This measure is also called "production" (PRO). Babe Ruth leads all other players with a career OPS of 1.164.

12.7 Power factor (PF)

$$PF = \frac{SLG}{BA} = \frac{TB}{H}$$

Comment: PF gives the number of bases per H. A lifetime PF > 2 has been attained very rarely in the annals of MLB.

12.8 Power-speed number (PSN)

$$PSN = \frac{2(HR)(SB)}{HR + BB}$$

Comment: Mathematically, this calculation is an example of a *harmonic mean*. The factor of "2" in the numerator guarantees that a player with 40 HR and 40 SB, for example, will score a PSN of 40. At 613.9, Barry Bonds has the highest PSN in history.

12.9 Runs created (RC)

$$RC = \frac{(H + BB)(TB)}{(AB + W)}$$

Comment: This metric, introduced by James, has a very close correlation to R.

12.10 Slugging times on base average (SLOB)

$$SLOB = (OBA)(SLG)$$

Comment: Since OPS is computed by adding two fractions with *different* denominators, thus making it mathematically suspect, the product of the two factors in the SLOB metric circumvents that annoyance.

12.11 Total power quotient (TPQ)

$$TPQ = \frac{TB + RBI + HR}{AB}$$

Comment: This metric combines SLG with HR frequency and RBI frequency.

12.12 Modified weighted pitcher's rating (MWPR)

$$MWPR = (IP)\left(\frac{W}{W + L} - \frac{W_{wo}}{W_{wo} + L_{wo}}\right)$$

Comment: W_{WO} and L_{WO} denote the team's wins and losses, respectively, minus the individual pitcher's wins and losses.

12.13 Pitching linear weights runs (PLWTS)

$$PR = (IP)\frac{League\ ERA}{9} - ER$$

Comment: This, too, is from Thorn and Palmer. The reader will find BSRs as well. Thorn and Palmer also discuss fielding runs (FRs). For a further discussion on FRs, the reader is advised to consult *The Hidden Game of Baseball* or *Practicing Sabermetrics* referenced in Section 11.6 of this book.

12.14 **Walks plus hits per innings pitched (WHIP)**

$$\text{WHIP} = \frac{\text{H} + \text{BB}}{\text{IP}}$$

Comment: A value of WHIP < 1 indicates a very good score in this metric.

We will now investigate the following sabermetric issues. Hints to solutions will be found in the back of the text.

12.1 What is sabermetrics and why is it important?

12.2 Where does Albert Pujols rate with the great first basemen in history?

12.3 Discuss ISO and PF.

12.4 Which year was Pedro Martinez's greatest year?

12.5 What was significant about Steve Carlton's record in 1972?

12.6 Is Joe DiMaggio's 56 consecutive game hitting streak likely to be broken?

12.7 Use the EC coefficient to predict Lou Gehrig's career HR totals based on 12,500 PA and a 4% decline.

12.8 Up through 2018, which season was Mike Trout's greatest?

12.9 For the season of the 2018 New York Yankees, compare R versus RC for MLB.

12.10 Using the seasons of 2013 through 2017, find the regression line that "best fits" the data, with R as the independent variable and W as the dependent variable for the Chicago Cubs.

12.11 Was Barry Bonds the greatest player ever?

12.12 What should be the criteria for the Most Valuable Player (MVP) Award?

12.13 What should be the criteria for the Cy Young Award?

12.14 Discuss the career of Hall of Famer Hank Greenberg.

12.15 Justin Verlander began his pitching career in 2005. What are his chances of getting into the Hall of Fame?

12.16 Should Pete Rose be enshrined into the Hall of Fame?

12.17 Discuss the Steroid Era.

12.18 Discuss SLOB.

12.19 Investigate WAR.

12.20 Critique FA and RF/G.

12.21 How great a player was Stan Musial?

12.22 Who was the greatest catcher ever?

12.23 Who was the greater centerfielder—Willie Mays or Tris Speaker?

12.24 How valuable are BSRs?

12.25 Give a sabermetric analysis of Babe Ruth.

The authors sincerely hope the reader has enjoyed this novel application to sabermetrics.

A word on technology

While it is true that much of the "number crunching" involved the basic operations of addition and multiplication, all would agree that much time could be consumed with these tasks.

As the second decade of this new millennium is coming to a close, we can take advantage of many tools. Calculators and computer algebra systems are ideal instruments that can be employed.

Note that we, as educators, are firm believers that students of mathematics, science, and engineering should first understand the *underlying fundamental concepts* involved with the topics presented in this text. However, once these ideas are mastered, a common sense approach would be appropriate regarding laborious numerical calculations.

That being said, we give a few suggestions below:

- TI-89 calculator produced by *Texas Instruments*
- Various *Mathematica* software packages
- Excel

Thank you for using our text.

Answers and hints to selected problems

CHAPTER 1
SECTION 1.1

1. $\mathbf{A}$ is 4×5, $\mathbf{B}$ is 3×3, $\mathbf{C}$ is 3×4, $\mathbf{D}$ is 4×4, $\mathbf{E}$ is 2×3, $\mathbf{F}$ is 5×1, $\mathbf{G}$ is 4×2, $\mathbf{H}$ is 2×2, $\mathbf{J}$ is 1×3.

2. $a_{13} = -2$, $a_{21} = 2$, $b_{13} = 3$, $b_{21} = 0$, $c_{13} = 3$, $c_{21} = 5$, $d_{13} = t^2$, $d_{21} = t - 2$, $e_{13} = \frac{1}{4}$, $e_{21} = \frac{2}{3}$, $f_{13} =$ does not exist, $f_{21} = 5$, $g_{13} =$ does not exist, $g_{21} = 2\pi$, $h_{13} =$ does not exist, $h_{21} = 0$, $j_{13} = -30$, j_{21} does not exist.

3. $a_{23} = -6$, $a_{32} = 3$, $b_{31} = 4$, $b_{32} = 3$, $c_{11} = 1$, $d = 22t^4$, $e_{13} = \frac{1}{4}$, $g_{22} = 18$, g_{23} and h_{32} do not exist.

4. $\mathbf{A} = \begin{bmatrix} 1 & -1 \\ -1 & 1 \end{bmatrix}$.

5. $\mathbf{A} = \begin{bmatrix} 1 & \dfrac{1}{2} & \dfrac{1}{3} \\ 2 & 1 & \dfrac{2}{3} \\ 3 & \dfrac{3}{2} & 1 \end{bmatrix}$.

6. $\mathbf{B} = \begin{bmatrix} 1 & 0 & -1 \\ 0 & -1 & -2 \\ -1 & -2 & -3 \end{bmatrix}$.

7. $\mathbf{C} = \begin{bmatrix} 1 & 1 & 1 & 1 \\ 1 & 2 & 3 & 4 \end{bmatrix}$.

8. $\mathbf{D} = \begin{bmatrix} 0 & -1 & -2 & -3 \\ 3 & 0 & -1 & -2 \\ 4 & 5 & 0 & -1 \end{bmatrix}$.

9. (a) $[9 \quad 15]$,
(b) $[12 \quad 0]$,
(c) $[13 \quad 30]$,
(d) $[21 \quad 15]$.

10. (a) $[7 \quad 4 \quad 1779]$,
(b) $[12 \quad 7 \quad 1941]$
(c) $[4 \quad 23 \quad 1809]$,
(d) $[10 \quad 31 \quad 1688]$.

11. $[950 \quad 1253 \quad 98]$.

12. $\begin{bmatrix} 3 & 5 & 3 & 4 \\ 0 & 2 & 9 & 5 \\ 4 & 2 & 0 & 0 \end{bmatrix}$.

13. $\begin{bmatrix} 72 & 12 & 16 \\ 45 & 32 & 16 \\ 81 & 10 & 35 \end{bmatrix}$.

14. $\begin{bmatrix} 100 & 150 & 50 & 500 \\ 27 & 45 & 116 & 2 \\ 29 & 41 & 116 & 3 \end{bmatrix}$.

15. (a) $\begin{bmatrix} 1000 & 2000 & 3000 \\ 0.07 & 0.075 & 0.0725 \end{bmatrix}$

(b) $\begin{bmatrix} 1070.00 & 2150.00 & 3217.50 \\ 0.075 & 0.08 & 0.0775 \end{bmatrix}$

16. $\begin{bmatrix} 0.95 & 0.05 \\ 0.01 & 0.99 \end{bmatrix}$.

17. $\begin{bmatrix} 0.6 & 0.4 \\ 0.7 & 03 \end{bmatrix}$.

18. $\begin{bmatrix} 0.10 & 0.50 & 0.40 \\ 0.20 & 0.60 & 0.20 \\ 0.25 & 0.65 & 0.10 \end{bmatrix}$.

19. $\begin{bmatrix} 0.80 & 0.15 & 0.05 \\ 0.10 & 0.88 & 0.02 \\ 0.25 & 0.30 & 0.45 \end{bmatrix}$.

SECTION 1.2

1. $\begin{bmatrix} 2 & 4 \\ 6 & 8 \end{bmatrix}$.

2. $\begin{bmatrix} -5 & -10 \\ -15 & -20 \end{bmatrix}$.

3. $\begin{bmatrix} 9 & 3 \\ -3 & 6 \\ 9 & -6 \\ 6 & 18 \end{bmatrix}$.

4. $\begin{bmatrix} -20 & 20 \\ 0 & -20 \\ 50 & -30 \\ 50 & 10 \end{bmatrix}$.

5. $\begin{bmatrix} 0 & -1 \\ 1 & 0 \\ 0 & 0 \\ -2 & -2 \end{bmatrix}$.

6. $\begin{bmatrix} 6 & 8 \\ 10 & 12 \end{bmatrix}$.

7. $\begin{bmatrix} 0 & 2 \\ 6 & 1 \end{bmatrix}$.

8. $\begin{bmatrix} 1 & 3 \\ -1 & 0 \\ 8 & -5 \\ 7 & 7 \end{bmatrix}$.

9. $\begin{bmatrix} 3 & 2 \\ -2 & 2 \\ 3 & -2 \\ 4 & 8 \end{bmatrix}$.

10. Does not exist.

11. $\begin{bmatrix} -4 & -4 \\ -4 & -4 \end{bmatrix}$.

12. $\begin{bmatrix} -2 & -2 \\ 0 & -7 \end{bmatrix}$.

13. $\begin{bmatrix} 5 & -1 \\ -1 & 4 \\ -2 & 1 \\ -3 & 5 \end{bmatrix}$.

14. $\begin{bmatrix} 3 & 0 \\ 0 & 2 \\ 3 & -2 \\ 0 & 4 \end{bmatrix}$.

15. $\begin{bmatrix} 17 & 22 \\ 27 & 32 \end{bmatrix}$.

16. $\begin{bmatrix} 5 & 6 \\ 3 & 18 \end{bmatrix}$.

17. $\begin{bmatrix} -0.1 & 0.2 \\ 0.9 & -0.2 \end{bmatrix}$.

18. $\begin{bmatrix} 4 & -3 \\ -1 & 4 \\ -10 & 6 \\ -8 & 0 \end{bmatrix}$.

19. $\mathbf{X} = \begin{bmatrix} 4 & 4 \\ 4 & 4 \end{bmatrix}$.

20. Y $= \begin{bmatrix} -11 & -12 \\ -11 & -19 \end{bmatrix}$.

21. X $= \begin{bmatrix} 11 & 1 \\ -3 & 8 \\ 4 & -3 \\ 1 & 17 \end{bmatrix}$.

22. Y $= \begin{bmatrix} -1.0 & 0.5 \\ 0.5 & -1.0 \\ 2.5 & -1.5 \\ 1.5 & -0.5 \end{bmatrix}$.

23. R $= \begin{bmatrix} -2.8 & -1.6 \\ 3.6 & -9.2 \end{bmatrix}$.

24. S $= \begin{bmatrix} -1.5 & 1.0 \\ -1.0 & -1.0 \\ -1.5 & 1.0 \\ 2.0 & 0 \end{bmatrix}$.

25. $\begin{bmatrix} 5 & 8 \\ 13 & 9 \end{bmatrix}$.

27. $\begin{bmatrix} -\theta^3 + \theta^2 + \theta & 6\theta - 6 \\ 21 & -\theta - 2\theta^2 - \theta + 6/2 \end{bmatrix}$.

32. (a) $\begin{bmatrix} 200 & 150 \end{bmatrix}$,
(b) $\begin{bmatrix} 600 & 450 \end{bmatrix}$,
(c) $\begin{bmatrix} 550 & 550 \end{bmatrix}$.

33. (b) $\begin{bmatrix} 11 & 2 & 6 & 3 \end{bmatrix}$,
(c) $\begin{bmatrix} 9 & 4 & 10 & 8 \end{bmatrix}$.

34. (b) $\begin{bmatrix} 10,500 & 6,000 & 4,500 \end{bmatrix}$,
(c) $\begin{bmatrix} 35,500 & 14,500 & 3,300 \end{bmatrix}$,

SECTION 1.3

1. (a) 2×2,
   (b) 4×4,
   (c) 2×1,
   (d) Not defined,
   (e) 4×2,
   (f) 2×4,
   (g) 4×2,
   (h) Not defined,
   (i) Not defined,
   (j) 1×4,
   (k) 4×4,
   (l) 4×2.

2. $\begin{bmatrix} 19 & 22 \\ 42 & 50 \end{bmatrix}.$

3. $\begin{bmatrix} 23 & 34 \\ 31 & 46 \end{bmatrix}.$

4. $\begin{bmatrix} 5 & -4 & 3 \\ 9 & -8 & 7 \end{bmatrix}.$

5. $\mathbf{A} = \begin{bmatrix} 13 & -12 & 11 \\ 17 & -16 & 15 \end{bmatrix}.$

6. Not defined.

7. $\begin{bmatrix} -5 & -6 \end{bmatrix}.$

8. $\begin{bmatrix} -9 & -10 \end{bmatrix}.$

9. $\begin{bmatrix} -7 & 4 & -1 \end{bmatrix}.$

10. Not defined.

11. $\begin{bmatrix} 1 & -3 \\ 7 & -3 \end{bmatrix}.$

12. $\begin{bmatrix} 2 & -2 & 2 \\ 7 & -4 & 1 \\ -8 & 4 & 0 \end{bmatrix}.$

13. $\begin{bmatrix} 1 & 3 \end{bmatrix}.$

14. Not defined.

15. Not defined.

16. Not defined.

17. $\begin{bmatrix} -1 & -2 & -1 \\ 1 & 0 & -3 \\ 1 & 3 & 5 \end{bmatrix}.$

18. $\begin{bmatrix} 2 & -2 & 1 \\ -2 & 0 & 0 \\ 1 & -2 & 2 \end{bmatrix}.$

19. $\begin{bmatrix} -1 & 1 & 5 \end{bmatrix}.$

22. $\begin{bmatrix} x + 2y \\ 3x + 4y \end{bmatrix}.$

23. $\begin{bmatrix} x - z \\ 3x + y + z \\ x + 3y \end{bmatrix}.$

24. $\begin{bmatrix} a_{11}x + a_{12}y \\ a_{21}x + a_{22}y \end{bmatrix}.$

25. $\begin{bmatrix} 2b_{11} - b_{12} + 3b_{13} \\ 2b_{21} - b_{22} + 3b_{23} \end{bmatrix}.$

26. $\begin{bmatrix} 0 & 0 \\ 0 & 0 \end{bmatrix}.$

27. $\begin{bmatrix} 0 & 40 \\ -16 & 8 \end{bmatrix}.$

28. $\begin{bmatrix} 0 & 0 & 0 \\ 0 & 0 & 0 \\ 0 & 0 & 0 \end{bmatrix}.$

29. $\begin{bmatrix} 7 & 5 \\ 11 & 10 \end{bmatrix}.$

32. $\begin{bmatrix} 2 & 3 \\ 4 & -5 \end{bmatrix} \begin{bmatrix} x \\ y \end{bmatrix} = \begin{bmatrix} 10 \\ 11 \end{bmatrix}$.

33. $\begin{bmatrix} 1 & 1 & 1 \\ 2 & 1 & 3 \\ 1 & 1 & 0 \end{bmatrix} \begin{bmatrix} x \\ y \\ z \end{bmatrix} = \begin{bmatrix} 2 \\ 4 \\ 0 \end{bmatrix}$.

34. $\begin{bmatrix} 5 & 3 & 2 & 4 \\ 1 & 1 & 0 & 1 \\ 3 & 2 & 2 & 0 \\ 1 & 1 & 2 & 3 \end{bmatrix} \begin{bmatrix} x \\ y \\ z \\ w \end{bmatrix} = \begin{bmatrix} 5 \\ 0 \\ -3 \\ 4 \end{bmatrix}$.

35. (a) **PN** = [38,000], which represents the total revenue for that flight.

(b) **NP** = $\begin{bmatrix} 26,5000 & 45,000 & 65,000 \\ 4,000 & 7,000 & 10,000 \\ 2,000 & 3,500 & 5,000 \end{bmatrix}$.

36. (a) **HP** = [9,625 9,762.50 9,887.50 10,100 9,887.50], which represents the portfolio value each day.
(b) **PH** does not exist.

37. **TW** = $[14.00\ 65.625\ 66.50]^{\mathrm{T}}$, which denotes the cost of producing each product.

38. **OTW** = [33,862.50], which denotes the cost of producing all items on order.

39. $\begin{bmatrix} 613 & 625 \\ 887 & 960 \\ 1870 & 1915 \end{bmatrix}$,

which represents the number of each sex in each state of sickness.

SECTION 1.4

1. $\begin{bmatrix} 7 & 4 & -1 \\ 6 & 1 & 0 \\ 2 & 2 & -6 \end{bmatrix}$.

2. $\begin{bmatrix} t^3 + 3t & 2t^2 + 3 & 3 \\ 2t^2 + t^2 & 4t^2 + t & t \\ t^4 + t^2 + t & 2t^3 + t + 1 & t + 1 \\ t^5 & 2t^4 & 0 \end{bmatrix}$.

3. (a) $\mathbf{BA}^T$,
  (b) $2\mathbf{A}^T + \mathbf{B}$,
  (c) $(\mathbf{B}^T + \mathbf{C})\mathbf{A} = \mathbf{B}^T\mathbf{A} + \mathbf{CA}$,
  (d) $\mathbf{AB} + \mathbf{C}^T$,
  (e) $\mathbf{A}^T\mathbf{A}^T + \mathbf{A}^T\mathbf{A} - \mathbf{AA}^T - \mathbf{AA}$.

4. $\mathbf{X}^T\mathbf{X} = [29]$, and $\mathbf{XX}^T = \begin{bmatrix} 4 & 6 & 8 \\ 6 & 9 & 12 \\ 8 & 12 & 16 \end{bmatrix}$.

5. $\mathbf{X}^T\mathbf{X} = \begin{bmatrix} 1 & -2 & 3 & -4 \\ -2 & 4 & -6 & 8 \\ 3 & -6 & 9 & -12 \\ -4 & 8 & -12 & 16 \end{bmatrix}$, and $\mathbf{XX}^T = [30]$.

6. $[2x^2 + 6xy + 4y^2]$.

7. **A, B, D, F, M, N, R**, and **T**.

8. **E, F, H, K, L, M, N, R**, and **T**.

9. Yes.

10. No, see **H** and **L** in Problem 7.

11. Yes, see **L** in Problem 7.

12. $\begin{bmatrix} -5 & 0 & 0 \\ 0 & 9 & 0 \\ 0 & 0 & 2 \end{bmatrix}$.

14. No.

19. $\mathbf{D}^2$ is a diagonal matrix with diagonal elements 4, 9, and 25; $\mathbf{D}^3$ is a diagonal matrix with diagonal elements 8, 27, and -125.

20. A diagonal matrix with diagonal elements 1, 8, 27.

23. A diagonal matrix with diagonal elements 4, 0, 10.

25. 4.

28. $\mathbf{A} = \mathbf{B} + \mathbf{C}$.

29.
$$
\begin{bmatrix}
1 & \frac{7}{2} & -\frac{1}{2} \\
\frac{7}{2} & 1 & 5 \\
-\frac{1}{2} & 5 & -8
\end{bmatrix}
+
\begin{bmatrix}
0 & \frac{3}{2} & -\frac{1}{2} \\
\frac{3}{2} & 0 & -2 \\
\frac{1}{2} & 2 & 0
\end{bmatrix}.
$$

30.
$$
\begin{bmatrix}
6 & \frac{3}{2} & 1 \\
\frac{3}{2} & 0 & -4 \\
1 & -4 & 2
\end{bmatrix}
+
\begin{bmatrix}
0 & \frac{1}{2} & 2 \\
\frac{1}{2} & 0 & 3 \\
-2 & -3 & 0
\end{bmatrix}.
$$

34. (a) $\mathbf{P}^2 = \begin{bmatrix} 0.37 & 0.63 \\ 0.28 & 0.72 \end{bmatrix}$ and $\mathbf{P}^2 = \begin{bmatrix} 0.289 & 0.711 \\ 0.316 & 0.684 \end{bmatrix}$.

(b) 0.37,
(c) 0.63,
(d) 0.711,
(e) 0.684.

35. $1 \to 1 \to 1 \to 1, 1 \to 1 \to 2 \to 1, 1 \to 2 \to 1 \to 1, 1 \to 2 \to 2 \to 1$.

36. (a) 0.097,
(b) 0.0194.

37. (a) 0.64,
(b) 0.636.

38. (a) 0.1,
(b) 0.21.

39. (a) 0.6675,
(b) 0.577,075,
(c) 0.267.

40.
$$\begin{bmatrix} 0 & 1 & 0 & 0 & 0 & 0 & 0 \\ 1 & 0 & 0 & 0 & 0 & 1 & 0 \\ 0 & 0 & 0 & 1 & 1 & 0 & 0 \\ 0 & 0 & 1 & 0 & 0 & 1 & 0 \\ 0 & 0 & 1 & 0 & 0 & 1 & 1 \\ 0 & 1 & 0 & 1 & 1 & 0 & 0 \\ 0 & 0 & 0 & 0 & 1 & 0 & 1 \end{bmatrix}.$$

41. (a) $\mathbf{M} = \begin{bmatrix} 0 & 2 & 0 & 1 & 0 \\ 2 & 0 & 0 & 2 & 1 \\ 0 & 0 & 0 & 1 & 0 \\ 1 & 2 & 1 & 0 & 1 \\ 0 & 1 & 0 & 1 & 0 \end{bmatrix}$,

(b) three paths consisting of two arcs connecting node 1 to node 5.

42. (a) $\mathbf{M} = \begin{bmatrix} 0 & 1 & 1 & 0 & 1 & 0 & 0 & 0 \\ 1 & 0 & 1 & 0 & 1 & 0 & 0 & 0 \\ 1 & 1 & 0 & 0 & 1 & 0 & 0 & 0 \\ 0 & 0 & 0 & 0 & 1 & 1 & 1 & 0 \\ 1 & 1 & 1 & 1 & 0 & 1 & 0 & 0 \\ 0 & 0 & 0 & 1 & 1 & 0 & 0 & 1 \\ 0 & 0 & 0 & 1 & 0 & 0 & 0 & 1 \\ 0 & 0 & 0 & 0 & 0 & 1 & 1 & 0 \end{bmatrix}$,

(b) $\mathbf{M}^3$ has a path from node 1 to node 7; it is the first integral power of $\mathbf{M}$ having m_{17} positive. The minimum number of *intermediate* cities is two.

SECTION 1.5

1. (a), (b), and (d) are submatrices.

3. $\begin{bmatrix} 4 & 5 & -1 & | & 9 \\ 15 & 10 & 4 & | & 22 \\ 1 & 1 & 5 & | & 9 \end{bmatrix}$.

4. Partition **A** and **B** into four 2×2 submatrices each. Then,

$$\mathbf{AB} = \left[\begin{array}{cc|cc} 11 & 9 & 0 & 0 \\ 4 & 6 & 0 & 0 \\ \hline 0 & 0 & 2 & 1 \\ 0 & 0 & 4 & -1 \end{array}\right].$$

5. $\left[\begin{array}{cc|cc} 18 & 6 & 0 & 0 \\ 12 & 6 & 0 & 0 \\ \hline 0 & 0 & 1 & 0 \\ 0 & 0 & 3 & 4 \end{array}\right].$

6. $\left[\begin{array}{cc|cc} 7 & 8 & 0 & 0 \\ -4 & -1 & 0 & 0 \\ \hline 0 & 0 & 5 & 1 \\ 0 & 0 & 1 & 2 \end{array}\right].$

7. $\mathbf{A}^2 = \left[\begin{array}{cc|cccc} 1 & 0 & 0 & 0 & 0 & 0 \\ 0 & 4 & 0 & 0 & 0 & 0 \\ \hline 0 & 0 & 0 & 0 & 1 & 0 \\ 0 & 0 & 0 & 0 & 0 & 1 \\ 0 & 0 & 0 & 0 & 0 & 0 \\ 0 & 0 & 0 & 0 & 0 & 0 \end{array}\right]. \ \mathbf{A}^3 = \left[\begin{array}{cc|cccc} 1 & 0 & 0 & 0 & 0 & 0 \\ 0 & 8 & 0 & 0 & 0 & 0 \\ \hline 0 & 0 & 0 & 0 & 0 & 1 \\ 0 & 0 & 0 & 0 & 0 & 0 \\ 0 & 0 & 0 & 0 & 0 & 0 \\ 0 & 0 & 0 & 0 & 0 & 0 \end{array}\right].$

8. $\mathbf{A}^n = \mathbf{A}$ when n is odd.

SECTION 1.6

1. $p = 1.$

2. $\left[\begin{array}{c} -4/3 \\ 1 \\ -8/3 \\ 1/3 \end{array}\right].$

3. $[1 \quad -0.4 \quad 1].$

4. (a) Not defined,

(b) $\begin{bmatrix} 6 & -3 & 12 & 3 \\ 2 & -1 & 4 & 1 \\ 12 & -6 & 24 & 6 \\ 0 & 0 & 0 & 0 \end{bmatrix}$,

(c) [29],

(d) [29].

5. (a) [4 −1 1],

(b) [−1],

(c) $\begin{bmatrix} 2 & 0 & -2 \\ -1 & 0 & 1 \\ 3 & 0 & -3 \end{bmatrix}$,

(d) $\begin{bmatrix} 1 & 0 & -1 \\ 0 & 0 & 0 \\ -1 & 0 & 1 \end{bmatrix}$.

6. (c), (d), (f), (g), (h), and (i).

7. (a) $\sqrt{2}$,

(b) 5,

(c) $\sqrt{3}$,

(d) $\frac{1}{2}\sqrt{3}$,

(e) $\sqrt{15}$,

(f) $\sqrt{39}$.

8. (a) $\sqrt{2}$,

(b) $\sqrt{5}$,

(c) $\sqrt{3}$,

(d) 2,

(e) $\sqrt{30}$,

(f) $\sqrt{2}$.

9. (a) $\sqrt{15}$,

(b) $\sqrt{39}$.

12. $x\begin{bmatrix} 2 \\ 4 \end{bmatrix} + y\begin{bmatrix} 3 \\ 5 \end{bmatrix} = \begin{bmatrix} 10 \\ 11 \end{bmatrix}$.

13. $x\begin{bmatrix} 3 \\ 0 \\ -1 \end{bmatrix} + y\begin{bmatrix} 4 \\ 1 \\ 1 \end{bmatrix} + z\begin{bmatrix} 5 \\ -2 \\ 2 \end{bmatrix} + w\begin{bmatrix} 6 \\ 8 \\ -1 \end{bmatrix} = \begin{bmatrix} 1 \\ 0 \\ 0 \end{bmatrix}$.

16. $[0.5 \quad 0.3 \quad 0.2]$.

17. (a) There is a 0.6 probability that an individual chosen at random initially will live in the city; thus, 60% of the population initially lives in the city, while 40% lives in the suburbs.

(b) $\mathbf{d}^{(1)} = [0.574 \quad 0.426]$.

(c) $\mathbf{d}^{(2)} = [0.54956 \quad 0.45044]$.

18. (a) 40% of customers now use brand X, 50% use brand Y, and 10% use other brands.

(b) $\mathbf{d}_1 = [0.395 \quad 0.530 \quad 0.075]$.

(c) $\mathbf{d}_2 = [0.38775 \quad 0.54815 \quad 0.06410]$.

19. (a) $\mathbf{d}^{(0)} = [0 \quad 1]$.

(b) $\mathbf{d}^{(1)} = [0.7 \quad 0.3]$.

20. (a) $\mathbf{d}^{(0)} = [1 \quad 0 \quad 0]$.

(b) $\mathbf{d}^{(2)} = [0.21 \quad 0.61 \quad 0.18]$. A probability of 0.18 that the harvest will be good in 2 years.

SECTION 1.7

1.

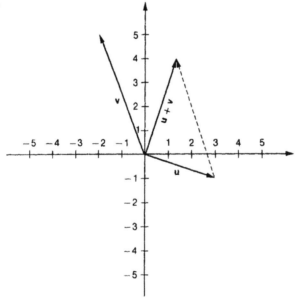

4.

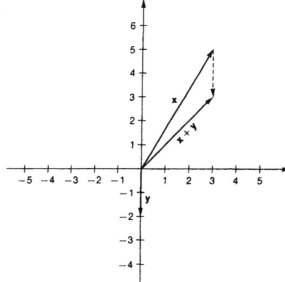

6.

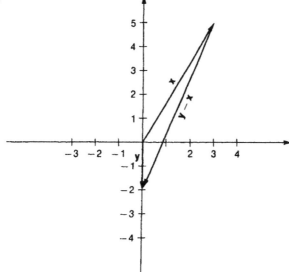

7.

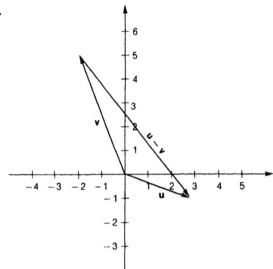

16.

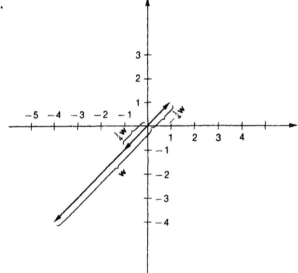

17. 341.57 degrees.

18. 111.80 degrees.

19. 225 degrees.

20. 59.04 degrees.

21. 270 degrees.

CHAPTER 2
SECTION 2.1

1. (a) No. (b) Yes.

2. (a) Yes. (b) No. (c) Yes.

3. No value of k will work.

4. $k = 1$.

5. $k = 1/12$.

6. k is arbitrary; any value will work.

7. No value of k will work.

8. $\begin{bmatrix} 3 & 5 \\ 2 & -7 \end{bmatrix} \begin{bmatrix} x \\ y \end{bmatrix} = \begin{bmatrix} 11 \\ -3 \end{bmatrix}$.

9. $\begin{bmatrix} 1 & 1 & 1 \\ 1 & -1 & -2 \\ 1 & 2 & 2 \end{bmatrix} \begin{bmatrix} x \\ y \\ z \end{bmatrix} = \begin{bmatrix} 2 \\ 0 \\ 1 \end{bmatrix}$.

10. $\begin{bmatrix} 1 & 2 & 3 \\ 1 & -3 & 2 \\ 3 & -4 & 7 \end{bmatrix} \begin{bmatrix} x \\ y \\ z \end{bmatrix} = \begin{bmatrix} 6 \\ 0 \\ 6 \end{bmatrix}$.

11. $\begin{bmatrix} 1 & 2 & 2 \\ 2 & 4 & 2 \\ -3 & -6 & -4 \end{bmatrix} \begin{bmatrix} x \\ y \\ z \end{bmatrix} = \begin{bmatrix} 0 \\ 0 \\ 0 \end{bmatrix}$.

12. $50r + 60s = 70{,}000$,
$30r + 40s = 45{,}000$.

13. $5d + 0.25b = 200$,
$10d + \quad b = 500$.

14. $8{,}000A + \quad 3{,}000B + \quad 1{,}000C = \quad 70{,}000$,
$5{,}000A + 12{,}000B + 10{,}000C = 181{,}000$,
$1{,}000A + \quad 3{,}000B + \quad 2{,}000C = \quad 41{,}000$.

15. $5A + \quad 4B + \quad 8C + 12D = \quad 80$,
$20A + 30B + 15C + \quad 5D = 200$,
$3A + \quad 3B + 10C + \quad 7D = \quad 50$.

16. $b + 0.05c + 0.05s = 20{,}000$,
$c \qquad = \quad 8{,}000$,
$0.03c + \quad s = 12{,}000$.

17. (a) $C = 800{,}000 + 30B$,

(b) Add the additional equation $S = C$.

18. $-0.60p_1 + 0.30p_2 + 0.50p_3 = 0$,

$\qquad 0.40p_1 - 0.75p_2 + 0.35p_3 = 0$,

$\qquad 0.20p_1 + 0.45p_2 - 0.85p_3 = 0$.

19. $-\dfrac{1}{2}p_1 + \dfrac{1}{3}p_2 + \dfrac{1}{6}p_3 = 0$,

$\qquad \dfrac{1}{4}p_1 - \dfrac{2}{3}p_2 + \dfrac{1}{3}p_3 = 0$,

$\qquad \dfrac{1}{4}p_1 + \dfrac{1}{3}p_2 - \dfrac{1}{2}p_3 = 0$.

20. $\quad -0.85p_1 + 0.10p_2 + \quad 0.15p_4 = 0$,

$\qquad 0.20p_1 - 0.60p_2 + \dfrac{1}{3}p_3 + 0.40p_4 = 0$,

$\qquad 0.30p_1 + 0.15p_2 - \dfrac{2}{3}p_3 + 0.45p_4 = 0$,

$\qquad 0.35p_1 + 0.35p_2 + \dfrac{1}{3}p_3 - p_4 = 0$.

22. $\mathbf{A} = \begin{bmatrix} \dfrac{1}{2} & \dfrac{1}{4} \\ \dfrac{1}{3} & 0 \end{bmatrix}$ and $\mathbf{d} = \begin{bmatrix} 20{,}000 \\ 30{,}000 \end{bmatrix}$.

23. $\mathbf{A} = \begin{bmatrix} 0 & 0.02 & 0.50 \\ 0.20 & 0 & 0.30 \\ 0.10 & 0.35 & 0.10 \end{bmatrix}$ and $\mathbf{d} = \begin{bmatrix} 50{,}000 \\ 80{,}000 \\ 30{,}000 \end{bmatrix}$.

24. $\mathbf{A} = \begin{bmatrix} 0.20 & 0.15 & 0.40 & 0.25 \\ 0 & 0.20 & 0 & 0 \\ 0.10 & 0.05 & 0 & 0.10 \\ 0.30 & 0.30 & 0.10 & 0.05 \end{bmatrix}$ and $\mathbf{d} = \begin{bmatrix} 0 \\ 5{,}000{,}000 \\ 0 \\ 0 \end{bmatrix}$.

SECTION 2.2

1. $x = 1, y = 1, z = 2.$

2. $x = -6z, y = 7z, z$ is arbitrary.

3. $x = y = 1.$

4. $r = t + 13/7, s = 2t + 15/7, t$ is arbitrary.

5. $l = \frac{1}{5}; (-n + 1), m = \frac{1}{5} (3n - 5p - 3), n$ and p are arbitrary.

6. $x = 0, y = 0, z = 0.$

7. $x = 2, y = 1, z = -1.$

8. $x = 1, y = 1, z = 0, w = 1.$

SECTION 2.3

1. $A^b = \begin{bmatrix} 1 & 2 & -3 \\ 3 & 1 & 1 \end{bmatrix}.$

2. $A^b = \begin{bmatrix} 1 & 2 & -1 & -1 \\ 2 & -3 & 2 & 4 \end{bmatrix}.$

3. $A^b = \begin{bmatrix} 1 & 2 & 5 \\ -3 & 1 & 13 \\ 4 & 3 & 0 \end{bmatrix}.$

4. $A^b = \begin{bmatrix} 2 & 4 & 0 & 2 \\ 3 & 2 & 1 & 8 \\ 5 & -3 & 7 & 15 \end{bmatrix}.$

5. $A^b = \begin{bmatrix} 2 & 3 & -4 & 12 \\ 3 & -2 & 0 & -1 \\ 8 & -1 & -4 & 10 \end{bmatrix}.$

6. $x + 2y = 5,$

$\qquad y = 8.$

7. $x - 2y + 3z = 10,$
$$y - 5z = -3,$$
$$z = \quad 4.$$

8. $r - 3s + 12t = \quad 40,$
$$s - \quad 6t = -200,$$
$$t = \quad 25.$$

9. $x + 3y \qquad = -8,$
$$y + 4z \quad = \quad 2,$$
$$0 \qquad = \quad 0.$$

10. $a - 7b + 2c = 0,$
$$b - \quad c = 0,$$
$$0 = 0.$$

11. $u - v \quad = -1,$
$$v - 2\omega \quad = \quad 2,$$
$$\omega \quad = -3,$$
$$0 \quad = \quad 1.$$

12. $x = -11,\ y = 8.$

13. $x = 32,\ y = 17,\ z = 4.$

14. $r = -410,\ s = -50,\ t = 25.$

15. $x = -14 + 12z,\ y = 2 - 4z,\ z$ is arbitrary.

16. $a = 5c,\ b = c,\ c$ is arbitrary.

17. No solution.

18. $\begin{bmatrix} 1 & -2 & 5 \\ 0 & 1 & 23 \end{bmatrix}.$

19. $\begin{bmatrix} 1 & 6 & 5 \\ 0 & 1 & 18 \end{bmatrix}.$

20. $\begin{bmatrix} 1 & 3.5 & -2.5 \\ 0 & 1 & -6 \end{bmatrix}$.

21. $\begin{bmatrix} 1 & 2 & 3 & 4 \\ 0 & 1 & 5 & 7 \\ 0 & 0 & 1 & 41/29 \end{bmatrix}$.

22. $\begin{bmatrix} 1 & 3 & 2 & 1 \\ 0 & 1 & -2 & 4 \\ 0 & 0 & 1 & -32/23 \end{bmatrix}$.

23. $\begin{bmatrix} 1 & 3 & 2 & 0 \\ 0 & 1 & -5 & 1 \\ 0 & 0 & 1 & -9/35 \\ 0 & 0 & 0 & 0 \end{bmatrix}$.

24. $\begin{bmatrix} 1 & 3/2 & 2 & 3 & 0 & 5 \\ 0 & 1 & -50 & -32 & -6 & -130 \\ 0 & 0 & 1 & 53/76 & 5/76 & 190/76 \end{bmatrix}$.

25. $x = 1, y = -2$.

26. $x = 5/7 - (1/7)z, y = -6/7 + (4/7)z$, z is arbitrary.

27. $a = -3, b = 4$.

28. $r = 13/3, s = t = -5/3$.

29. $r = \frac{1}{13}(21 + 8t), s = \frac{1}{13}(38 + 12t)$, t is arbitrary.

30. $x = 1, y = 1, z = 2$.

31. $x = -6z, y = 7z$, z is arbitrary.

32. $x = y = 1$.

33. $r = t + 13/7, s = 2t + 15/7$, t is arbitrary.

34. $l = \frac{1}{5}(-n + 1), m = \frac{1}{5}(3n - 5p - 3)$, n and p are arbitrary.

35. $r = 500$, $s = 750$.

36. $d = 30$, $b = 200$.

37. $A = 5$, $B = 8$, $C = 6$.

38. $A = 19.759 - 4.145D$, $B = -7.108 + 2.735D$, $C = 1.205 - 0.277D$, D is arbitrary.

39. $b = \$19,012$.

40. 80,000 barrels.

41. $p_1 = (48/33)p_3$, $p_2 = (41/33)p_3$, p_3 is arbitrary.

42. $p_1 = (8/9)p_3$, $p_2 = (5/6)p_3$, p_3 is arbitrary.

43. $p_1 = 0.3435p_4$, $p_2 = 1.4195p_4$, $p_3 = 1.1489p_4$, p_4 is arbitrary.

44. $x_1 = \$66,000$; $x_2 = \$52,000$.

45. To construct an elementary matrix that will interchange the ith and jth rows, simply interchange those rows in the identity matrix of appropriate order.

46. To construct an elementary matrix that will multiply the ith row of a matrix by the scalar r, simply replace the unity element in the $i-i$ position of an identity matrix of appropriate order by r.

47. To construct an elementary matrix that will add r times the ith row to the jth row, simply do the identical process to an identity matrix of appropriate order.

48. $\mathbf{x}^{(0)} = \begin{bmatrix} 40,000 \\ 60,000 \end{bmatrix}$, $\mathbf{x}^{(1)} = \begin{bmatrix} 55,000 \\ 43,333 \end{bmatrix}$, $\mathbf{x}^{(2)} \begin{bmatrix} 58,333 \\ 48,333 \end{bmatrix}$.

49. $\mathbf{x}^{(0)} = \begin{bmatrix} 100,000 \\ 160,000 \\ 60,000 \end{bmatrix}$, $\mathbf{x}^{(1)} = \begin{bmatrix} 83,200 \\ 118,000 \\ 102,000 \end{bmatrix}$, $\mathbf{x}^{(2)} = \begin{bmatrix} 103,360 \\ 127,240 \\ 89,820 \end{bmatrix}$.

The solution is $x_1 = \$99,702$; $x_2 = \$128,223$; and $x_3 = \$94,276$, rounded to the nearest dollar.

50. $\mathbf{x}^{(0)} = \begin{bmatrix} 0 \\ 10,000,000 \\ 0 \\ 0 \end{bmatrix}$, $\quad \mathbf{x}^{(1)} = \begin{bmatrix} 1,500,000 \\ 7,000,000 \\ 500,000 \\ 3,000,000 \end{bmatrix}$, $\quad \mathbf{x}^{(2)} = \begin{bmatrix} 2,300,000 \\ 6,400,000 \\ 800,000 \\ 2,750,000 \end{bmatrix}.$

The solution is: energy $=$ \$2,484,488; tourism $=$ \$6,250,000; transportation $=$ \$845,677; and construction $=$ \$2,847,278, all rounded to the nearest dollar.

SECTION 2.4

1. (a) 4,
(b) 4,
(c) 8.

2. (a) 5,
(b) 5,
(c) 5.

3. (a) 3,
(b) 3,
(c) 8.

4. (a) 4,
(b) -3,
(c) 8.

5. (a) 9,
(b) 9,
(c) 11.

6. (a) 4,
(b) 1,
(c) 10.

7. $a = -3$, $b = 4$.

8. $r = 13/3$, $s = t = -5/3$.

9. Depending on the roundoff procedure used, the last equation may not be $0 = 0$, but rather numbers very close to zero. Then only one answer is obtained.

SECTION 2.5

1. Independent.

2. Independent.

3. Dependent.

4. Dependent.

5. Independent.

6. Dependent.

7. Independent.

8. Dependent.

9. Dependent.

10. Dependent.

11. Independent.

12. Dependent.

13. Independent.

14. Independent.

15. Dependent.

16. Independent.

17. Dependent.

18. Dependent.

19. Dependent.

20. $\begin{bmatrix} 2 \\ 1 \\ 2 \end{bmatrix} = (-2)\begin{bmatrix} 1 \\ 1 \\ 0 \end{bmatrix} + (1)\begin{bmatrix} 1 \\ 0 \\ -1 \end{bmatrix} + (3)\begin{bmatrix} 1 \\ 1 \\ 1 \end{bmatrix}.$

21. (a) $[2 \quad 3] = 2[1 \quad 0] + 3[0 \quad 1]$, (b) $[2 \quad 3] = \frac{5}{2}[1 \quad 1] + \left(-\frac{1}{2}\right)[1 \quad -1]$,

  (c) No.

22. (a) $\begin{bmatrix} 1 \\ 1 \\ 1 \end{bmatrix} = \left(\frac{1}{2}\right)\begin{bmatrix} 1 \\ 0 \\ 1 \end{bmatrix} + \left(\frac{1}{2}\right)\begin{bmatrix} 1 \\ 1 \\ 0 \end{bmatrix} + \left(\frac{1}{2}\right)\begin{bmatrix} 0 \\ 1 \\ 1 \end{bmatrix}$,   **(b) No,**

(c) $\begin{bmatrix} 1 \\ 1 \\ 1 \end{bmatrix} = (0)\begin{bmatrix} 1 \\ 0 \\ 1 \end{bmatrix} + (1)\begin{bmatrix} 1 \\ 1 \\ 1 \end{bmatrix} + (0)\begin{bmatrix} 1 \\ -1 \\ 1 \end{bmatrix}.$

23. $\begin{bmatrix} 2 \\ 0 \\ 3 \end{bmatrix} = (1)\begin{bmatrix} 1 \\ 0 \\ 1 \end{bmatrix} + (1)\begin{bmatrix} 1 \\ 0 \\ 2 \end{bmatrix} + (0)\begin{bmatrix} 2 \\ 0 \\ 1 \end{bmatrix}.$

24. $[a \ \ b] = (a)[1 \ \ 0] + (b)[0 \ \ 1].$

25. $[a \ \ b] = \left(\frac{a+b}{2}\right)[1 \ \ 1] + \left(\frac{a-b}{2}\right)[1 \ \ -1].$

26. $[1 \ \ 0]$ cannot be written as a linear combination of these vectors.

27. $[a \ \ -2a] = (a/2)[2 \ \ -4] + (0)[-3 \ \ 6].$

28. $[a \ \ b] = \left(\dfrac{a+2b}{7}\right)[1 \ \ 3] + \left(\dfrac{3a-b}{7}\right)[2 \ \ -1] + (0)[1 \ \ 1].$

29. $\begin{bmatrix} a \\ b \\ c \end{bmatrix} = \left(\dfrac{a-b+c}{2}\right)\begin{bmatrix} 1 \\ 0 \\ 1 \end{bmatrix} + \left(\dfrac{a+b-c}{2}\right)\begin{bmatrix} 1 \\ 1 \\ 0 \end{bmatrix} + \left(\dfrac{-a+b+c}{2}\right)\begin{bmatrix} 0 \\ 1 \\ 1 \end{bmatrix}.$

30. No, impossible to write any vector with a nonzero second component as a linear combination of these vectors.

31. $\begin{bmatrix} a \\ 0 \\ a \end{bmatrix} = (a)\begin{bmatrix} 1 \\ 0 \\ 1 \end{bmatrix} + (0)\begin{bmatrix} 1 \\ 0 \\ 2 \end{bmatrix} + (0)\begin{bmatrix} 2 \\ 0 \\ 1 \end{bmatrix}.$

32. 1 and 2 are bases.

33. 7 and 11 are bases.

39. $(-k)\, x + (1)k\mathbf{x} = \mathbf{0}.$

42. $\mathbf{0} = \mathbf{A0} = \mathbf{A}(c_1\mathbf{x}_1 + c_2\mathbf{x}_2 + \cdots + c_k\mathbf{x}_k) = c_1\mathbf{Ax}_1 + c_2\mathbf{Ax}_2 + \cdots + c_k\mathbf{Ax}_k$
$$= c_1\mathbf{y}_1 + c_2\mathbf{y}_2 + \cdots + c_k\mathbf{y}_k).$$

SECTION 2.6

1. 2.

2. 2.

3. 1.

4. 2.

5. 3.

6. Independent.

7. Independent.

8. Dependent.

9. Dependent.

10. Independent.

11. Dependent.

12. Independent.

13. Dependent.

14. Dependent.

15. Dependent.

16. Independent.

17. Dependent.

18. Independent.

19. Dependent.

20. Independent.

21. Dependent.

22. Dependent.

23. (a) Yes,
    (b) Yes,
    (c) No.

24. (a) Yes,
    (b) No,
    (c) Yes.

25. Yes.

26. Yes.

27. No.

28. First two.

29. First two.

30. First and third.

31. 0.

SECTION 2.7

1. Consistent with no arbitrary unknowns; $x = 2/3$, $y = 1/3$.

2. Inconsistent.

3. Consistent with one arbitrary unknown; $x = (1/2) \, (3 - 2z)$, $y = -1/2$.

4. Consistent with two arbitrary unknowns; $x = (1/7) \, (11 - 5z - 2w)$, $y = (1/7) \, (1 - 3z + 3w)$.

5. Consistent with no arbitrary unknowns; $x = y = 1$, $z = -1$.

6. Consistent with no arbitrary unknowns; $x = y = 0$.

7. Consistent with no arbitrary unknowns; $x = y = z = 0$.

8. Consistent with no arbitrary unknowns; $x = y = z = 0$.

9. Consistent with two arbitrary unknowns; $x = z - 7w$, $y = 2z - 2w$.

CHAPTER 3
SECTION 3.1

1. (c).

2. None.

3. $\begin{bmatrix} \dfrac{3}{14} & \dfrac{-2}{14} \\[2ex] \dfrac{-5}{14} & \dfrac{8}{14} \end{bmatrix}$.

4. $\begin{bmatrix} -\dfrac{1}{3} & \dfrac{2}{3} \\[2ex] \dfrac{2}{3} & -\dfrac{1}{3} \end{bmatrix}$.

5. D has no inverse.

7. $\begin{bmatrix} 4 & -1 \\ -3 & 1 \end{bmatrix}$.

8. $\begin{bmatrix} \dfrac{3}{2} & -\dfrac{1}{2} \\[2ex] -2 & 1 \end{bmatrix}$.

9. $\begin{bmatrix} 4 & -6 \\ -6 & 12 \end{bmatrix}$.

10. $\begin{bmatrix} \dfrac{-1}{5} & \dfrac{1}{10} \\[2ex] \dfrac{3}{20} & \dfrac{-1}{20} \end{bmatrix}$.

11. $\begin{bmatrix} 0 & 1 \\ 1 & 0 \end{bmatrix}$.

12. $\begin{bmatrix} 3 & 0 \\ 0 & 1 \end{bmatrix}.$

13. $\begin{bmatrix} 1 & 0 \\ 0 & -5 \end{bmatrix}.$

14. $\begin{bmatrix} 1 & 0 & 0 \\ 0 & -5 & 0 \\ 0 & 0 & 1 \end{bmatrix}.$

15. $\begin{bmatrix} 1 & 0 \\ 3 & 1 \end{bmatrix}.$

16. $\begin{bmatrix} 1 & 3 \\ 0 & 1 \end{bmatrix}.$

17. $\begin{bmatrix} 1 & 0 & 0 \\ 0 & 1 & 3 \\ 0 & 0 & 1 \end{bmatrix}.$

18. $\begin{bmatrix} 1 & 0 & 0 \\ 0 & 1 & 0 \\ 5 & 0 & 1 \end{bmatrix}.$

19. $\begin{bmatrix} 1 & 0 & 0 & 0 \\ 0 & 1 & 0 & 8 \\ 0 & 0 & 1 & 0 \\ 0 & 0 & 0 & 1 \end{bmatrix}.$

20. $\begin{bmatrix} 1 & 0 & 0 & 0 & 0 \\ 0 & 1 & 0 & 0 & 0 \\ 0 & 0 & 1 & 0 & 0 \\ -2 & 0 & 0 & 1 & 0 \\ 0 & 0 & 0 & 0 & 1 \end{bmatrix}.$

21. $\begin{bmatrix} 1 & 0 & 0 & 0 \\ 0 & 0 & 0 & 1 \\ 0 & 0 & 1 & 0 \\ 0 & 1 & 0 & 0 \end{bmatrix}.$

22.
$$\begin{bmatrix} 1 & 0 & 0 & 0 \\ 0 & 0 & 0 & 1 \\ 0 & 0 & 1 & 0 \\ 0 & 1 & 0 & 0 \end{bmatrix}.$$

23.
$$\begin{bmatrix} 1 & 0 & 0 & 0 & 0 & 0 \\ 0 & 0 & 0 & 1 & 0 & 0 \\ 0 & 0 & 1 & 0 & 0 & 0 \\ 0 & 1 & 0 & 0 & 0 & 0 \\ 0 & 0 & 0 & 0 & 1 & 0 \\ 0 & 0 & 0 & 0 & 0 & 0 \end{bmatrix}.$$

24.
$$\begin{bmatrix} 1 & 0 \\ 0 & 7 \end{bmatrix}.$$

25.
$$\begin{bmatrix} 1 & 0 & 0 & 0 & 0 \\ 0 & 1 & 0 & 0 & 0 \\ 0 & 0 & 7 & 0 & 0 \\ 0 & 0 & 0 & 1 & 0 \\ 0 & 0 & 0 & 0 & 1 \end{bmatrix}.$$

26.
$$\begin{bmatrix} 1 & 0 & 0 \\ 0 & -0.2 & 0 \\ 0 & 0 & 1 \end{bmatrix}.$$

27.
$$\begin{bmatrix} \dfrac{1}{2} & 0 \\ 0 & 1 \end{bmatrix}.$$

28.
$$\begin{bmatrix} 1 & -2 \\ 0 & 1 \end{bmatrix}.$$

29.
$$\begin{bmatrix} 1 & 0 \\ 3 & 1 \end{bmatrix}.$$

30. $\begin{bmatrix} 1 & 0 \\ -1 & 0 \end{bmatrix}$.

31. $\begin{bmatrix} 1 & 0 & 0 \\ 0 & \dfrac{1}{2} & 0 \\ 0 & 0 & 1 \end{bmatrix}$.

32. $\begin{bmatrix} 0 & 1 & 0 \\ 1 & 0 & 0 \\ 0 & 0 & 1 \end{bmatrix}$.

33. $\begin{bmatrix} 1 & 0 & 0 \\ 0 & 1 & 0 \\ -3 & 0 & 1 \end{bmatrix}$.

34. $\begin{bmatrix} 1 & 0 & -3 \\ 0 & 1 & 0 \\ 0 & 0 & 1 \end{bmatrix}$.

35. $\begin{bmatrix} 1 & 0 & 0 \\ 0 & 1 & 2 \\ 0 & 0 & 1 \end{bmatrix}$.

36. $\begin{bmatrix} 1 & 0 & 0 \\ 0 & 1 & 0 \\ 0 & 0 & -\dfrac{1}{4} \end{bmatrix}$.

37. $\begin{bmatrix} 1 & 0 & 0 & 0 \\ 0 & 1 & 0 & 0 \\ 0 & 0 & 0 & 1 \\ 0 & 0 & 1 & 0 \end{bmatrix}$.

38. $\begin{bmatrix} 1 & 0 & 0 & 0 \\ 0 & 1 & 0 & -7 \\ 0 & 0 & 1 & 0 \\ 0 & 0 & 0 & 1 \end{bmatrix}$.

39. $\begin{bmatrix} 1 & 0 & 0 & 0 \\ 0 & 1 & 0 & 0 \\ 3 & 0 & 1 & 0 \\ 0 & 0 & 0 & 1 \end{bmatrix}.$

40. $\begin{bmatrix} 0 & 0 & 0 & 1 \\ 0 & 1 & 0 & 0 \\ 0 & 0 & 1 & 0 \\ 1 & 0 & 0 & 0 \end{bmatrix}.$

41. $\begin{bmatrix} 1 & 0 & 0 & 0 \\ 0 & 1 & 0 & 0 \\ 0 & -1 & 1 & 0 \\ 0 & 0 & 0 & 1 \end{bmatrix}.$

42. $\begin{bmatrix} 1 & 0 & 0 & 0 \\ 0 & 1 & 0 & 0 \\ 0 & 0 & -2 & 0 \\ 0 & 0 & 0 & 1 \end{bmatrix}.$

43. $\begin{bmatrix} \dfrac{1}{2} & 0 \\ 0 & \dfrac{1}{3} \end{bmatrix}.$

44. No inverse.

45. $\begin{bmatrix} \dfrac{1}{3} & 0 \\ 0 & -\dfrac{1}{3} \end{bmatrix}.$

46. $\begin{bmatrix} 2 & 0 \\ 0 & -\dfrac{3}{2} \end{bmatrix}.$

47. $\begin{bmatrix} \dfrac{1}{10} & 0 & 0 \\[2mm] 0 & \dfrac{1}{5} & 0 \\[2mm] 0 & 0 & \dfrac{1}{5} \end{bmatrix}.$

48. $\begin{bmatrix} 1 & -1 & 0 \\ 0 & 1 & 0 \\ 0 & 0 & -1 \end{bmatrix}.$

49. $\begin{bmatrix} -\dfrac{1}{4} & 0 & 0 \\[2mm] 0 & -\dfrac{1}{2} & 0 \\[2mm] 0 & 0 & \dfrac{5}{3} \end{bmatrix}.$

50. $\begin{bmatrix} 1 & -2 & 0 & 0 \\ 0 & 1 & 0 & 0 \\ 0 & 0 & 1 & 0 \\ 0 & 0 & -2 & 1 \end{bmatrix}.$

51. $\begin{bmatrix} \dfrac{1}{2} & 0 & 0 & 0 \\[2mm] 0 & \dfrac{1}{3} & 0 & 0 \\[2mm] 0 & 0 & 1 & 3 \\[1mm] 0 & 0 & 0 & 1 \end{bmatrix}.$

52. $\begin{bmatrix} \dfrac{1}{4} & 0 & 0 & 0 \\[2mm] 0 & \dfrac{1}{5} & 0 & 0 \\[2mm] 0 & 0 & \dfrac{1}{6} & 0 \\[2mm] 0 & 0 & 0 & 1 \end{bmatrix}.$

53. $\begin{bmatrix} 0 & 1 & 0 & 0 \\ 1 & 0 & 0 & 0 \\ 0 & 0 & 0 & 1 \\ 0 & 0 & 1 & 0 \end{bmatrix}.$

54. $\begin{bmatrix} 0 & 0 & 1 & 0 \\ 0 & 1 & 0 & 0 \\ 1 & 0 & 0 & 0 \\ 0 & 0 & 0 & \frac{1}{7} \end{bmatrix}.$

55. $\begin{bmatrix} \frac{1}{4} & 0 & 0 & 0 \\ 0 & \frac{1}{5} & 0 & 0 \\ 0 & 0 & 1 & -6 \\ 0 & 0 & 0 & 1 \end{bmatrix}.$

SECTION 3.2

1. $\begin{bmatrix} 4 & -1 \\ -3 & 1 \end{bmatrix}.$

2. $\frac{1}{3}\begin{bmatrix} 2 & -1 \\ -1 & 2 \end{bmatrix}.$

3. Does not exist.

4. $\frac{1}{11}\begin{bmatrix} 4 & 1 \\ -3 & 2 \end{bmatrix}.$

5. $\begin{bmatrix} 2 & -3 \\ -5 & 8 \end{bmatrix}.$

6. $\begin{bmatrix} 4 & -6 \\ -6 & 12 \end{bmatrix}.$

7. $\frac{1}{2}\begin{bmatrix} 1 & 1 & -1 \\ 1 & -1 & 1 \\ -1 & 1 & 1 \end{bmatrix}.$

8. $\begin{bmatrix} 0 & 1 & 0 \\ 0 & 0 & 1 \\ 1 & 0 & 0 \end{bmatrix}$.

9. $\begin{bmatrix} -1 & -1 & 1 \\ 6 & 5 & -4 \\ -3 & -2 & 2 \end{bmatrix}$.

10. Does not exist.

11. $\dfrac{1}{2} \begin{bmatrix} 1 & 0 & 0 \\ -5 & 2 & 0 \\ 1 & -2 & 2 \end{bmatrix}$.

12. $\dfrac{1}{6} \begin{bmatrix} 3 & -1 & -8 \\ 0 & 2 & 1 \\ 0 & 0 & 3 \end{bmatrix}$.

13. $\begin{bmatrix} 9 & -5 & -2 \\ 5 & -3 & -1 \\ -36 & 21 & 8 \end{bmatrix}$.

14. $\dfrac{1}{17} \begin{bmatrix} 1 & 7 & -2 \\ 7 & -2 & 3 \\ -2 & 3 & 4 \end{bmatrix}$.

15. $\dfrac{1}{17} \begin{bmatrix} 14 & 5 & -6 \\ -5 & -3 & 7 \\ 13 & 1 & -8 \end{bmatrix}$.

16. Does not exist.

17. $\dfrac{1}{33} \begin{bmatrix} 5 & 3 & 1 \\ -6 & 3 & 12 \\ -8 & 15 & 5 \end{bmatrix}$.

18. $\dfrac{1}{4} \begin{bmatrix} 0 & -4 & 4 \\ 1 & 5 & -4 \\ 3 & 7 & -8 \end{bmatrix}$.

19. $\dfrac{1}{4}\begin{bmatrix} 4 & -4 & -4 & -4 \\ 0 & 4 & 2 & 5 \\ 0 & 0 & 2 & 3 \\ 0 & 0 & 0 & -2 \end{bmatrix}.$

20. $\begin{bmatrix} 1 & 0 & 0 & 0 \\ 2 & -1 & 0 & 0 \\ -8 & 3 & \dfrac{1}{2} & 0 \\ -25 & 10 & 2 & -1 \end{bmatrix}.$

21. Inverse of a nonsingular lower triangular matrix is lower triangular.

22. Inverse of a nonsingular upper triangular matrix is upper triangular.

23. 35 62 5 10 47 75 2 3 38 57 15 25 18 36.

24. 14 116 10 20 −39 131 −3 5 −57 95 −5 45 36 72.

25. 3 5 48 81 14 28 47 75 2 3 28 42 27 41 5 10.

26. HI THERE.

27. THIS IS FUN.

28. 24 13 27 19 28 9 0 1 1 24 10 24 10 0 18.

SECTION 3.3

1. $x = 1, y = -2.$

2. $a = -3, b = 4.$

3. $x = 2, y = -1.$

4. $l = 1, p = 3.$

5. Not possible; A is singular.

6. $x = -8, y = 5, z = 3.$

7. $x = y = z = 1.$

8. $l = 1, m = -2, n = 0.$

9. $r = 4.333, s = t = -1.667.$

10. $r = 3.767$, $s = -1.133$, $t = -1.033$.

11. Not possible; A is singular.

12. $x = y = 1$, $z = 2$.

13. $r = 500$, $s = 750$.

14. $d = 30$, $b = 200$.

15. $A = 5$, $B = 8$, $C = 6$.

16. $B = \$19{,}012$.

17. 80,000 barrels.

18. $x_1 = 66{,}000$; $x_2 = 52{,}000$.

19. $x_1 = 99{,}702$; $x_2 = 128{,}223$; $x_3 = 94{,}276$.

SECTION 3.4

11. $\mathbf{A}^{-2} = \begin{bmatrix} 11 & -4 \\ -8 & 3 \end{bmatrix}$, $\mathbf{B}^{-2} = \begin{bmatrix} 9 & -20 \\ -4 & 9 \end{bmatrix}$.

12. $\mathbf{A}^{-3} = \begin{bmatrix} 41 & -15 \\ -30 & 11 \end{bmatrix}$, $\mathbf{B}^{-3} = \begin{bmatrix} -38 & 85 \\ 17 & -38 \end{bmatrix}$.

13. $\mathbf{A}^{-2} = \frac{1}{4}\begin{bmatrix} 22 & -10 \\ -15 & 7 \end{bmatrix}$, $\mathbf{B}^{-4} = \frac{1}{512}\begin{bmatrix} 47 & 15 \\ -45 & -13 \end{bmatrix}$.

14. $\mathbf{A}^{-2} = \begin{bmatrix} 1 & -2 & 1 \\ 0 & 1 & -2 \\ 0 & 0 & 1 \end{bmatrix}$, $\mathbf{B}^{-2} = \begin{bmatrix} 1 & -4 & 4 \\ 0 & 1 & 2 \\ 0 & 0 & 1 \end{bmatrix}$.

15. $\mathbf{A}^{-3} = \begin{bmatrix} 1 & -3 & 3 \\ 0 & 1 & -3 \\ 0 & 0 & 1 \end{bmatrix}$, $\mathbf{B}^{-3} = \begin{bmatrix} 1 & -6 & -9 \\ 0 & 1 & 3 \\ 0 & 0 & 1 \end{bmatrix}$.

16. $\dfrac{1}{125} = \begin{bmatrix} -11 & -2 \\ 2 & -11 \end{bmatrix}$.

17. First show that $(\mathbf{BA}^{-1})^{\mathrm{T}} = \mathbf{A}^{-1}\,\mathbf{B}^{\mathrm{T}}$ and that $(\mathbf{A}^{-1}\,\mathbf{B}^{\mathrm{T}})^{-1} = (\mathbf{B}^{\mathrm{T}})^{-1}\,\mathbf{A}$.

SECTION 3.5

1. $\begin{bmatrix} 1 & 0 \\ 3 & 1 \end{bmatrix} \begin{bmatrix} 1 & 1 \\ 0 & 1 \end{bmatrix}$, $\mathbf{x} = \begin{bmatrix} 10 \\ -9 \end{bmatrix}$.

2. $\begin{bmatrix} 1 & 0 \\ 0.5 & 1 \end{bmatrix} \begin{bmatrix} 2 & 1 \\ 0 & 1.5 \end{bmatrix}$, $\mathbf{x} = \begin{bmatrix} 8 \\ -5 \end{bmatrix}$.

3. $\begin{bmatrix} 1 & 0 \\ 0.625 & 1 \end{bmatrix} \begin{bmatrix} 8 & 3 \\ 0 & 0.125 \end{bmatrix}$, $\mathbf{x} = \begin{bmatrix} -400 \\ 1275 \end{bmatrix}$.

4. $\begin{bmatrix} 1 & 0 & 0 \\ 1 & 1 & 0 \\ 0 & -1 & 1 \end{bmatrix} \begin{bmatrix} 1 & 1 & 0 \\ 0 & -1 & 1 \\ 0 & 0 & 2 \end{bmatrix}$, $\mathbf{x} \begin{bmatrix} 3 \\ 1 \\ -2 \end{bmatrix}$.

5. $\begin{bmatrix} 1 & 0 & 0 \\ -1 & 1 & 0 \\ -2 & -2 & 1 \end{bmatrix} \begin{bmatrix} -1 & 2 & 0 \\ 0 & -1 & 1 \\ 0 & 0 & 5 \end{bmatrix}$, $\mathbf{x} = \begin{bmatrix} 5 \\ 2 \\ -1 \end{bmatrix}$.

6. $\begin{bmatrix} 1 & 0 & 0 \\ 2 & 1 & 0 \\ -1 & 0 & 0 \end{bmatrix} \begin{bmatrix} 2 & 1 & 3 \\ 0 & -1 & -6 \\ 0 & 0 & 1 \end{bmatrix}$, $\mathbf{x} = \begin{bmatrix} -10 \\ 0 \\ 10 \end{bmatrix}$.

7. $\begin{bmatrix} 1 & 0 & 0 \\ \dfrac{4}{3} & 1 & 0 \\ 1 & -\dfrac{21}{8} & 1 \end{bmatrix} \begin{bmatrix} 3 & 2 & 1 \\ 0 & -\dfrac{8}{3} & -\dfrac{1}{3} \\ 0 & 0 & \dfrac{1}{8} \end{bmatrix}$, $\mathbf{x} = \begin{bmatrix} 10 \\ -10 \\ 40 \end{bmatrix}$.

8. $\begin{bmatrix} 1 & 0 & 0 \\ 2 & 1 & 0 \\ -1 & -0.75 & 1 \end{bmatrix} \begin{bmatrix} 1 & 2 & -1 \\ 0 & -4 & 3 \\ 0 & 0 & 4.25 \end{bmatrix}$, $\mathbf{x} = \begin{bmatrix} 79 \\ 1 \\ 1 \end{bmatrix}$.

9. $\begin{bmatrix} 1 & 0 & 0 \\ 0 & 1 & 0 \\ 0 & 0 & 1 \end{bmatrix} \begin{bmatrix} 1 & 2 & -1 \\ 0 & 2 & 1 \\ 0 & 0 & 1 \end{bmatrix}$, $\mathbf{x} = \begin{bmatrix} 19 \\ -3 \\ 5 \end{bmatrix}$.

10. $\begin{bmatrix} 1 & 0 & 0 \\ 3 & 1 & 0 \\ 1 & \dfrac{1}{2} & 1 \end{bmatrix} \begin{bmatrix} 1 & 0 & 0 \\ 0 & 2 & 0 \\ 0 & 0 & 2 \end{bmatrix}$, $\mathbf{x} = \begin{bmatrix} 2 \\ -1 \\ \dfrac{1}{2} \end{bmatrix}$.

11.
$$\begin{bmatrix} 1 & 0 & 0 & 0 \\ 1 & 1 & 0 & 0 \\ 1 & 1 & 1 & 0 \\ 0 & 1 & 2 & 1 \end{bmatrix} \begin{bmatrix} 1 & 0 & 1 & 1 \\ 0 & 1 & -1 & 0 \\ 0 & 0 & 1 & -1 \\ 0 & 0 & 0 & 3 \end{bmatrix}, \quad \mathbf{x} = \begin{bmatrix} 1 \\ -5 \\ 2 \\ 1 \end{bmatrix}.$$

12.
$$\begin{bmatrix} 1 & 0 & 0 & 0 \\ \dfrac{1}{2} & 1 & 0 & 0 \\ 0 & 0 & 1 & 0 \\ 0 & \dfrac{2}{7} & \dfrac{5}{7} & 1 \end{bmatrix} \begin{bmatrix} 2 & 1 & -1 & 3 \\ 0 & \dfrac{7}{2} & \dfrac{5}{2} & -\dfrac{1}{2} \\ 0 & 0 & -1 & 1 \\ 0 & 0 & 0 & \dfrac{3}{7} \end{bmatrix}, \quad \mathbf{x} = \begin{bmatrix} 266.67 \\ -166.67 \\ 166.67 \\ 266.67 \end{bmatrix}.$$

13.
$$\begin{bmatrix} 1 & 0 & 0 & 0 \\ 1 & 1 & 0 & 0 \\ 1 & 1 & 1 & 0 \\ 0 & -1 & -2 & 1 \end{bmatrix} \begin{bmatrix} 1 & 2 & 1 & 1 \\ 0 & -1 & 1 & 0 \\ 0 & 0 & -1 & 1 \\ 0 & 0 & 0 & 3 \end{bmatrix}, \quad \mathbf{x} = \begin{bmatrix} 10 \\ 10 \\ 10 \\ -10 \end{bmatrix}.$$

14.
$$\begin{bmatrix} 1 & 0 & 0 & 0 \\ 1 & 1 & 0 & 0 \\ -2 & 1.5 & 1 & 0 \\ 0.5 & 0 & 0.25 & 1 \end{bmatrix} \begin{bmatrix} 2 & 0 & 2 & 0 \\ 0 & 2 & -2 & 6 \\ 0 & 0 & 8 & -8 \\ 0 & 0 & 0 & 3 \end{bmatrix}, \quad \mathbf{x} = \begin{bmatrix} -2.5 \\ -1.5 \\ 1.5 \\ 2.0 \end{bmatrix}.$$

15. (a) $x = 5, y = -2$;   (b) $x = -5/7, y = 1/7$.

16. (a) $x = 1, y = 0, z = 2$;   (b) $x = 140, y = -50, z = -20$.

17. (a) $\begin{bmatrix} 8 \\ -3 \\ -1 \end{bmatrix}$, (b) $\begin{bmatrix} 2 \\ 0 \\ 0 \end{bmatrix}$, (c) $\begin{bmatrix} 35 \\ 5 \\ 15 \end{bmatrix}$, (d) $\begin{bmatrix} -0.5 \\ 1.5 \\ 1.5 \end{bmatrix}$.

18. (a) $\begin{bmatrix} -1 \\ -1 \\ 1 \\ 1 \end{bmatrix}$, (b) $\begin{bmatrix} 0 \\ 0 \\ 0 \\ 0 \end{bmatrix}$, (c) $\begin{bmatrix} 80 \\ 50 \\ -10 \\ 20 \end{bmatrix}$, (d) $\begin{bmatrix} -\dfrac{1}{3} \\ \dfrac{1}{3} \\ \dfrac{1}{3} \\ \dfrac{1}{3} \end{bmatrix}$.

21. (d) $\mathbf{A}$ is singular.

CHAPTER 4
SECTION 4.1

1.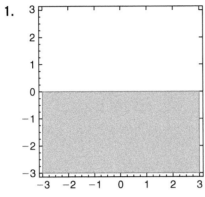

2.

3.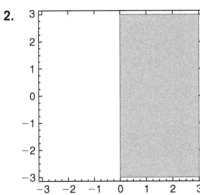

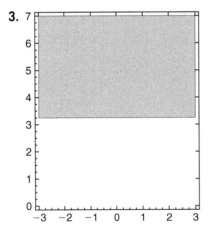

4.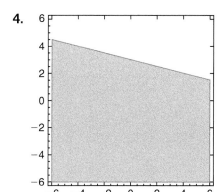

5.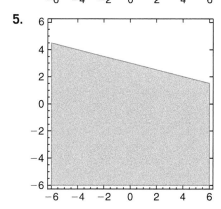

6.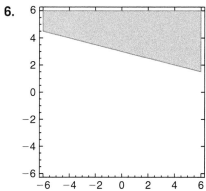

7.

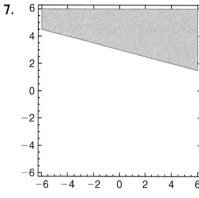

8.

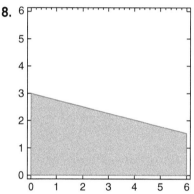

9.

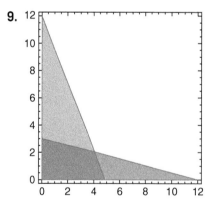

10.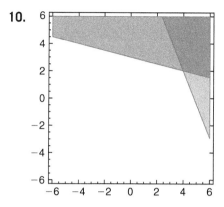

11.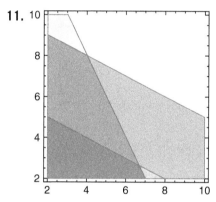

12.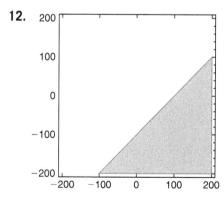

13.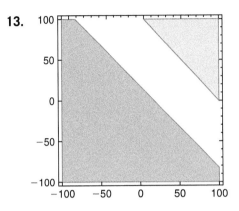

14.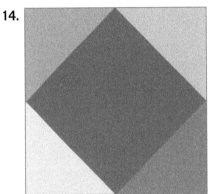

SECTION 4.2

Note: Assume all variables are nonnegative for (1) through (8).

1. Let $x =$ the number of trucks of wheat; $y =$ the number of trucks of corn.
$2x + 3y \le 23$, $3x + y \le 17$.

The objective function is $5000x + 6000y$.

2. The objective function is $8000x + 5000y$.

3. Let $x =$ the number of units of X; $y =$ the number of units of Y. $2x + 3y \ge 180$, $3x + 2y \ge 240$. The objective function is $500x + 750y$.

4. The objective function is $750x + 500y$.

5. Add the third constraint $10x + 10y \ge 210$.

6. Let $x =$ the number of ounces of zinc and $y =$ the number of ounces of calcium. $2x + y \geq 10$, $x + 4y \geq 15$. The objective function is $.04x + 0.05y$.

7. Add the third constraint $3x + 2y \geq 12$.

8. The objective function is $.07x + 0.08y$.

9. The *Richard Nardone Emporium* needs at least 1800 cases of regular scotch and at least 750 cases of premium scotch. Each foreign shipment from distributor "x" can deliver two cases of the former and three cases of the latter, while distributor "y" can produce nine cases of the former and one case of the latter for each foreign shipment. Minimize the cost if each "x" shipment costs \$400 and each "y" shipment costs \$1100. Note that the units for $K (x, y)$ are in \$100s.
   (g) Three components are required to produce a special force (in pounds): mechanical, chemical, and electrical. The following constraints are imposed:
   - Every x force requires one mechanical unit, two chemical units, and one electrical unit;
   - Every y force needs one mechanical unit, one chemical unit, and three electrical units;
   - Every z force requires two mechanical units, one chemical unit, and one electrical unit.

The limits on these components are 12, 14, and 15 units, respectively. The *Cafone Force Machine* uses $2x$ plus $3y$ plus $4z$ pounds of force; maximize the sum of these forces.

SECTION 4.3

1. \$50,000.

2. \$57,000.

3. \$45,000. Note that the minimum occurs at every point on the line segment connecting (72,12) and (90,0).

4. \$60,000. Note that the minimum occurs at every point on the line segment connecting (72,12) and (0,120).

5. $X = 72$, $Y = 12$ is one solution, $X = 90$, $Y = 0$ is another solution.

6. About 29 cents.

9. 400.

12. 3280.

14. 60,468.8.

15. 3018.8.

SECTION 4.4

1. $50,000.

2. $57,000.

3. 30.

4. 20.

5. 72.

7.
$$\begin{array}{ccccc} x_1 & x_2 & s_1 & s_2 & z \\ \left[\begin{array}{ccccc|c} 2 & 5 & 1 & 0 & 0 & 10 \\ 3 & 4 & 0 & 1 & 0 & 12 \\ \hline -100 & -55 & 0 & 0.5 & 1 & 0 \end{array}\right]. \end{array}$$

CHAPTER 5
SECTION 5.1

1. -2.

2. 38.

3. 38.

4. -2.

5. 82.

6. -82.

7. 9.

8. -20.

9. 21.

10. 2.

11. 20.

12. 0.

13. 0.

14. 0.

15. -93.

16. $4t - 6$.

17. $2t^2 + 6$.

18. $5t^2$.

19. 0 and 2.

20. -1 and 4.

21. 2 and 3.

22. $\pm\sqrt{6}$.

23. $\lambda^2 - 9\lambda - 2$.

24. $\lambda^2 - 9\lambda + 38$.

25. $\lambda^2 - 13\lambda - 2$.

26. $\lambda^2 - 8\lambda + 9$.

27. $|\mathbf{A}||\mathbf{B}| = |\mathbf{AB}|$.

28. They differ by a sign.

29. The new determinants are the chosen constant times the old determinants, respectively.

30. No change.

31. Zero.

32. Identical.

33. Zero.

SECTION 5.2

1. -6.

2. 22.

3. 0.

4. -9.

5. -33.

6. 15.

7. -5.

8. -10.

9. 0.

10. 0.

11. 0.

12. 119.

13. -8.

14. 22.

15. -7.

16. -40.

17. 52.

18. 25.

19. 0.

20. 0.

21. -11.

22. 0.

23. Product of diagonal elements.

24. Always zero.

25. $-\lambda^3 + 7\lambda + 22$.

26. $-\lambda^3 + 4\lambda^2 - 17\lambda$.

27. $-\lambda^3 + 6\lambda - 9$.

28. $-\lambda^3 + 10\lambda^2 - 22\lambda - 33$.

SECTION 5.3

2. For an upper triangular matrix, expand by the first column at each step.

3. Use the third column to simplify both the first and second columns.

6. Factor the numbers $-1, 2, 2$, and 3 from the third row, second row, first column, and second column, respectively.

7. Factor a 5 from the third row. Then use this new third row to simplify the second row and the new second row to simplify the first row.

8. Interchange the second and third rows, and then transpose.

9. Multiply the first row by 2, the second row by -1, and the second column by 2.

10. Apply the third elementary row operation with the third row to make the first two rows identical.

11. Multiply the first column by 1/2, the second column by 1/3, to obtain identical columns.

13. $1 = \det(\mathbf{I}) = \det(\mathbf{A}\mathbf{A}^{-1}) = \det(\mathbf{A}) \det(\mathbf{A}^{-1})$.

SECTION 5.4

1. -1.

2. 0.

3. -311.

4. -10.

5. 0.

6. -5.

7. 0.

8. 0.

9. 119.

10. -9.

11. -33.

12. 15.

13. 2187.

14. 52.

15. 25.

16. 0.

17. 0.

18. 152.

19. 0.

20. 0.

SECTION 5.5

1. Does not exist.

2. $\begin{bmatrix} 4 & -1 \\ -3 & 1 \end{bmatrix}.$

3. $\begin{bmatrix} 4 & -6 \\ -6 & 12 \end{bmatrix}.$

4. $\dfrac{1}{11}\begin{bmatrix} 4 & 1 \\ -3 & 2 \end{bmatrix}.$

5. $\begin{bmatrix} 2 & -3 \\ -5 & 8 \end{bmatrix}.$

6. Does not exist.

7. $\dfrac{1}{2}\begin{bmatrix} 1 & 1 & -1 \\ 1 & -1 & 1 \\ -1 & 1 & 1 \end{bmatrix}.$

8. $\begin{bmatrix} 0 & 1 & 0 \\ 0 & 0 & 1 \\ 1 & 0 & 0 \end{bmatrix}.$

9. $\begin{bmatrix} -1 & -1 & 1 \\ 6 & 5 & -4 \\ -3 & -2 & 2 \end{bmatrix}$.

10. Does not exist.

11. $\dfrac{1}{2} \begin{bmatrix} 1 & 0 & 0 \\ -5 & 2 & 0 \\ 1 & -2 & 2 \end{bmatrix}$.

12. $\dfrac{1}{17} \begin{bmatrix} 14 & 5 & -6 \\ -5 & -3 & 7 \\ 13 & 1 & -8 \end{bmatrix}$.

13. Does not exist.

14. $\dfrac{1}{33} \begin{bmatrix} 5 & 3 & 1 \\ -6 & 3 & 12 \\ -8 & 15 & 5 \end{bmatrix}$.

15. $\dfrac{1}{4} \begin{bmatrix} 0 & -4 & 4 \\ 1 & 5 & -4 \\ 3 & 7 & -8 \end{bmatrix}$.

16. $\dfrac{1}{ad - bc} \begin{bmatrix} d & -b \\ -c & a \end{bmatrix}$.

17. $\det(\mathbf{AB}) = \det(\mathbf{A}) \det(\mathbf{B})$.

19. Equals the number of rows in the matrix.

SECTION 5.6

1. $x = 1, y = -2$.

2. $x = 3, y = -3$.

3. $a = 10/11, b = -20/11$.

4. $s = 50, t = 30$.

5. Determinant of coefficient matrix is zero.

6. System is not square.

7. $x = 10, y = z = 5$.

8. $x = 1, y = -4, z = 5.$

9. $x = y = 1, z = 2.$

10. $a = b = c = 1.$

11. Determinant of coefficient matrix is zero.

12. $r = 3, s = -2, t = 3.$

13. $x = 1, y = 2, z = 5, w = -3.$

CHAPTER 6
SECTION 6.1

1. (a), (d), (e), (f), and (h).

2. (a) 3
   (d) 5
   (e) 3
   (f) 3
   (h) 5.

3. (c), (e), (f), and (g).

4. (c) 0
   (e) 0
   (f) −4
   (g) −4.

5. (b), (c), (d), (e), and (g).

6. (b) 2
   (c) 1
   (d) 1
   (e) 3
   (g) 3.

7. (a), (b), and (d).

8. (a) −2
   (b) −1
   (d) 2.

SECTION 6.2

1. 2, 3.
2. 1, 4.
3. 0, 8.
4. -3, 12.
5. 3, 3.
6. 3, -3.
7. $\pm\sqrt{34}$.
8. $\pm 4i$.
9. $\pm i$.

10. 1,1.

11. 0, 0.

12. 0, 0.

13. $\pm\sqrt{2}$.

14. 10, -11.

15. -10, 11.

16. t, $-2t$.

17. $2t$, $2t$.

18. 2θ, 3θ.

19. 2, 4, -2.

20. 1, 2, 3.

21. 1, 1, 3.

22. 0, 2, 2.

23. 2, 3, 9.

24. 1, -2, 5.

25. 2, 3, 6.

26. 0, 0, 14.

27. 0, 10, 14.

28. 2, 2, 5.

29. 0, 0, 6.

30. 3, 3, 9.

31. 3, $\pm 2i$.

32. 0, $\pm i$.

33. 3, 3, 3.

34. 2, 4, 1, $\pm i\sqrt{5}$.

35. 1, 1, 2, 2.

SECTION 6.3

1. $\begin{bmatrix} 2 \\ 1 \end{bmatrix}, \begin{bmatrix} 1 \\ 1 \end{bmatrix}$.

2. $\begin{bmatrix} 1 \\ -1 \end{bmatrix}, \begin{bmatrix} 1 \\ 2 \end{bmatrix}$.

3. $\begin{bmatrix} 3 \\ -2 \end{bmatrix}, \begin{bmatrix} 1 \\ 2 \end{bmatrix}$.

4. $\begin{bmatrix} 1 \\ -1 \end{bmatrix}, \begin{bmatrix} 2 \\ 3 \end{bmatrix}$.

5. $\begin{bmatrix} 1 \\ 1 \end{bmatrix}, \begin{bmatrix} 1 \\ -2 \end{bmatrix}$.

6. $\begin{bmatrix} -5 \\ 3 - \sqrt{34} \end{bmatrix}, \begin{bmatrix} -5 \\ 3 + \sqrt{34} \end{bmatrix}$.

7. $\begin{bmatrix} -5 \\ 3 - 4i \end{bmatrix}, \begin{bmatrix} -5 \\ 3 + 4i \end{bmatrix}$.

8. $\begin{bmatrix} -5 \\ 2 - i \end{bmatrix}, \begin{bmatrix} -5 \\ 2 + i \end{bmatrix}$.

9. $\begin{bmatrix} -2 - \sqrt{2} \\ 1 \end{bmatrix}, \begin{bmatrix} -2 + \sqrt{2} \\ 1 \end{bmatrix}$.

10. $\begin{bmatrix} 5 \\ 3 \end{bmatrix}, \begin{bmatrix} 2 \\ -3 \end{bmatrix}$.

11. $\begin{bmatrix} 1 \\ 1 \end{bmatrix}, \begin{bmatrix} -1 \\ 2 \end{bmatrix}$.

12. $\begin{bmatrix} -1 \\ 1 \end{bmatrix}, \begin{bmatrix} -2 \\ 1 \end{bmatrix}$.

13. $\begin{bmatrix} 0 \\ 1 \\ 0 \end{bmatrix}, \begin{bmatrix} 1 \\ 1 \\ 1 \end{bmatrix}, \begin{bmatrix} -1 \\ 0 \\ 1 \end{bmatrix}$.

14. $\begin{bmatrix} 1 \\ -4 \\ 1 \end{bmatrix}, \begin{bmatrix} 0 \\ 1 \\ 0 \end{bmatrix}, \begin{bmatrix} -1 \\ 0 \\ 1 \end{bmatrix}$.

15. $\begin{bmatrix} 1 \\ -4 \\ 1 \end{bmatrix}, \begin{bmatrix} 0 \\ 1 \\ 0 \end{bmatrix}, \begin{bmatrix} -1 \\ 0 \\ 1 \end{bmatrix}$.

16. $\begin{bmatrix} 0 \\ -1 \\ 1 \end{bmatrix}, \begin{bmatrix} 3 \\ -2 \\ 0 \end{bmatrix}, \begin{bmatrix} 0 \\ 4 \\ 3 \end{bmatrix}$.

17. $\begin{bmatrix} 0 \\ 1 \\ 1 \end{bmatrix}, \begin{bmatrix} -1 \\ 0 \\ 1 \end{bmatrix}, \begin{bmatrix} 1 \\ 1 \\ 1 \end{bmatrix}$.

18. $\begin{bmatrix} 1 \\ -1 \\ 0 \end{bmatrix}, \begin{bmatrix} 1 \\ 1 \\ 1 \end{bmatrix}, \begin{bmatrix} 1 \\ 1 \\ -2 \end{bmatrix}$.

19. $\begin{bmatrix} 9 \\ 1 \\ 13 \end{bmatrix}, \begin{bmatrix} 5 \\ -1 + 2i \\ 0 \end{bmatrix}, \begin{bmatrix} 5 \\ -1 - 2i \\ 0 \end{bmatrix}$.

20. $\begin{bmatrix} 1 \\ 0 \\ 0 \end{bmatrix}, \begin{bmatrix} -1 \\ -i \\ 1 \end{bmatrix}, \begin{bmatrix} -1 \\ i \\ 1 \end{bmatrix}.$

21. $\begin{bmatrix} 1 \\ 0 \\ 0 \end{bmatrix}, \begin{bmatrix} -1 \\ -1 \\ 1 \end{bmatrix}, \begin{bmatrix} 1 \\ 0 \\ 2 \end{bmatrix}.$

22. $\begin{bmatrix} -1 \\ 1 \\ 0 \\ 0 \end{bmatrix}, \begin{bmatrix} -1 \\ 3 \\ 0 \\ 0 \end{bmatrix}, \begin{bmatrix} 0 \\ 0 \\ 2 \\ 1 \end{bmatrix}, \begin{bmatrix} 0 \\ 0 \\ -2 \\ 1 \end{bmatrix}.$

23. $\begin{bmatrix} 10 \\ -6 \\ 11 \\ 4 \end{bmatrix}, \begin{bmatrix} 1 \\ 0 \\ 0 \\ 0 \end{bmatrix}, \begin{bmatrix} 2 \\ 0 \\ 1 \\ 0 \end{bmatrix}, \begin{bmatrix} 0 \\ 0 \\ 1 \\ -1 \end{bmatrix}.$

24. $\begin{bmatrix} 2/\sqrt{5} \\ 1/\sqrt{5} \end{bmatrix}, \begin{bmatrix} 1/\sqrt{5} \\ 1/\sqrt{5} \end{bmatrix}.$

25. $\begin{bmatrix} 1/\sqrt{2} \\ -1/\sqrt{2} \end{bmatrix}, \begin{bmatrix} 1/\sqrt{5} \\ 2/\sqrt{5} \end{bmatrix}.$

26. $\begin{bmatrix} 3/\sqrt{13} \\ -2/\sqrt{13} \end{bmatrix}, \begin{bmatrix} 1/\sqrt{5} \\ 2/\sqrt{5} \end{bmatrix}.$

27. $\begin{bmatrix} 0 \\ 1 \\ 0 \end{bmatrix}, \begin{bmatrix} 1/\sqrt{3} \\ 1/\sqrt{3} \\ 1/\sqrt{3} \end{bmatrix}, \begin{bmatrix} -1/\sqrt{2} \\ 0 \\ 1/\sqrt{2} \end{bmatrix}.$

28. $\begin{bmatrix} 1/\sqrt{18} \\ -4/\sqrt{18} \\ 1/\sqrt{18} \end{bmatrix}, \begin{bmatrix} 0 \\ 1 \\ 0 \end{bmatrix}, \begin{bmatrix} -1/\sqrt{2} \\ 0 \\ 1/\sqrt{2} \end{bmatrix}.$

29. $\begin{bmatrix} 0 \\ -1/\sqrt{2} \\ 1/\sqrt{2} \end{bmatrix}, \begin{bmatrix} 3/\sqrt{2} \\ -2/\sqrt{13} \\ 0 \end{bmatrix}, \begin{bmatrix} 0 \\ 4/5 \\ 3/5 \end{bmatrix}.$

30. $[1 \ {-}1], [-1 \ 2].$

31. $[-2 \ 1], [1 \ 1].$

32. $[-2 \ 1], [2 \ 3].$

33. $[-3 \ 2], [1 \ 1].$

34. $[1{-}2 \ 1], [1 \ 0 \ 1], [-1 \ 0 \ 1].$

35. $[1 \ 0 \ 1], [2 \ 1 \ 2], [-1 \ 0 \ 1].$

36. $[-2 \ {-}3 \ 4], [1 \ 0 \ 0], [2 \ 3 \ 3].$

37. $[1 \ {-}1 \ 0], [1 \ 1 \ 1], [1 \ 1{-}2].$

38. $\mathbf{A}\mathbf{x} = \lambda\mathbf{x}$, so $(\mathbf{A}\mathbf{x})^{\mathrm{T}} = (\lambda\mathbf{x})^{\mathrm{T}}$, and $\mathbf{x}^{\mathrm{T}}\mathbf{A} = \lambda\mathbf{x}^{\mathrm{T}}$.

39. $\begin{bmatrix} 1 & 1 \\ 2 & 2 \end{bmatrix}.$

40. $\begin{bmatrix} 2 & 3 \\ 5 & 5 \end{bmatrix}.$

41. $\begin{bmatrix} 1 & 2 & 5 \\ 8 & 8 & 8 \end{bmatrix}.$

42. (a) $\begin{bmatrix} 1 & 5 \\ 6 & 6 \end{bmatrix}.$ (b) $\frac{1}{6}.$

43. $[7/11 \ 4/11]$; probability of having a Republican is $7/11 = 0.636.$

44. [23/120 71/120 26/120]; probability of a good harvest is 26/120 = 0.217.

45. [40/111 65/111 6/111]; probability of a person using brand Y is 65/111 = 0.586

SECTION 6.4

1. 9.

2. 9.2426.

3. $5 + 8 + \lambda = -4, \lambda = -17$.

4. $(5)(8)\lambda = -4, \lambda = -0.1$.

5. Their product is -24.

6. (a) $-6, 8$;
(b) $-15, 20$;
(c) $-6, 1$;
(d) $1, 8$.

7. (a) $4, 4, 16$;
(b) $-8, 8, 64$;
(c) $6, -6, -12$;
(d) $1, 5, 7$.

8. (a) $2\mathbf{A}$,
(b) $5\mathbf{A}$,
(c) $\mathbf{A}^2$,
(d) $\mathbf{A} + 3\mathbf{I}$.

9. (a) $2\mathbf{A}$,
(b) $\mathbf{A}^2$,
(c) $\mathbf{A}^3$,
(d) $\mathbf{A} - 2\mathbf{I}$.

SECTION 6.5

1. $\begin{bmatrix} 1 \\ -1 \end{bmatrix}$.

2. $\begin{bmatrix} 1 \\ 0 \end{bmatrix}$.

3. $\begin{bmatrix} 1 \\ 0 \end{bmatrix}, \begin{bmatrix} 0 \\ 1 \end{bmatrix}.$

4. $\begin{bmatrix} 1 \\ 0 \\ -1 \end{bmatrix}, \begin{bmatrix} 1 \\ -1 \\ 0 \end{bmatrix}, \begin{bmatrix} 1 \\ 0 \\ 1 \end{bmatrix}.$

5. $\begin{bmatrix} 1 \\ 0 \\ -1 \end{bmatrix}, \begin{bmatrix} 1 \\ 0 \\ 1 \end{bmatrix}.$

6. $\begin{bmatrix} 0 \\ 1 \\ 0 \end{bmatrix}, \begin{bmatrix} 1 \\ 0 \\ 1 \end{bmatrix}, \begin{bmatrix} 1 \\ 2 \\ -1 \end{bmatrix}.$

7. $\begin{bmatrix} 5 \\ -4 \\ 1 \end{bmatrix}, \begin{bmatrix} -1 \\ 0 \\ 1 \end{bmatrix}.$

8. $\begin{bmatrix} 3 \\ 0 \\ -1 \end{bmatrix}, \begin{bmatrix} -1 \\ 5 \\ -3 \end{bmatrix}, \begin{bmatrix} 1 \\ 2 \\ 3 \end{bmatrix}.$

9. $\begin{bmatrix} 1 \\ 0 \\ -1 \end{bmatrix}, \begin{bmatrix} 1 \\ 2 \\ 1 \end{bmatrix}, \begin{bmatrix} 1 \\ -1 \\ 1 \end{bmatrix}.$

10. $\begin{bmatrix} 1 \\ 3 \\ 9 \end{bmatrix}.$

11. $\begin{bmatrix} 1 \\ 1 \\ 1 \end{bmatrix}.$

12. $\begin{bmatrix} 1 \\ 0 \\ -1 \end{bmatrix}, \begin{bmatrix} 1 \\ 2 \\ 1 \end{bmatrix}, \begin{bmatrix} 1 \\ -1 \\ 1 \end{bmatrix}.$

13. $\begin{bmatrix} 1 \\ 1 \\ 1 \\ 1 \end{bmatrix}.$

14. $\begin{bmatrix} 1 \\ 0 \\ 0 \\ 0 \end{bmatrix}, \begin{bmatrix} 0 \\ 1 \\ 1 \\ 1 \end{bmatrix}.$

15. $\begin{bmatrix} -1 \\ 0 \\ 0 \\ 1 \end{bmatrix}, \begin{bmatrix} -1 \\ 0 \\ 1 \\ 0 \end{bmatrix}, \begin{bmatrix} -1 \\ 0 \\ 0 \\ 1 \end{bmatrix}, \begin{bmatrix} 0 \\ 1 \\ 1 \\ 1 \end{bmatrix}.$

16. $\begin{bmatrix} 1 \\ 0 \\ 0 \\ 0 \end{bmatrix}, \begin{bmatrix} 0 \\ -1 \\ 1 \\ 0 \end{bmatrix}, \begin{bmatrix} -1 \\ -1 \\ 0 \\ 1 \end{bmatrix}.$

SECTION 6.6

1.

Iteration	Eigenvector components		Eigenvalue
0	1.0000	1.0000	
1	0.6000	1.0000	5.0000
2	0.5238	1.0000	4.2000
3	0.5059	1.0000	4.0476
4	0.5015	1.0000	4.0118
5	0.5004	1.0000	4.0029

2.

Iteration	Eigenvector components		Eigenvalue
0	1.0000	1.0000	
1	0.5000	1.0000	10.0000
2	0.5000	1.0000	8.0000
3	0.5000	1.0000	8.0000

3.

Iteration	Eigenvector components		Eigenvalue
0	1.0000	1.0000	
1	0.6000	1.0000	15.0000
2	0.6842	1.0000	11.4000
3	0.6623	1.0000	12.1579
4	0.6678	1.0000	11.9610
5	0.6664	1.0000	12.0098

4.

Iteration	Eigenvector components		Eigenvalue
0	1.0000	1.0000	
1	0.5000	1.0000	2.0000
2	0.2500	1.0000	4.0000
3	0.2000	1.0000	5.0000
4	0.1923	1.0000	5.2000
5	0.1912	1.0000	5.2308

5.

Iteration	Eigenvector components		Eigenvalue
0	1.0000	1.0000	
1	1.0000	0.6000	10.0000
2	1.0000	0.5217	9.2000
3	1.0000	0.5048	9.0435
4	1.0000	0.5011	9.0096
5	1.0000	0.5002	9.0021

6.

Iteration	Eigenvector components		Eigenvalue
0	1.0000	1.0000	
1	1.0000	0.4545	11.0000
2	1.0000	0.4175	9.3636
3	1.0000	0.4145	9.2524
4	1.0000	0.4142	9.2434
5	1.0000	0.4142	9.2427

7.

Iteration	Eigenvector components			Eigenvalue
0	1.0000	1.0000	1.0000	
1	0.2500	1.0000	0.8333	12.0000
2	0.0763	1.0000	0.7797	9.8333
3	0.0247	1.0000	0.7605	9.2712
4	0.0081	1.0000	0.7537	9.0914
5	0.0027	1.0000	0.7513	9.0310

8.

Iteration	Eigenvector components			Eigenvalue
0	1.0000	1.0000	1.0000	
1	0.6923	0.6923	1.0000	13.0000
2	0.5586	0.7241	1.0000	11.1538
3	0.4723	0.6912	1.0000	11.3448
4	0.4206	0.6850	1.0000	11.1471
5	0.3883	0.6774	1.0000	11.1101

9.

Iteration	Eigenvector components			Eigenvalue
0	1.0000	1.0000	1.0000	
1	0.4000	0.7000	1.0000	20.0000
2	0.3415	0.6707	1.0000	16.4000
3	0.3343	0.6672	1.0000	16.0488
4	0.3335	0.6667	1.0000	16.0061
5	0.3333	0.6667	1.0000	16.0008

10.

Iteration	Eigenvector components			Eigenvalue
0	1.0000	1.0000	1.0000	
1	0.4000	1.0000	0.3000	−20.0000
2	1.0000	0.7447	0.0284	−14.1000
3	0.5244	1.0000	−0.3683	−19.9504
4	1.0000	0.7168	−0.5303	−18.5293
5	0.6814	1.0000	−0.7423	−20.3976

11. $\begin{bmatrix} 1 \\ 1 \\ 1 \end{bmatrix}$ is a linear combination of $\begin{bmatrix} 1 \\ -4 \\ 1 \end{bmatrix}$ and $\begin{bmatrix} 0 \\ 1 \\ 0 \end{bmatrix}$, which are eigenvectors

corresponding to $\lambda = 1$ and $\lambda = 2$, not $\lambda = 3$. Thus, the power method converges to $\lambda = 2$.

12. There is no single dominant eigenvalue. Here, $|\lambda_1| = |\lambda_2| = \sqrt{34}$.

13. Shift by $\lambda = 4$. Power method on $\mathbf{A} = \begin{bmatrix} -2 & 1 \\ 2 & -1 \end{bmatrix}$ converges after three

iterations to $\mu = -3$. $\lambda + \mu = 1$.

14. Shift by $\lambda = 16$. Power method on $\mathbf{A} = \begin{bmatrix} -13 & 2 & 3 \\ 2 & -10 & 6 \\ 3 & 6 & -5 \end{bmatrix}$ converges after

three iterations to $\mu = -14$. $\lambda = \mu = 2$.

15.

Iteration	Eigenvector components		Eigenvalue
0	1.0000	1.0000	
1	−0.3333	1.0000	0.6000
2	1.0000	−0.7778	0.6000
3	−0.9535	1.0000	0.9556
4	1.0000	0.9904	0.9721
5	−0.9981	1.0000	0.9981

16.

Iteration	Eigenvector components		Eigenvalue
0	1.0000	−0.5000	
1	−0.8571	1.0000	0.2917
2	1.0000	−0.9615	0.3095
3	−0.9903	1.0000	0.3301
4	1.0000	0.9976	0.3317
5	−0.9994	1.0000	0.3331

17.

Iteration	Eigenvector components		Eigenvalue
0	1.0000	1.0000	
1	0.2000	1.0000	0.2778
2	−0.1892	1.0000	0.4111
3	−0.2997	1.0000	0.4760
4	−0.3258	1.0000	0.4944
5	−0.3316	1.0000	0.4987

18.

Iteration	Eigenvector components		Eigenvalue
0	1.0000	1.0000	
1	−0.2000	1.0000	0.7143
2	−0.3953	1.0000	1.2286
3	−0.4127	1.0000	1.3123
4	−0.4141	1.0000	1.3197
5	−0.4142	1.0000	1.3203

19.

Iteration	Eigenvector components			Eigenvalue
0	1.0000	1.0000	1.0000	
1	1.0000	0.4000	−0.2000	0.3125
2	1.0000	0.2703	−0.4595	0.4625
3	1.0000	0.2526	−0.4949	0.4949
4	1.0000	0.2503	−0.4994	0.4994
5	1.0000	0.2500	−0.4999	0.4999

20.

Iteration	Eigenvector components			Eigenvalue
0	1.0000	1.0000	1.0000	
1	0.3846	1.0000	0.9487	−0.1043
2	0.5004	0.7042	1.0000	−0.0969
3	0.3296	0.7720	1.0000	−0.0916
4	0.3857	0.6633	1.0000	−0.0940
5	0.3244	0.7002	1.0000	−0.0907

21.

Iteration	Eigenvector components			Eigenvalue
0	1.0000	1.0000	1.0000	
1	−0.6667	1.0000	−0.6667	−1.5000
2	−0.3636	1.0000	−0.3636	1.8333
3	−0.2963	1.0000	−0.2963	1.2273
4	−0.2712	1.0000	−0.2712	1.0926
5	−0.2602	1.0000	−0.2602	1.0424

22. Cannot construct an **LU** decomposition. Shift as explained in Problem 13.

23. Cannot solve $\mathbf{Lx}_1 = \mathbf{y}$ uniquely for $\mathbf{x}_1$ because one eigenvalue is zero. Shift as explained in Problem 13.

24. Yes, on occasion.

25. Inverse power method applied to $\mathbf{A} = \begin{bmatrix} -7 & 2 & 3 \\ 2 & -4 & 6 \\ 3 & 6 & 1 \end{bmatrix}$ converges to $\mu = 1/6$.

$\lambda + 1/\mu = 10 + 6 = 16$.

26. Inverse power method applied to $\mathbf{A} = \begin{bmatrix} 27 & -17 & 7 \\ -17 & 21 & 1 \\ 7 & 1 & 11 \end{bmatrix}$ converges to

$\mu = 1/3$. $\lambda + 1/\mu = -25 + 3 = -22$.

CHAPTER 7
SECTION 7.1

1. (a) $\begin{bmatrix} 0 & -4 & 8 \\ 0 & 4 & -8 \\ 0 & 0 & 0 \end{bmatrix}, \begin{bmatrix} 0 & 8 & -16 \\ 0 & -8 & 16 \\ 0 & 0 & 0 \end{bmatrix}$;

(b) $\begin{bmatrix} 57 & 78 \\ 117 & 174 \end{bmatrix}, \begin{bmatrix} 234 & 348 \\ 522 & 756 \end{bmatrix}$.

2. $p_k(\mathbf{A}) = \begin{bmatrix} p_k(\lambda_1) & 0 & 0 \\ 0 & p_k(\lambda_2) & 0 \\ 0 & & p_k(\lambda_3) \end{bmatrix}$.

4. In general, $\mathbf{AB} \neq \mathbf{BA}$.

5. Yes.

6. $\begin{bmatrix} 0 & 2 \\ 3 & 0 \end{bmatrix}$.

7. $\begin{bmatrix} 2 \\ 0 \\ 3/2 \end{bmatrix}$.

8. 2−2 element tends to ∞, so limit diverges.

9. a, b, d, and f.

10. f.

11. All except c.

13. $\begin{bmatrix} e & 0 \\ 0 & e^2 \end{bmatrix}$.

14. $\begin{bmatrix} e^{-1} & 0 \\ 0 & e^{28} \end{bmatrix}$.

15. $\begin{bmatrix} e^2 & 0 & 0 \\ 0 & e^{-2} & 0 \\ 0 & 0 & 1 \end{bmatrix}$.

16. $\sin(\mathbf{A}) = \begin{bmatrix} \sin(\lambda_1) & 0 & \cdots & 0 \\ 0 & \sin(\lambda_2) & \cdots & 0 \\ \vdots & \vdots & \cdots & \vdots \\ 0 & 0 & \cdots & \sin(\lambda_n) \end{bmatrix}$.

17. $\begin{bmatrix} \sin(1) & 0 \\ 0 & \sin(2) \end{bmatrix}$.

18. $\begin{bmatrix} \sin(-1) & 0 \\ 0 & \sin(28) \end{bmatrix}$.

19. $\cos \mathbf{A} = \sum_{k=0}^{\infty} \frac{(-1)^k \mathbf{A}^{2k}}{(2k)!}$, $\cos \begin{bmatrix} 1 & 0 \\ 0 & 2 \end{bmatrix} = \begin{bmatrix} \cos(1) & 0 \\ 0 & \cos(2) \end{bmatrix}$.

20. $\begin{bmatrix} \cos(2) & 0 & 0 \\ 0 & \cos(-2) & 0 \\ 0 & 0 & 1 \end{bmatrix}$.

SECTION 7.2

1. $\mathbf{A}^{-1} = \begin{bmatrix} -2 & 1 \\ 3/2 & -1/2 \end{bmatrix}$.

2. Since $\alpha_0 = 0$, the inverse does not exist.

3. Since $\alpha_0 = 0$, the inverse does not exist.

4. $\mathbf{A}^{-1} \begin{bmatrix} -1/3 & -1/3 & 2/3 \\ -1/3 & 1/6 & 1/6 \\ 1/2 & 1/4 & -1/4 \end{bmatrix}$.

5. $\mathbf{A}^{-1} = \begin{bmatrix} 1 & 0 & 0 & 0 \\ 0 & -1 & 0 & 0 \\ 0 & 0 & -1 & 0 \\ 0 & 0 & 0 & 1 \end{bmatrix}$.

SECTION 7.3

1. $\begin{matrix} 1 = \alpha_1 + \alpha_0, \\ -1 = -\alpha_1 + \alpha_0; \end{matrix}$ $\begin{bmatrix} -2 & 3 \\ -1 & 2 \end{bmatrix}$.

2. $\begin{bmatrix} 1 & 0 \\ 0 & 1 \end{bmatrix}$.

3. $\begin{matrix} 0 = \alpha_0, \\ -1 = -\alpha_1 + \alpha_0; \end{matrix}$ $\begin{bmatrix} 0 & 1 \\ 0 & -1 \end{bmatrix}$.

4. $\begin{matrix} 0 = \alpha_0, \\ 1 = -\alpha_1 + \alpha_0; \end{matrix}$ $\begin{bmatrix} 3 & -6 \\ 1 & -2 \end{bmatrix}$.

5. $\begin{bmatrix} -3 & 6 \\ -1 & 2 \end{bmatrix}$.

6. $\begin{bmatrix} 0 & -1 \\ 0 & 1 \end{bmatrix}.$

7. $3^{78} = 3\alpha_1 + \alpha_0,$

$\qquad 4^{78} = 4^{78} = 4\alpha_1 + \alpha_0;$ $\quad \begin{bmatrix} -4^{78} + 2(3^{78}) & -4^{78} + 3^{78} \\ 2(4^{78}) - 2(3^{78}) & 2(4^{78}) - 3^{78} \end{bmatrix}.$

8. $\begin{bmatrix} -4^{41} + 2(3^{41}) & -4^{41} + 3^{41} \\ 2(4^{41}) - 2(3^{41}) & 2(4^{41}) - 3^{41} \end{bmatrix}.$

9. $\quad 1 = \alpha_2 + \alpha_1 + \alpha_0,$

$\qquad\quad 1 = \alpha_2 - \alpha_1 + \alpha_0,$ $\quad \begin{bmatrix} 1 & 0 & (-4 + 4(2^{222}))/3 \\ 0 & 1 & (-2 + 2(2^{222}))/3 \\ 0 & 0 & 2^{222} \end{bmatrix}.$

$\qquad 2^{222} = 4\alpha_2 + 2\alpha_1 + \alpha_0,$

10. $\quad 3^{17} = \quad 9\alpha_2 + \quad 3\alpha_1 + \alpha_0,$

$\qquad\quad 5^{17} = \quad 25\alpha_2 + \quad 5\alpha_1 + \alpha_0,$

$\qquad 10^{17} = 100\alpha_2 + 10\alpha_1 + \alpha_0.$

11. $\qquad 2^{25} = \quad 8\alpha_3 + \quad 4\alpha_2 + 2\alpha_1 + \alpha_0,$

$\qquad (-2)^{25} = -8\alpha_3 + \quad 4\alpha_2 - 2\alpha_1 + \alpha_0,$

$\qquad\quad 3^{25} = 27\alpha_3 + \quad 9\alpha_2 + 3\alpha_1 + \alpha_0,$

$\qquad\quad 4^{25} = 64\alpha_3 + 16\alpha_2 + 4\alpha_1 + \alpha_0.$

12. $\qquad\quad 1 = \qquad \alpha_3 + \quad \alpha_2 + \quad \alpha_1 + \alpha_0,$

$\qquad (-2)^{25} = \quad -8\alpha_3 + \quad 4\alpha_2 - 2\alpha_1 + \alpha_0,$

$\qquad\quad 3^{25} = \quad 27\alpha_3 + \quad 9\alpha_2 + 3\alpha_1 + \alpha_0,$

$\qquad (-4^{25}) = \quad -64\alpha_3 + 16\alpha_2 - 4\alpha_1 + \alpha_0.$

13. $\qquad\quad 1 = \quad \alpha_4 + \quad \alpha_3 + \quad \alpha_2 + \quad \alpha_1 + \alpha_0,$

$\qquad\qquad 1 = \quad \alpha_4 - \quad \alpha_3 + \quad \alpha_2 - \quad \alpha_1 + \alpha_0,$

$\qquad\quad 256 = 16\alpha_4 + \quad 8\alpha_3 + 4\alpha_2 + 2\alpha_1 + \alpha_0,$

$\qquad\quad 256 = 16\alpha_4 - \quad 8\alpha_3 + 4\alpha_2 - 2\alpha_1 + \alpha_0,$

$\qquad 6,561 = 81\alpha_4 + 27\alpha_3 + 9\alpha_2 + 3\alpha_1 + \alpha_0.$

14.
$$5,837 = 9\alpha_2 + 3\alpha_1 + \alpha_0,$$
$$381,255 = 25\alpha_2 + 5\alpha_1 + \alpha_0,$$
$$10^8 - 3(10)^5 + 5 = 100\alpha_2 + 10\alpha_1 + \alpha_0.$$

15.
$$165 = 8\alpha_3 + 4\alpha_2 + 2\alpha_1 + \alpha_0,$$
$$357 = -8\alpha_3 + 4\alpha_2 - 2\alpha_1 + \alpha_0,$$
$$5,837 = 27\alpha_3 + 9\alpha_2 + 3\alpha_1 + \alpha_0,$$
$$62,469 = 64\alpha_3 + 16\alpha_2 + 4\alpha_1 + \alpha_0.$$

16.
$$3 = \alpha_3 + \alpha_2 + \alpha_1 + \alpha_0,$$
$$357 = -8\alpha_3 + 4\alpha_2 - 2\alpha_1 + \alpha_0,$$
$$5,837 = 27\alpha_3 + 9\alpha_2 + 3\alpha_1 + \alpha_0,$$
$$68,613 = -64\alpha_3 + 16\alpha_2 - 4\alpha_1 + \alpha_0.$$

17.
$$15 = \alpha_3 + \alpha_2 + \alpha_1 + \alpha_0,$$
$$960 = -8\alpha_3 + 4\alpha_2 - 2\alpha_1 + \alpha_0,$$
$$59,235 = 27\alpha_3 + 9\alpha_2 + 3\alpha_1 + \alpha_0,$$
$$1,048,160 = -64\alpha_3 + 16\alpha_2 - 4\alpha_1 + \alpha_0.$$

18.
$$15 = \alpha_4 + \alpha_3 + \alpha_2 + \alpha_1 + \alpha_0,$$
$$-13 = \alpha_4 - \alpha_3 + \alpha_2 - \alpha_1 + \alpha_0,$$
$$1,088 = 16\alpha_4 + 8\alpha_3 + 4\alpha_2 + 2\alpha_1 + \alpha_0,$$
$$960 = 16\alpha_4 - 8\alpha_3 + 4\alpha_2 - 2\alpha_1 + \alpha_0,$$
$$59,235 = 81\alpha_4 + 27\alpha_3 + 9\alpha_2 + 3\alpha_1 + \alpha_0.$$

19. $\begin{bmatrix} 9 & -9 \\ 3 & -3 \end{bmatrix}.$

20. $\begin{bmatrix} 6 & -9 \\ 3 & -6 \end{bmatrix}.$

21. $\begin{bmatrix} -50,801 & -56,632 \\ 113,264 & 119,095 \end{bmatrix}$.

22. $\begin{bmatrix} 3,007 & -5,120 \\ 1,024 & -3,067 \end{bmatrix}$.

23. $\begin{bmatrix} 938 & 160 \\ -32 & 1130 \end{bmatrix}$.

24. $\begin{bmatrix} 2 & -4 & -3 \\ 0 & 0 & 0 \\ 1 & -5 & -2 \end{bmatrix}$.

25. $2,569 = 4\alpha_2 + 2\alpha_1 + \alpha_0,$ $\begin{bmatrix} -339 & -766 & 1110 \\ -4440 & 4101 & 344 \\ -1376 & -3064 & 4445 \end{bmatrix}$.
$5,633 = 4\alpha_2 - 2\alpha_1 + \alpha_0,$
$5 = \alpha_2 + \alpha_1 + \alpha_0.$

26. $0.814453 = 0.25\alpha_2 + 0.5\alpha_1 + \alpha_0,$ $\begin{bmatrix} 1.045578 & 0.003906 & -0.932312 \\ 0.058270 & 0.812500 & -0.229172 \\ 0.014323 & 0.000977 & 0.755207 \end{bmatrix}$.
$0.810547 = 0.25\alpha_2 - 0.5\alpha_1 + \alpha_0,$
$0.988285 = 0.0625\alpha_2 + 0.25\alpha_1 + \alpha_0.$

SECTION 7.4

1. $128 = 2\alpha_1 + \alpha_0,$
$448 = \alpha_1.$

2. $128 = 4\alpha_2 + 2\alpha_1 + \alpha_0,$
$448 = 4\alpha_1 + \alpha_1,$
$1,344 = 2\alpha_2.$

3. $128 = 4\alpha_2 + 2\alpha_1 + \alpha_0,$
$448 = 4\alpha_2 + \alpha_1,$
$1 = \alpha_2 + \alpha 1 + \alpha_0.$

4. $59,049 = 3\alpha_1 + \alpha_0,$
$196,830 = \alpha_1.$

5. $59,049 = 9\alpha_2 + 3\alpha_1 + \alpha_0,$

   $196,830 = 6\alpha_2 + \alpha_1,$

   $590,490 = 2\alpha_2.$

6. $59,049 = 27\alpha_3 + 9\alpha_2 + 3\alpha_1 + \alpha_0,$

   $196,830 = 27\alpha_3 + 6\alpha_2 + \alpha_1,$

   $590,490 = 18\alpha_3 + 2\alpha_2,$

   $1,574,640 = 6\alpha_3.$

7. $512 = 8\alpha_3 + 4\alpha_2 + 2\alpha_1 + \alpha_0,$

   $2,304 = 12\alpha_3 + 4\alpha_2 + \alpha_1,$

   $9,216 = 12\alpha_3 + 2\alpha_2,$

   $32,256 = 6\alpha_3.$

8. $512 = 8\alpha_3 + 4\alpha_2 + 2\alpha_1 + \alpha_0,$

   $2,304 = 12\alpha_3 + 4\alpha_2 + \alpha_1,$

   $9,216 = 12\alpha_3 + 2\alpha_2,$

   $1 = \alpha_3 + \alpha_2 + \alpha_1 + \alpha_0.$

9. $512 = 8\alpha_3 + 4\alpha_2 + 2\alpha_1 + \alpha_0,$

   $2,304 = 12\alpha_3 + 4\alpha_2 + \alpha_1,$

   $1 = \alpha_3 + \alpha_2 + \alpha_1 + \alpha_0,$

   $9 = 3\alpha_3 + 2\alpha_2 + \alpha_1.$

10. $(5)^{10} - 3(5)^5 = \alpha_5(5)^5 + \alpha_4(5)^4 + \alpha_3(5)^3 + \alpha_2(5)^2 + \alpha_1(5) + \alpha_0,$

   $10(5)^9 - 15(5)^4 = 5\alpha_5(5)^4 + 4\alpha_4(5)^3 + 3\alpha_3(5)^2 + 2\alpha_2(5) + \alpha_1,$

   $90(5)^8 - 60(5)^3 = 20\alpha_5(5)^3 + 12\alpha_4(5)^2 + 6\alpha_3(5) + 2\alpha_2,$

   $720(5)^7 - 180(5)^2 = 60\alpha_5(5)^2 + 24\alpha_4(5) + 6\alpha_3,$

   $(2)^{10} - 3(2)^5 = \alpha_5(2)^5 + \alpha_4(2)^4 + \alpha_3(2)^3 + \alpha_2(2)^2 + \alpha_1(2) + \alpha_0,$

   $10(2)^9 - 15(2)^4 = 5\alpha_5(2)^4 + 4\alpha_4(2)^3 + 3\alpha_3(2)^2 + 2\alpha_2(2) + \alpha_1.$

11. $\begin{bmatrix} 729 & 0 \\ 0 & 729 \end{bmatrix}.$

12. $\begin{bmatrix} 4 & 1 & -3 \\ 0 & -1 & 0 \\ 5 & 1 & -4 \end{bmatrix}$.

13. $\begin{bmatrix} 0 & 0 & 0 \\ 0 & 0 & 0 \\ 0 & 0 & 0 \end{bmatrix}$.

SECTION 7.5

1. $e = \alpha_1 + \alpha_0,$

$e^2 = 2\alpha_1 + \alpha_0.$

2. $e^2 = 2\alpha_1 + \alpha_0,$

$e^2 = \alpha 1.$

3. $e^2 = 4\alpha_2 + 2\alpha_1 + \alpha_0,$

$e^2 = 4\alpha_2 + \alpha_1,$

$e^2 = 2\alpha_2.$

4. $e^1 = \alpha_2 + \alpha_1 + \alpha_0,$

$e^{-2} = 4\alpha_2 - 2\alpha_1 + \alpha_0,$

$e^3 = 9\alpha_2 + 3\alpha_1 + \alpha_0.$

5. $e^{-2} = 4\alpha_2 - 2\alpha_1 + \alpha_0,$

$e^{-2} = -4\alpha_2 + \alpha_1,$

$e^1 = \alpha_2 + \alpha_1 + \alpha_0.$

6. $\sin(1) = \alpha_2 + \alpha_1 + \alpha_0,$

$\sin(2) = 4\alpha_2 + 2\alpha_1 + \alpha_0,$

$\sin(3) = 9\alpha_2 + 3\alpha_1 + \alpha_0.$

7. $\sin(-2) = 4\alpha_2 - 2\alpha_1 + \alpha_0,$

$\sin(-2) = -4\alpha_2 + \alpha_1,$

$\sin(1) = \alpha_2 + \alpha_1 + \alpha_0.$

8. $e^2 = 8\alpha_3 + 4\alpha_2 + 2\alpha_1 + \alpha_0,$

$e^2 = 12\alpha_3 + 4\alpha_2 + \alpha_1,$

$e^2 = 12\alpha_3 + 2\alpha_2,$

$e^2 = 6\alpha_3.$

9. $e^2 = 8\alpha_3 + 4\alpha_2 + 2\alpha_1 + \alpha_0,$

$e^2 = 12\alpha_3 + 4\alpha_2 + \alpha_1,$

$e^{-2} = -8\alpha_3 + 4\alpha_2 - 2\alpha_1 + \alpha_0,$

$e^{-2} = 12\alpha_3 - 4\alpha_2 + \alpha_1.$

10. $\sin(2) = 8\alpha_3 + 4\alpha_2 + 2\alpha_1 + \alpha_0,$

$\cos(2) = 12\alpha_3 + 4\alpha_2 + \alpha_1,$

$\sin(-2) = -8\alpha_3 + 4\alpha_2 - 2\alpha_1 + \alpha_0,$

$\cos(-2) = 12\alpha_3 - 4\alpha_2 + \alpha_1.$

11. $e^3 = 27\alpha_3 + 9\alpha_2 + 3\alpha_1 + \alpha_0,$

$e^3 = 27\alpha_3 + 6\alpha_2 + \alpha_1,$

$e^3 = 18\alpha_3 + 2\alpha_2,$

$e^{-1} = -\alpha_3 + \alpha_2 - \alpha_1 + \alpha_0.$

12. $\cos(3) = 27\alpha_3 + 9\alpha_2 + 3\alpha_1 + \alpha_0,$

$-\sin(3) = 27\alpha_3 + 6\alpha_2 + \alpha_1,$

$-\cos(3) = 18\alpha_3 + 2\alpha_2,$

$\cos(-1) = -\alpha_3 + \alpha_2 - \alpha_1 + \alpha_0.$

13. $\dfrac{1}{7} \begin{bmatrix} 3e^5 + 4e^{-2} & 3e^5 - 3e^{-2} \\ 4e^5 - 4e^{-2} & 4e^5 + 3e^{-2} \end{bmatrix}.$

14. $e^3 \begin{bmatrix} 2 & -1 \\ 1 & 0 \end{bmatrix}.$

15. $e^2 \begin{bmatrix} 0 & 1 & 3 \\ -1 & 2 & 5 \\ 0 & 0 & 1 \end{bmatrix}.$

16. $\frac{1}{16} \begin{bmatrix} 12e^2 + 4e^{-2} & 4e^2 - 4e^{-2} & 38e^2 + 2e^{-2} \\ 12e^2 - 12e^{-2} & 4e^2 + 12e^{-2} & 46e^2 - 6e^{-2} \\ 0 & 0 & 16e^2 \end{bmatrix}.$

17. $\frac{1}{5} \begin{bmatrix} -1 & 6 \\ 4 & 1 \end{bmatrix}.$

18. (a) $\begin{bmatrix} \log(3/2) & \log(3/2) - \log(1/2) \\ 0 & \log(1/2) \end{bmatrix}.$

(b) and (c) are not defined since they possess eigenvalues having absolute value greater than 1.

(d) $\begin{bmatrix} 0 & 0 \\ 0 & 0 \end{bmatrix}.$

SECTION 7.6

1. $1/7 \begin{bmatrix} 3e^{8t} + 4e^t & 4e^{8t} - 4e^t \\ 3e^{8t} - 3e^t & 4e^{8t} + 3e^t \end{bmatrix}.$

2. $\begin{bmatrix} \left(2/\sqrt{2}\right)\sin h\sqrt{3}t + \cos h\sqrt{3}t & \left(1/\sqrt{3}\right)\sin h\sqrt{3}t \\ \left(-1/\sqrt{3}\right)\sin h\sqrt{3}t & \left(-2/\sqrt{3}\right)\sin h\sqrt{3}t + \cos h\sqrt{3}t \end{bmatrix}.$

Note:

$$\sin h\sqrt{3}t = \frac{e^{\sqrt{3}t} - e^{-\sqrt{3}t}}{2} \quad \text{and} \quad \cos h\sqrt{3}t = \frac{e^{\sqrt{3}t} + e^{-\sqrt{3}t}}{2}.$$

3. $e^{3t} \begin{bmatrix} 1+t & t \\ -t & 1-t \end{bmatrix}.$

4. $\begin{bmatrix} 1.4e^{-2t} - 0.4e^{-7t} & 0.2e^{-2t} - 0.2e^{-7t} \\ -2.8e^{-2t} + 2.8e^{-7t} & -0.4e^{-2t} + 1.4e^{-7t} \end{bmatrix}.$

5. $\begin{bmatrix} 0.8e^{-2t} + 0.2e^{-7t} & 0.4e^{-2t} - 0.4e^{-7t} \\ 0.4e^{-2t} - 0.4e^{-7t} & 0.2e^{-2t} + 0.8e^{-7t} \end{bmatrix}.$

6. $\begin{array}{cc} 0.5e^{-4t} + 0.5e^{-16t} & 0.5e^{-4t} - 0.5e^{-16t} \\ 0.5e^{-4t} - 0.5e^{-16t} & 0.5e^{-4t} + 0.5e^{-16t} \end{array}$

7. $\begin{bmatrix} 1 & t & t^2/2 \\ 0 & 1 & t \\ 0 & 0 & 1 \end{bmatrix}.$

8. $\dfrac{1}{12} \begin{bmatrix} 12e^t & 0 & 0 \\ -9e^t + 14e^{3t} - 5e^{-3t} & 8e^{3t} + 4e^{-3t} & 4e^{3t} - 4e^{-3t} \\ -24e^t + 14e^{3t} + 10e^{-3t} & 8e^{3t} - 8e^{-3t} & 4e^{3t} + 8e^{-3t} \end{bmatrix}.$

SECTION 7.7

1. $\begin{bmatrix} (1/2)\sin 2t + \cos 2t & (-1/2)\sin 2t \\ (5/2)\sin 2t & (-1/2)\sin 2t + \cos 2t \end{bmatrix}.$

2. $\begin{bmatrix} \sqrt{2}\sin\sqrt{2}t + \cos\sqrt{2}t & -\sqrt{2}\sin\sqrt{2}t \\ \left(3/\sqrt{2}\right)\sin\sqrt{2}t & -\sqrt{2}\sin\sqrt{2}t + \cos\sqrt{2}t \end{bmatrix}.$

3. $\begin{bmatrix} \cos(8t) & \dfrac{1}{8}\sin(8t) \\ -8\sin(8t) & \cos(8t) \end{bmatrix}.$

4. $\dfrac{1}{4} \begin{bmatrix} 2\sin(8t) + 4\cos(8t) & -4\sin(8t) \\ 5\sin(8t) & -2\sin(8t) + 4\cos(8t) \end{bmatrix}.$

5. $\begin{bmatrix} 2\sin(t) + \cos(t) & 5\sin(t) \\ -\sin(t) & -2\sin(t) + \cos(t) \end{bmatrix}.$

6. $\dfrac{1}{3}e^{-4t} \begin{bmatrix} 4\sin(3t) + 3\cos(3t) & \sin(3t) \\ -25\sin(3t) & -4\sin(3t) + 3\cos(3t) \end{bmatrix}.$

7. $e^{4t} \begin{bmatrix} -\sin(t) + \cos t & \sin t \\ -2\sin t & \sin t + \cos t \end{bmatrix}.$

8. $\begin{bmatrix} 1 & -2 + 2\cos(t) + \sin(t) & -5 + 5\cos(t) \\ 0 & \cos(t) - 2\sin(t) & -5\sin(t) \\ 0 & \sin(t) & \cos(t) + 2\sin(t) \end{bmatrix}.$

SECTION 7.8

3. $\mathbf{A}$ does not have an inverse.

8. $e^{\mathbf{A}} = \begin{bmatrix} e & e-1 \\ 0 & 1 \end{bmatrix}$, $\quad e^{\mathbf{B}} = \begin{bmatrix} 1 & e-1 \\ 0 & 1 \end{bmatrix}$, $\quad e^{\mathbf{A}}e^{\mathbf{B}} = \begin{bmatrix} e & 2e^2 - 2e \\ 0 & e \end{bmatrix}$,

$e^{\mathbf{B}}e^{\mathbf{A}} = \begin{bmatrix} e & 2e-2 \\ 0 & e \end{bmatrix}$, $\quad e^{\mathbf{A}+\mathbf{B}} = \begin{bmatrix} e & 2e \\ 0 & e \end{bmatrix}$.

9. $\mathbf{A} = \begin{bmatrix} 1 & 0 \\ 0 & 2 \end{bmatrix}$, $\quad \mathbf{B} = \begin{bmatrix} 3 & 0 \\ 0 & 4 \end{bmatrix}$. Also see Problem 10.

11. First show that for any integer n, $(\mathbf{P}^{-1}\mathbf{B}\mathbf{P})^n = \mathbf{P}^{-1}\mathbf{B}^n\mathbf{P}$, and then use Eq. (6) directly.

SECTION 7.9

1. (a) $\begin{bmatrix} -\sin t & 2t \\ 2 & e^{(t-1)} \end{bmatrix}$. (b) $\begin{bmatrix} 6t^2 e^{t^3} & 2t-1 & 0 \\ 2t+3 & 2\cos 2t & 1 \\ -18t\cos^2(3t^2)\sin(3t^2) & 0 & 1/t \end{bmatrix}$.

4. $\begin{bmatrix} \sin t + c_1 & \frac{1}{3}t^3 - t + c_2 \\ t^2 + c_3 & e^{(t-1)} + c_4 \end{bmatrix}$.

CHAPTER 8
SECTION 8.1

1. $\mathbf{x}(t) = \begin{bmatrix} x(t) \\ \dot{x}(t) \end{bmatrix}$, $\quad \mathbf{A}(t) = \begin{bmatrix} 2 & 3 \\ 4 & 5 \end{bmatrix}$, $\quad \mathbf{f}(t) = \begin{bmatrix} 0 \\ 0 \end{bmatrix}$, $\quad \mathbf{c} = \begin{bmatrix} 6 \\ 7 \end{bmatrix}$, $\quad t_0 = 0$.

2. $\mathbf{x}(t) = \begin{bmatrix} y(t) \\ z(t) \end{bmatrix}$, $\quad \mathbf{A}(t) = \begin{bmatrix} 3 & 2 \\ 4 & 1 \end{bmatrix}$, $\quad \mathbf{f}(t) = \begin{bmatrix} 0 \\ 0 \end{bmatrix}$, $\quad \mathbf{c} = \begin{bmatrix} 1 \\ 1 \end{bmatrix}$, $\quad t_0 = 0$.

3. $\mathbf{x}(t) = \begin{bmatrix} x(t) \\ y(t) \end{bmatrix}$, $\quad \mathbf{A}(t) = \begin{bmatrix} -3 & 3 \\ 4 & -4 \end{bmatrix}$, $\quad \mathbf{f}(t) = \begin{bmatrix} 1 \\ -1 \end{bmatrix}$, $\quad \mathbf{c} = \begin{bmatrix} 0 \\ 0 \end{bmatrix}$, $\quad t_0 = 0$.

4. $\mathbf{x}(t) = \begin{bmatrix} x(t) \\ y(t) \end{bmatrix}$, $\mathbf{A}(t) = \begin{bmatrix} 3 & 0 \\ 2 & 0 \end{bmatrix}$, $\mathbf{f}(t) = \begin{bmatrix} t \\ t+1 \end{bmatrix}$, $\mathbf{c} = \begin{bmatrix} 1 \\ -1 \end{bmatrix}$, $t_0 = 0$.

5. $\mathbf{x}(t) = \begin{bmatrix} x(t) \\ y(t) \end{bmatrix}$, $\mathbf{A}(t) = \begin{bmatrix} 3t^2 & 7 \\ 1 & t \end{bmatrix}$, $\mathbf{f}(t) = \begin{bmatrix} 2 \\ 2t \end{bmatrix}$, $\mathbf{c} = \begin{bmatrix} 2 \\ -3 \end{bmatrix}$, $t_0 = 0$.

6. $\mathbf{x}(t) = \begin{bmatrix} u(t) \\ v(t) \\ w(t) \end{bmatrix}$, $\mathbf{A}(t) = \begin{bmatrix} e^t & t & 1 \\ t^2 & -3 & t+1 \\ 0 & 1 & e^{t^2} \end{bmatrix}$, $\mathbf{f}(t) = \begin{bmatrix} 0 \\ 0 \\ 0 \end{bmatrix}$, $\mathbf{c} = \begin{bmatrix} 0 \\ 1 \\ -1 \end{bmatrix}$, $t_0 = 4$.

7. $\mathbf{x}(t) = \begin{bmatrix} x(t) \\ y(t) \\ z(t) \end{bmatrix}$, $\mathbf{A}(t) = \begin{bmatrix} 0 & 6 & 1 \\ 1 & 0 & -3 \\ 0 & -2 & 0 \end{bmatrix}$, $\mathbf{f}(t) = \begin{bmatrix} 0 \\ 0 \\ 0 \end{bmatrix}$, $\mathbf{c} = \begin{bmatrix} 10 \\ 10 \\ 20 \end{bmatrix}$, $t_0 = 0$.

8.

$\mathbf{x}(t) = \begin{bmatrix} r(t) \\ s(t) \\ u(t) \end{bmatrix}$, $\mathbf{A}(t) = \begin{bmatrix} t^2 & -3 & -\sin t \\ 1 & -1 & 0 \\ 2 & e^t & t^2-1 \end{bmatrix}$, $\mathbf{f}(t) = \begin{bmatrix} \sin t \\ t^2-1 \\ \cos t \end{bmatrix}$,

$\mathbf{c} = \begin{bmatrix} 4 \\ -2 \\ 5 \end{bmatrix}$, $t_0 = 1$.

9. Only (c).

10. Only (c).

11. Only (b).

SECTION 8.2

1. $\mathbf{x}(t) = \begin{bmatrix} x_1(t) \\ x_2(t) \end{bmatrix}$, $\mathbf{A}(t) = \begin{bmatrix} 0 & 1 \\ 3 & 2 \end{bmatrix}$, $\mathbf{f}(t) = \begin{bmatrix} 0 \\ 0 \end{bmatrix}$, $\mathbf{c} = \begin{bmatrix} 4 \\ 5 \end{bmatrix}$, $t_0 = 0$.

2. $\mathbf{x}(t) = \begin{bmatrix} x_1(t) \\ x_2(t) \end{bmatrix}$, $\mathbf{A}(t) = \begin{bmatrix} 0 & 1 \\ t & -e^t \end{bmatrix}$, $\mathbf{f}(t) = \begin{bmatrix} 0 \\ 0 \end{bmatrix}$, $\mathbf{c} = \begin{bmatrix} 2 \\ 0 \end{bmatrix}$, $t_0 = 1$.

3. $\mathbf{x}(t) = \begin{bmatrix} x_1(t) \\ x_2(t) \end{bmatrix}$, $\mathbf{A}(t) = \begin{bmatrix} 0 & 1 \\ 1 & 0 \end{bmatrix}$, $\mathbf{f}(t) = \begin{bmatrix} 0 \\ t^2 \end{bmatrix}$, $\mathbf{c} = \begin{bmatrix} -3 \\ 3 \end{bmatrix}$, $t_0 = 0$.

4. $\mathbf{x}(t) = \begin{bmatrix} x_1(t) \\ x_2(t) \end{bmatrix}$, $\mathbf{A}(t) = \begin{bmatrix} 0 & 1 \\ 3 & 2e^t \end{bmatrix}$, $\mathbf{f}(t) = \begin{bmatrix} 0 \\ 2e^{-1} \end{bmatrix}$, $\mathbf{c} = \begin{bmatrix} 0 \\ 0 \end{bmatrix}$, $t_0 = 0$.

5. $\mathbf{x}(t) = \begin{bmatrix} x_1(t) \\ x_2(t) \end{bmatrix}$, $\mathbf{A}(t) = \begin{bmatrix} 0 & 1 \\ -2 & 3 \end{bmatrix}$, $\mathbf{f}(t) = \begin{bmatrix} 0 \\ e^{-1} \end{bmatrix}$, $\mathbf{c} = \begin{bmatrix} 2 \\ 2 \end{bmatrix}$, $t_0 = 1$.

6.

$\mathbf{x}(t) = \begin{bmatrix} x_1(t) \\ x_2(t) \\ x_3(t) \end{bmatrix}$, $\mathbf{A}(t) = \begin{bmatrix} 0 & 1 & 0 \\ 0 & 0 & 1 \\ 1/4 & 0 & -t/4 \end{bmatrix}$, $\mathbf{f}(t) = \begin{bmatrix} 0 \\ 0 \\ 0 \end{bmatrix}$, $\mathbf{c} = \begin{bmatrix} 2 \\ 1 \\ -205 \end{bmatrix}$, $t_0 = -1$.

7.

$\mathbf{x}(t) = \begin{bmatrix} x_1(t) \\ x_2(t) \\ x_3(t) \\ x_4(t) \end{bmatrix}$, $\mathbf{A}(t) = \begin{bmatrix} 0 & 1 & 0 & 0 \\ 0 & 0 & 1 & 0 \\ 0 & 0 & 0 & 1 \\ 0 & e^{-t} & -et^{-t} & 0 \end{bmatrix}$, $\mathbf{f}(t) = \begin{bmatrix} 0 \\ 0 \\ 0 \\ e^{-1} \end{bmatrix}$, $\mathbf{c} = \begin{bmatrix} 1 \\ 2 \\ \pi \\ e^3 \end{bmatrix}$, $t_0 = 0$.

8.

$\mathbf{x}(t) = \begin{bmatrix} x_1(t) \\ x_2(t) \\ x_3(t) \\ x_4(t) \\ x_5(t) \\ x_6(t) \end{bmatrix}$, $\mathbf{A}(t) = \begin{bmatrix} 0 & 1 & 0 & 0 & 0 & 0 \\ 0 & 0 & 1 & 0 & 0 & 0 \\ 0 & 0 & 0 & 1 & 0 & 0 \\ 0 & 0 & 0 & 0 & 1 & 0 \\ 0 & 0 & 0 & 0 & 0 & 1 \\ 0 & 0 & 0 & 0 & -4 & 0 \end{bmatrix}$, $\mathbf{f}(t) = \begin{bmatrix} 0 \\ 0 \\ 0 \\ 0 \\ 0 \\ t^2 - 1 \end{bmatrix}$, $\mathbf{c} = \begin{bmatrix} 2 \\ 1 \\ 0 \\ 2 \\ 1 \\ 0 \end{bmatrix}$, $t_0 = \pi$.

SECTION 8.3

1. $\mathbf{x}(t) = \begin{bmatrix} x_1(t) \\ x_2(t) \\ y_1(t) \end{bmatrix}$, $\mathbf{A}(t) = \begin{bmatrix} 0 & 1 & 0 \\ 3 & 2 & 4 \\ 5 & 0 & -6 \end{bmatrix}$, $\mathbf{f}(t) = \begin{bmatrix} 0 \\ 0 \\ 0 \end{bmatrix}$, $\mathbf{c} = \begin{bmatrix} 7 \\ 8 \\ 9 \end{bmatrix}$, $t_0 = 0$.

2. $\mathbf{x}(t) = \begin{bmatrix} x_1(t) \\ x_2(t) \\ y_1(t) \\ y_2(t) \end{bmatrix}$, $\mathbf{A}(t) = \begin{bmatrix} 0 & 1 & 0 & 0 \\ 0 & 1 & 0 & 1 \\ 0 & 0 & 0 & 1 \\ 0 & -1 & 0 & 1 \end{bmatrix}$, $\mathbf{f}(t) = \begin{bmatrix} 0 \\ 0 \\ 0 \\ 0 \end{bmatrix}$,

$\mathbf{c} = \begin{bmatrix} 2 \\ 3 \\ 4 \\ 4 \end{bmatrix}$, $t_0 = 0$.

3. $\mathbf{x}(t) = \begin{bmatrix} x_1(t) \\ y_1(t) \\ y_2(t) \end{bmatrix}$, $\mathbf{A}(t) = \begin{bmatrix} -4 & 0 & t^2 \\ 0 & 0 & 1 \\ t^2 & t & 0 \end{bmatrix}$, $\mathbf{f}(t) = \begin{bmatrix} 0 \\ 0 \\ 0 \end{bmatrix}$,

$\mathbf{c} = \begin{bmatrix} -1 \\ 0 \\ 0 \end{bmatrix}$, $t_0 = 2$.

4. $\mathbf{x}(t) = \begin{bmatrix} x_1(t) \\ x_1(t) \\ y_2(t) \end{bmatrix}$, $\mathbf{A}(t) = \begin{bmatrix} -4 & 0 & 2 \\ 0 & 0 & 1 \\ 3 & t & 0 \end{bmatrix}$, $\mathbf{f}(t) = \begin{bmatrix} t \\ 0 \\ -1 \end{bmatrix}$,

$\mathbf{c} = \begin{bmatrix} 0 \\ 0 \\ 0 \end{bmatrix}$, $t_0 = 3$.

5. $\mathbf{x}(t) = \begin{bmatrix} x_1(t) \\ x_2(t) \\ y_1(t) \\ y_2(t) \\ y_3(t) \\ y_4(t) \end{bmatrix}$, $\mathbf{A}(t) = \begin{bmatrix} 0 & 1 & 0 & 0 & 0 & 0 \\ 0 & 2 & 0 & 0 & 0 & 1 \\ 0 & 0 & 0 & 1 & 0 & 0 \\ 0 & 0 & 0 & 0 & 1 & 0 \\ 0 & 0 & 0 & 0 & 0 & 1 \\ t & 0 & -t & 0 & 1 & 0 \end{bmatrix}$, $\mathbf{f}(t) = \begin{bmatrix} 0 \\ -t \\ 0 \\ 0 \\ 0 \\ -e^t \end{bmatrix}$,

$\mathbf{c} = \begin{bmatrix} 2 \\ 0 \\ 0 \\ 3 \\ 9 \\ 4 \end{bmatrix}$, $t_0 = -1$.

6. $\mathbf{x}(t) = \begin{bmatrix} x_1(t) \\ x_2(t) \\ x_3(t) \\ y_1(t) \\ y_2(t) \end{bmatrix}$, $\mathbf{A}(t) = \begin{bmatrix} 0 & 1 & 0 & 0 & 0 \\ 0 & 0 & 1 & 0 & 0 \\ 1 & 0 & 0 & -1 & 1 \\ 0 & 0 & 0 & 0 & 1 \\ -1 & 0 & 1 & 0 & 2 \end{bmatrix}$, $\mathbf{f}(t) = \begin{bmatrix} 0 \\ 0 \\ 0 \\ 0 \\ 0 \end{bmatrix}$,

$\mathbf{c} = \begin{bmatrix} 21 \\ 4 \\ -5 \\ 5 \\ 7 \end{bmatrix}$, $t_0 = -0$.

7. $\mathbf{x}(t) = \begin{bmatrix} x_1(t) \\ y_1(t) \\ y_1(t) \\ z_1(t) \end{bmatrix}$, $\quad \mathbf{A}(t) = \begin{bmatrix} 0 & 1 & 0 & 0 \\ 0 & 0 & 1 & 0 \\ 0 & 0 & 0 & 1 \\ 1 & 1 & 0 & 0 \end{bmatrix}$, $\quad \mathbf{f}(t) = \begin{bmatrix} -2 \\ 0 \\ -2 \\ 0 \end{bmatrix}$,

$\mathbf{c} = \begin{bmatrix} 1 \\ 2 \\ 17 \\ 0 \end{bmatrix}$, $\quad t_0 = \pi.$

8. $\mathbf{x}(t) = \begin{bmatrix} x_1(t) \\ x_2(t) \\ y_1(t) \\ y_2(t) \\ z_1(t) \\ z_2(t) \end{bmatrix}$, $\quad \mathbf{A}(t) = \begin{bmatrix} 0 & 1 & 0 & 0 & 0 & 0 \\ 0 & 0 & 1 & 0 & 1 & 0 \\ 0 & 0 & 0 & 1 & 0 & 0 \\ 0 & 0 & 1 & 0 & 0 & 0 \\ 0 & 0 & 0 & 0 & 0 & 1 \\ 1 & 0 & 0 & 0 & -1 & 0 \end{bmatrix}$, $\quad \mathbf{f}(t) = \begin{bmatrix} 0 \\ 2 \\ 0 \\ -1 \\ 0 \\ 1 \end{bmatrix}$,

$\mathbf{c} = \begin{bmatrix} 4 \\ -4 \\ 5 \\ -5 \\ 9 \\ -9 \end{bmatrix}$, $\quad t_0 = 20.$

SECTION 8.4

3. (a) $e^{3t} \begin{bmatrix} 1 & -t & t^2/2 \\ 0 & 0 & -1 \\ 0 & 0 & - \end{bmatrix}$,

(b) $e^{3(t-2)} \begin{bmatrix} 1 & (t-2) & (t-2)^2/2 \\ 0 & 1 & (t-2) \\ 0 & 0 & 1 \end{bmatrix}$,

(c) $e^{3(t-s)} \begin{bmatrix} 1 & (t-2) & (t-s)^2/2 \\ 0 & 1 & (t-s) \\ 0 & 0 & 1 \end{bmatrix}$,

(d) $e^{-3(t-2)} \begin{bmatrix} 1 & -(t-2) & (t-2)^2/2 \\ 0 & 1 & -(t-s) \\ 0 & 0 & 1 \end{bmatrix}$,

5. (a) $\dfrac{1}{6}\begin{bmatrix} 2e^{-5t}+4e^t & 2e^{-5t}-2e^t \\ 4e^{5t}-4e^t & 4e^{-5t}+2e^t \end{bmatrix}$,

(b) $\dfrac{1}{6}\begin{bmatrix} 2e^{-5s}+4e^s & 2e^{-5s}-2e^s \\ 4e^{5s}-4e^s & 4e^{-5s}+2e^s \end{bmatrix}$,

(c) $\dfrac{1}{6}\begin{bmatrix} 2e^{5(t-3)}+4e^{-(t-3)} & 2e^{5(t-3)}-2e^{-(t-3)} \\ 4e^{5(t-3)}-4e^{-(t-3)} & 4e^{5(t-3)}+2e^{-(t-3)} \end{bmatrix}$.

6. (a) $\dfrac{1}{3}\begin{bmatrix} \sin 3t+3\cos 3t & -5\sin 3t \\ 2\sin 3t & -\sin 3t+3\cos 3t \end{bmatrix}$,

(b) $\dfrac{1}{3}\begin{bmatrix} \sin 3s+3\cos 3s & -5\sin 3s \\ 2\sin 3s & -\sin 3t+3\cos 3s \end{bmatrix}$,

(c) $\dfrac{1}{3}\begin{bmatrix} \sin 3(t-s)+3\cos 3(t-s) & -5\sin 3(t-s) \\ 2\sin 3(t-s) & -\sin 3(t-s)+3\cos 3(t-s)t \end{bmatrix}$,

7. $x(t)=5e^{(t-2)}-3e^{-(t-2)}$, $y(t)=5e^{(t-2)}-e^{-(t-2)}$.

8. $x(t)=2e^{(t-1)}-1$, $y(t)=2e^{(t-1)}-1$.

9. $x(t)=k_3e^t+3k_4e^t$, $y(t)=k_3e^t+k_4e^{-t}$.

10. $x(t)=k_3e^t+3k_4e^{-t}-1$, $y(t)=k_3e^t+k_4e^{-t}-1$.

11. $x(t)=\cos 2t-(1/6)\sin 2t+(1/3)\sin t$.

12. $x(t)=t^4/24+(5/4)t^2-(2/3)t+3/8$.

13. $x(t)=(4/9)\,e^{2t}+(5/9)\,e^{-1t}-(1/3)\,te^{-1t}$.

14. $x(t)=-8\cos t-6\sin t+8+6t$, $y(t)=4\cos t-2\sin t-3$.

SECTION 8.5

4. First show that

$$\boldsymbol{\Phi}^{\mathrm{T}}(t_1,t_0)\left[\int_{t_0}^{t_1}\boldsymbol{\Phi}(t_1,s)\ \boldsymbol{\Phi}^{\mathrm{T}}(t_1,s)ds\right]^{-1}\boldsymbol{\Phi}(t_1,t_0)$$

$$=\left[\boldsymbol{\Phi}(t_0,t_1)\int_{t_0}^{t_1}\boldsymbol{\Phi}(t_1,s)\ \boldsymbol{\Phi}'(t_1,s)ds\boldsymbol{\Phi}^{\mathrm{T}}(t_0,t_1)\right]^{-1}$$

$$=\left[\int_{t_0}^{t_1}\boldsymbol{\Phi}(t_0,t_1)\boldsymbol{\Phi}(t_1,s)[\boldsymbol{\Phi}(t_0,t_1)\boldsymbol{\Phi}(t_1,s)]^{\mathrm{T}}ds\right]^{-1}.$$

CHAPTER 9
SECTION 9.1

1. (a) The English alphabet: *a*, *b*, *c*, … *x*, *y*, *z*. 26. 5/26.
   (b) The 366 days designated by a 2008 calendar, ranging from January 1 through December 31. 366. 1/366.
   (c) A list of all 43 United States presidents. 43. 1/43.
   (d) Same as (c). 43. 2/43 (Grover Cleveland was both the 22nd and 24th president).
   (e) Regular deck of 52 cards. 52. 1/52.
   (f) Pinochle deck of 48 cards. 48. 2/48.
   (g) See Fig. 9.1 of Chapter 9. 36. 1/36.
   (h) Same as (g).
   (i) Same as (g). 5/36.
   (j) Same as (g). 2/36.
   (k) Same as (g). 18/36.
   (l) Same as (g). 7/36.
   (m) Same as (g). 5/36.
   (n) Same as (g). 12/36.
   (o) Same as (n).
   (p) Same as (g). 0.

2. The sample space would consist of all 216 possibilities, ranging from rolling a "3" to tossing an "18".

3. 90.

4. 1950.

SECTION 9.2

1. (a) 8/52.
   (b) 16/52.
   (c) 28/52.
   (d) 2/52.
   (e) 28/52.
   (f) 26/52.
   (g) 39/52.
   (h) 48/52.
   (i) 36/52.

2. (a) 18/36.
   (b) 15/36.
   (c) 10/36.
   (d) 30/36.
   (e) 26/36.
   (f) 1.

3. (a) 108/216.
   (b) 1/216.
   (c) 1/216.
   (d) 3/216.
   (e) 3/216.
   (f) 0.
   (g) 213/216.
   (h) 210/216.
   (i) 206/216.

4. 0.75.

5. 0.4.

6. $P(A \cup B \cup C \cup D) = P(A) + P(B) + P(C) + P(D) - P(A \cap B) - P(A \cap C) - P(A \cap D) - P(B \cap C) - P(B \cap D) - P(C \cap D) + P(A \cap B \cap C) + P(A \cap B \cap D) + P(A \cap C \cap D) + P(B \cap C \cap D) - P(A \cap B \cap C \cap D).$

SECTION 9.3

1. (a) 15.
   (b) 7.
   (c) 56.
   (d) 190.
   (e) 190.
   (f) 1.
   (g) 1.
   (h) 100.
   (i) 1000.
   (j) 1.

2. 2,042,975.

3. 5005.

4. (a) Approximately 0.372.

(b) Approximately 0.104.

(c) Approximately 0.135.

(d) Approximately 0.969.

(e) Approximately 0.767.

5. (a) $\binom{500}{123}\ (.65)^{123}(.35)^{377}.$

(b) $\binom{500}{485}\ (.65)^{485}(.35)^{15}.$

(c) $\binom{500}{497}\ (.65)^{497}(.35)^{3} + \binom{500}{498}\ (.65)^{498}(.35)^{2}$

$+ \binom{500}{499}\ (.65)^{499}(.35)^{1} + \binom{500}{500}\ (.65)^{500}(.35)^{0}.$

(d) $1 - \binom{500}{498}\ (.65)^{498}(.35)^{2} - \binom{500}{499}\ (.65)^{499}(.35)^{1}$

$- \binom{500}{500}\ (.65)^{500}(.35)^{0}.$

(e) $\binom{500}{100}\ (.65)^{100}(.35)^{400} + \binom{500}{200}\ (.65)^{200}(.35)^{300}$

$+ \binom{500}{300}\ (.65)^{300}(.35)^{200} + \binom{500}{400}\ (.65)^{400}(.35)^{100}$

$+ \binom{500}{500}\ (.65)^{500}(.35)^{0}.$

6. Approximately 0.267.

7. Approximately 0.267.

SECTION 9.4

1. (a) There is a negative element in the second row.
   (b) The first row does not add to 1.
   (c) The third row does not add to 1.
   (d) It is not a square matrix.

2. (a) If it is sunny today, there is a probability of 0.5 that it will be sunny tomorrow and a 0.5 probability that it will rain tomorrow. If it rains today, there is a 0.7 probability that it will be sunny tomorrow and a 0.3 chance that it will rain tomorrow.
   (b) If a parking meter works today, there is a probability of 0.95 that it will work tomorrow with a 0.05 probability that it will not work tomorrow. If the parking meter is inoperative today, there is a probability of 0.02 that it will be fixed tomorrow and a 0.98 probability that it will not be fixed tomorrow.
   (c) Any scenario has a "50–50" chance at any stage.
   (d) What is "good" stays "good"; what is "bad" stays "bad".
   (e) What is "good" today is "bad" tomorrow; what is "bad" today is "good" tomorrow.
   (f) See Example 2 in Section 9.4 and use Tinker, Evers, and Chance for Moe, Curly, and Larry and instead of visiting or staying home use "borrowing a car" or "not borrowing a car".

3. Clearly if we raise either matrix to any power, we obtain the original matrix.

4. The even powers produce $\begin{bmatrix} 1 & 0 \\ 0 & 1 \end{bmatrix}$ and the odd powers give back the original matrix. And the situation repeats itself after an even number of time periods.

5. $p^{(2)}_{11} = 0.4$, $p^{(2)}_{21} = 0.15$, $p^{(3)}_{12} = 0.7$, $p^{(3)}_{22} = 0.825$.

6. (a) $\begin{bmatrix} .7 & .3 \\ .4 & .6 \end{bmatrix}$.
   (b) Approximately 0.5725.
   (c) Approximately 0.5717.

CHAPTER 10
SECTION 10.1

1. 11, 5.
2. 8, 4.

3. −50, 74.

4. 63, 205.

5. 64, 68.

6. 6, 5.

7. 26, 24.

8. −30, 38.

9. 5/6, 7/18.

10. $5/\sqrt{6}$, 1.

11. 7/24, 1/3.

12. 0, 1400.

13. 2, 3.

14. 1, 1.

15. −19, 147.

16. −1/5, 1/5.

17. Undefined, 6.

18. $\begin{bmatrix} 3/5 \\ 4/5 \end{bmatrix}.$

19. $\begin{bmatrix} 4/\sqrt{41} \\ -5/\sqrt{41} \end{bmatrix}.$

20. $\begin{bmatrix} 7/\sqrt{58} & 3/\sqrt{58} \end{bmatrix}.$

21. $\begin{bmatrix} -4/\sqrt{34} \\ 3/\sqrt{34} \\ -3/\sqrt{34} \end{bmatrix}.$

22. $\begin{bmatrix} 3/\sqrt{17} \\ -2/\sqrt{17} \\ -2/\sqrt{17} \end{bmatrix}.$

23. $\left[2\big/\sqrt{21} \quad 4\big/\sqrt{21} \quad 1\big/\sqrt{21} \right]$.

24.
$$
\begin{bmatrix}
4\big/\sqrt{197} \\
-6\big/\sqrt{197} \\
-9\big/\sqrt{197} \\
8\big/\sqrt{197}
\end{bmatrix}.
$$

25. $\left[1\big/\sqrt{55} \quad 2\big/\sqrt{55} \quad -3\big/\sqrt{55} \quad 4\big/\sqrt{55} \quad -5\big/\sqrt{55} \right]$.

26. $\left[-3\big/\sqrt{259} \quad 8\big/\sqrt{259} \quad 11\big/\sqrt{259} \quad -4\big/\sqrt{259} \quad 7\big/\sqrt{259} \right]$.

27. No vector **x** exists.

28. Yes, see Problem 12.

33. $\begin{aligned}\|\mathbf{x} + \mathbf{y}\|^2 &= \langle \mathbf{x} + \mathbf{y}, \, \mathbf{x} + \mathbf{y} \rangle = \langle \mathbf{x}, \, \mathbf{x} \rangle + 2\langle \mathbf{x}, \, \mathbf{y} \rangle + \langle \mathbf{y}, \, \mathbf{y} \rangle \\ &= \|\mathbf{x}\|^2 + 2\langle \mathbf{x}, \mathbf{y} \rangle + \|\mathbf{y}\|^2. \end{aligned}$

34. Show that $\|\mathbf{x} - \mathbf{y}\|^2 = \|\mathbf{x}\|^2 - 2\langle \mathbf{x}, \mathbf{y} \rangle + \|\mathbf{y}\|^2$. And then use Problem 33.

37. Note that $\langle \mathbf{x}, \, \mathbf{y} \rangle \le \ |\langle \mathbf{x}, \, \mathbf{y} \rangle|$.

38. $\langle \mathbf{x}, \mathbf{y} \rangle = \det(\mathbf{x}^{\mathsf{T}}\mathbf{y})$.

40. 145.

41. 27.

42. 32.

SECTION 10.2

1. **x** and **y**, **x** and **u**, **y** and **v**, **u** and **v**.
2. **x** and **z**, **x** and **u**, **y** and **u**, **z** and **u**, **y** and **v**.
3. $-20/3$.
4. -4.
5. 0.5.

6. $x = -3y$.

7. $x = 1, y = -2$.

8. $x = y = -z$.

9. $x = y = -z;\ z = \pm 1/\sqrt{3}$.

10. $\begin{bmatrix} 1/\sqrt{5} \\ 2/\sqrt{5} \end{bmatrix}, \begin{bmatrix} 2/\sqrt{5} \\ -1/\sqrt{5} \end{bmatrix}$.

11. $\begin{bmatrix} 1/\sqrt{2} \\ 1/\sqrt{2} \end{bmatrix}, \begin{bmatrix} -1/\sqrt{2} \\ 1/\sqrt{2} \end{bmatrix}$.

12. $\begin{bmatrix} 3/\sqrt{15} \\ -2/\sqrt{15} \end{bmatrix}, \begin{bmatrix} 2/\sqrt{13} \\ 3/\sqrt{13} \end{bmatrix}$.

13. $\begin{bmatrix} 1/\sqrt{6} \\ 2/\sqrt{6} \\ 1/\sqrt{6} \end{bmatrix}, \begin{bmatrix} 1/\sqrt{3} \\ -1/\sqrt{3} \\ 1/\sqrt{3} \end{bmatrix}, \begin{bmatrix} -1/\sqrt{2} \\ 0 \\ 1/\sqrt{2} \end{bmatrix}$.

14. $\begin{bmatrix} 2/\sqrt{2} \\ 1/\sqrt{2} \\ 0 \end{bmatrix}, \begin{bmatrix} -2/\sqrt{45} \\ 4/\sqrt{45} \\ 5/\sqrt{45} \end{bmatrix}, \begin{bmatrix} 1/3 \\ -2/3 \\ 2/3 \end{bmatrix}$.

15. $\begin{bmatrix} 1/\sqrt{2} \\ 1/\sqrt{2} \\ 0 \end{bmatrix}, \begin{bmatrix} 1/\sqrt{3} \\ -1/\sqrt{3} \\ 1/\sqrt{3} \end{bmatrix}, \begin{bmatrix} -1/\sqrt{6} \\ 1/\sqrt{6} \\ 2/\sqrt{6} \end{bmatrix}$.

16. $\begin{bmatrix} 0 \\ 3/5 \\ 4/5 \end{bmatrix}, \begin{bmatrix} 3/5 \\ 16/25 \\ -12/25 \end{bmatrix}, \begin{bmatrix} 4/5 \\ -12/25 \\ 9/25 \end{bmatrix}$.

17. $\begin{bmatrix} 0 \\ 1/\sqrt{3} \\ 1/\sqrt{3} \\ 1/\sqrt{3} \end{bmatrix}, \begin{bmatrix} 3/\sqrt{15} \\ -2/\sqrt{15} \\ 1/\sqrt{15} \\ 1/\sqrt{15} \end{bmatrix}, \begin{bmatrix} 3/\sqrt{35} \\ 3/\sqrt{35} \\ -4/\sqrt{35} \\ 1/\sqrt{35} \end{bmatrix}, \begin{bmatrix} 1/\sqrt{7} \\ 1/\sqrt{7} \\ 1/\sqrt{7} \\ -2/\sqrt{7} \end{bmatrix}.$

18. $\begin{bmatrix} 1/\sqrt{2} \\ 1/\sqrt{2} \\ 0 \\ 0 \end{bmatrix}, \begin{bmatrix} -1/\sqrt{6} \\ 1/\sqrt{6} \\ -2/\sqrt{6} \\ 0 \end{bmatrix}, \begin{bmatrix} 1/\sqrt{3} \\ -1/\sqrt{3} \\ -1/\sqrt{3} \\ 0 \end{bmatrix}, \begin{bmatrix} 0 \\ 0 \\ 0 \\ -1 \end{bmatrix}.$

23. $\|\mathbf{x} - \mathbf{y}\|^2 = \langle \mathbf{x} - \mathbf{y}, \mathbf{x} - \mathbf{y} \rangle = \|\mathbf{x}\|^2 - 2\langle \mathbf{x}, \mathbf{y} \rangle + \|\mathbf{y}\|^2.$

24. $\|s\mathbf{x} + t\mathbf{y}\|^2 = \langle s\mathbf{x} - t\mathbf{y}, s\mathbf{x} - t\mathbf{y} \rangle = \|s\mathbf{x}\|^2 - 2st\langle \mathbf{x}, \mathbf{y} \rangle + \|t\mathbf{y}\|^2.$

25. I.

26. Set $\mathbf{y} = \mathbf{x}$ and use Property (**I1**) of Section 10.1.

28. Denote the columns of $\mathbf{A}$ as $\mathbf{A}_1, \mathbf{A}_2, \dots, \mathbf{A}_n$, and the elements of $\mathbf{y}$ as $y_1, y_2, \dots, y_n$, respectively. Then, $\mathbf{A}\mathbf{y} = \mathbf{A}_1 y_1 + \mathbf{A}_2 y_2 + \dots + \mathbf{A}_n y^n$ and $\langle \mathbf{A}\mathbf{y}, \mathbf{p} \rangle = y^1 \langle \mathbf{A}^1, \mathbf{p} \rangle + y^2 \langle \mathbf{A}^2, \mathbf{p} \rangle + \dots + y^n \langle \mathbf{A}^n, \mathbf{p} \rangle$

SECTION 10.3

1. (a) $\theta = 36.9$ degrees,

(b) $\begin{bmatrix} 1.6 \\ 0.8 \end{bmatrix}$,

(c) $\begin{bmatrix} -0.6 \\ 1.2 \end{bmatrix}$.

2. (a) $\theta = 14.0$ degrees,

(b) $\begin{bmatrix} 0.7059 \\ 1.1765 \end{bmatrix}$,

(c) $\begin{bmatrix} -0.2941 \\ -0.1765 \end{bmatrix}$.

3. (a) $\theta = 78.7$ degrees,

(b) $\begin{bmatrix} 0.5 \\ 0.5 \end{bmatrix}$,

(c) $\begin{bmatrix} 2.5 \\ -2.5 \end{bmatrix}$.

4. (a) $\theta = 90$ degrees,

(b) $\begin{bmatrix} 0 \\ 0 \end{bmatrix}$,

(c) $\begin{bmatrix} 4 \\ -1 \end{bmatrix}$.

5. (a) $\theta = 118.5$ degrees,

(b) $\begin{bmatrix} -0.7529 \\ 3.3882 \end{bmatrix}$,

(c) $\begin{bmatrix} -6.2471 \\ 1.3882 \end{bmatrix}$.

6. (a) $\theta = 50.8$ degrees,

(b) $\begin{bmatrix} 1 \\ 0 \\ 1 \end{bmatrix}$,

(c) $\begin{bmatrix} 1 \\ 1 \\ -1 \end{bmatrix}$.

7. (a) $\theta = 19.5$ degrees,

(b) $\begin{bmatrix} 8/9 \\ 8/9 \\ 4/9 \end{bmatrix}$,

(c) $\begin{bmatrix} 1/9 \\ 1/9 \\ -4/9 \end{bmatrix}$.

8. (a) $\theta = 17.7$ degrees,

(b) $\begin{bmatrix} 1.2963 \\ 3.2407 \\ 3.2407 \end{bmatrix}$,

(c) $\begin{bmatrix} -1.2963 \\ -0.2407 \\ 0.7593 \end{bmatrix}$.

9. (a) $\theta = 48.2$ degrees,

(b) $\begin{bmatrix} 2/3 \\ 2/3 \\ 2/3 \\ 0 \end{bmatrix}$,

(c) $\begin{bmatrix} -2/3 \\ 1/3 \\ 1/3 \\ 1 \end{bmatrix}$.

10. (a) $\theta = 121.4$ degrees,

(b) $\begin{bmatrix} -7/6 \\ 7/3 \\ 0 \\ 7/6 \end{bmatrix}$,

(c) $\begin{bmatrix} -13/6 \\ -1/3 \\ 3 \\ 17/6 \end{bmatrix}$.

11. $\begin{bmatrix} 0.4472 & 0.8944 \\ 0.8944 & -0.4472 \end{bmatrix} \begin{bmatrix} 2.2361 & 1.7889 \\ 0.000 & 1.3416 \end{bmatrix}$.

12. $\begin{bmatrix} 0.7071 & -0.7071 \\ 0.7071 & 0.7071 \end{bmatrix} \begin{bmatrix} 1.4142 & 5.6569 \\ 0.0000 & 1.4142 \end{bmatrix}$.

13. $\begin{bmatrix} 0.8321 & 0.5547 \\ -0.5547 & 0.8321 \end{bmatrix} \begin{bmatrix} 3.6056 & 0.8321 \\ 0.0000 & 4.1603 \end{bmatrix}$.

14. $\begin{bmatrix} 0.3333 & 0.8085 \\ 0.6667 & 0.1617 \\ 0.6667 & -0.5659 \end{bmatrix} \begin{bmatrix} 3.0000 & 2.6667 \\ 0.0000 & 1.3744 \end{bmatrix}$.

15. $\begin{bmatrix} 0.3015 & -0.2752 \\ 0.3015 & 0.2752 \\ 0.9045 & 0.3853 \end{bmatrix} \begin{bmatrix} 3.3166 & 4.8242 \\ 0.0000 & 1.6514 \end{bmatrix}$.

16. $\begin{bmatrix} 0.7746 & 0.4043 \\ -0.5164 & 0.5714 \\ 0.2582 & 0.4706 \\ -0.2582 & 0.5378 \end{bmatrix} \begin{bmatrix} 3.8730 & 0.2582 \\ 0.0000 & 1.9833 \end{bmatrix}$.

17. $\begin{bmatrix} 0.8944 & -0.2981 & 0.3333 \\ 0.4472 & 0.5963 & -0.6667 \\ 0.0000 & 0.7454 & 0.6667 \end{bmatrix} \begin{bmatrix} 2.2361 & 0.4472 & 1.7889 \\ 0.0000 & 1.3416 & 0.8944 \\ 0.0000 & 0.0000 & 2.0000 \end{bmatrix}$.

18. $\begin{bmatrix} 0.7071 & -0.5774 & -0.4082 \\ 0.7071 & -0.5774 & 0.4082 \\ 0.0000 & 0.5774 & 0.8165 \end{bmatrix} \begin{bmatrix} 1.4142 & 1.4142 & 2.8284 \\ 0.0000 & 1.7321 & 0.5774 \\ 0.0000 & 0.0000 & 0.8165 \end{bmatrix}$.

19. $\begin{bmatrix} 0.00 & 0.60 & 0.80 \\ 0.60 & 0.64 & -0.48 \\ 0.80 & -0.48 & 0.36 \end{bmatrix} \begin{bmatrix} 5 & 3 & 07 \\ 0 & 5 & 2 \\ 0 & 0 & 1 \end{bmatrix}$.

20. $\begin{bmatrix} 0.0000 & 0.7746 & 0.5071 \\ 0.5774 & -0.5164 & 0.5071 \\ 0.5774 & 0.2582 & 0.6761 \\ 0.5774 & 0.2582 & 0.1690 \end{bmatrix} \begin{bmatrix} 1.7321 & 1.1547 & 1.1547 \\ 0.0000 & 1.2910 & 0.5164 \\ 0.0000 & 0.0000 & 1.1832 \end{bmatrix}$.

21. $\begin{bmatrix} 0.7071 & -0.4082 & 0.5774 \\ 0.7071 & 0.4082 & -0.5774 \\ 0.0000 & -0.8165 & -0.5774 \\ 0.0000 & 0.0000 & 0.0000 \end{bmatrix} \begin{bmatrix} 1.4142 & 0.7071 & 0.7071 \\ 0.0000 & 1.2247 & 0.4082 \\ 0.0000 & 0.0000 & 1.1547 \end{bmatrix}$.

24. QR $\neq$ **A**.

SECTION 10.4

1. $\mathbf{A}_1 = \mathbf{R}_0 \mathbf{Q}_0 + 7\mathbf{I}$

$$= \begin{bmatrix} 19.3132 & -1.2945 & 0.0000 \\ 0.0000 & 7.0231 & -0.9967 \\ 0.0000 & 0.0000 & 0.0811 \end{bmatrix} \begin{bmatrix} -0.3624 & 0.07565 & 0.9289 \\ 0.0000 & -0.9967 & 0.0811 \\ 0.9320 & 0.0294 & 0.3613 \end{bmatrix}$$

$$+ 7 \begin{bmatrix} 1 & 0 & 0 \\ 0 & 1 & 0 \\ 0 & 0 & 1 \end{bmatrix} = \begin{bmatrix} 0.0000 & 2.7499 & 17.8357 \\ -0.9289 & -0.0293 & 0.2095 \\ 0.0756 & 0.0024 & 7.0293 \end{bmatrix}.$$

2. $\mathbf{A}_1 = \mathbf{R}_0\mathbf{Q}_0 - 14\mathbf{I}$

$$= \begin{bmatrix} 24.3721 & -17.8483 & 3.8979 \\ 0.0000 & 8.4522 & -4.6650 \\ 0.0000 & 0.0000 & 3.6117 \end{bmatrix} \begin{bmatrix} 0.6565 & -0.6250 & 0.4233 \\ -0.6975 & -0.2898 & 0.6553 \\ 0.2872 & 0.7248 & 0.6262 \end{bmatrix}$$

$$-14 \begin{bmatrix} 1 & 0 & 0 \\ 0 & 1 & 0 \\ 0 & 0 & 1 \end{bmatrix} = \begin{bmatrix} 15.5690 & -7.2354 & 1.0373 \\ -7.2354 & -19.8307 & 2.6178 \\ 1.0373 & 2.6178 & -11.7382 \end{bmatrix}.$$

3. Shift by 4.

$$\mathbf{R}_0 = \begin{bmatrix} 4.1231 & -0.9701 & 0.0000 & 13.5820 \\ 0.0000 & 4.0073 & -0.9982 & -4.1982 \\ 0.0000 & 0.0000 & 4.0005 & 12.9509 \\ 0.0000 & 0.0000 & 0.0000 & 3.3435 \end{bmatrix},$$

$$\mathbf{Q}_0 = \begin{bmatrix} -0.9701 & -0.2349 & -0.0586 & -0.0151 \\ 0.2425 & -0.9395 & -0.2344 & -0.0605 \\ 0.0000 & 0.2495 & -0.9376 & -0.2421 \\ 0.0000 & 0.0000 & 0.2500 & -0.9683 \end{bmatrix}.$$

$$\mathbf{A}_1 = \mathbf{R}_0\mathbf{Q}_0 + 4\mathbf{I} = \begin{bmatrix} -0.2353 & -0.0570 & 3.3809 & -13.1546 \\ 0.9719 & -0.0138 & -1.0529 & 4.0640 \\ 0.0000 & 0.9983 & 3.4864 & -13.5081 \\ 0.0000 & 0.0000 & 0.8358 & 0.7626 \end{bmatrix}.$$

4. $7.2077, \quad -0.1039 \pm 1.5769i.$

5. $-11, \quad -22, \quad 17.$

6. $2, \ 3, \ 9.$

7. Method fails. $\mathbf{A}_0 - 7\mathbf{I}$ does not have linearly independent columns, so no $\mathbf{QR}$-decomposition is possible.

8. $2, \ 2, \ 16.$

9. $1, \ 3, \ 3.$

10. $2, \ 3 \pm i$.

11. $1, \ \pm i$.

12. $\pm i, \ 2 \pm 3i$.

13. $3.1265 \ \pm 1.2638i, \ -2.6265 \pm 0.7590i$.

14. $0.0102, \ 0.8431, \ 3.8581, \ 30.887$.

SECTION 10.5

1. $x = 2.225, y = 1.464$.
2. $x = 3.171, y = 2.286$.
3. $x = 9.879, y = 18.398$.
4. $x = -1.174, y = 8.105$.
5. $x = 1.512, y = 0.639, z = 0.945$.
6. $x = 7.845, y = 1.548, z = 5.190$.
7. $x = 81.003, y = 50.870, z = 38.801$.
8. $x = 2.818, y = -0.364, z = -1.364$.
9. 2 and 4.

10. (b) $y = 2.3x + 8.1$,
   (c) 21.9.

11. (b) $y = -2.6x + 54.4$,
   (c) 31 in week 9, 28 in week 10.

12. (b) $y = 0.27x + 10.24$,
   (c) 12.4.

13. $m = \dfrac{N\sum\limits_{i=1}^{N} x_i y_i - \sum\limits_{i=1}^{N} x_i \sum\limits_{i=1}^{N} y_i}{N\sum\limits_{i=1}^{N} x_i^2 - \left(\sum\limits_{i=1}^{N} x_i\right)^2}, c = \dfrac{\sum\limits_{i=1}^{N} y_i \sum\limits_{i=1}^{N} x_i^2 - \sum\limits_{i=1}^{N} x_i \sum\limits_{i=1}^{N} x_i y_i}{N\sum\limits_{i=1}^{N} x_i^2 - \left(\sum\limits_{i=1}^{N} x_i\right)^2}.$

If $N\sum_{i=1}^{N} x_i^2$ is near $\left(\sum_{i=1}^{N} x_i\right)^2$, then the denominator is near zero.

14. $\sum_{i=1}^{N} x'_i = 0$, so the denominator for **m** and **c** as suggested in Problem 13 is simply $N\sum_{i=1}^{N}(x'_i)^2$.

15. $y = 2.3x + 15$.

16. $y = -2.6x + 4s2.9.$

17. (a) $y = -0.198x' + 21.18,$
   (b) Year 2000 is coded as $x' = 30$; $y(30) = 15.2.$

23. $\mathbf{E} = \begin{bmatrix} 0.841 \\ 0.210 \\ -2.312 \end{bmatrix}.$

24. $\mathbf{E} = \begin{bmatrix} 0.160 \\ 0.069 \\ -0.042 \\ -0.173 \end{bmatrix}.$

CHAPTER 11

Note: In this chapter all statistics are taken from https://www.baseball-reference.com/

1. **1.** Carefully *identify* the question(s) or problem(s) to be discussed.
   2. Get all *relevant* information.
   3. Use as many *measures* and *instruments* as possible.
   4. Use *relativity* and *normalization* arguments when comparing players of different eras. The effects of specific ballparks should be included in this part of the analysis as well.
   5. Exercise *care* with regard to:
   - Labels (e.g., HR/AB vs. AB/HR)
   - Definitions (e.g., ratio vs. differential)
   - Interpolation versus extrapolation
   - Projections
   - Etc.
   6. *Interpret* the results. Do various approaches seem to *converge* to an objective and plausible conclusion? Can these results be *extended*?
   7. Consider the historical context with regard to technological developments, medical advances, etc.
   8. Consider any nonsabermetric factors germane to the analysis.
   9. Be open to further discussions and developments.
   10. Realize the limitations of a "sabermetric proof".

2. The ranking will probably be: Robinson, Killebrew, Colavito, and Cash. Note that Robinson leads in HR (586), RBI (1812), and TB (5373).

3. Lou Gehrig has a lifetime MSLG of 0.6857, one of the highest in history. Foxx and Greenberg would be slightly less, while McCovey and Stargell would be a notch or so under Foxx and Greenberg.

4. The article can be accessed at http://www.jstor.org/stable/167156

5. Over their careers, Walter Johnson and Lefty Grove put up some great numbers, including W, SHO, PR, and MWPR. Sandy Koufax was brilliant for a 5–6 year period during the early 1960s. And Pedro Martinez had a spectacular hurling record during the steroid era. All things considered, it seems that Johnson or Grove would get the nod over all others.

6. One of the three great scandals in baseball history, along with the color barrier and the steroid era. Question: Should Joe Jackson be enshrined in the Hall of Fame?

7. In his career, Donnie Baseball had 588 BB and 7003 AB, for a BB + AB total of 7591 plate appearances. He slugged 222 HR. So, to give Mattingly 12,000 plate appearances in the same proportion of AB to BB, we prorate to determine that 4068 additional AB are required. Since we are assuming Mattingly would be on a 5% decline, his EC becomes:

$$1 + (4068/7003)(.95) = 1.55.$$

Hence, his projected HR total is: $1.55(222) = 345$ HR.

8. Care must be taken when comparing a scale that ranges from 0.000 to 1.000 as BA does, vis-à-vis the SLG scale, which runs a gamut from 0.000 to 4.000. Nevertheless, it seems that Ruth was a greater slugger than Cobb was as a hitter. For example, in 1920, Ruth slugged 0.847, while the league slugged 0.387, giving Ruth a ratio of 2.189, which could be interpreted as Ruth being better than "twice the average slugger" for 1920 ALers. Cobb's highest "batting ratio" was 1.584, when he hit 0.385 and the league hit 0.243. By the way, Ruth's lowest title ratio was 1.688 in 1922, when he slugged only 0.672.

9. Using Excel, we can easily construct a scatter plot, and determine that the straight line that "best fits" is given by: RBI $= 1.9887$ HR $+ 34.589$, with an R^2 value of 0.7484 (which very roughly indicates a $3/4$ fit).

10. Through 2018, Albert Pujols put up some remarkable numbers including approximately 630 HR and 1980 RBI. However, Mike Trout accumulated 240 HR and nearly 650 RBI through 2018, at the age of 26 years old. If Trout continues to perform at this same playing level for the next 10 years or so, and is relatively injury free, he may surpass even Prince Albert.

11. The 2018 NL Cy Young Award winner had a fantastic season last year, and has a lifetime W-L record of only 57-43 but will turn 31 years old this year. He is certainly a top-notch pitcher but has to put up a least a few more very good years to rank with the greatest hurlers.

12. Absolutely. Hodges retired with 370 HR. At the time of his retirement in 1963, this number was the most ever by a right-handed hitter in the NL, and second best for a right-handed slugger (only to Jimmie Foxx's 534 circuit clouts). Hodges anchored a great Brooklyn Dodger infield in the 1950s and was a top RBI man. When one considers Hodges's record as a manager, first with the Washington Senators and then with the 1969 Miracle Mets, it is difficult to understand why this individual has not been admitted into the Hall of Fame at Cooperstown.

13. Because Williams was younger, and seemed to be less fragile, it would seem that the Red Sox would have received the better of the deal. This is reinforced by the fact that Williams retired in 1960, 9 years after DiMaggio.

14. (a) Consider Barry Bonds, Babe Ruth, Stan Musial, and Hank Aaron.
(b) Consider Steve Carlton, Walter Johnson, Whitey Ford, and Pedro Martinez.
(c) Consider Mark McGwire, Babe Ruth, and Mike Schmidt.

15. (a) Consider Ed Walsh, Christy Mathewson, Grover Cleveland Alexander, Mariano Rivera, Clayton Kershaw, and Pedro Martinez.
(b) Consider Babe Ruth, Jimmie Foxx, Lou Gehrig, and Barry Bonds.
(c) Consider Babe Ruth and Ted Williams.

16. Honus Wagner ranks among the 10 greatest position players ever, perhaps in the top five. It is no insult to Jeter to consider him as the second greatest of all shortstops.

17. The Gray Eagle leads all MLers in lifetime doubles with 792. In the "Dead Ball" era, Speaker would on occasion be involved in unassisted double plays. With more than 3500 H and a 0.345 BA, he certainly was a great all-around player. Speaker was also the playing-manager for the 1920 World Champion Cleveland Indians.

The Yankee Clipper played in a different era. He exhibited more power than Speaker but had far fewer stolen bases (30–436). Both were unequivocally the "captains" of their respective outfields.

One is tempted to call it a standoff, while adding Willie Mays into the mix, settling for the assertion that they were the three greatest defensive centerfielders in the Hall of Fame.

18. After all is said and done, the sabermetric community generally agrees that one of the following two candidates must prevail: Ted Williams or Babe Ruth. The nod goes to the Sultan of Swat due to his transcendence of the game and to his records that still endure. For example, in 1921, Ruth played in 152 G, scored 177 R, had 119 extra-base hits and 457 TB. These last three statistics are still ML records.

    It has been pointed out that Williams lost valuable time during two tours of military duty. Indeed, Williams logged about 7700 career AB. However, Ruth only had around 700 additional AB in his career (due to his years as being primarily a pitcher, illness, or suspensions). The fact that Williams lost *prime* years has been taken into consideration with the development of the *equivalence coefficient*.

    By the way, Ruth and Williams are first and second in lifetime OPS (1.164 and 1.116).

    The reader is referred to Issue 23 of *Reasoning with Sabermetrics*, Costa, Huber, and Saccoman (978-0-7864-6071-7) and to inning 3 of *Understanding Sabermetrics*, Costa, Huber, and Saccoman (978-0-7864-3388-9).

19. Babe Ruth was also a great pitcher. He won 23 + G twice and held the World Series consecutive scoreless inning streak (29 + IP) for 40 years. In 10 seasons of pitching, Ruth had 10 winning seasons.

    Ruth was an excellent fielder (see Tris Speaker's comment on Ruth's fielding on page 282 of *The Year Babe Ruth Hit 104 Home Runs*, Jenkinson (978-0-78,671-906-8)) and stole Home 10 times in his career.

    Regarding his hitting, see answer 18 above.

    There is a saying in analytics, which may be considered as the Fundamental Theorem of Sabermetrics: *"If one does an analysis and Babe Ruth does not come out on top, there is something wrong with the analysis."*

    Who was baseball's second greatest player ever? Some candidates are: Willie Mays, Joe DiMaggio, Henry Aaron, Honus Wagner, Ty Cobb … Mike Trout(?).

20. Sabermetrics gives one plausibility to compare player with player, era with era. In baseball, unlike football or basketball, the biggest, fastest, or strongest players do not always win. Before the advent of sabermetrics, there was a static—frozen—aspect to the statistics. For example, Ty Cobb hit 0.366 lifetime, so he *had* to be the greatest hitter … "Not so!" proclaims the sabermetrician. "Let's look deeper … how about Bases on Balls? Etc."

    And let's have fun!

CHAPTER 12

Note: In this chapter all statistics are taken from https://www.baseball-reference.com/

1. Sabermetrics gives plausibility to any discussion regarding the game of baseball. It seeks objectivity and attempts to bring all factors into the discussion. It has taken hold on many different levels: in scouting, during contract negotiations, and in sportscasting.

Furthermore, it is gaining solid academic footing in universities all over the United States.

2. Albert Pujols will rate slightly below the likes of Lou Gehrig and Jimmie Foxx. His SLG has dropped over the years. Still, he will most probably finish up at or about 700 HR and with over 2000 RBI.

3. $ISO = SLG - BA = (TB - H)/AB$. This is a measure of power beyond first base. At 0.348, Babe Ruth has the highest ISO of all time.

$PF = SLG/BA = TB/H$. It measures the total bases per hit … kind of an "average hit". This statistic almost "rewards" the batter for a low BA.

A $PF > 2.000$ is an excellent accomplishment.

4. This Hall of Famer had a number of sterling seasons, but his best year may have been in 1999. As a Red Sox pitcher, Martinez had 23 W and 4 L, for a winning percentage of 0.852. He struck out 313 batters, while walking only 37 hitters. He had a league-leading ERA of 2.07 and was the unanimous choice for the Cy Young Award.

5. In 1972, the Philadelphia Phillies had an abysmal season, winning just 59 G and posting 97 L. Their ace, Steve Carlton, led the league with a 27-10 record and an ERA of 1.97. He also paced the circuit with 346.1 IP, 41 games started and 30 complete games. His MWPR was:

$$\text{MWPR} = (\text{IP})\left(\frac{W}{W+L} - \frac{W_{wo}}{W_{wo}+L_{wo}}\right) = (346.1)\left(\frac{27}{37} - \frac{32}{119}\right) = 160,$$

which ranks with the all-time best.

6. No. DiMaggio rarely struck out and did not receive a great number of walks (13 and 76, respectively, in 1941 when he set the record). To break the mark, a player would have to be a contact hitter who used the entire field. Speed would also be an asset.

7. The Iron Horse 8001 AB and 1508 BB, for a BB + AB total of 9509 plate appearances. He slugged 493 HR. So, to give Mattingly 12,500 plate appearances in the same proportion of AB to BB, we prorate to determine that 2517 additional AB are required. Since we are assuming Mattingly would be on a 4% decline, his EC becomes:

$$1 + (2517/8001)(.96) = 1.302.$$

Hence, his projected HR total is: $1.302(493) = 642$ HR.

8. Trout is arguably the best player in the game today. He certainly looks like he will be a Hall of Famer. Perhaps his best season was last year, 2018, even though he only played in 140 G. Trout hit 0.312, drove in 79 runs, while scoring 101 R. He had 24 SB and a 0.628 SLG. His OPS was an amazing 1.088. He also had 24 SB.

9. In 2018, the Yankees scored 851 R. They accumulated 1374 H, 625 BB, and 2490 TB in 5515 AB. Their RC is computed to be:

$$RC = \frac{(H + BB)(TB)}{(AB + W)} = \frac{(1374 + 625)(2490)}{(5515 + 625)} = 811$$

10. ■

11. Bonds has put up Ruthian numbers, but only toward the end of his career. Bonds has hit 50 + HR in a season just once (albeit it was 73) nor is he a 0.300 lifetime hitter. Controversies have followed him for the latter part of his career and it is doubtful that he will ever get into Cooperstown.
As of now, it is difficult to put him ahead of the likes of Willie Mays and, of course, Babe Ruth.

12. Excellent hitting and fielding … leadership … having that sense of the "intangible" à la Derek Jeter … where the team finishes in the pennant race. Still, this will always be one of the "debatable" questions.

13. See 12 above. Excellent pitching record … leadership … being a "stopper" when the team is going bad. Dominance over the league.
This, too, will always be one of the "debatable" questions.

14. Henry Benjamin Greenberg was a superior power hitter. Rivaling both Lou Gehrig and Jimmie Foxx as baseball's premier first baseman in the 1930s; Greenberg hit 58 HR in 1938.

Greenberg lost three prime years due to World War II.
By using the metric of the equivalence coefficient, and bringing his 5193 AB total to 9000 or 10,000, it is plausible to conjecture that Greenberg would have hit between 500 and 550 career HR.

15. Up through 2018 Verlander had 204 W and only 123 L. His lifetime ERA is about 3.40, and he has a WHIP of 1.16 with a K/BB ratio of about 3.36. As he entered the 2019 campaign, he was 36 years old. Should he have another 3−4 "good" years, Verlander should be a strong candidate for Cooperstown.

16.

YEAR	RUNS	WINS
2013	602	66
2014	614	73
2015	689	97
2016	808	103
2017	822	93

Using Excel®, we can easily construct a scatter plot and determine that the straight line that "best fits" is given by: $W = 0.1297\,R - 5.3127$, with an R^2 value of 0.7114 (which very roughly indicates a 70% fit).

17. This, along with the Black Sox World Series (1919) and the barring of players of color until 1947, is one of the great scandals in the history of the game.
There are medical and legal issues; economics has also been greatly affected by this menace. Youngsters who look up to ballplayers have doubtlessly been aversely influenced by HGH, PEDs, etc.
Baseball records have been skewed, making heretofore comparisons much more difficult.
The "jury" is still out on this issue, and this whole fiasco is not close to being ended.
To be continued.

18. SLOB = (OBA) (SLG). Babe Ruth has the highest statistic in history.

19. WAR (wins above replacement) is the measure that has enjoyed a lot of popularity. It seems, though, that the definition (or definitions of variations of WAR) of this metric, is not very simple. Various "adjustments" are taken into consideration, and to some it seems to have grown out of the "linear weights" school of Thorn and Palmer.

At this point it seems best to let WAR "settle" a bit and look into how this instrument will be modified in the future.

The following site may prove helpful:

https://www.baseball-reference.com/about/war_explained.shtml

20. Fielding, in general, is the least "measurable" baseball statistic. For example, a player may have very little range, but have sure hands, catching everything that comes to him. He might have a near perfect FA. Yet, is he more valuable than someone else with less sure hands, but a greater range? RF/G tries to take this into consideration. In fact, RF/inning is an even better statistic.

As time goes on, the suspicion is that even better instruments will emerge regarding Fielding.

21. Stan "The Man" Musial must be considered as one of the 10 greatest position players of all time. He had over 3600 H, hit 475 HR, and batted 0.331 lifetime. A perennial All-Star, Musial was universally admired by his teammates and opponents.

22. Mike Piazza was probably the greatest hitting catcher in history, although Johnny Bench was close and a much better defensive player. Other great all-around receivers were Yogi Berra, Bill Dickey, Mickey Cochrane, and Roy Campanella. Defensively, Ivan Rodriquez was excellent.

Given his penchant for winning (10 World Series rings), many would pick Yogi Berra over everyone else.

23. This is much like the Joe DiMaggio/Tris Speaker question from the previous chapter (#17).

It would seem that the consensus would select Mays over Speaker, yet if Joe DiMaggio is also taken into consideration, we are back to square one.

To hedge a bit, if one is considering total *career* value, then Speaker and Mays would take precedence over Joe DiMaggio (even though DiMaggio won eight World Championships).

For *peak* value, however, a young Joe DiMaggio would probably outperform both Mays and Speaker.

This is one of those debates that can go on and on.

24. Very little. From the formula, one can easily deduce that if SB are not at least twice CS, there is a *negative* contribution from BSR.

"Outs" are the currency. You only get 27 of them in a nine-inning game. So care must be exercised when considering a sacrifice, etc.

25. Please see answers to #18 and #19 in the previous chapter.

The Ruth is Mighty and Shall Prevail!

Index

Note: Page numbers followed by "t" indicate tables "f" indicate figures.